BARRON'S

THE TRUSTED NAME IN TEST PREP

2025

AP® Precalculus

PREMIUM

Christina Pawlowski-Polanish, M.S.

AP® is a registered trademark of the College Board, which is not affiliated with Barron's and was not involved in the production of, and does not endorse, this product.

Dedication

To my Cailey, continue looking up to the sky at the stars, and you can reach them! Love, Mommy

To my family, thank you for your time and support. It wouldn't have been possible without you!

AP® is a registered trademark of the College Board, which is not affiliated with Barron's and was not involved in the production of, and does not endorse, this product.

© Copyright 2024, 2023 by Kaplan North America, LLC, d/b/a Barron's Educational Series

All rights reserved.
No part of this publication may be reproduced in any form, or by any means, without the written permission of the copyright owner.

Published by Kaplan North America, LLC, d/b/a Barron's Educational Series
1515 West Cypress Creek Road
Fort Lauderdale, FL 33309
www.barronseduc.com

ISBN: 978-1-5062-9203-8

10 9 8 7 6 5 4 3 2 1

Kaplan North America, LLC, d/b/a Barron's Educational Series print books are available at special quantity discounts to use for sales promotions, employee premiums, or educational purposes. For more information or to purchase books, please call the Simon & Schuster special sales department at 866-506-1949.

About the Author

Christina Pawlowski-Polanish has taught all levels of math, including 10 years of AP Calculus BC at Commack High School in Commack, NY, and College Precalculus for 7 years. She earned a B.A. in mathematics and secondary education and an M.S. in mathematics from Hofstra University. She frequently presents on mathematics literacy at local, state, and national conferences and leads workshops in her district and at colleges. Christina is a member of many organizations, including the National Council of Teachers of Mathematics, Suffolk County Math Teachers' Association, and the Association of Mathematics Teachers of New York State, where she served as recording secretary and is presently a county chair. Christina was accepted into the New York State (NYS) Master Teacher Program and the NYS Academy of Master Teachers. She was chosen as Commack School District's Teacher of the Year and was a finalist for the 2018 NYS Teacher of the Year award. Christina joined the National Network of State Teachers of the Year and has been published on their blog. She was honored by the governor as a recipient of the NYS Empire State Excellence in Teaching Award in 2018. Most recently, Christina was named the 2019 New York State winner of the Presidential Award for Excellence in Mathematics and Science Teaching and was honored in Washington, D.C.

Table of Contents

UNIT 2: EXPONENTIAL AND LOGARITHMIC FUNCTIONS

UNIT 3: TRIGONOMETRIC AND POLAR FUNCTIONS

Visit Barron's Online Learning Hub for an additional full-length practice test.

How to Use This Book

This book provides comprehensive review and extensive practice for AP Precalculus. It is based on the latest AP Precalculus Course and Exam Description published by the College Board.

Introduction

Start with the Introduction, which contains an overview of the AP Precalculus course and exam. Learn the topics covered within each unit, familiarize yourself with the exam format, and review the guidelines for calculator use on test day. Then read through the Prior Knowledge section, which covers important tools and skills (that you may have learned in previous math classes) that can be applied to solve AP Precalculus problems. This section features key formulas, laws, and steps as well as carefully worked-out examples.

Review and Practice

Study Chapters 1 through 12, which are organized according to the units of AP Precalculus. Every chapter includes Learning Objectives that will be covered, a review of each topic, dozens of tables and figures that illustrate key concepts, numerous examples with clear solutions, and a set of multiple-choice practice questions (with detailed answer explanations) to check your progress. Throughout these chapters, you'll also find helpful tips and advice to keep in mind for test day.

Note: The last three review chapters in this book focus on Unit 4 topics. While Unit 4 topics will not be tested on the AP Precalculus exam, they may be part of your school's course curriculum. Furthermore, reviewing these chapters may help build your understanding of topics that you will encounter on the exam.

Practice Tests

This book concludes with two full-length practice tests that mirror the actual exam in format, content, and level of difficulty. Each test is followed by detailed answer explanations for all questions.

Glossary of Key Terms

Be sure to consult the Glossary of Key Terms at the end of this book. AP Precalculus is a very vocabulary-based subject, so a strong command of the most frequently used precalculus terms will further prepare you for success on the exam.

Online Practice

There is also an additional full-length practice test online. You may take this test in practice (untimed) mode or in timed mode. All questions are answered and explained.

For Students

Students who use this book independently will improve their performance by studying the examples carefully and trying to complete the practice problems before referring to the answer explanations. You are strongly encouraged to read the book from start to finish since each topic builds on the previous ones. However, it is possible to use this book as a reference for specific topics that need review.

For Teachers

The teacher who uses this book with a class can do so in several ways. If the book is used throughout the year, you can assign all or part of each set of multiple-choice questions after the topic has been covered. These sets can also be used for review purposes shortly before exam time. The practice tests will also be helpful in reviewing toward the end of the year.

BARRON'S ESSENTIAL 5

As you review the content in this book to work toward earning that **5** on your AP PRECALCULUS exam, here are five things that you **MUST** know above everything else:

Know the mathematical content. It is much better to understand the different mathematical concepts rather than memorizing properties and formulas. By understanding why the properties and formulas work, you will have a firm grasp of the topic, which will enable you to solve the problems. No two math questions are alike, and so it is impossible to memorize your way through a math course, especially an AP math course. This book often goes through the derivations of formulas and gives background explanations of the concepts so that you will understand how the math works. With deeper understanding comes true ownership of the material.

Build your vocabulary. Math is its own language with its own notations and terms. Often terms in math can have multiple meanings within the context of the problem. For example:

- In trigonometry, *tangent* refers to the ratio of sides of a right triangle or the ratio of the trigonometric functions, $\frac{\sin\theta}{\cos\theta}$.
- In geometry, *tangent* refers to a line that intersects a circle in only one point.
- In calculus, *tangent* refers to a line that intersects a curve at a point, matching the curve's slope there.

Make sure you understand the specific mathematical meanings of each term, especially those in the Glossary at the end of this book. Most of the multiple-choice questions on the exam measure your knowledge and ability to apply mathematical vocabulary.

Mark your model. Since math is its own language, the way you read and approach a problem should be different from how you read in other subject areas. Many problems in math require visualization. If a model is provided, it is important to mark the model with any given information. The markings in the model may tell you about a relationship in the figure that will lead to an equation. If a verbal description is provided, begin by creating a model that can be marked.

4

Pause for punctuation. When you are reading a problem and see a comma, period, semicolon, colon, or the words *and* or *is*, you should stop and ask yourself, "Why might this break in the problem be important?" The answer to this question may tell you about some equation to focus on, information that will allow you to create an equation, or some important information that will lead to a solution to the problem. Think of these pauses as a way for the question writer to signal to you a list of important givens in an organized manner.

5

Improve your study habits. Reaching this level in math means that you have been a student for many years and are not afraid to take on a challenging course. You might already have your own study strategies in place or may have found math very easy up until now. Studying for the AP Precalculus exam is an opportunity for you to enhance your study strategies. Study strategies can include:

- Spacing your studying over a span of weeks as opposed to cramming the days before an exam.
- Reflecting on topics you struggled with, reading the review material on that section, and answering the practice questions. Repeating a question is a great learning tool to ensure you understand the concept.
- Using the practice questions at the end of each chapter to test your knowledge.
- Taking all of the practice tests to become familiar with the testing format and using them to understand which topics you need to repeat and practice more.
- Using flashcards to review important vocabulary or equations that need to be better understood. The terms you select to make cards on should be ones that are seen often or ones that you are not familiar with.

Introduction

AP Precalculus is a full-year course designed to prepare you for the math you will encounter at the college level. This course should be taken after you have completed Geometry and Algebra II or once you have completed Integrated Math 3. AP Precalculus contains content that is already present in standard high school precalculus courses. This course will develop your ability to express functions, concepts, problems, and conclusions analytically, graphically, numerically, and verbally and to understand how these are related. You will also learn how to use a graphing calculator as a tool for mathematical investigations and problem solving.

Unit Outline

The course is divided into four units. The first three units are assessed on the AP Precalculus exam. The fourth unit includes additional topics available to schools but that are **not** included on the AP Precalculus exam.

Unit 1: Polynomial and Rational Functions

Unit 1 covers the following topics:

- Rates of change
- Linear functions
- Quadratic functions
- Complex zeros of polynomial functions
- End behavior of polynomial functions
- Zeros of rational functions
- Vertical asymptotes of rational functions
- Holes of rational functions, long division
- Transformations of functions, modeling

Unit 2: Exponential and Logarithmic Functions

Unit 2 covers the following topics:

- Arithmetic and geometric sequences
- Exponential functions
- Exponential function manipulation
- Data modeling
- Composition of functions
- Inverse functions
- Logarithmic expressions
- Inverses of exponential functions
- Logarithmic functions
- Logarithmic function manipulation
- Exponential and logarithmic equations
- Exponential and logarithmic inequalities, semi-log plots

Unit 3: Trigonometric and Polar Functions

Unit 3 covers the following topics:

- Periodic phenomena
- Sine function
- Cosine function
- Tangent function
- Graphs of trigonometric functions
- Transformations of sinusoidal functions
- Reciprocal trigonometric functions
- Solving trigonometric equations
- Trigonometric identity proofs
- Trigonometry and polar coordinates
- Polar function graphs

Unit 4: Functions Involving Parameters, Vectors, and Matrices

Unit 4 covers the following topics:

These Unit 4 topics are **not** assessed on the AP exam.

- Parametric functions
- Rectilinear motion
- Implicit functions
- Conic sections, vectors
- Vector operations
- Vector-valued functions
- Matrices, matrix operations
- Linear transformations and matrices
- Matrices as functions
- Markov chains

Exam Format

The AP Precalculus exam is 3 hours long and is comprised of the following sections.

Section and Question Type		Number of Questions	Time	Exam Weighting
I (Multiple-Choice)	Part A—No Calculator	28	80 min	43.75%
	Part B—Calculator Active*	12	40 min	18.75%
II (Free-Response)	Part A—Calculator Active**	2	30 min	18.75%
	Part B—No Calculator	2	30 min	18.75%

*Some of these questions require the use of a graphing calculator.
**Some of these questions require the use of a graphing calculator. Note that, after 30 minutes, *you will no longer be permitted to use a calculator.*

Scoring of the Exam

Each completed AP exam paper receives a grade according to the following five-point scale:

AP Grade	Qualification
5	Extremely well qualified
4	Well qualified
3	Qualified
2	Possibly qualified
1	No recommendation

Many colleges and universities accept a grade of 3 or better for college credit or advanced placement. Be sure to check the AP credit policies on individual colleges' websites.

Using Your Graphing Calculator on the AP Exam

Each student is expected to bring a graphing calculator to the AP exam. Different models of calculators vary in their features and capabilities; however, there are three procedures you must be able to perform on your calculator:

1. Produce the graph of a function within an arbitrary viewing window.
2. Solve an equation numerically.
3. Compute regression models numerically.

Guidelines for Calculator Use

1. For the multiple-choice questions in Section I: Part B, you may use any feature or program on your calculator.
2. On the free-response questions in Section II: Part A, remember the following:
   - You may use the calculator to perform any of the three listed procedures. When you do, you need to write only the equation or setup that will produce the solution and then the calculator result to the required degree of accuracy (three places after the decimal point unless otherwise specified).
   - For a solution for which you use a calculator capability other than the three listed above, you must write down the mathematical steps that yield the answer. A correct answer alone will not earn full credit.
   - When asked to "justify," you must provide mathematical reasoning to support your answer. Calculator results alone will not be sufficient.

WARNING

Don't rely on your graphing calculator too much. Only a few of these questions require a calculator, and in some cases, it may be an inefficient use of time.

Additional Calculator Notes

- Learn the proper syntax for your calculator: the correct way to enter operations, functions, and other commands. Parentheses, commas, variables, or parameters that are missing or entered in the wrong order can produce error messages or yield wrong answers.
- Keep your calculator set in radian mode. Almost all questions about angles and trigonometric functions use radians. If you ever need to change to degrees for a specific calculation, return the calculator to radian mode as soon as that calculation is complete.
- Many calculators do not have keys for the secant, cosecant, or cotangent functions. To obtain these functions, use their reciprocals. For example, $\sec\left(\frac{\pi}{8}\right) = \frac{1}{\cos\left(\frac{\pi}{8}\right)}$.

- Evaluate inverse functions such as arcsin, arccos, and arctan on your calculator. Those function keys are usually denoted as $\sin^{-1}$, $\cos^{-1}$, and $\tan^{-1}$.
- To achieve three-place accuracy in a final answer, do *not* round off variables at intermediate steps since this is likely to produce error accumulations. Round off only after your calculator produces the final answer.
- Although final answers expressed in this book are in simplest form, this is not necessary on Section II questions of the AP exam. According to the directions printed on the exam, "unless otherwise specified" (1) you need not simplify algebraic or numerical answers and (2) answers involving decimals should be correct to three places after the decimal point. However, be aware that if you try to simplify, you must do so correctly or you will lose credit.

Prior Knowledge

Math is very much like a staircase; each course is a stair you climb, providing you with the skills and techniques to guide you upward to the next course. This section will give a brief review of important tools that are used to help solve topics found in this course.

Factoring

Factoring is the process of breaking down a polynomial expression or equation and finding its factors that, when multiplied together, return to the original polynomial. A few different factoring techniques will be reviewed.

Greatest Common Factor (GCF)

Greatest common factor factoring refers to dividing out the greatest like term in a polynomial. The greatest common factor can be a number, a variable, or both.

Example

Factor each of the following polynomials.

1. $3x^4 - x$
2. $2x^3 + 6x^4 - 4x^5$
3. $(x + 1)^3 - 4(x + 1)^2$

Solution

1. There are two terms in this polynomial, $3x^4$ and x. Their greatest common factor is x. Factoring out x gives $3x^4 - x = x(3x^3 - 1)$.
2. There are three terms in this polynomial, $2x^3$, $6x^4$, and $4x^5$. Their greatest common factor is $2x^3$. Factoring out $2x^3$ gives $2x^3 + 6x^4 - 4x^5 = 2x^3(1 + 3x - 2x^2)$.
3. There are two terms in this polynomial, $(x + 1)^3$ and $4(x + 1)^2$. Their greatest common factor is $(x + 1)^2$. Factoring out $(x + 1)^2$ gives $(x + 1)^3 - 4(x + 1)^2 = (x + 1)^2((x + 1) - 4) = (x + 1)^2(x - 3)$.

Difference of Two Squares

Difference of two squares factoring refers to binomials of the form $x^2 - p^2$. To factor a difference of two squares, $x^2 - p^2 = (x - p)(x + p)$. The factors are known as conjugate pairs because each factor has the same terms, x and p, but one factor is a sum and the other factor is a difference.

Example

Factor each of the following polynomials.

1. $16x^2 - 9$
2. $(y - 2)^2 - 25$
3. $49 - 4t^2$
4. $x^2 - 5$

Solution

1. Two terms are being subtracted, $16x^2$ and 9. Factoring using the difference of two squares gives $16x^2 - 9 = (4x - 3)(4x + 3)$.
2. Two terms are being subtracted, $(y - 2)^2$ and 25. Factoring using the difference of two squares gives $(y - 2)^2 - 25 = ((y - 2) - 5)((y - 2) + 5) = (y - 7)(y + 3)$.
3. Two terms are being subtracted, 49 and $4t^2$. Factoring using the difference of two squares gives $49 - 4t^2 = (7 - 2t)(7 + 2t)$.
4. Two terms are being subtracted, x^2 and 5. Factoring using the difference of two squares gives $x^2 - 5 = (x - \sqrt{5})(x + \sqrt{5})$.

Factoring Trinomials

To factor a quadratic of the form $x^2 + bx + c$, look for two real numbers, p and q, that satisfy $p + q = b$ and $pq = c$.

Example

Factor each of the following polynomials.

1. $y^2 - 24y + 144$
2. $x^2 + 6x + 8$
3. $z^2 - z - 20$

Solution

1. Two numbers that multiply to $c = 144$ are -12 and -12, and these same two numbers add to $b = -24$. Therefore, $y^2 - 24y + 144 = (y - 12)(y - 12)$.
2. Two numbers that multiply to $c = 8$ are 4 and 2, and these same two numbers add to $b = 6$. Therefore, $x^2 + 6x + 8 = (x + 4)(x + 2)$.
3. Two numbers that multiply to $c = -20$ are -5 and 4, and these same two numbers add to $b = -1$. Therefore, $z^2 - z - 20 = (z - 5)(z + 4)$.

To factor a quadratic of the form $ax^2 + bx + c$ where $a \neq 0$ and $a \neq 1$, you can either guess and check or use the split the middle method.

Steps for Factoring Using Split the Middle

1. Find the product *ac*.
2. Determine numbers whose product is *ac* and whose sum is *b*.
3. Split up the middle term *bx*, and write it as a sum of the two numbers determined in step 2.
4. The trinomial is written as an equivalent polynomial with four terms. Group together the first two terms, group together the last two terms, and factor out a common factor from each.
5. A common factor for the resulting two terms should appear that can then be factored out. Doing this will reveal the two factors of the original trinomial.

Example

Factor each of the following quadratics.

1. $25t^2 + 10t + 1$
2. $3a^2 - 10a + 8$

Solution

1. Since this trinomial has an *a*-value other than 1, factor using the split the middle method. The product of *a* and *c* is 25. Two numbers that multiply to 25 are 5 and 5, and they also add to the *b*-value of 10.

   Split the middle term:

   $$25t^2 + 5t + 5t + 1$$

   Factor by grouping:

   $$(25t^2 + 5t) + (5t + 1)$$

   Take out a common factor from each set of parentheses:

   $$5t(5t + 1) + 1(5t + 1)$$

   Factor out the common factor $(5t + 1)$ from each term:

   $$(5t + 1)(5t + 1)$$

2. Since this trinomial has an *a*-value other than 1, factor using split the middle. The product of *a* and *c* is 24. Two numbers that multiply to 24 are -6 and -4, and they also add to the *b*-value of -10.

   Split the middle term:

   $$3a^2 - 6a - 4a + 8$$

   Factor by grouping:

   $$(3a^2 - 6a) + (-4a + 8)$$

   Take out a common factor from each set of parentheses:

   $$3a(a - 2) - 4(a - 2)$$

   Factor out the common factor $(a - 2)$ from each term:

   $$(3a - 4)(a - 2)$$

Sum or Difference of Cubes

To factor a sum or difference of cubes, use the factoring formulas below:

$$x^3 + p^3 = (x + p)(x^2 - px + p^2)$$

$$x^3 - p^3 = (x - p)(x^2 + px + p^2)$$

The trinomial obtained by factoring a sum or difference of cubes does not factor further.

Example

Factor each of the following polynomials.

1. $x^3 - 8$
2. $1 + 1000x^3$

Solution

1. The polynomial is an example of a difference of cubes where $p = 2$ since $2^3 = 8$. Use the factoring formula:

$$x^3 - 8 = (x - 2)(x^2 + 2x + 2^2) = (x - 2)(x^2 + 2x + 4)$$

2. The polynomial is an example of a sum of cubes where $x = 1$ since $1^3 = 1$ and $p = 10x$ since $(10x)^3 = 1000x^3$. Use the factoring formula:

$$1 + 1000x^3 = (1 + 10x)(1^2 - 10x \cdot 1 + (10x)^2) = (1 + 10x)(1 - 10x + 100x^2)$$

Quadratic Formula

When asked to solve a quadratic equation, one method is to use factoring and set each factor equal to zero. Sometimes it is impossible to factor a quadratic equation. In those cases, the solutions can be found using the quadratic formula.

The Quadratic Formula

Given a quadratic equation in standard form, $ax^2 + bx + c = 0$, the equation below can find the solutions to the quadratic equation:

$$x = \frac{-b \pm \sqrt{b^2 - 4ac}}{2a}$$

Example

Solve the following quadratic equations, and write your answer in simplest radical form.

1. $2x^2 - 5x - 7 = 0$
2. $4t^2 = 16t + 20$

Solution

1. The quadratic equation is in standard form with $a = 2$, $b = -5$, and $c = -7$. Substitute the values into the quadratic formula:

$$x = \frac{-(-5) \pm \sqrt{(-5)^2 - 4(2)(-7)}}{2(2)}$$

$$x = \frac{5 \pm \sqrt{25 + 56}}{4}$$

$$x = \frac{5 \pm \sqrt{81}}{4}$$

$$x = \frac{5 \pm 9}{4}$$

$$x = \frac{5 + 9}{4}, x = \frac{5 - 9}{4}$$

The solutions are $x = \frac{7}{2}$, $x = -1$.

2. The quadratic equation needs to be written in standard form first, $-4t^2 + 16t + 20 = 0$. The equation is in standard form with $a = -4$, $b = 16$, $c = 20$. Substitute the values into the quadratic formula:

$$x = \frac{-(16) \pm \sqrt{(16)^2 - 4(-4)(20)}}{2(-4)}$$

$$x = \frac{-16 \pm \sqrt{256 + 320}}{-8}$$

$$x = \frac{-16 \pm \sqrt{576}}{-8}$$

$$x = \frac{-16 \pm 24}{-8}$$

$$x = \frac{-16 + 24}{-8}, x = \frac{-16 - 24}{-8}$$

The solutions are $x = -1$, $x = 5$.

Imaginary Numbers

There are many different sets of numbers, such as integers, natural numbers, rational numbers, and real numbers. There is also the set of imaginary numbers. This occurs when there is a negative number under a square root.

The most basic imaginary number is $\sqrt{-1}$ and is notated as i. In other words, $i = \sqrt{-1}$. Operations using imaginary numbers are much like operations in the real number system.

Addition: $2i + 3i = 5i$

Subtraction: $7i - 3i = 4i$

When multiplying imaginary numbers, the exponents of the imaginary number are added together and their coefficients are multiplied.

Multiplication: $2i \cdot 5i = 10i^2$

Special properties are associated with the powers of i, as noted in the sidebar to the right.

Powers of *i*

$i = \sqrt{-1}$

$i^2 = (\sqrt{-1})^2 = -1$

$i^3 = i^2 \cdot i = -1 \cdot i = -i$

$i^4 = i^2 \cdot i^2 = -1 \cdot -1 = 1$

The powers continue to cycle following this pattern.

When dividing imaginary numbers, the exponents of the imaginary numbers are subtracted and their coefficients are divided.

$$\text{Division: } \frac{49i^5}{7i^3} = 7i^2 = 7(-1) = -7$$

Imaginary numbers can appear when working with square roots. A problem may ask for a solution to be written in simplest $a + bi$ form, which is a complex number. The a represents the real portion of the number, and the bi represents the imaginary portion of the number.

To simplify a complex number, first rewrite all radicals with a negative radicand as imaginary numbers by removing the negative sign and replacing it with the imaginary number i. Finally, simplify the radical.

Example

Solve the quadratic equation, and write your solution in simplest $a + bi$ form.

$$x^2 + 6x + 13 = 0$$

Solution

The quadratic equation is written in standard form with $a = 1$, $b = 6$, $c = 13$. Use the quadratic formula:

$$x = \frac{-(6) \pm \sqrt{(6)^2 - 4(1)(13)}}{2(1)}$$

$$x = \frac{-6 \pm \sqrt{36 - 52}}{2}$$

$$x = \frac{-6 \pm \sqrt{-16}}{2}$$

$$x = \frac{-6 \pm i\sqrt{16}}{2}$$

$$x = \frac{-6 \pm 4i}{2}$$

$$x = \frac{-6 + 4i}{2},\ x = \frac{-6 - 4i}{2}$$

The solutions are $x = -3 + 2i$, $x = -3 - 2i$.

Right Triangle Trigonometry

Right triangles are made up of a right angle and two acute angles. The sum of the 3 angles is 180°. The sides opposite the acute angles are called legs, and the side opposite the right angle is called the hypotenuse. The lengths of the sides of a right triangle must satisfy the Pythagorean theorem, $a^2 + b^2 = c^2$, where a and b are the legs and c is the hypotenuse.

The relationship between the angle measures and the side lengths is known as trigonometry. There are three trigonometric ratios of sides that correspond to an acute angle measure in a right triangle.

Three Trigonometric Ratios

Given the right triangle ABC, where angle A is the right angle shown in the figure, the sides are labeled based upon their relationship to angle θ.

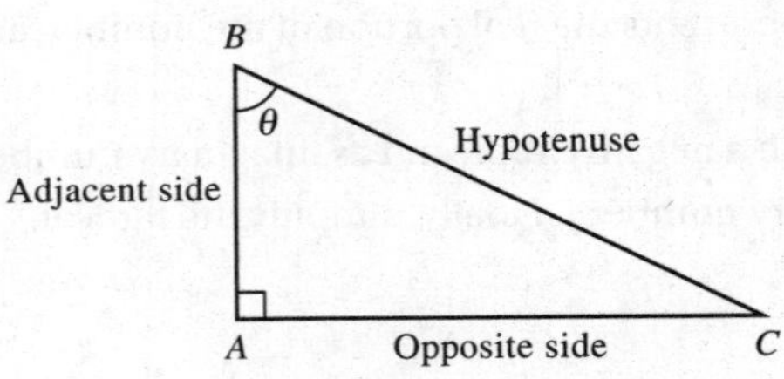

Trigonometric Ratios of Angle θ

$$\sin\theta = \frac{\text{opposite}}{\text{hypotenuse}} = \frac{AC}{BC}$$

$$\cos\theta = \frac{\text{adjacent}}{\text{hypotenuse}} = \frac{AB}{BC}$$

$$\tan\theta = \frac{\text{opposite}}{\text{adjacent}} = \frac{AC}{AB}$$

The ratio of sides in the box above can be remembered using the following mnemonic device:

$$S\frac{O}{H} \quad C\frac{A}{H} \quad T\frac{O}{A}$$

Special Right Triangles

Two triangles have known angle measures and corresponding side lengths. These are known as the special right triangles.

30°–60°–90° Triangle

In this triangle, the two acute angles measure 30° and 60° and the third angle is the right angle, 90°. Their corresponding side lengths are in the ratio $1{:}\sqrt{3}{:}2$ as shown in the figure below.

As a result, below are six known trigonometric values that correspond to this special right triangle:

$$\sin 30^\circ = \frac{1}{2} \qquad \cos 30^\circ = \frac{\sqrt{3}}{2}$$

$$\tan 30^\circ = \frac{1}{\sqrt{3}} = \frac{\sqrt{3}}{3} \qquad \sin 60^\circ = \frac{\sqrt{3}}{2}$$

$$\cos 60^\circ = \frac{1}{2} \qquad \tan 60^\circ = \frac{\sqrt{3}}{1} = \sqrt{3}$$

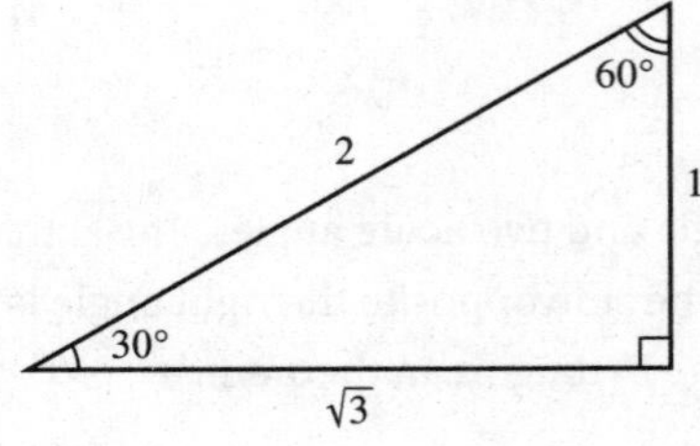

45°–45°–90° Triangle

In this triangle, the two acute angles measure 45° and 45° and the third angle is the right angle, 90°. Their corresponding side lengths are in the ratio 1:1:$\sqrt{2}$ as shown in the figure.

As a result, below are three known trigonometric values that correspond to this special right triangle:

$$\sin 45° = \frac{1}{\sqrt{2}} = \frac{\sqrt{2}}{2}$$

$$\cos 45° = \frac{1}{\sqrt{2}} = \frac{\sqrt{2}}{2}$$

$$\tan 45° = \frac{1}{1} = 1$$

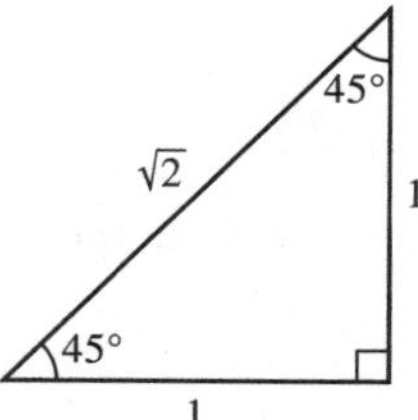

The trigonometric values of angles other than 30°, 45°, and 60° can be measured using calculators. In this course, you will learn that the trigonometric values of angles that measure more than 90° can also be evaluated.

Law of Sines and Law of Cosines

Two other equations help determine the lengths and angles of triangles that don't necessarily have to be right triangles. They are known as the law of sines and the law of cosines.

Law of Sines

The law of sines relates the lengths of two sides of a triangle to the sines of the angles opposite those sides.

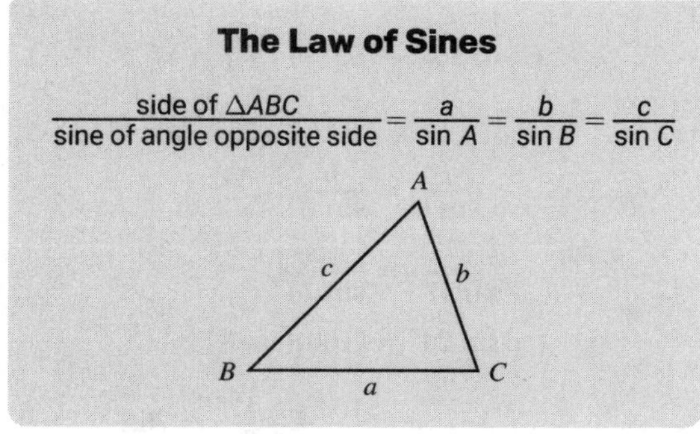

Example

In ΔABC, $a = 12$, $\sin A = 0.6$, and $\sin B = 0.4$. What is the length of side b?

Solution

Since the sine of two angles and the length of a side are given, solve for b by using the law of sines.

$$\frac{a}{\sin A} = \frac{b}{\sin B}$$

$$\frac{12}{0.6} = \frac{b}{0.4}$$

$$0.6b = (0.4)(12)$$

$$b = \frac{4.8}{0.6} = 8$$

Example

In ΔABC, $\angle A = 59°$, $\angle B = 74°$, and $b = 100$ meters. Find c to the nearest tenth of a meter.

Solution

It is helpful to make a model of the given information to determine the necessary equation needed to solve the problem.

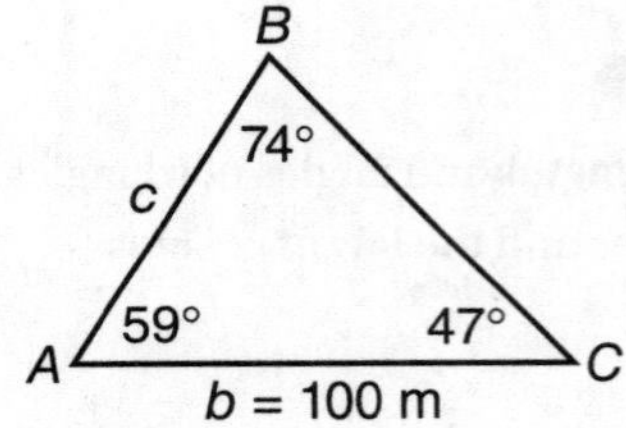

First find the measure of $\angle C$, which is the third angle of the triangle:

$$\angle C = 180° - (59° + 74°) = 47°$$

Now find c using the law of sines.

$$\frac{c}{\sin C} = \frac{b}{\sin B}$$

$$\frac{c}{\sin 47°} = \frac{100}{\sin 74°}$$

$$c \sin 74° = (100)(\sin 47°)$$

$$c = \frac{100 \sin 47°}{\sin 74°} = 76.1$$

Law of Cosines

The law of cosines relates the lengths of the three sides of a triangle to the cosine of the angle opposite one of the sides.

The Law of Cosines

$a^2 = b^2 + c^2 - 2bc \cos A$

$b^2 = a^2 + c^2 - 2ac \cos B$

$c^2 = a^2 + b^2 - 2ab \cos C$

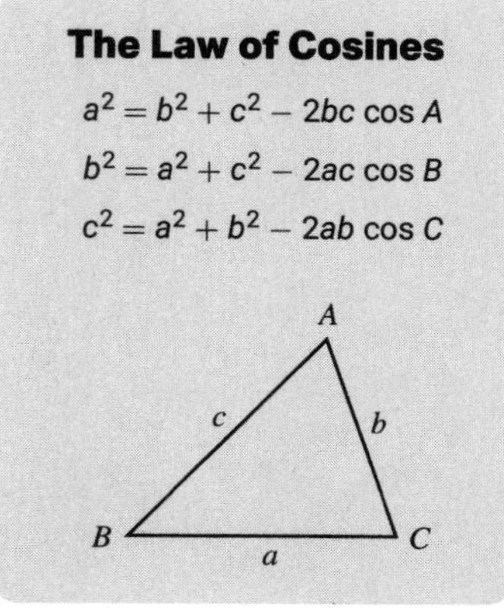

Example

In $\triangle ABC$, $a = 6$, $b = 10$, and $\angle C = 120°$. What is the length of side c?

Solution

It is helpful to make a model of the given information to determine the necessary equation needed to solve the problem.

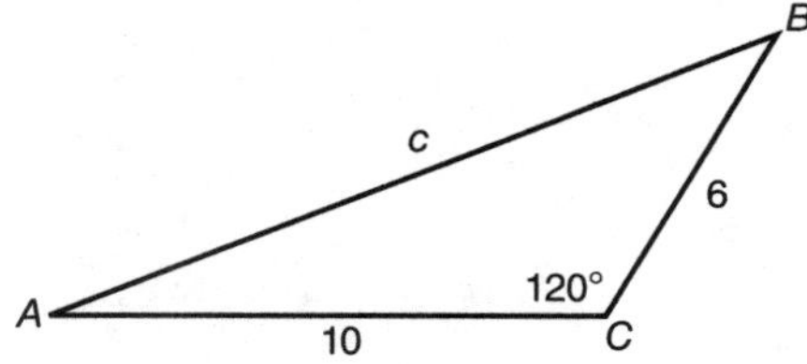

Since three sides and one angle are marked in the model, use the law of cosines to find the length of side c.

$$c^2 = a^2 + b^2 - 2ab \cos C$$

$$c^2 = (6)^2 + (10)^2 - 2(6)(10) \cos 120°$$

Use your calculator.

$$c^2 = 196$$

$$c = \pm\sqrt{196} = \pm 14$$

Since c represents the length of a side, reject the negative length. Therefore, $c = 14$.

UNIT 1
Polynomial and Rational Functions

Rates of Change

Learning Objectives

In this chapter, you will learn:

- ➔ The difference between a relation and a function
- ➔ How to calculate the rate of change and its meaning
- ➔ How to calculate the rate of change of different functions

1.1 Relations and Functions

Relations

Most mathematical studies are based upon two different quantities and how they relate to one another. This is known as a relation. A relation can be represented by a diagram, an equation, or a list that specifies a specific relationship between groups of elements.

For example, consider relation r shown in Figure 1.1 as both a diagram and a list of ordered pairs.

r:

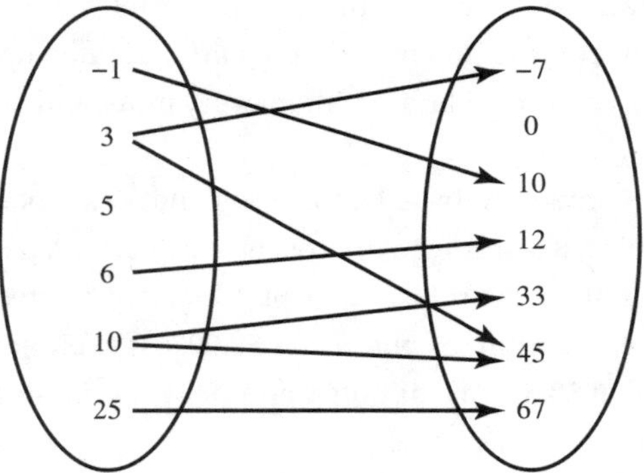

Figure 1.1

The elements in the oval on the left are known as the set of input values, which is also called the domain. The elements in the oval on the right are known as the set of output values, which is also called the range. The arrows lead from the inputs and end at their corresponding outputs.

Following the arrows helps to write the relation as a set of ordered pairs as shown below:

$$r: \{(-1, 10), (3, -7), (3, 45), (6, 12), (10, 33), (10, 45), (25, 67)\}$$

There are two things to notice about relation r:

1. Not all of the elements in the domain were used: 5 did not have a corresponding output.
2. Some inputs had multiple outputs: 3 mapped to -7 and 45, and 10 mapped to 33 and 45.

Functions

A relation that maps a set of input values to a set of output values where each input value is mapped to one and only one output value is defined as a function. For relation r as a function, 5 would need to be removed from the domain, and both 3 and 10 should be mapped to only one output each. This is shown in Figure 1.2.

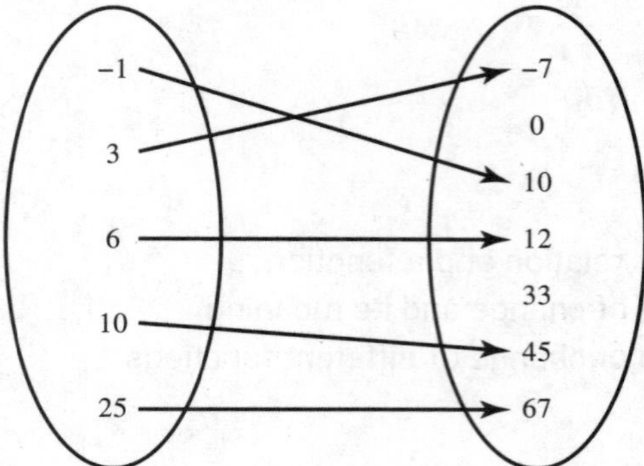

Figure 1.2

An example of a real-life model of a function is the cost of a taxi. Taxis calculate the cost of a ride based on the amount of time a passenger spends in the vehicle. Therefore, the cost of a particular ride is a function of the length of the ride. A passenger cannot have two charges for the same duration. In this example, time is the domain and cost is the range.

Functions are powerful mathematical tools for describing how one variable quantity depends on another. A function can be represented verbally, numerically, algebraically, or graphically. Variables are often used to represent the infinitely many possibilities for the domain and range of a function. The variable representing input values is called the independent variable, and the variable representing output values is called the dependent variable. Since functions, like relations, can be represented by ordered pairs, x is often used to represent the independent variable and y often represents the dependent variable. Then every input and its corresponding output can be represented by the ordered pair (x, y).

Often functions can be written as equations based upon a rule that describes the relationship between the independent and dependent variables. Returning to the cost of the taxi, for a passenger to ride, it costs $2.00 to enter the car and $1.50 for every minute of the ride. A table of values can be created to list the different corresponding inputs and outputs. For example, at 0 minutes, the cost is $2.00 just to enter the taxi. After a minute of driving, $1.50 is added to that cost for a total of $3.50. At 2 minutes, another $1.50 is added for a total cost of $5.00. This is illustrated in Table 1.1.

Table 1.1

x = time (minutes)	y = cost (dollars)
0	2.00
1	3.50
2	5.00
3	6.50
4	8.00

Since there are infinitely many values that time can take on, listing all the corresponding pairs is tedious. It is more efficient to write the cost of a ride as an equation dependent upon time. Since the independent variable is time, let x represent the time in minutes and y represent the cost in dollars. The cost equation could be written as $y = 2.00 + 1.50x$.

There is another way to notate functions as equations. If f represents a function, then $f(x)$, read as "f of x," represents the value of the function when x takes on a specific value. In other words, $y = f(x)$. The benefit of this notation is that it allows you to see the two variables and clearly identify the independent and dependent variables. It is also a nice way to identify specific values. For example, to find the cost of the taxi ride after 5 minutes, we would input 5 for x to find the cost. This substitution would look like $f(5) = 2.00 + 1.50(5) = 9.50 = y$. After 5 minutes, the cost of the taxi would be \$9.50.

Different letters can be used to fit what the variables represent. For example, t can stand for time and c for cost. Our taxi equation can be rewritten as $c(t) = 2.00 + 1.50t$.

› Example

If $f(x) = 10x - 2$, find the value of $f(3)$.

✓ Solution

To find the value of $f(3)$, the input is $x = 3$. This value is substituted in for x into the equation.

$$f(3) = 10(3) - 2 = 30 - 2 = 28$$

When $x = 3$, $f(3) = 28 = y$. The ordered pair that represents this solution is (3, 28).

› Example

Given $g = \{(-1, 1), (0, 3), (4, 7), (5, 1), (8, -10)\}$, find $g(0)$.

✓ Solution

To find the value of $g(0)$, the input is $x = 0$. Find the ordered pair whose x-value is 0. The y-value will represent the function value. The ordered pair is (0, 3), so $g(0) = 3$.

› Example

If $v(t) = t^2 + 5$, what is $v(c + 2)$?

✓ Solution

To evaluate $v(c + 2)$, the input is $t = c + 2$. This input is substituted for t in the equation.

$$v(c + 2) = (c + 2)^2 + 5 = (c + 2)(c + 2) + 5 = c^2 + 2c + 2c + 4 + 5 = c^2 + 4c + 9$$

Another way functions can be represented is by their graphs. To graph a function, plot its coordinate pairs as points on an xy-plane. The xy-plane is a coordinate grid with a horizontal line that represents the x-axis and a vertical line that represents the y-axis. Coordinate pairs are plotted by moving left or right on the x-axis and then moving down or up on the y-axis. If the function is continuous over its domain, the points of the pairs can be connected to reveal the graph. The graph provides a nice visual and can show different behaviors of the graph that can lead to important conclusions about the function.

Returning to the taxi cost function, $f(x) = 2.00 + 1.50x$, graph the function on the coordinate plane. Begin by plotting the points from Table 1.2 of values on the coordinate plane as shown in Figure 1.3.

Table 1.2

x = time (minutes)	y = cost (dollars)
0	2.00
1	3.50
2	5.00
3	6.50
4	8.00

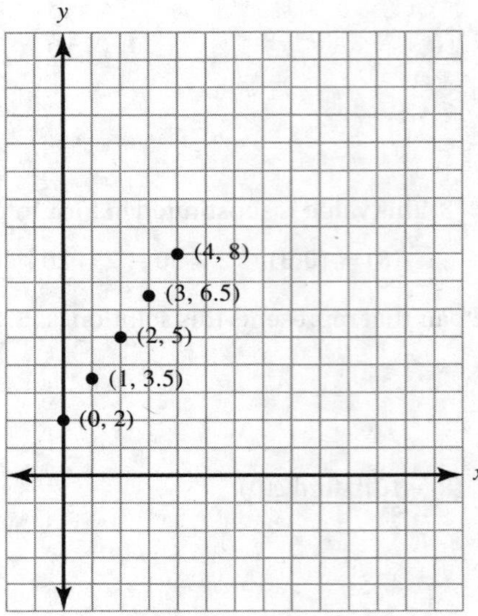

Figure 1.3

Since the domain is time, assume the inputs are all real numbers greater than zero. The points can be connected to create a continuous graph as shown in Figure 1.4.

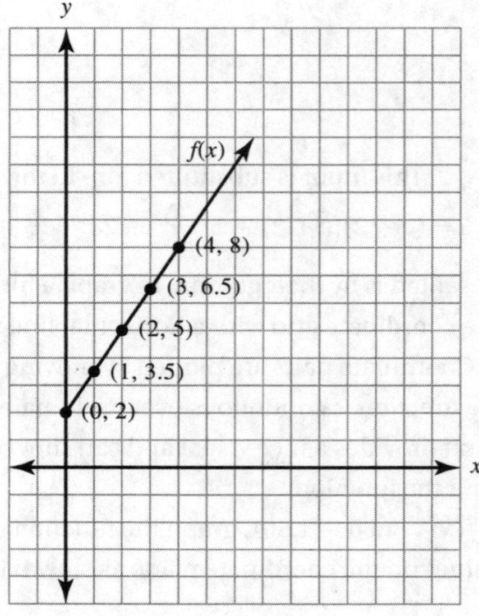

Figure 1.4

This is a linear function because its graph is a line. Note that the axes can change names, too, depending on the independent and dependent variables being used. If we graphed $c(t) = 2.00 + 1.50t$, the x-axis could be renamed as t and the y-axis could be renamed as either c or $c(t)$.

Various types of functions will be explored in later chapters of this book. Many have general concepts that can be applied, including domain, range, x- and y-intercepts, and graph behaviors.

Domain and Range

The domain of a function, $f(x)$, is the largest possible set of numbers x for which $f(x)$ is a real number. To find the domain graphically, look along the x-axis from left to right. The x-value of the point farthest to the left represents the lower bound of the domain, and the x-value of the point farthest to the right represents the upper bound of the domain.

The range of a function, $f(x)$, is the set of all values that y can have as x takes on each of its possible values. Figure 1.5 is the graph of function $g(x)$.

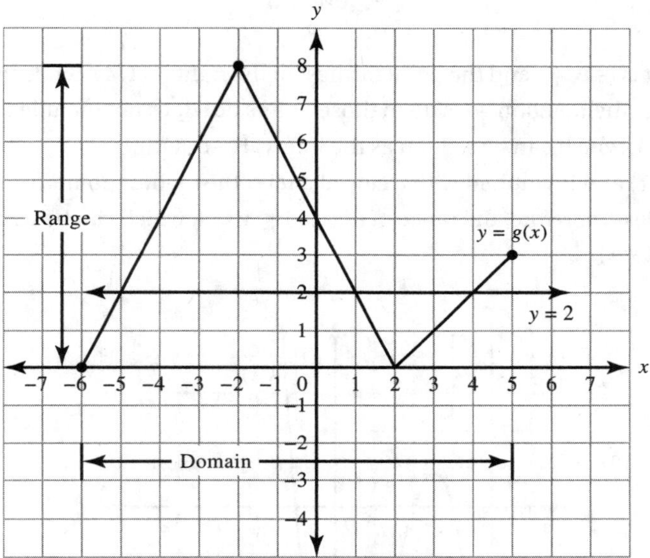

Figure 1.5

The point farthest to the left is $(-6, 0)$, and the point farthest to the right is $(5, 3)$. Therefore, the domain of $g(x)$ is $-6 \leq x \leq 5$, which can be written as the interval $[-6, 5]$.

The lowest points are $(-6, 0)$ and $(2, 0)$. The highest point is $(-2, 8)$. Therefore, the range of $g(x)$ is $0 \leq y \leq 8$, which can be written as the interval $[0, 8]$.

Not all graphs have closed endpoints such as in the last example.

Figure 1.6 is the graph of function $h(x)$.

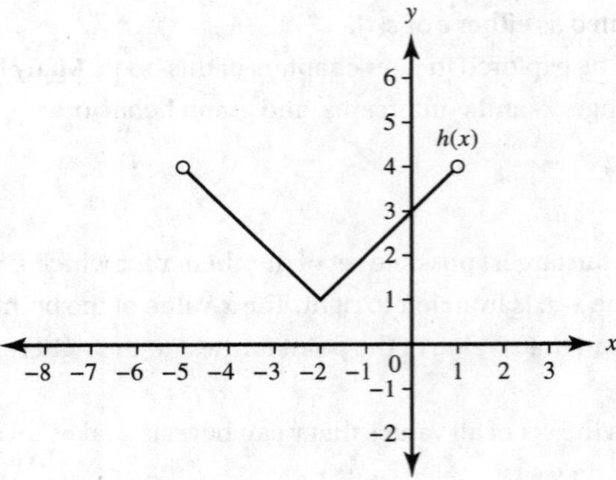

Figure 1.6

The point farthest to the left is $(-5, 4)$, and the point farthest to the right is $(1, 4)$. Since these points have open circles, it means the graph of the function goes up to those points but does not include those values. Therefore, the domain of $h(x)$ is $-5 < x < 1$, which can be written as the interval $(-5, 1)$.

The lowest point is $(-2, 1)$ and is graphed with a closed circle. The highest points are $(-5, 4)$ and $(1, 4)$, which are graphed with open circles. Therefore, the range of $h(x)$ is $1 \leq y < 4$, which can be written as the interval $[1, 4)$.

Figure 1.7 is the graph of $k(x)$.

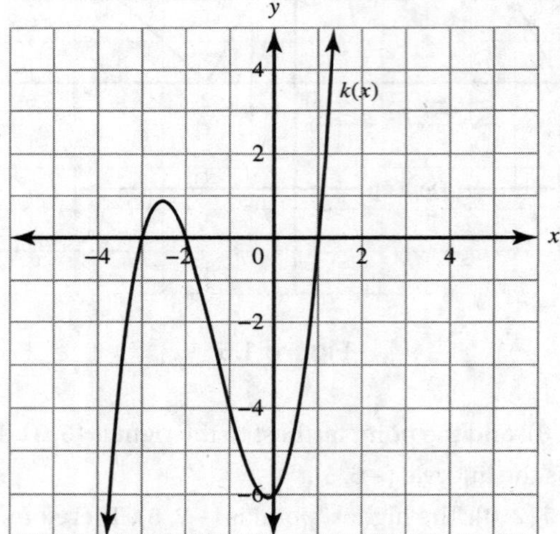

Figure 1.7

There are arrowheads on both endpoints. This means the graph continues in those directions forever. Since there is no left or right endpoint, the domain is all real numbers, which can be written in interval notation as $(-\infty, \infty)$. Since there is no lowest or highest endpoint, the range is all real numbers, which is written in interval notation as $(-\infty, \infty)$.

Example

State the domain and range of each of the functions found in Figures 1.8 and 1.9.

1.

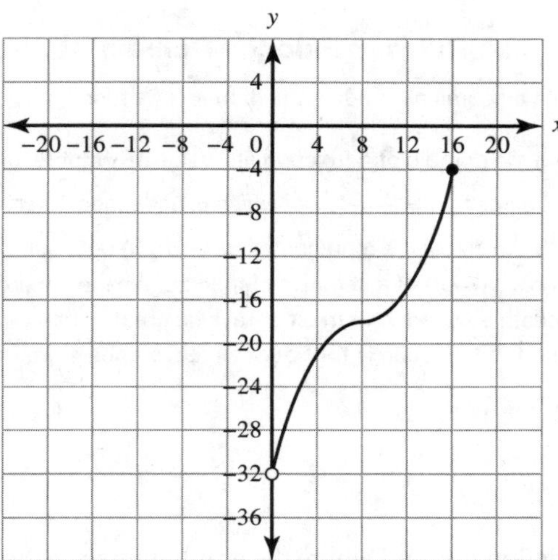

Figure 1.8

2.

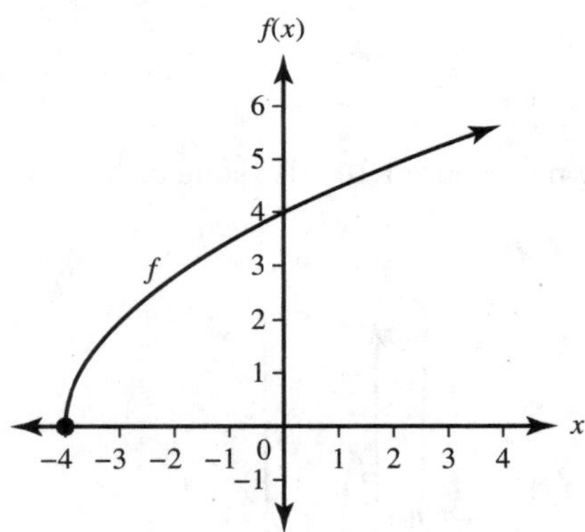

Figure 1.9

Solution

1. The point farthest to the left is (0, −32) and has an open circle. The point farthest to the right is (16, −4) and has a closed circle. Therefore, the domain is $0 < x \leq 16$, which can be written as the interval (0, 16].

   The lowest point is (0, −32) and has an open circle. The highest point is (16, −4) and has a closed circle. Therefore, the range is $-32 < y \leq -4$, which can be written as the interval (−32, −4].

2. The point farthest to the left is (−4, 0) and has a closed circle. There is no point farthest to the right since there is an arrowhead on the right side of the graph. Therefore, the domain is $-4 \leq x$ or $x \geq -4$, which can be written as the interval $[-4, \infty)$.

The lowest point is $(-4, 0)$ and has a closed circle. There is no highest point since there is an arrowhead on the right side of the graph. Therefore, the range is $0 \le y$ or $y \ge 0$, which can be written as the interval $[0, \infty)$.

Algebraically, the domain can be calculated by considering the input values that would make the function not exist or be undefined.

Rules to Find the Domain

There are two key rules to follow when algebraically finding the domain of a function.

1. Do not divide by zero. Exclude from the domain of a function any value of x that results in a 0 in the denominator. For example, if $f(x) = \frac{x+4}{x-3}$, then $x \neq 3$ since $f(3) = \frac{3+4}{3-3} = \frac{7}{0}$, which is undefined. Therefore, the domain of the function is the set of all real numbers except 3, which can be represented as $(-\infty, 3) \cup (3, \infty)$.
2. Do not take the square root (or any even root) of a negative number. Since the square root of a negative number is not a real number, the quantity underneath a square root must always be greater than or equal to 0. If $f(x) = \sqrt{x+5}$, then $x + 5 \ge 0$ and $x \ge -5$. The domain of the function is the set of all real numbers greater than or equal to -5, which can be represented as $[-5, \infty)$.

Intercepts

The x-intercept of the graph of a function is found where the graph intersects the x-axis. At this point, the output value is zero and the corresponding input values are said to be the zeros of the function.

When graphically finding the zeros of a function, look to see where the graph intersects the x-axis.

The y-intercept of the graph of a function is found where the graph intersects the y-axis. At this point, the input value is zero.

Example

Find the zeros of the function $f(x)$ shown in Figure 1.10. Also state the coordinate of the y-intercept.

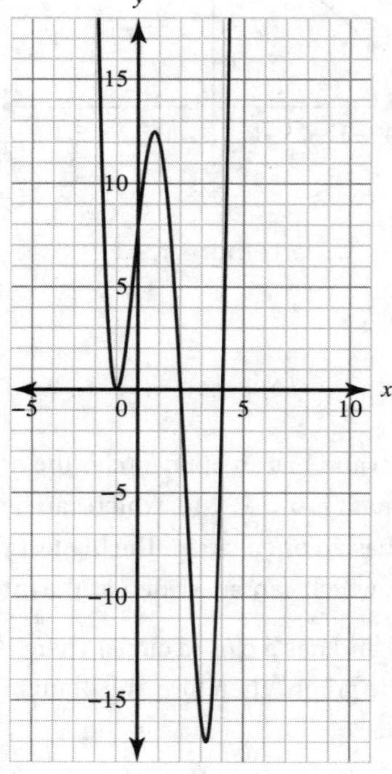

Figure 1.10

Solution

To find the zeros, locate the x-intercepts. The x-intercepts are the points $(-1, 0)$, $(2, 0)$, and $(4, 0)$. Therefore, the zeros are $x = -1$, 2, and 4.

The y-intercept is the point $(0, 8)$.

To find the zeros of a function algebraically, set the equation equal to zero and solve for x. The x-values will represent the zeros of the function.

To find the y-intercept of a function algebraically, substitute 0 for the input value and solve for y. The y-value will represent the y-coordinate of the y-intercept.

Example

Find the zeros of each of the following functions.

1. $f(x) = 2x - 48$
2. $g(x) = (x - 3)(x + 5)(x + 13)$

Solution

1. To find the zeros of the function, first set the equation equal to 0. Solve for x by adding 48 to both sides of the equation and then dividing both sides by 2.

$$2x - 48 = 0$$
$$2x = 48$$
$$x = 24$$

2. First set the equation equal to 0. To solve for x, set each of the factors equal to 0 since the only way a product can equal 0 is if at least one of the terms in the parentheses is 0.

$$(x - 3)(x + 5)(x + 13) = 0$$

$$(x - 3) = 0 \qquad (x + 5) = 0 \qquad (x + 13) = 0$$
$$x = 3 \qquad x = -5 \qquad x = -13$$

Example

Find the y-intercept of each of the following functions.

1. $f(x) = 2x - 48$
2. $g(x) = (x - 3)(x + 5)(x + 13)$

Solution

1. To find the y-intercept of a function, substitute 0 for the input value and solve for y.

$$f(0) = 2(0) - 48 = 0 - 48 = -48 = y$$

The y-intercept has the coordinates $(0, -48)$.

2. Substitute 0 for the input value, and solve for y.

$$g(0) = (0 - 3)(0 + 5)(0 + 13) = (-3)(5)(13) = -195 = y$$

The y-intercept has the coordinates $(0, -195)$.

Notice in the previous two examples that a function can have multiple zeros or *x*-intercepts but only one *y*-intercept. This is true for every function since a function can have one and only one output for any input. Since the input for a *y*-intercept is always 0, that 0 can correspond to only one *y*-value in order to remain a function.

Remember that a function can have many *x*-intercepts or zeros but can have only one *y*-intercept.

This rule can easily be visualized using a graph and is called the vertical line test. The vertical line test states that an equation represents a function if no vertical line intersects the graph of the equation in more than one point.

The graph in Figure 1.11 represents a function since any vertical line that can be drawn intersects the graph in at most one point. The graph in Figure 1.12 is not a function since it fails the vertical line test.

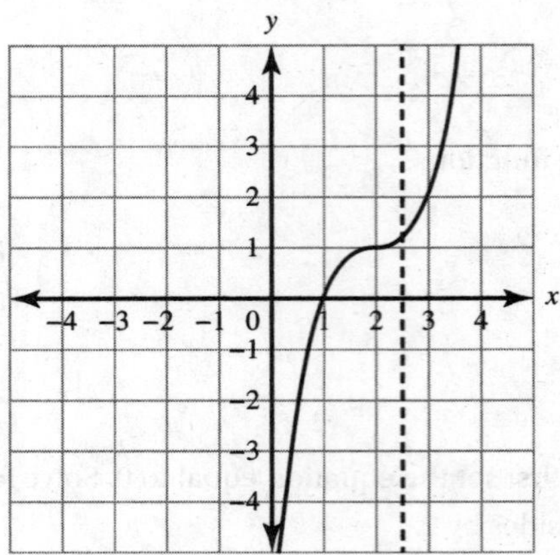

Figure 1.11

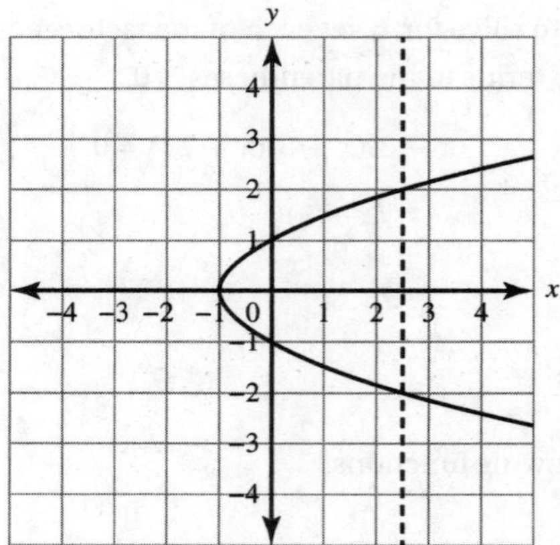

Figure 1.12

Additional Behaviors of Functions

The last general behaviors that will be discussed about functions can be easily viewed from a function's graph and help describe its shape. These behaviors can also be tested algebraically and are a major topic of conversation in calculus.

Increasing/Decreasing

A function is increasing over an interval of its domain if, as the input values increase, the output values always increase. In other words, for all a and b in the interval of the domain, if $a < b$, then $f(a) < f(b)$.

A perfect example of an increasing function is the taxi cost equation, $f(x) = 2.00 + 1.50x$. As time increases, the cost of the taxi continues to increase. As shown in Figure 1.4, as you follow the path of the graph from left to right, the graph continues to travel up.

Alternatively, a function is decreasing over an interval of its domain if, as the input values increase, the output values always decrease. In other words, for all a and b in the interval, if $a < b$, then $f(a) > f(b)$.

A real-world example that models a decreasing function is the value of a new car. After you buy a car, as time increases, the value of the car continues to decrease as the car is said to depreciate.

Figure 1.13 shows the graph of an increasing function, and Figure 1.14 shows the graph of a decreasing function.

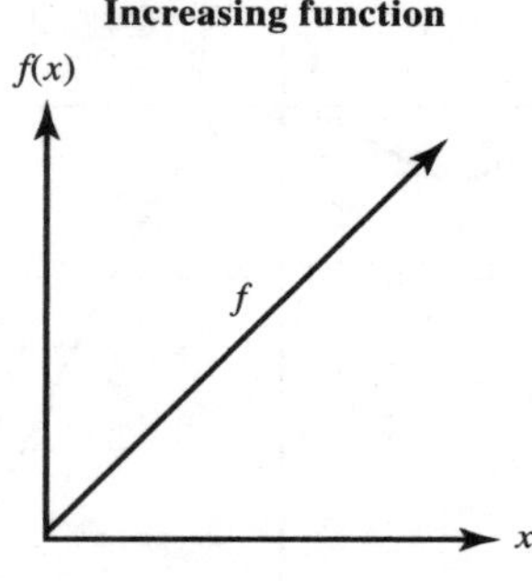

Figure 1.13

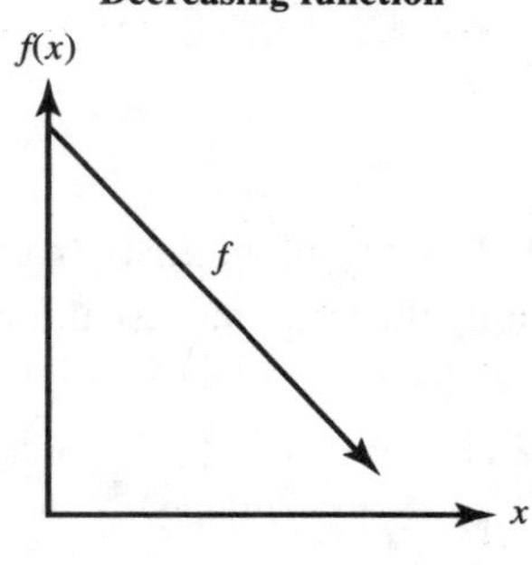

Figure 1.14

Note that a graph of a function can have both increasing and decreasing intervals over the domain.

Example

Using the graph of the function shown in Figure 1.15, identify the intervals where the function is increasing and the intervals where it is decreasing.

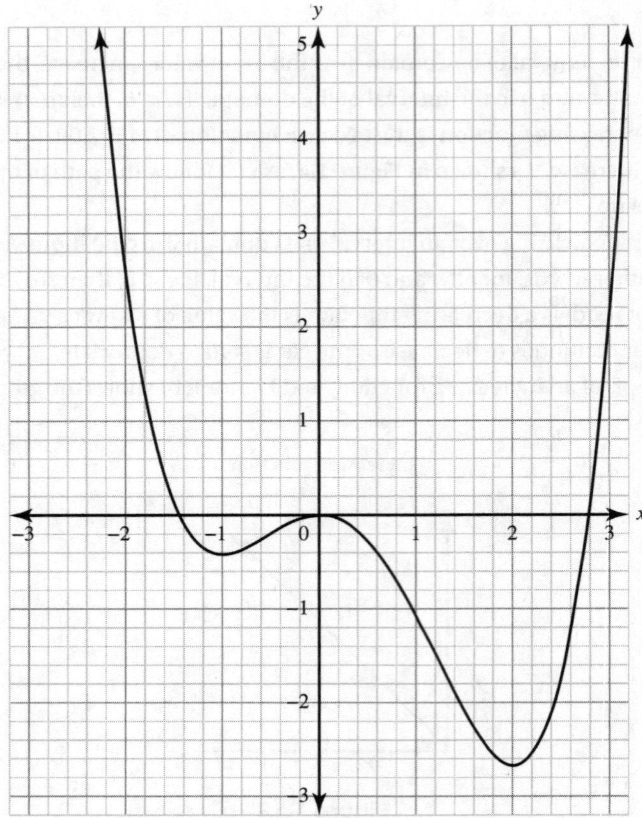

Figure 1.15

✓ Solution

Begin at the left endpoint of the graph, and follow its path to the right endpoint of the graph. Note that the function changes behavior more than once. When stating the intervals, use the x-values since they represent the independent variable.

The function is decreasing over the intervals $(-\infty, -1] \cup (0, 2)$.

The function is increasing over the intervals $(-1, 0) \cup (2, \infty)$.

Concavity

Unlike linear functions, the graphs of other functions that will be discussed are not only continuous but are also smooth and appear to bend, sometimes upward or downward. This is known as the concavity of a graph, as shown in Figure 1.16.

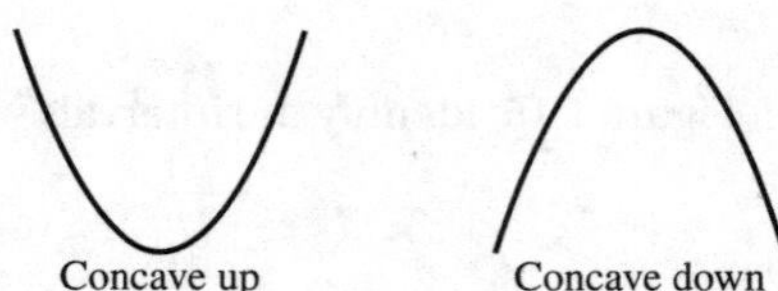

Figure 1.16

A more formal definition of concavity now leads to a discussion of rates of change.

1.2 Rates of Change

A rate is a comparison between two different quantities.

Average Rate of Change

The average rate of change is the change in the value of one quantity divided by the change in the value of another quantity. Common examples of average rate of change include finding the average speed $= \frac{\text{change in distance}}{\text{change in time}}$ and the slope of a line segment $= \frac{\Delta y}{\Delta x} = \frac{y_2 - y_1}{x_2 - x_1}$.

To find the average rate of change, either two points or two sets of inputs with their corresponding outputs are needed. Then the differences can be calculated.

Finding the average rate of change between an initial point and a final point is equivalent to finding the slope of the segment (or secant) between those points.

Example

A car travels between two cities according to the distance-time graph shown in Figure 1.17.

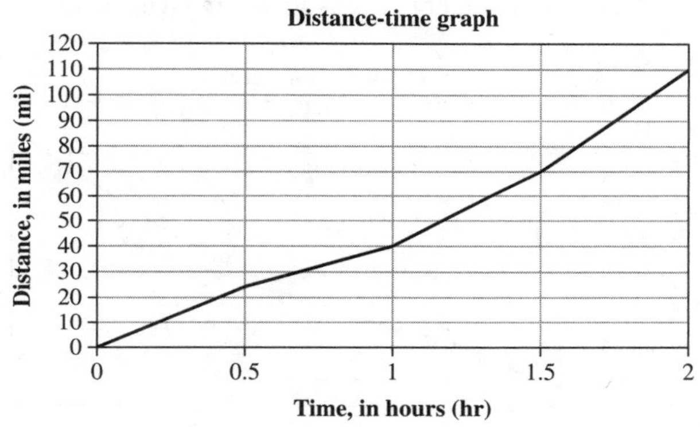

Figure 1.17

(a) What is the car's average speed for the first hour?
(b) What is the car's average speed for the entire trip?
(c) How does the average speed in part (a) compare with the average speed in part (b)?

Solution

(a) Initially, the car has traveled 0 miles in 0 hours. After 1 hour, the car has traveled 40 miles. The average speed for the first hour is

$$\frac{40 - 0}{1 - 0} = 40 \text{ mph}$$

(b) Initially, the car has traveled 0 miles in 0 hours. By the end of the trip, the car has traveled 110 miles in 2 hours. The average speed for the entire trip is

$$\frac{110 - 0}{2 - 0} = 55 \text{ mph}$$

(c) Figure 1.18 shows the two secants that represent the solutions to parts (a) and (b). The slope of the secant for part (b) is steeper than the slope of the secant for part (a). In other words, over the course of the entire trip, more distance was gained after the initial hour.

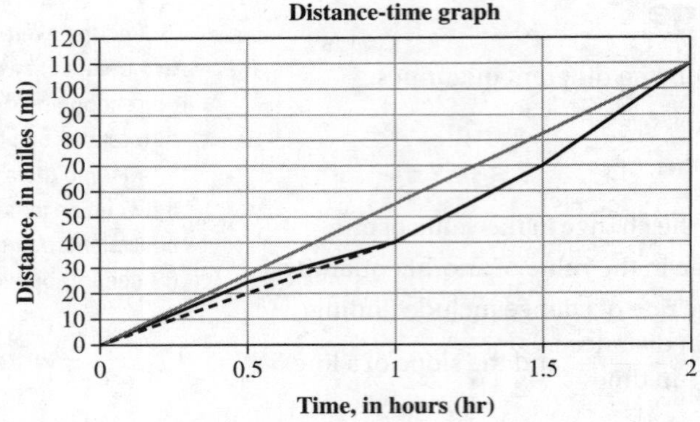

Figure 1.18

The average speed for the entire trip does not give all the details for the speeds along the way. The average speed is just that, an average or an overall idea of the rate of change between a starting point and an ending point.

Finding the speed at a specific instant in time is something entirely different. Graphically at one point, such as $t = 1$ hour, the secant line would change to a tangent line, as shown in Figure 1.19.

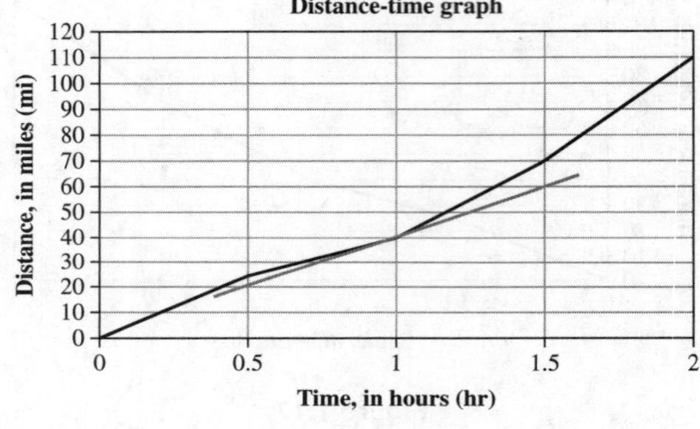

Figure 1.19

Finding the rate of change at this value is an issue since the previous slope formula needs two points and only one point is known for the tangent line.

The rate of change at a point can be approximated by the average rate of change of the function over small intervals containing the point, if such values exist.

Example

Given the graph of function $f(x) = x^2$ shown in Figure 1.20, find the average rate of change at point $P(4, 16)$.

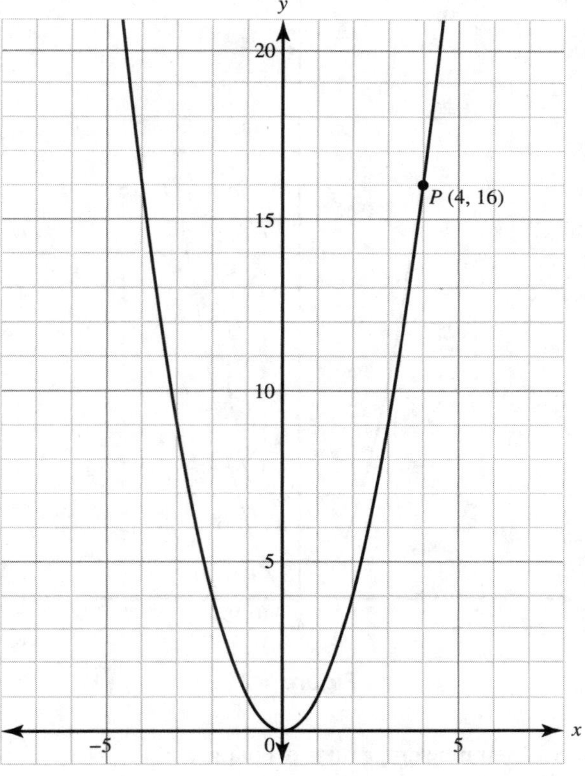

Figure 1.20

Solution

It is impossible to find the average rate of change at one point since two points are needed. To approximate, we will choose points to serve as endpoints with point P to create secant lines. As the endpoints get closer and closer to point P, the slopes of the secant lines should approach one value. This value will be the approximation of the average rate of change at point P. As seen in Figure 1.21, as the endpoints get closer to point P, the secant lines begin to look the same.

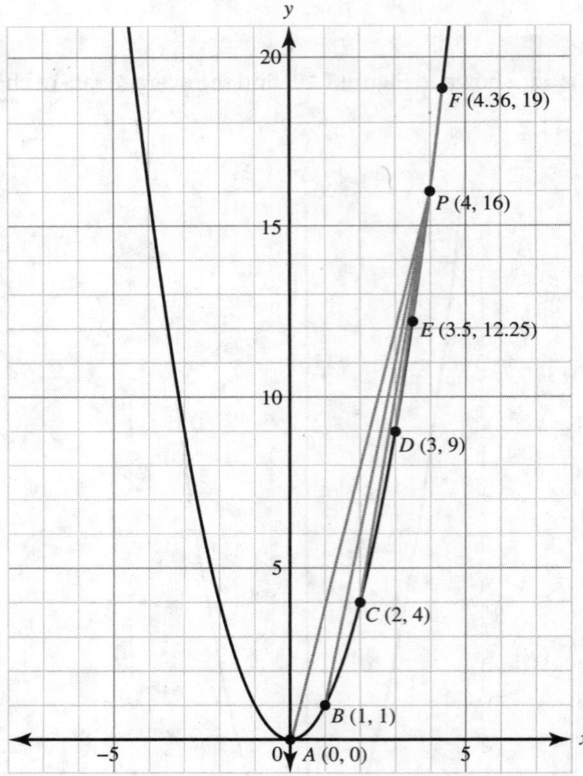

Figure 1.21

Table 1.3 will help to organize the different average rates of change.

Table 1.3

Point	P(4, 16)	Average Rate of Change
$A(0, 0)$	$P(4, 16)$	$\frac{16-0}{4-0}=\frac{16}{4}=4$
$B(1, 1)$	$P(4, 16)$	$\frac{16-1}{4-1}=\frac{15}{3}=5$
$C(2, 4)$	$P(4, 16)$	$\frac{16-4}{4-2}=\frac{12}{2}=6$
$D(3, 9)$	$P(4, 16)$	$\frac{16-9}{4-3}=\frac{7}{1}=7$
$E(3.5, 12.25)$	$P(4, 16)$	$\frac{16-12.25}{4-3.5}=\frac{3.75}{0.5}=7.5$
$F(4.36, 19)$	$P(4, 16)$	$\frac{16-19}{4-4.36}=\frac{-3}{-0.36}=8.\overline{333}$

For x-values smaller than 4, the slopes approach 7.5. For x-values greater than 4, the values approach $8.\overline{333}$.

Since the equation of the function is given, x-values that are even closer to 4 from both the left and right directions can be substituted into the function as shown in Table 1.4.

Table 1.4

Point	P(4, 16)	Average Rate of Change
$G(3.9, 15.21)$	$P(4, 16)$	$\frac{16-15.21}{4-3.9}=\frac{0.79}{0.1}=7.9$
$H(3.95, 15.6025)$	$P(4, 16)$	$\frac{16-15.6025}{4-3.95}=\frac{0.3975}{0.05}=7.95$
$I(3.99, 15.9201)$	$P(4, 16)$	$\frac{16-15.9201}{4-3.99}=\frac{0.0799}{0.01}=7.99$
$J(4.001, 16.008001)$	$P(4, 16)$	$\frac{16-16.008001}{4-4.001}=\frac{-0.008001}{-0.001}=8.001$
$K(4.01, 16.0801)$	$P(4, 16)$	$\frac{16-16.0801}{4-4.01}=\frac{-0.0801}{-0.01}=8.01$
$L(4.1, 16.81)$	$P(4, 16)$	$\frac{16-16.81}{4-4.1}=\frac{-0.81}{-0.1}=8.1$

Coming in from the left and getting closer to point $P(4, 16)$, points G, H, and I have slopes that approach 7.99. Coming in from the right and getting closer to point $P(4, 16)$, points J, K, and L have slopes that approach 8.001. Overall, it seems that the slopes are approaching the value of 8. Therefore, the approximate average rate of change at point P is 8.

> The rates of change at two points can be approximated by calculating average rate of change approximations over sufficiently small intervals containing each point, if such values exist.

Example

A small local restaurant sponsored a contest to name their new special dish. Patrons submitted their entries between noon ($t = 0$) and 10 P.M. ($t = 10$). The number of entries t hours after noon is modeled by the function N for $0 \leq t \leq 10$. Values of $N(t)$, at various times t, are shown in Table 1.5. Use the data in the table to approximate the rate at which entries were being deposited at time $t = 8$.

Table 1.5

t (hours)	0	3	6	7	9	10
$N(t)$ (number of entries)	0	25	43	50	79	125

Solution

Since $t = 8$ is between $t = 7$ and $t = 9$, which are included in the table, find the average rate of change between these two points to use as an approximate rate of change for the entries being deposited at time $t = 8$.

Calculate the approximate average rate of change at $t = 8$:

$$\frac{N(9)-N(7)}{9-7}=\frac{79-50}{2}=\frac{29}{2}=14.5$$

As demonstrated in the previous examples, a rate of change quantifies how two quantities vary together.

A positive rate of change indicates that as one quantity increases or decreases, the other quantity does the same. Algebraically, when calculating the average rate of change, the value simplifies to a positive number. In the previous example as time increased from $t = 0$ to $t = 10$, the number of entries continuously increased from $N = 0$ to $N = 125$. In other words, as the day went by, the number of entries increased as more were collected. A positive rate of change can be seen on the graph shown in Figure 1.22.

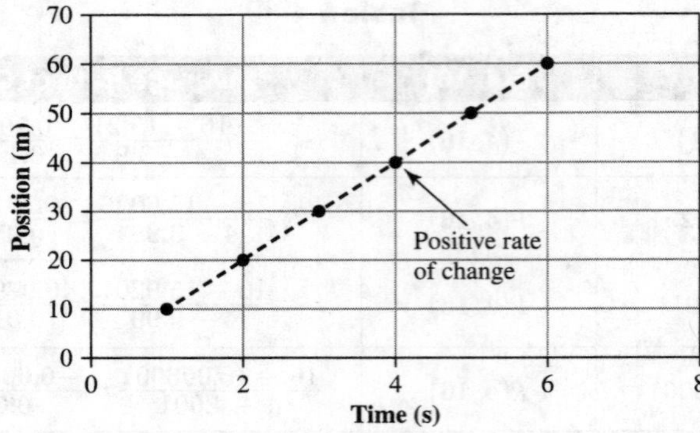

Figure 1.22

A negative rate of change indicates that as one quantity increases, the other decreases. For example, a freshman in college begins the semester with \$1,000 in their savings account. Each week they take out \$75 to pay for their expenses. After one week, the student has \$925. The average rate of change is $\frac{925-1000}{1-0}=\frac{-75}{1}=-75$. Graphically, a negative rate of change can be seen in Figure 1.23.

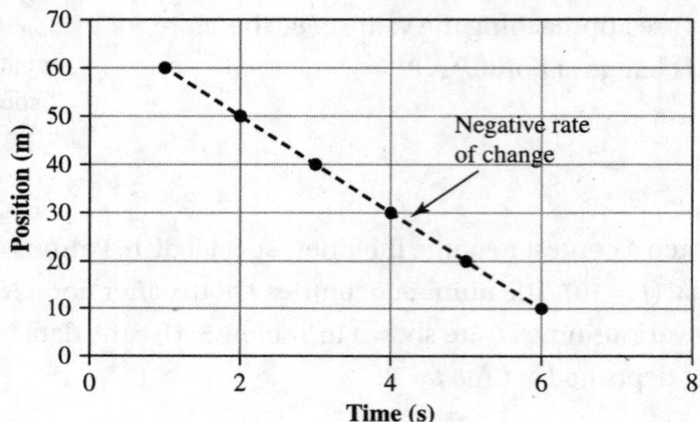

Figure 1.23

Unlike Figures 1.22 and 1.23, not all graphs are linear. Each type of function that will be discussed in later chapters has its own unique graphical behaviors. One such behavior is known as concavity. The concavity of a function describes how the graph of the function curves and bends as shown in Figure 1.24.

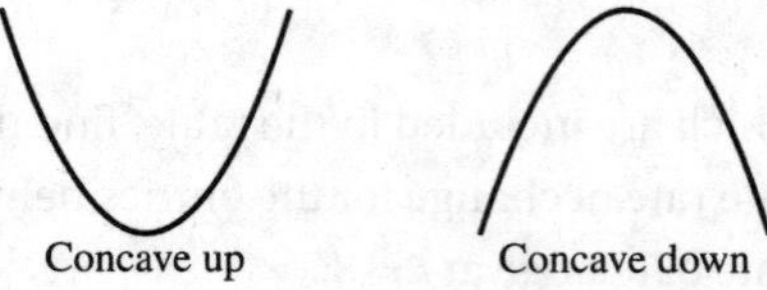

Figure 1.24

More formally, the graph of a function is concave up on intervals in which the rate of change is increasing. To illustrate this, consider the function $f(x) = (x - 4)^2 + 2$, as shown in Figure 1.25 and Table 1.6.

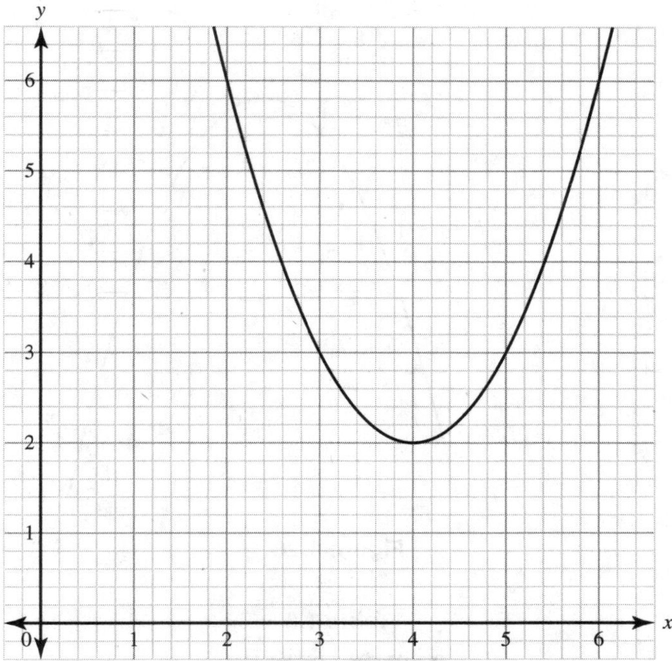

Figure 1.25

Table 1.6

x	$f(x)$
2	6
3	3
4	2
5	3
6	6

Table 1.7 calculates the average rates of change between the pairs of points.

Table 1.7

Point 1	Point 2	Average Rate of Change
(2, 6)	(3, 3)	−3
(3, 3)	(4, 2)	−1
(4, 2)	(5, 3)	1
(5, 3)	(6, 6)	3

As x is increasing, the average rate of change is also increasing. This illustrates why the graph of the function is concave up in Figure 1.25.

Alternatively, the graph of a function is concave down on intervals in which the rate of change is decreasing. To illustrate this, consider the function $f(x) = -(x - 4)^2 + 2$, as shown in Figure 1.26 and Table 1.8.

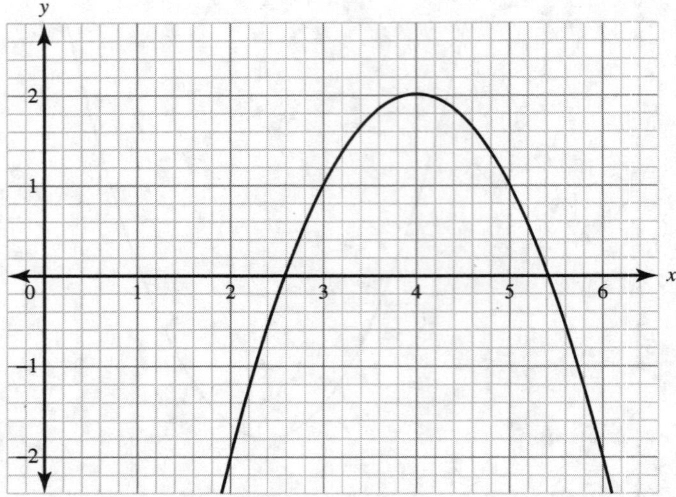

Figure 1.26

Table 1.8

x	$f(x)$
2	−2
3	1
4	2
5	1
6	−2

Table 1.9 calculates the average rates of change between the pairs of points.

Table 1.9

Point 1	Point 2	Average Rate of Change
(2, −2)	(3, 1)	3
(3, 1)	(4, 2)	1
(4, 2)	(5, 1)	−1
(5, 1)	(6, −2)	−3

As x is increasing, the average rate of change is decreasing. This illustrates why the graph of the function is concave down in Figure 1.26.

1.3 Rates of Change in Linear Functions

Many different types of functions will be discussed and analyzed in this course. Calculating the average rate of change can be used to help identify some of them. Consider Table 1.10, which shows the ordered pairs of the linear function $y = f(x) = -4x + 7$.

Table 1.10

x	$y = f(x)$	Average Rate of Change $= \frac{\Delta y}{\Delta x}$
−2	15	
		$= \frac{11 - 15}{-1 - (-2)} = \frac{-4}{1} = -4$
−1	11	
		$= \frac{7 - 11}{0 - (-1)} = \frac{-4}{1} = -4$
0	7	
		$= \frac{3 - 7}{1 - 0} = \frac{-4}{1} = -4$
1	3	
		$= \frac{-1 - 3}{2 - 1} = \frac{-4}{1} = -4$
2	−1	
		$= \frac{-5 - (-1)}{3 - 2} = \frac{-4}{1} = -4$
3	−5	

As the value of x increases by 1, the value of y decreases by 4. In other words, for every consecutive integer ordered pair, the average rate of change is −4. This would also be true over any interval length. Consider the ordered pairs (−2, 15) and (3, −5). The average rate of change is

$$\frac{\Delta y}{\Delta x} = \frac{-5 - 15}{3 - (-2)} = \frac{-20}{5} = -4$$

It is not a coincidence that the average rate of change of the linear function $f(x) = -4x + 7$ is the slope of the linear function.

Rate of Change

In general, all linear functions have a constant rate of change, $m = \frac{\Delta y}{\Delta x}$, and can be written in the slope-intercept form $y = f(x) = mx + b$, where b is the y-coordinate of the y-intercept. Only linear functions have a constant average rate of change. The average rate of change can be used to determine whether a particular function is linear or not. Functions that are not linear are said to be nonlinear.

Example

A strain of bacteria is placed into a Petri dish and allowed to grow. The data collected are shown in Table 1.11. The population is measured in kilograms and the time in hours. The data in Table 1.12 represent the maximum number of hours an individual should work during the week. For both tables, use the average rate of change to determine whether the function is linear. Graph each to check.

(a)

Table 1.11

Time (hours), x	Population (kg), y
0	0.09
1	0.12
2	0.16
3	0.22
4	0.29
5	0.39

(b)

Table 1.12

Age, x	Time (hours), y
20	50
30	47.5
40	45
50	42.5
60	40
70	37.5

✓ **Solution**

(a) Find the average rate of change for each pair of consecutive ordered pairs to see if it is constant, as shown in Table 1.13.

Table 1.13

Time (hours), x	Population (kg), y	Average Rate of Change $= \frac{\Delta y}{\Delta x}$
0	0.09	
		$= \frac{0.12 - 0.09}{1 - 0} = \frac{0.03}{1} = 0.03$
1	0.12	
		$= 0.04$
2	0.16	
		$= 0.06$
3	0.22	
		$= 0.07$
4	0.29	
		$= 0.10$
5	0.39	

Since the average rate of change is not constant, we know that the function is not linear for the data in Table 1.11. Because the average rate of change is increasing as x increases, we say that the function is increasing at an increasing rate. In other words, not only is the population increasing over time, but it is also growing more rapidly as time passes. We also know that when we look at the graph of this function in Figure 1.27, the graph will be concave up.

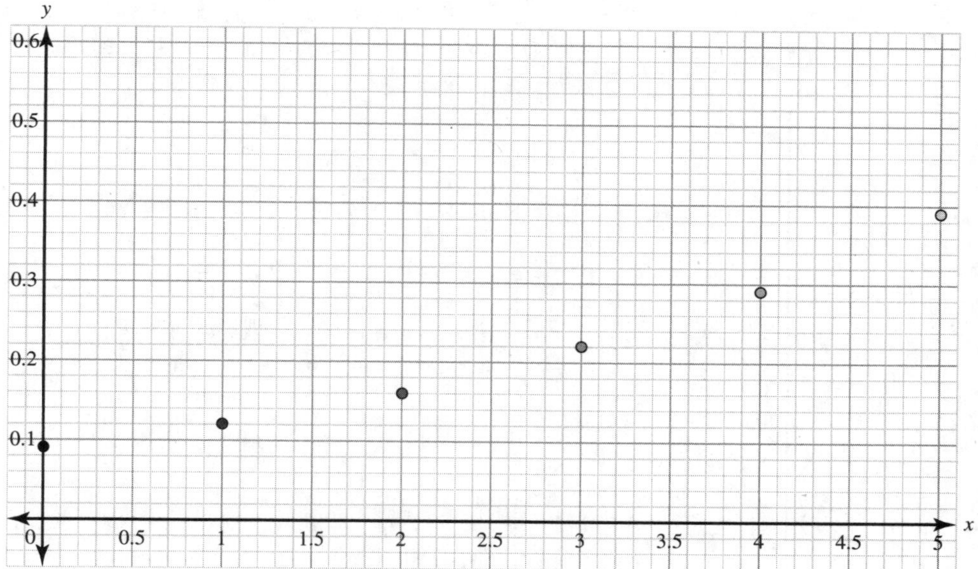

Figure 1.27

(b) Find the average rate of change for each pair of consecutive ordered pairs to see if it is constant, as shown in Table 1.14.

Table 1.14

Age, x	Time (hours), y	Average Rate of Change $= \frac{\Delta y}{\Delta x}$
20	50	
		$= \frac{47.5 - 50}{30 - 20} = \frac{-2.5}{10} = -0.25$
30	47.5	
		$= -0.25$
40	45	
		$= -0.25$
50	42.5	
		$= -0.25$
60	40	
		$= -0.25$
70	37.5	

Since the average rate of change is constant, we know that the function is linear for the data in Table 1.12. When the data are graphed as in Figure 1.28, we can see that it resembles a straight line.

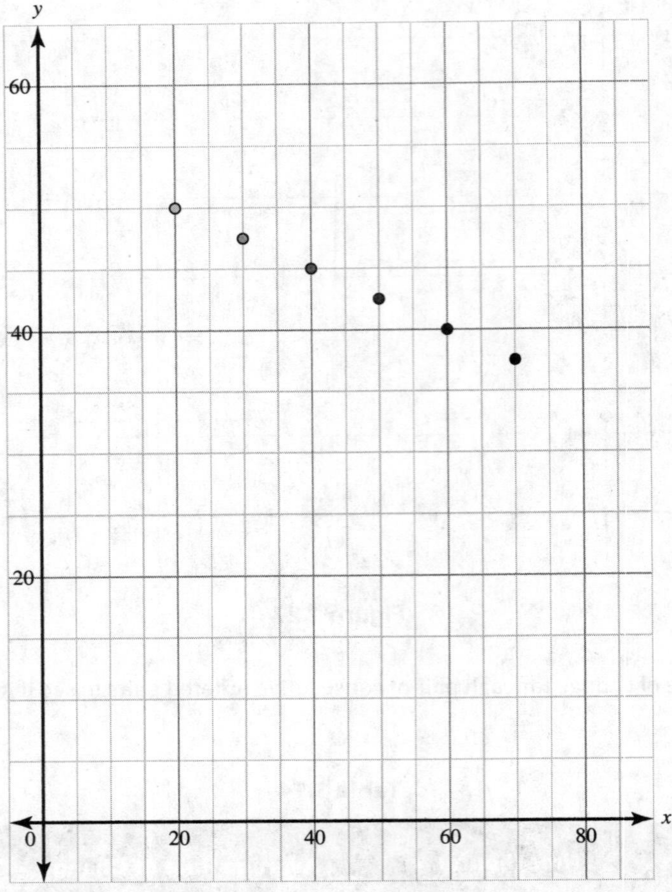

Figure 1.28

We can find the equation of the linear function using the point-slope formula $(y - y_1) = m(x - x_1)$ and then solving for y. Since the average rate of change is -0.25, the slope is $m = -0.25$. Choose any point to substitute in for x_1 and y_1:

$$y - 50 = -0.25(x - 20)$$
$$y = -0.25x + 55$$

Since the average rate of change of the linear function in Table 1.12 is constant, the rate of change of the average rates of change of a linear function is zero. That is why the graph is neither concave up nor concave down.

Over any length interval, the average rate of change for a linear function is constant. The rate of change of the average rates of change of a linear function is zero.

1.4 Rates of Change in Quadratic Functions

A quadratic function is a function of the form $f(x) = ax^2 + bx + c$, where a, b, and c are real numbers and $a \neq 0$. The domain of a quadratic function is the set of all real numbers.

As seen in the previous section, the rate of change of a linear function is constant. The rate of change of a quadratic function is not constant; there are no straight-line segments on a parabola, which is the graph of a quadratic function. To calculate the rate of change, determine the average rate of change by finding the slope of the segment that connects two points on the parabola. This segment is called the secant line.

Example

Consider the quadratic function $y = x^2$ and its graph shown in Figure 1.29.

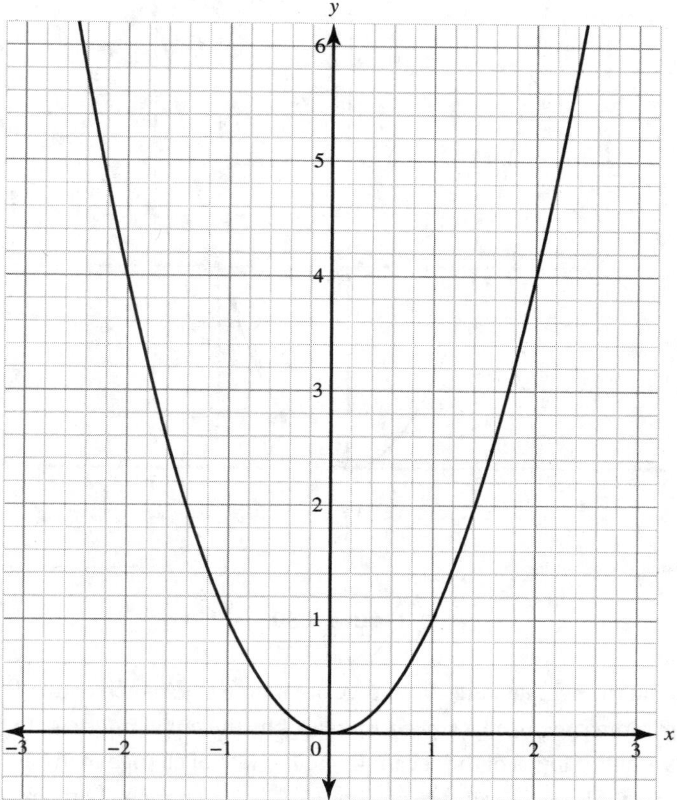

Figure 1.29

Table 1.15 shows the average rate of change between points that have consecutive, even-length input values.

Table 1.15

Points	Average Rate of Change = Slope of Secant Line
(0, 0) and (1, 1)	$= \frac{1-0}{1-0} = 1$
(1, 1) and (2, 4)	$= \frac{4-1}{2-1} = 3$
(2, 4) and (3, 9)	$= \frac{9-4}{3-2} = 5$

The graph in Figure 1.30 gives a visual representation of the slope of the secant lines.

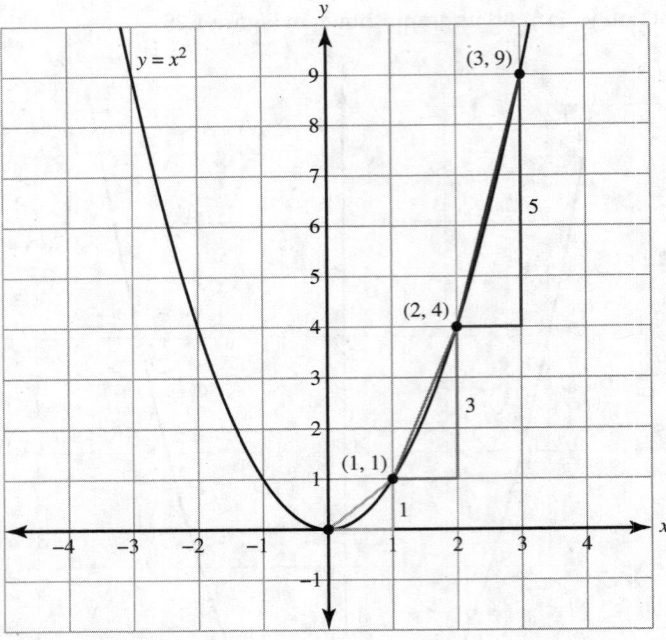

Figure 1.30

As x increases, we can see that the average rates of change are increasing, which coincides with the graph of the parabola being concave up. More specifically, since x increases by the equal interval of 1, we can see a more specific pattern emerging with the slopes of the secant lines. The rate of change of the average rates of change of this quadratic function is constant, specifically the constant 2. This will happen with any quadratic function.

Example

Which of the following functions represents a quadratic equation? State if the graph of the parabola would be concave up or concave down.

(a)

Table 1.16

x	$f(x)$
−3	−30
−2	−14
−1	−4
0	0
1	−2
2	−10

(b)

Table 1.17

x	$g(x)$
−3	76
−2	55
−1	36
0	19
1	4
2	−9

(c)

Table 1.18

x	$h(x)$
−3	81
−2	16
−1	1
0	0
1	1
2	16

✓ **Solution**

Since all the x-values in all three tables increase by 1, find the average rates of change between each pair of points.

(a)

Pair of Points from $f(x)$	Average Rate of Change
(−3, −30) and (−2, −14)	16
(−2, −14) and (−1, −4)	10
(−1, −4) and (0, 0)	4
(0, 0) and (1, −2)	−2
(1, −2) and (2, −10)	−8

The average rates of change are not constant, but the rate of change between each is a constant −6. Therefore, $f(x)$ is a quadratic function, and its parabola would be concave down.

(b)

Pair of Points from $g(x)$	Average Rate of Change
$(-3, 76)$ and $(-2, 55)$	-21
$(-2, 55)$ and $(-1, 36)$	-19
$(-1, 36)$ and $(0, 19)$	-17
$(0, 19)$ and $(1, 4)$	-15
$(1, 4)$ and $(2, -9)$	-13

The average rates of change are not constant, but the rate of change between each is a constant 2. Therefore, $g(x)$ is a quadratic function, and its parabola would be concave up.

(c)

Pair of Points from $h(x)$	Average Rate of Change
$(-3, 81)$ and $(-2, 16)$	-65
$(-2, 16)$ and $(-1, 1)$	-15
$(-1, 1)$ and $(0, 0)$	-1
$(0, 0)$ and $(1, 1)$	1
$(1, 1)$ and $(2, 16)$	15

The average rates of change are not constant, and the rate of change between each is also not constant. Therefore, $h(x)$ is neither a linear nor a quadratic function.

Multiple-Choice Questions

1. Which of the following functions has a domain of $x > 0$?

   (A) $f(x) = x^2$
   (B) $f(x) = \sqrt{x}$
   (C) $f(x) = \log x$
   (D) $f(x) = e^x$

2. Which of the following are functions?

   I. $y - x = 17 + 2x$
   II. $x^2 + y^2 = 64$
   III. $x - y^2 = 81$
   IV. $x^2 + y = 4$

   (A) I only
   (B) I and III only
   (C) I and IV only
   (D) I, II, III, and IV

3. Find the average rate of change of $f(x) = 3x^2 - 2$ from $x = 2$ to $x = 6$.

   (A) 24
   (B) 18
   (C) $\frac{1}{18}$
   (D) $\frac{1}{24}$

4. Which of the following linear functions is increasing?

   (A) $s(t) = 4 - \frac{1}{2}t$
   (B) $h(z) = 10$
   (C) $g(x) = -3x + 2$
   (D) $f(x) = -7 + 0.9x$

5. If a represents the zero of the function $f(x) = (x - 3)^2$, evaluate $g(a)$ if $g(t) = 10 - 2(t - 1)^3$.

   (A) −6
   (B) −3
   (C) 1
   (D) 3

6. As a pot of boiling water cools, the temperature of the water is recorded and placed in the table below, where time t is measured in minutes and temperature $F(t)$ is measured in degrees Fahrenheit. Using the data in the table, which of the following best approximates the rate at which the temperature of the water is changing at time $t = 3.5$?

t (minutes)	0	2	5	9	10
$F(t)$ (degrees Fahrenheit)	150.8	140	125.6	111.2	109.4

   (A) −5.04°F/min
   (B) −4.8°F/min
   (C) −4.4°F/min
   (D) −4.14°F/min

7. Given the table of values below, which of the following best describes the behavior of the graph of the function in the interval $1 \leq x \leq 6$?

x	1	2	3	4	5	6
y	11	6	−5	−22	−45	−74

   (A) Increasing, concave down
   (B) Increasing, concave up
   (C) Decreasing, concave down
   (D) Decreasing, concave up

8. Cailey loves to bake and decides to open her own cookie stand, named Cailey's Cookies. Her daily fixed costs are \$120 a day, and each cookie costs \$0.99 to bake. Which of the following represents a model that expresses the cost, C, of baking m cookies in a day?

   (A) $C(m) = 0.99m + 120$
   (B) $C(m) = 0.99x + 120$
   (C) $C(m) = 120m + 0.99$
   (D) $C(m) = 0.04125m + 5$

9. For the function $f(x) = x^2 - 4x + 1$, evaluate $f(x + h)$.

   (A) $x^2 - 4x + 1 + h$
   (B) $h^2 - 4h + 1$
   (C) $x^2 + h^2 + 2xh - 4x - 4h + 1$
   (D) $x^2 + h^2 - 4x - 4h + 1$

10. Which of the following functions has the domain $(-\infty, \infty)$, $x \neq -9$, $x \neq 4$?

   (A) $f(x) = \sqrt{x - 36}$
   (B) $g(x) = \dfrac{x - 3}{x^2 + 5x - 36}$
   (C) $h(x) = \dfrac{x}{(x + 4)(x - 9)}$
   (D) $i(x) = x^2 + 5x - 36$

Answer Explanations

1. **(C)** The domain of each of the functions can be seen from their graphs.

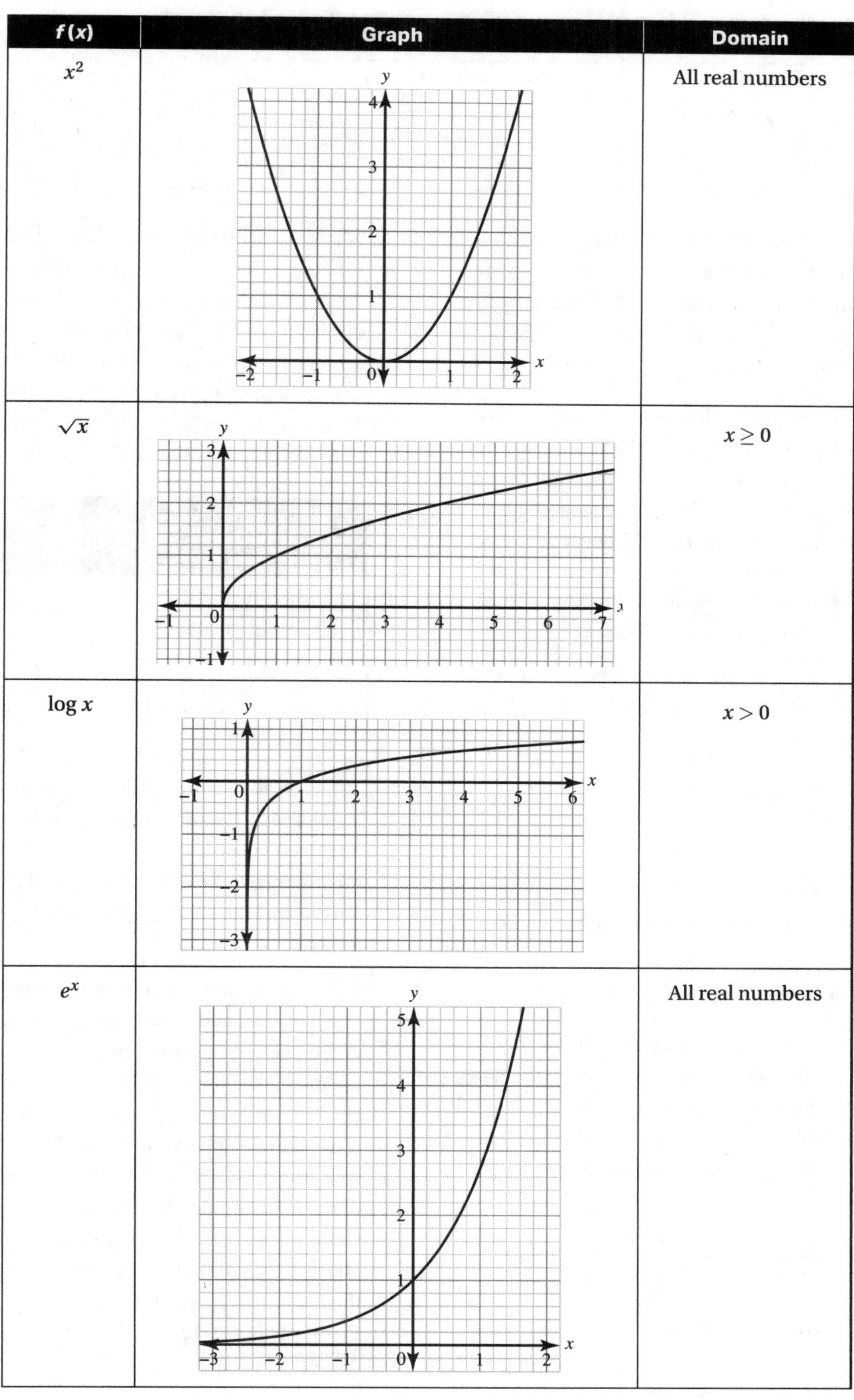

$f(x)$	Graph	Domain
x^2		All real numbers
$\sqrt{x}$		$x \geq 0$
$\log x$		$x > 0$
e^x		All real numbers

2. **(C)** For an equation to be a function, every x can correspond to one and only one y. This can be verified graphically by seeing if the graph passes the vertical line test or algebraically by solving for y and seeing if each y has only one solution. The only issues are II and III since the y is squared. When isolating the y, we extract the square root. This would give two solutions, one positive and one negative.

I. $y - x = 17 + 2x$	II. $x^2 + y^2 = 64$	III. $x - y^2 = 81$	IV. $x^2 + y = 4$
$y = 3x + 17$	$y^2 = 64 - x^2$ $y = \pm\sqrt{64 - x^2}$	$-y^2 = 81 - x$ $y^2 = -81 + x$ $y = \pm\sqrt{-81 + x}$	$y = 4 - x^2$

3. **(A)** To find the average rate of change of a quadratic function, calculate the slope of the secant line of the endpoints provided. At $x = 2$, $y = 10$, and at $x = 6$, $y = 106$. The average rate of change $=$ slope of secant line $= \dfrac{\Delta y}{\Delta x} = \dfrac{106 - 10}{6 - 2} = 24.$

4. **(D)** To determine if a linear function is increasing, decreasing, or constant, determine the slope. A linear function is increasing if the slope is positive, decreasing if the slope is negative, and constant if the slope is zero. All the choices are written in the form $y = mx + b$, where m represents the slope. The only linear function with a positive slope is $f(x) = -7 + 0.9x$.

5. **(A)** First find the value of a. To find a zero, set the equation equal to zero and solve for the independent variable:

$$0 = (x - 3)^2$$
$$\sqrt{0} = \sqrt{(x - 3)^2}$$
$$0 = x - 3$$
$$x = 3 = a$$

To evaluate $g(a)$, substitute $a = 3$ for x into the function $g(x)$:

$$g(a) = g(3) = 10 - 2(3 - 1)^3 = 10 - 2(2)^3 = -6$$

6. **(B)** The rate of change at a point can be approximated by the average rates of change of the function over small intervals containing the point. The smallest interval containing $t = 3.5$ in the table is from $t = 2$ to $t = 5$. Calculate the average rate of change:

$$\frac{125.6 - 140}{5 - 2} = \frac{-14.4}{3} = -4.8°\text{F/min}$$

7. **(C)** Since the y-values are getting smaller as x increases over the interval, the function is decreasing. One way to describe the concavity of the function is to calculate and interpret the average rate of change between intervals of equal length. Since the given table has x increase by 1 each time, calculate the average rate of change between points as shown in the table.

Points Used	Average Rate of Change $= \dfrac{\Delta y}{\Delta x}$
(1, 11) and (2, 6)	−5
(2, 6) and (3, −5)	−11
(3, −5) and (4, −22)	−17
(4, −22) and (5, −45)	−23
(5, −45) and (6, −74)	−29

The average rate of change is not constant, so it is not linear. The rate of change of the average rate of change is a constant −6, so the function is quadratic. Since the average rates of change are decreasing, the parabola is concave down over the interval.

8. **(A)** Since we are asked to write a model for only one day, the fixed costs of \$120 occur only once and represent the initial cost even if zero cookies are made. The total cost depends on how many cookies are baked. The number of cookies baked is the independent variable, m. The \$0.99 represents the cost per cookie and reflects the slope of the linear model. Since it is a constant rate of change, the cost to bake the cookie is not going to change. Using the slope-intercept form of a line, the linear cost model is written as $C(m) = 0.99m + 120$.

9. **(C)** To evaluate $f(x+h)$, substitute $(x+h)$ for x into the given function $f(x) = x^2 - 4x + 1$:

$$\begin{aligned} f(x+h) &= (x+h)^2 - 4(x+h) + 1 \\ &= (x+h)(x+h) - 4x - 4h + 1 \\ &= x^2 + xh + xh + h^2 - 4x - 4h + 1 \\ &= x^2 + 2xh + h^2 - 4x - 4h + 1 \end{aligned}$$

10. **(B)** The domain of a function is the set of input values for which the function is defined and has a corresponding output value. In Choice (A), a radical cannot be a negative number. So the domain for $f(x)$ is $x \geq 36$. For functions like Choices (B) and (C), there cannot be a zero in the denominator. Set each denominator equal to zero and solve for x. Then restrict these values from the domain. In Choice (B), $x^2 + 5x - 36 = 0$, by factoring $(x+9)(x-4) = 0$, and therefore $x \neq -9$, $x \neq 4$. In Choice (C), $(x+4)(x-9) = 0$, which means $x \neq -4$, $x \neq 9$. Choice (D) is a quadratic equation, so its domain is all real numbers.

2

Polynomial Functions

Learning Objectives

In this chapter, you will learn:

- → The different properties of polynomial functions
- → How to graph polynomial functions
- → How to calculate the rate of change of different functions
- → How to write equivalent forms of polynomial functions
- → How to solve polynomial inequalities

A polynomial function, p, is a function that can be written in the form $p(x) = a_n x^n + a_{n-1}x^{n-1} + a_{n-2}x^{n-2} + \ldots + a_1 x + a_0$, where $a_n \neq 0$ and n is a positive integer called the degree of the polynomial. Each member of the sum is a term of the polynomial. The number constants $a_n, a_{n-1}, a_{n-2}, \ldots, a_1, a_0$ are coefficients. The leading term is $a_n x^n$, the leading coefficient is a_n, and the constant term is a_0.

Polynomial functions should be written in descending order, starting with the term of the highest power. Any term missing in the polynomial means that 0 is the coefficient of that term. A constant is considered a polynomial function of degree 0.

Example

Which of the following functions are considered polynomial functions?

1. $f(x) = x^5 + \sqrt{x}$
2. $g(x) = x^3 - \frac{1}{x}$
3. $h(x) = 7x^4 - \frac{1}{2}x^3 + x - 10$

Solution

1. This is not a polynomial function since it contains $\sqrt{x} = x^{1/2}$.
2. This is not a polynomial function since it contains $\frac{1}{x} = x^{-1}$.
3. This is a polynomial function since each of the exponents in $h(x) = 7x^4 - \frac{1}{2}x^3 + x - 10$ is a whole number.

2.1 Polynomial Functions and Rates of Change

In Chapter 1, calculating the rate of change helped to distinguish between linear and quadratic functions and helped discover the concavity of quadratic functions. Since linear functions (degree 1) and quadratic functions (degree 2) are both types of polynomial functions, we will now discover how the rate of change can be applied to polynomial functions of all degrees.

Relative Extrema

A relative extremum (also known as a local extremum) means the existence of a relative maximum or minimum point on a graph. This is a point that looks like it is either the top or bottom of a hill. More formally, where a polynomial function switches between increasing and decreasing, the polynomial function will have a local (relative) maximum or minimum output value.

This behavior can easily be identified in our basic quadratic functions, $f(x) = x^2$ and $g(x) = -x^2$.

For $f(x) = x^2$, shown in Figure 2.1, the graph is decreasing when $x < 0$ and increasing when $x > 0$. At $x = 0$ there is a local minimum.

For $g(x) = -x^2$, shown in Figure 2.2, the graph is increasing when $x < 0$ and decreasing when $x > 0$. At $x = 0$, there is a local maximum.

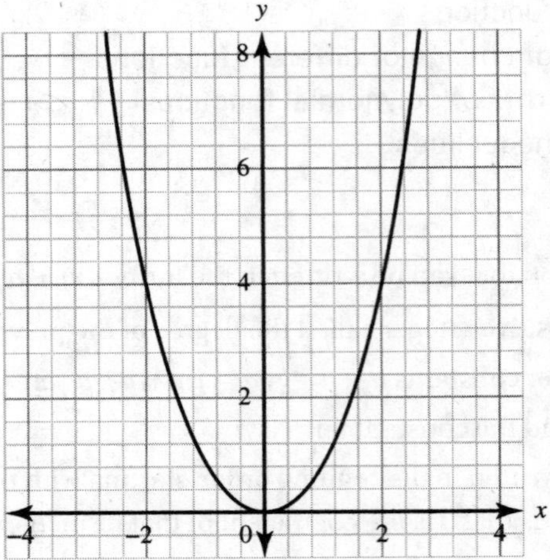

Figure 2.1

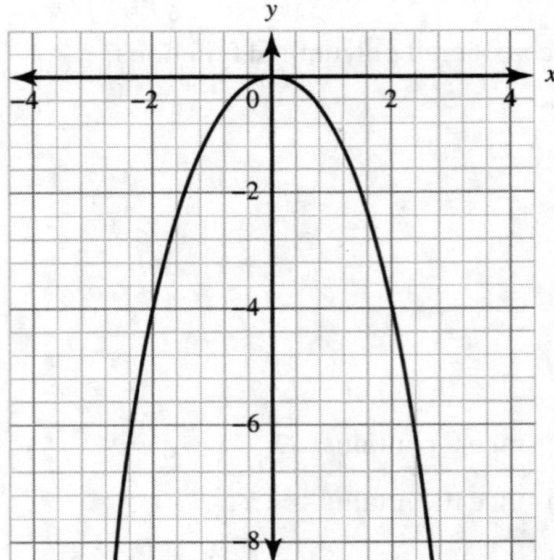

Figure 2.2

If a polynomial function has a restricted domain, at the included endpoints the polynomial will also have a local extremum value.

Consider the linear function $f(x) = x + 5$ over the domain $[-2, 4]$. This linear function is of degree 1 and is considered a polynomial. By restricting the domain, the graph of $f(x)$ is shown in Figure 2.3.

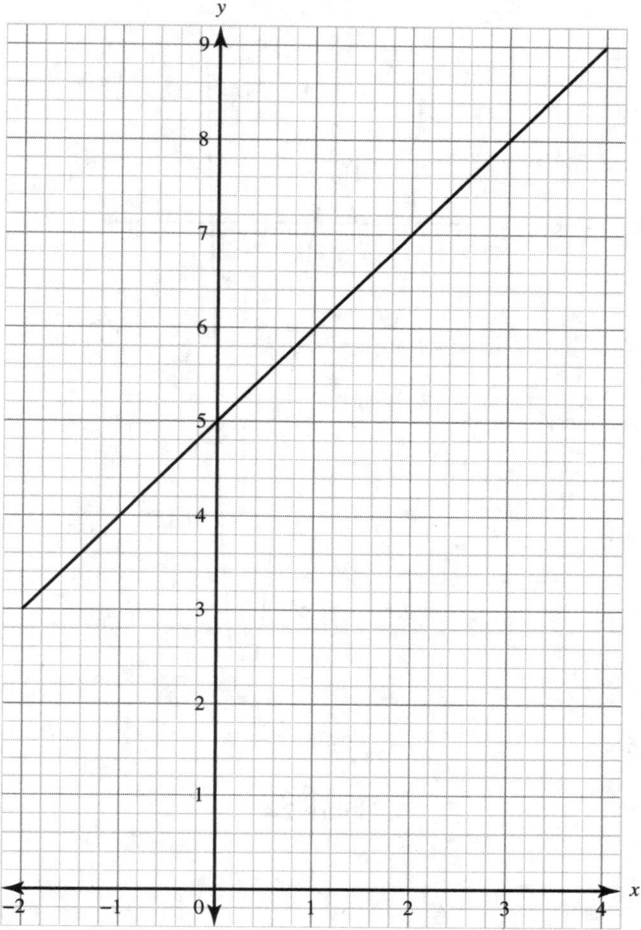

Figure 2.3

At the endpoint $(-2, 3)$, there is a local minimum. At the endpoint $(4, 9)$, there is a local maximum.

This leads to two important theorems, the extreme value theorem and the local extrema theorem, as defined on the right.

Extreme Value Theorem

If a polynomial function f is on a closed interval $[a, b]$, then f has both a maximum and a minimum value on $[a, b]$.

Local Extrema Theorem

A polynomial function of degree n has at most $n - 1$ relative maxima/minima.

Example

State the number of possible relative maxima/minima of $f(x) = x^4 - x^3 - 5x^2 + 1$. Confirm your answer graphically.

Solution

The degree of $f(x)$ is 4, meaning there are at most $4 - 1 = 3$ relative extrema. It can be expected that there will be 3 or less possible relative maximum or relative minimum points on the graph of $f(x)$.

Figure 2.4 is the graph of $f(x) = x^4 - x^3 - 5x^2 + 1$. The graph has two relative minima at $x = -1.25$ and $x = 2$ and one relative maximum at $x = 0$.

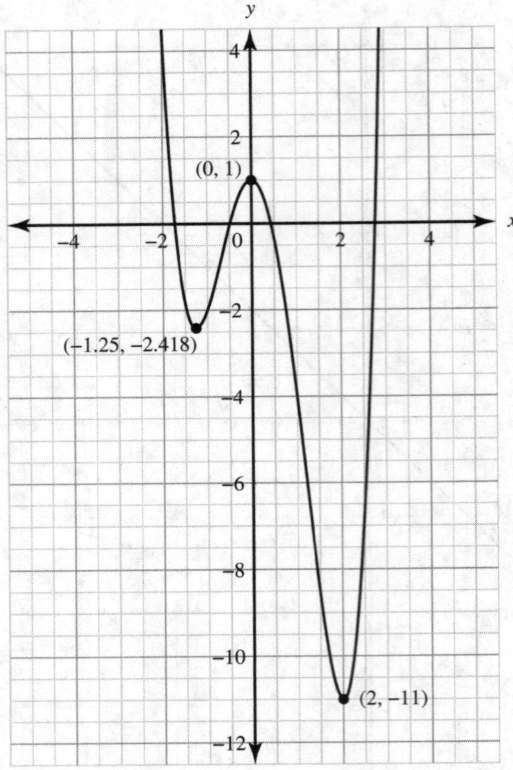

Figure 2.4

Absolute Extrema

Over a closed interval, the greatest of all the local maxima is called the global, or absolute, maximum. Likewise, the least of all local minima is called the global, or absolute, minimum.

Absolute extrema can occur at either relative extrema or at the endpoints of the closed interval.

The graphs in Table 2.1 discuss the existence of absolute extrema.

Table 2.1

Graph	Absolute Extrema
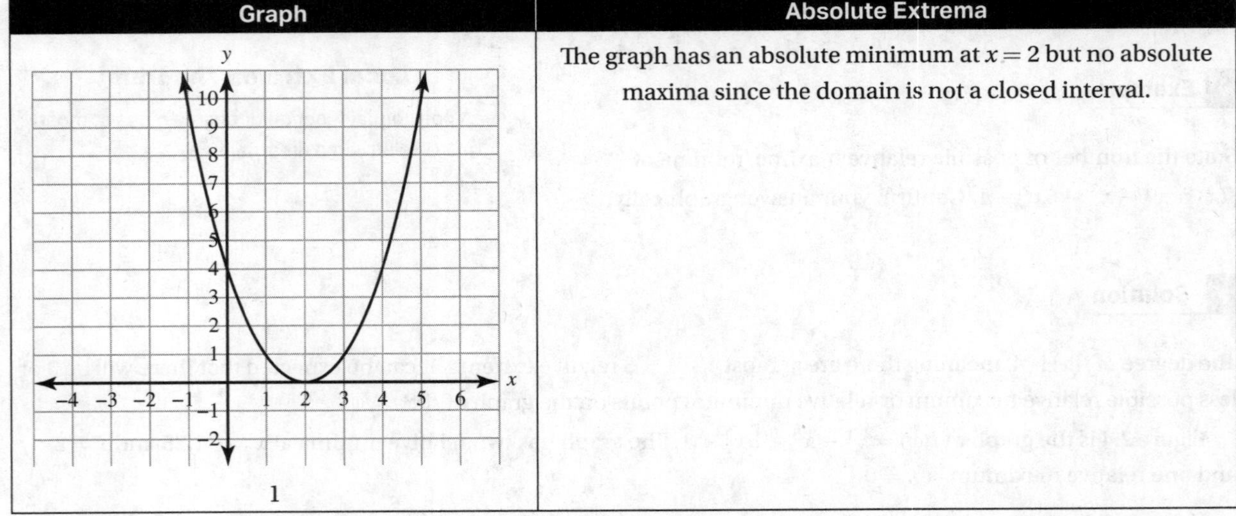 1	The graph has an absolute minimum at $x = 2$ but no absolute maxima since the domain is not a closed interval.

(*Continued*)

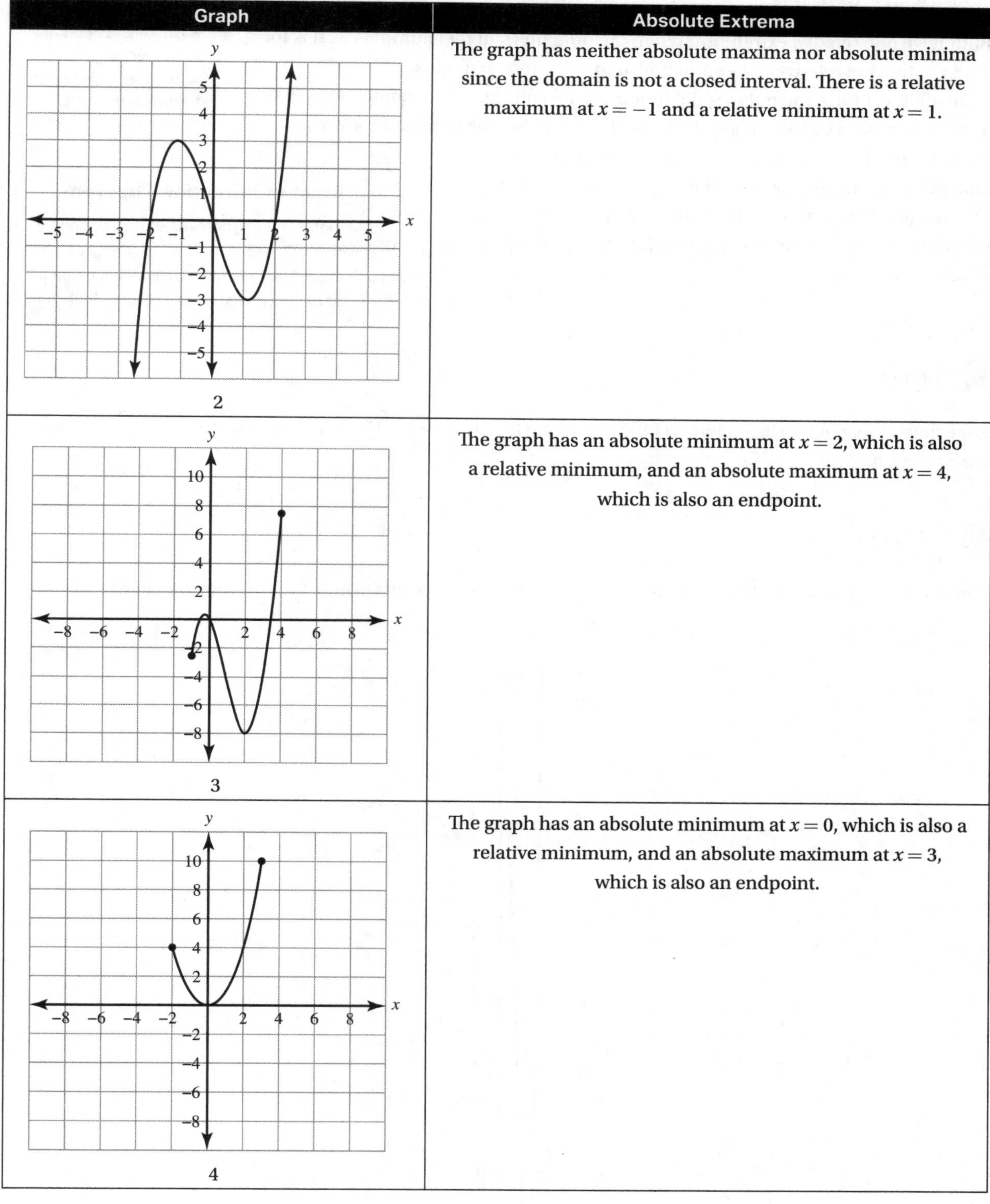

Graph	Absolute Extrema
2	The graph has neither absolute maxima nor absolute minima since the domain is not a closed interval. There is a relative maximum at $x = -1$ and a relative minimum at $x = 1$.
3	The graph has an absolute minimum at $x = 2$, which is also a relative minimum, and an absolute maximum at $x = 4$, which is also an endpoint.
4	The graph has an absolute minimum at $x = 0$, which is also a relative minimum, and an absolute maximum at $x = 3$, which is also an endpoint.

In graphs 2 and 3, the polynomials have zeros where the graphs intersect the x-axis. It can be observed that between every two distinct real zeros of a nonconstant polynomial function, there must be at least one input value corresponding to a local maximum or local minimum. In graph 2, there are zeros at $x = -2$ and $x = 0$ and a relative maximum at $x = -1$. There is also a set of zeros at $x = 0$ and $x = 2$ and a relative minimum at $x = 1$.

Graphs 1 and 4 are parabolas, meaning they are a type of quadratic equation. Both have one relative extremum. When the domain is all real numbers as it is in graph 1, the one relative extremum is also the absolute extremum.

In all the graphs, the polynomial functions are continuous, meaning no breaks, and have smooth curves. In graphs 1 and 4, we see that the graphs are always concave up. However, in graphs 2 and 3, the graphs go from concave down to concave up. The point where the concavity changes is known as the point of inflection. For graph 2, the point of inflection is at $x = 0$, and for graph 3, the point of inflection is at $x = 1$.

Polynomial functions of an even degree have either a global maximum or a global minimum.

Point of Inflection Theorem

The graph of a polynomial function of degree n, with $n \geq 2$, has at most $n - 2$ points of inflection. The graph of a polynomial function of odd degree has at least 1 point of inflection.

Example

State the number of possible points of inflection of $p(x) = 3x^4 + 5x^3 - 17x^2 - 14x + 6$. Use a graph to estimate their location.

Solution

The degree of the polynomial is 4. There will be at most $4 - 2 = 2$ points of inflection, as shown in Figure 2.5.

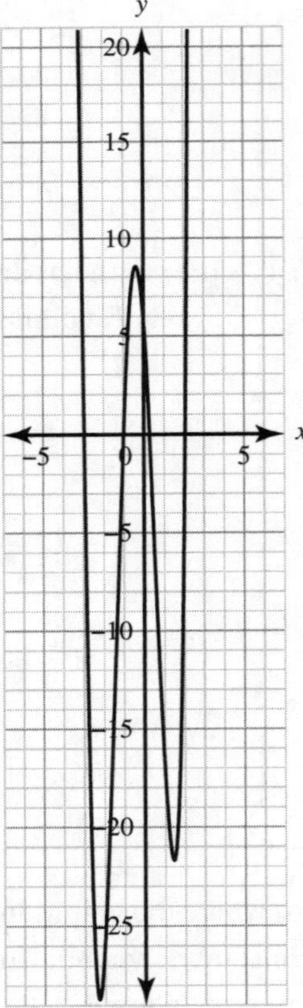

Figure 2.5

The graph changes from concave up to concave down to concave up. With the two changes there are two points of inflection. The points of inflection are at approximately $x = -1$ and $x = \frac{1}{2}$.

2.2 Real and Complex Zeros of Polynomial Functions

There are different ways to find the zeros of a polynomial function. Graphically, the zeros are located at the x-intercepts. Algebraically, the equation is set to 0 and solved for the independent variable. As the degree of the polynomial function increases and becomes larger than 2, it is common that the equation will have to be factored more than once or a graphing calculator will have to be used to locate the x-intercepts.

The degree of the polynomial states how many zeros the function has, both real and imaginary. For the polynomial $p(x) = a_n x^n + a_{n-1}x^{n-1} + a_{n-2}x^{n-2} + \ldots + a_1 x + a_0$, where $a_n \neq 0$, there will be n complex zeros. A complex number is of the form $a + bi$, where $i = \sqrt{-1}$ and where a and b are real numbers. The real component of the number is a, and bi is the imaginary component. While a polynomial function of degree n will have exactly n complex zeros, it doesn't necessarily mean that it will have exactly n real zeros. This means that the graph of a polynomial function will not always have the same number of x-intercepts as its degree. This may be due to repeated zeros or imaginary zeros.

Example

Find the zeros of each of the polynomial functions. Confirm your answer graphically.

1. $f(x) = x^3 - 3x^2 - 4x + 12$
2. $g(x) = x^4 - 5x^2 - 36$

Solution

To algebraically find the zeros, set each equation equal to 0 and factor to solve for x.

1. Factor this equation by grouping. Group together the first two terms and the last two terms. Then factor out their respective greatest common factors:

$$
\begin{aligned}
x^3 - 3x^2 - 4x + 12 &= 0 \\
(x^3 - 3x^2) + (-4x + 12) &= 0 \\
x^2(x - 3) - 4(x - 3) &= 0 \\
(x^2 - 4)(x - 3) &= 0
\end{aligned}
$$

The first factor can be factored again since it is a difference of two perfect squares:

$$(x + 2)(x - 2)(x - 3) = 0$$

Set each factor equal to zero and solve for x:

$$x = -2, x = 2, x = 3$$

There are three real zeros of this polynomial function.

Consider the graph of $f(x)$ in Figure 2.6.

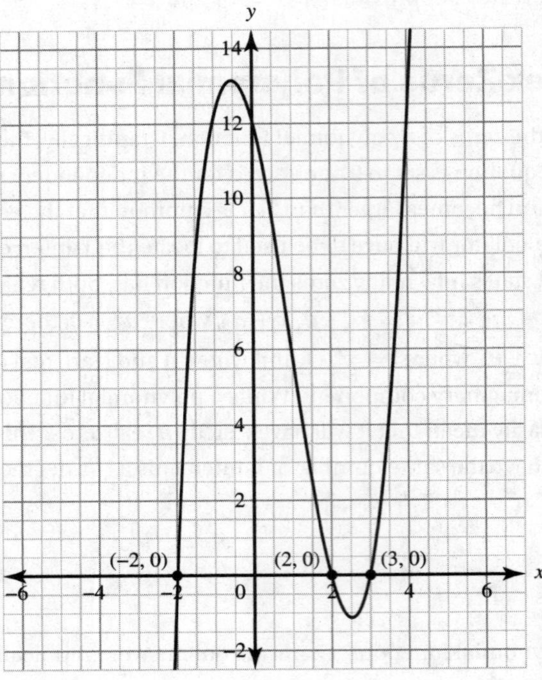

Figure 2.6

There are three x-intercepts, $(-2, 0)$, $(2, 0)$, and $(3, 0)$. As expected, the x-coordinates match the zeros found algebraically.

This was a third-degree polynomial that had three real zeros and three x-intercepts.

2. Factor this equation using product-sum:

$$x^4 - 5x^2 - 36 = 0$$
$$(x^2 - 9)(x^2 + 4) = 0$$

The first factor can be factored again since it is a difference of two perfect squares. The second factor is a sum of two perfect squares and cannot be factored with real numbers:

$$(x + 3)(x - 3)(x^2 + 4) = 0$$

Set each factor equal to zero and solve for x:

$$x = -3, x = 3, x^2 = -4$$

To solve for x for the last factor, extract the square root:

$$\sqrt{x^2} = \sqrt{-4}$$
$$x = \pm\sqrt{-4}$$

Simplify the square root using imaginary numbers:

$$x = 2i, x = -2i$$

There are four complex zeros: $x = 3$, $x = -3$, $x = 2i$, $x = -2i$. There are two real zeros and two imaginary zeros.

Consider the graph of $g(x)$ in Figure 2.7.

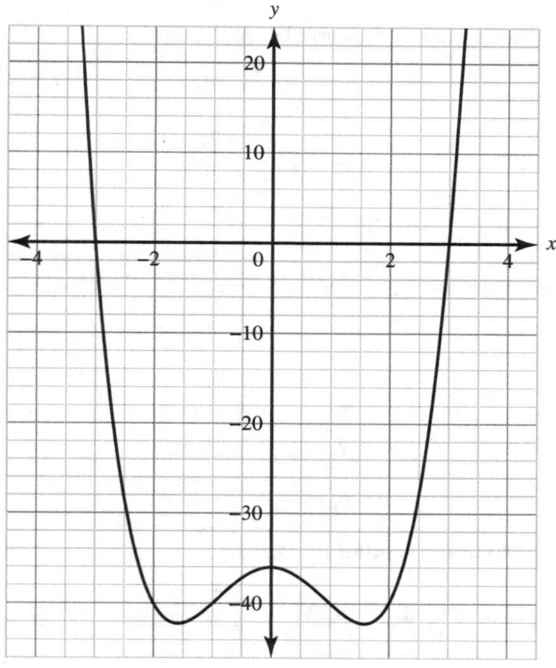

Figure 2.7

There are two x-intercepts, $(-3, 0)$ and $(3, 0)$. Only the real zeros are x-intercepts for the graph. Since the coordinate plane graphs only real numbers, the imaginary zeros do not appear on the graph.

A few key properties were demonstrated in the previous examples.

Key Properties of the Polynomial Function $p(x)$

1. If a is a complex number and $p(a) = 0$, then a is called a zero of p, or a root of the polynomial function p.
2. If a is a real number, then $(x - a)$ is a linear factor of p if and only if a is a zero of p.
3. If a is a real zero of a polynomial function p, the graph of $y = p(x)$ has an x-intercept at the point $(a, 0)$.
4. If $a + bi$ is a nonreal zero of a polynomial p, its conjugate $a - bi$ is also a zero of p.

Example

Find the zeros of each function.

1. $p(x) = x^4 - 25$
2. $q(x) = x^3 - 2x^2 + 5x$

Solution

Set each equation equal to 0, and factor each polynomial completely.

1.

$$x^4 - 25 = 0$$
$$(x^2 - 5)(x^2 + 5) = 0$$
$$(x - \sqrt{5})(x + \sqrt{5})(x - i\sqrt{5})(x + i\sqrt{5}) = 0$$

The zeros of the polynomial are $x = \sqrt{5}, -\sqrt{5}, i\sqrt{5}, -i\sqrt{5}$. The irrational zeros occur in conjugate pairs, as do the imaginary zeros.

2.

$$x^3 - 2x^2 + 5x = 0$$
$$x(x^2 - 2x + 5) = 0$$
$$x = 0,\ x^2 - 2x + 5 = 0$$

Use the quadratic formula to find the other two zeros.

$$x = \frac{-(-2) \pm \sqrt{(-2)^2 - 4(1)(5)}}{2(1)} = \frac{2 \pm \sqrt{-16}}{2} = \frac{2 \pm 4i}{2} = 1 \pm 2i$$

The zeros of the polynomial are $x = 0, 1 + 2i, 1 - 2i$. There is one real zero, and the two imaginary zeros occur in conjugate pairs.

Example

Find a polynomial function $f(x)$ with real coefficients of degree 4 that has the zeros -2, 3, and $-1 + i$ and whose leading coefficient is 1.

Solution

Since it is given that $f(x)$ is degree 4, there should be exactly four complex zeros. Two of the three given zeros are real, and the third zero is $-1 + i$. That means that the fourth zero of $f(x)$ is its conjugate pair, $-1 - i$. Now that all four zeros have been found, their linear factors can be written, as shown in Table 2.2.

Table 2.2

Zero	$x = -2$	$x = 3$	$x = -1 + i$	$x = -1 - i$
Factor	$(x + 2)$	$(x - 3)$	$(x - (-1 + i))$	$(x - (-1 - i))$

Since the leading coefficient was given as 1, to find the polynomial, multiply the factors together:

$$f(x) = (x + 2)(x - 3)(x - (-1 + i))(x - (-1 - i))$$
$$f(x) = (x^2 - 3x + 2x - 6)(x^2 - (-1 - i)x - (-1 + i)x + (-1 + i)(-1 - i))$$
$$f(x) = (x^2 - x - 6)(x^2 + 1x + ix + 1x - ix + 1 + i - i - i^2)$$
$$f(x) = (x^2 - x - 6)(x^2 + 2x + 2)$$
$$f(x) = x^4 + x^3 - 6x^2 - 14x - 12$$

Counting Repeated Zeros

There is no guarantee that linear factors of a polynomial are all different. For example, $p(x) = x^2 - 8x + 16 = (x - 4)(x - 4) = (x - 4)^2$. Since the degree is 2, one might assume there would be two zeros. However, $p(x)$ has two equal zeros, $x = 4$. Since $x - 4$ is repeated two times when $p(x)$ is written as a product of linear factors, it is said that "4 is a zero of multiplicity two."

Repeated zeros have an interesting relationship when the polynomial is graphed, as shown in Figure 2.8.

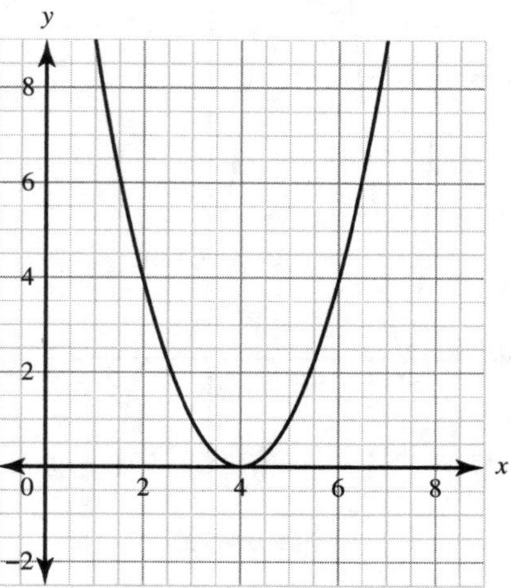

Figure 2.8

For the graph of $p(x) = (x - 4)^2$, at the zero $x = 4$, the graph has an x-intercept at (4, 0) but does not pass through the x-axis. Instead, the graph seems to bounce off. This is known as the graph being tangent to the x-axis at $x = 4$. It is also important to note that the x-values close to $x = 4$, both on the left and right side, have corresponding y-values that are positive.

When polynomial functions are written as a product of their linear factors, the sum of the multiplicities of each factor represents the degree of the polynomial.

Example

Given: $f(x) = (1 - x)^3(4 + x)^2$

(a) Find the degree of the polynomial.

(b) Discuss the behavior of the graph of $f(x)$ at the x-intercepts.

✓ Solution

(a) When all the factors of the polynomial function are multiplied together, the term with the greatest power will be $(x^3)(x^2) = x^5$. So the degree of $f(x)$ is 5.

The sum of the multiplicities of the factors is $3 + 2 = 5$, confirming the degree of $f(x)$. Since $f(x)$ is a fifth-degree polynomial, it has a total of five complex zeros. Specifically, 1 is a zero of multiplicity 3, and -4 is a zero of multiplicity 2.

(b) To see the behavior of the graph at both an even and odd multiplicity, consider the graph of $f(x)$ shown in Figure 2.9.

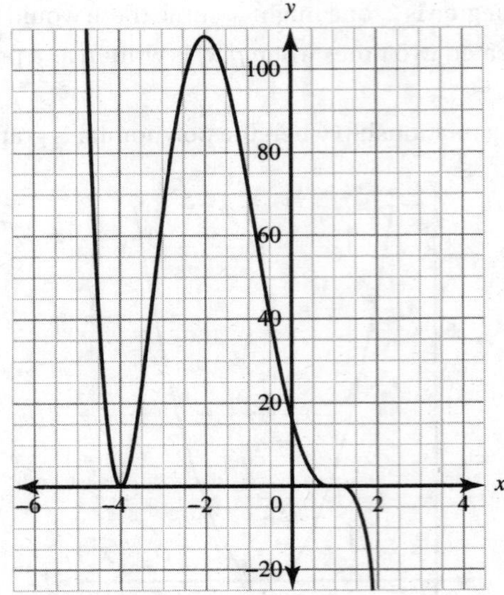

Figure 2.9

The behaviors at each of the x-intercepts are different. At $(1, 0)$ the graph crosses the x-axis, but at $(-4, 0)$ the graph is tangent to the x-axis.

Remember

When counting possible zeros from the graph of a function, we cannot assume that an x-intercept corresponds to exactly one zero. The following generalizations exist:

1. A polynomial function of degree n has exactly n complex zeros, provided that repeated zeros are counted the number of times they occur as factors when the function is written as a product of linear factors.
2. If the graph of a polynomial function $y = f(x)$ crosses the x-axis at $(k, 0)$, then k occurs as a zero an odd number of times.
3. If the graph of a polynomial function $y = f(x)$ is tangent to the x-axis at $(k, 0)$, then k occurs as a zero an even number of times.

Example

Find a polynomial function f for which 2 is a zero of multiplicity 2 and $\sqrt{5}$ is also a zero.

Solution

Since 2 is a zero of multiplicity 2 for function f, $(x-2)^2$ is a factor of the function. Since $\sqrt{5}$ is a zero, then its conjugate, $-\sqrt{5}$, is also a zero. Therefore, $(x-\sqrt{5})$ and $(x+\sqrt{5})$ are also factors of function f.

Multiply to find one possible polynomial function for $f(x)$:

$$f(x) = (x-2)^2(x-\sqrt{5})(x+\sqrt{5}) = (x^2-4x+4)(x^2-5)$$

$$f(x) = x^4 - 4x^3 - x^2 + 20x - 20$$

Example

Find a polynomial function f that would match the graph of the polynomial function in Figure 2.10.

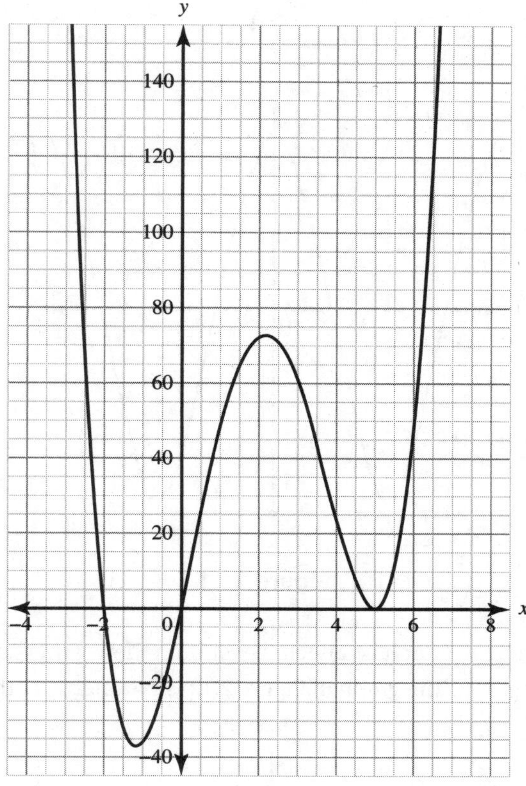

Figure 2.10

Solution

Table 2.3 shows a way to organize the given and needed information.

Table 2.3

Coordinates of the x-intercept	$(-2, 0)$	$(0, 0)$	$(5, 0)$
Zero	$x = -2$	$x = 0$	$x = 5$
Graph behavior at the x-intercept	Crosses x-axis	Crosses x-axis	Tangent to the x-axis
Multiplicity	Odd	Odd	Even
Possible factor	$(x + 2)$	x	$(x - 5)^2$

Assuming the leading coefficient is 1, one possible function is $f(x) = x(x + 2)(x - 5)^2$.

Line and Origin Symmetry

When working with graphs of functions, it is helpful to know whether the function has any special symmetry.

A graph has line symmetry if the graph can be mapped onto itself by a reflection over a line. Figure 2.11 shows a graph with vertical line symmetry.

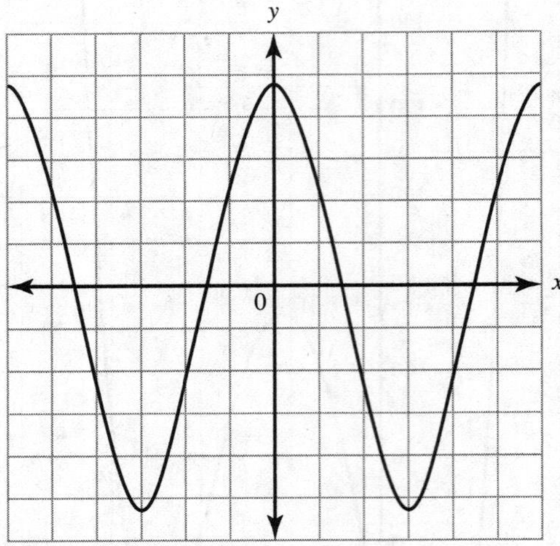

Figure 2.11

A graph is symmetric to the origin if, after it is reflected over both the x- and y-axes, in either order, the new graph coincides with the original graph. The graph in Figure 2.12 is symmetric to the origin.

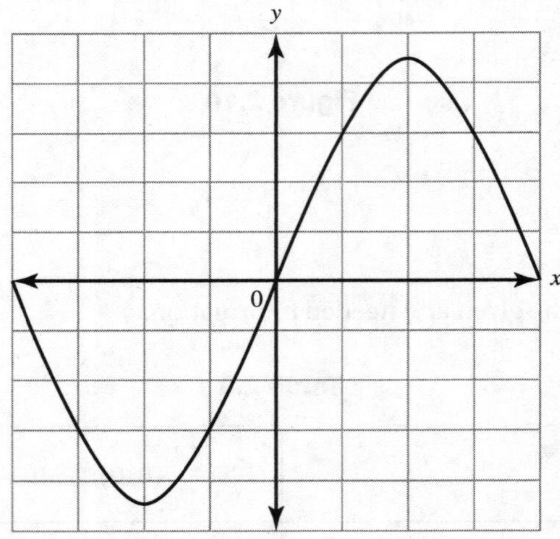

Figure 2.12

Even and Odd Functions

Functions that have y-axis symmetry or origin symmetry are given special names.

A function f is an even function if its equation remains unchanged when x is replaced with $-x$, that is, $f(-x) = f(x)$ for all x. Even functions are symmetric to the y-axis. For example, the function $f(x) = x^2$ is an even function since replacing x with $-x$ gives $f(-x) = (-x)^2 = x^2 = f(x)$. The graph in Figure 2.11 describes an even function since it has y-axis symmetry.

A function f is an odd function if replacing x with $-x$ changes the sign of each term of the equation to its opposite, that is, when $f(-x) = -f(x)$ for all x. The graph of an odd function is symmetric to the origin. For example, the function $f(x) = x^3 - x$ is an odd function since replacing x with $-x$ gives $f(-x) = (-x)^3 - (-x) = -x^3 + x = -f(x)$. The graph in Figure 2.12 describes an odd function since it has origin symmetry.

Example

The function $f(x)$ over the domain $-10 \leq x \leq 10$ is represented for selected values in Table 2.4. Given that $f(x)$ is an even function, complete the rest of the table.

Table 2.4

x	-10	-4	-1	0	1	4	10
$f(x)$	18		2	0		-3	

Solution

Since the function is even, the graph is symmetric with respect to the y-axis, meaning that the y-value is the same for every positive and negative x-value. This is shown in Table 2.5.

Table 2.5

x	-10	-4	-1	0	1	4	10
$f(x)$	18	**-3**	2	0	**2**	-3	**18**

2.3 Polynomial Functions and End Behavior

The graphs of all polynomial functions are continuous because there are no values of x for which the function would be undefined. That means that there are no breaks, holes, jumps, or asymptotes in the graph.

The discussion of end behavior of the graph of a polynomial function is an interesting one since the domain of a polynomial function is all real numbers, meaning the domain does not end. To really examine the end behavior, the concept of a limit must be introduced.

Definition of a Limit

Given a function $f(x)$, the limit of $f(x)$ as x approaches c is a real number L if $f(x)$ can be made arbitrarily close to L by having x approach close to c but not be equal to c. Table 2.6 shows limit notation and how to read it.

Table 2.6

Notation	Read As
$\lim_{x \to c} f(x) = L$	The limit as x approaches c of $f(x)$ equals L

If the limit equals a value L, it is said that the limit exists. It is also understood that the values of $f(x)$ approach L as x approaches c from both the left and the right.

Our focus is on end behavior, so our limit won't be approaching a value. Instead for x to go all the way to the end on the left, it will be going toward negative infinity, and on the right, it will be going toward positive infinity.

The symbol ∞ is used to represent infinity. Infinity is the idea of something unlimited, without bounds. Infinity can appear in limits in two ways: as what x is approaching or as what the function is approaching.

To test for end behavior, first determine if the leading coefficient of the polynomial is either positive or negative. Then determine if the degree is either even or odd. This leads to four possible scenarios as explained in Table 2.7 for the polynomial $p(x)$.

Table 2.7

Leading Coefficient	Degree	End Behavior	Limit Notation
Positive	Even	Graph rises on the left and right 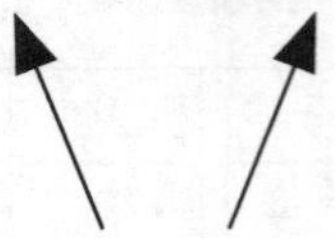	$\lim_{x\to-\infty} p(x) = +\infty$ $\lim_{x\to\infty} p(x) = +\infty$
Positive	Odd	Graph falls on the left and rises on the right 	$\lim_{x\to-\infty} p(x) = -\infty$ $\lim_{x\to\infty} p(x) = +\infty$
Negative	Even	Graph falls on the left and right 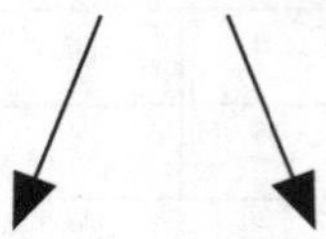	$\lim_{x\to-\infty} p(x) = -\infty$ $\lim_{x\to\infty} p(x) = -\infty$
Negative	Odd	Graph rises on the left and falls on the right 	$\lim_{x\to-\infty} p(x) = +\infty$ $\lim_{x\to\infty} p(x) = -\infty$

This behavior can be easily identified in two basic polynomial functions, $f(x) = x^2$ and $g(x) = x^3$. For $f(x) = x^2$, the leading coefficient (1) is positive, and the degree (2) is even. In Figure 2.13, it can be verified that the graph rises on the left, $\lim_{x\to-\infty} f(x) = +\infty$, and rises on the right, $\lim_{x\to\infty} f(x) = +\infty$.

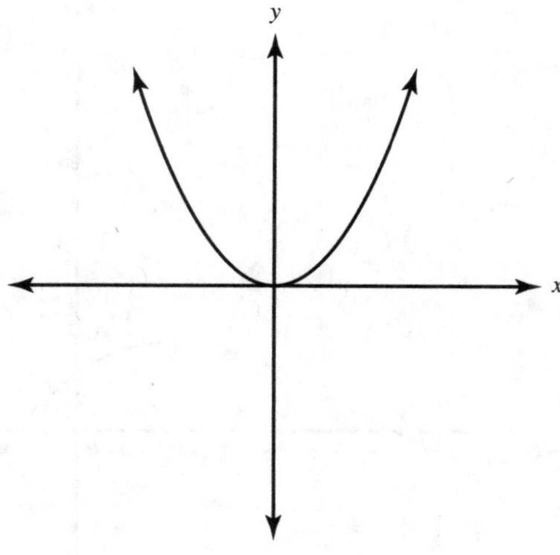

Figure 2.13

For $g(x) = x^3$, the leading coefficient (1) is positive, and the degree (3) is odd. In Figure 2.14, it can be verified that the graph falls on the left, $\lim_{x \to -\infty} g(x) = -\infty$, and rises on the right, $\lim_{x \to \infty} g(x) = +\infty$.

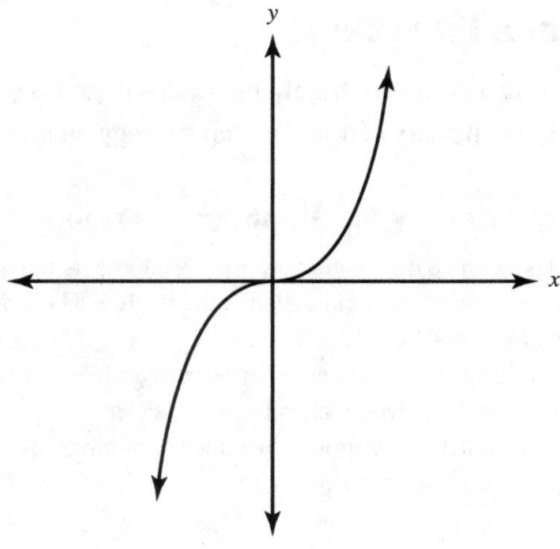

Figure 2.14

Example

State the left and right end behavior of $h(x) = -4x^3 - 17x^2 - 14x + 6$ using limit notation. Confirm your answer graphically.

✓ Solution

The leading coefficient of $h(x)$ is -4, which is negative, and the degree of $h(x)$ is 3, which is odd. Therefore, $\lim_{x \to -\infty} h(x) = +\infty$ and $\lim_{x \to \infty} h(x) = -\infty$. This is confirmed by the graph in Figure 2.15, which rises on the left and falls on the right.

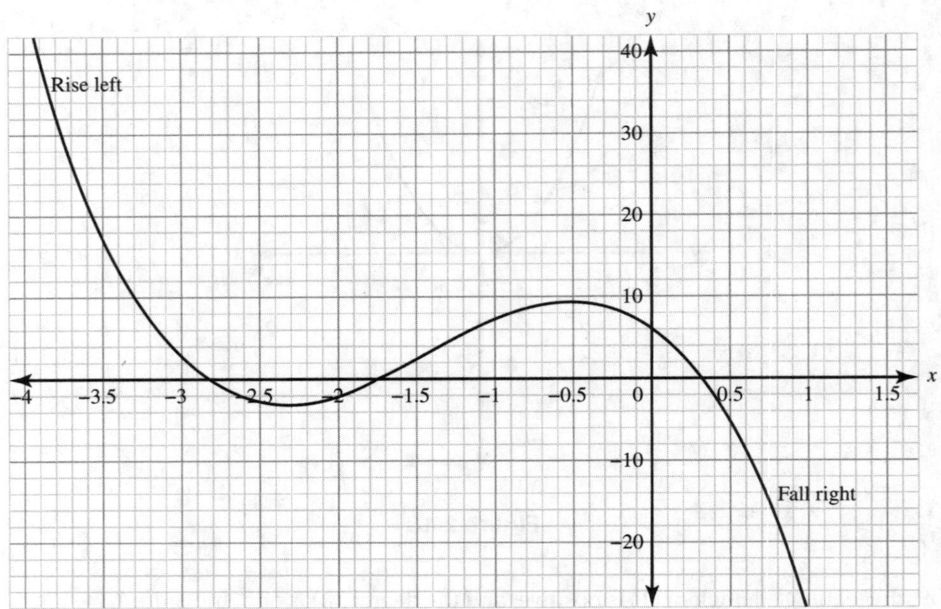

Figure 2.15

2.4 Graphs of Polynomial Functions

Using information from all the previous sections in this chapter helps to graph polynomial functions without the use of a graphing calculator. Below is a summary of a seven-step strategy for graphing a polynomial function.

Seven-Step Strategy for Graphing a Polynomial Function

1. State the degree of the function: this will help determine how many x-intercepts the graph may have.
2. State the number of possible relative extrema and points of inflection: this will help determine how the graph will bend and if there will be any relative minima or maxima.
3. Test for end behavior: examine the leading coefficient and degree to determine the end behavior of the graph.
4. Find the x-intercept(s): completely factor the polynomial and find its zeros.
5. State the multiplicity of each zero: this will help determine the behavior of the graph at each x-intercept.
6. Find the y-intercept: evaluate the polynomial function at $x = 0$.
7. Sketch the graph using the above steps.

Example

Sketch the graph of the polynomial function $f(x) = 2x^3 + 2x^2 - 8x - 8$.

Solution

Use the seven-step strategy.

STEP 1 State the degree

Degree $= 3$, the function will have at most 3 real zeros so at most 3 x-intercepts

STEP 2 Relative minima/maxima and points of inflection

$3 - 1 = 2$, there will be at most 2 relative minima/maxima

$3 - 2 = 1$, there will be at most 1 point of inflection

STEP 3 Test for end behavior

Degree = 3, odd

Leading coefficient = 2, positive

$\lim_{x \to -\infty} f(x) = -\infty$: graph will fall on the left

$\lim_{x \to \infty} f(x) = +\infty$: graph will rise on the right

STEP 4 Find the x-intercept(s)

Set the equation equal to 0 and factor completely. Then solve for x.

$$2x^3 + 2x^2 - 8x - 8 = 0$$
$$2(x^3 + x^2 - 4x - 4) = 0$$
$$2[x^2(x + 1) - 4(x + 1)] = 0$$
$$2(x^2 - 4)(x + 1) = 0$$
$$2(x - 2)(x + 2)(x + 1) = 0$$

Divide by 2 on both sides of the equation:

$$(x - 2)(x + 2)(x + 1) = 0$$

The zeros are $x = -2, 2, -1$.

The x-intercepts are $(-2, 0)$, $(2, 0)$, $(-1, 0)$.

STEP 5 State the multiplicity of each zero

Using Step 4, $f(x)$ is written as a product of linear factors. Each factor has a power of 1, so each zero has a multiplicity of 1. Therefore, the graph of $f(x)$ crosses through the x-axis at each x-intercept.

STEP 6 Find the y-intercept

$$f(0) = 2(0)^3 + 2(0)^2 - 8(0) - 8 = -8$$

The y-intercept is at the point $(0, -8)$.

STEP 7 Graph

First graph the information found in Steps 1–6 on a set of axes as shown in Figure 2.16.

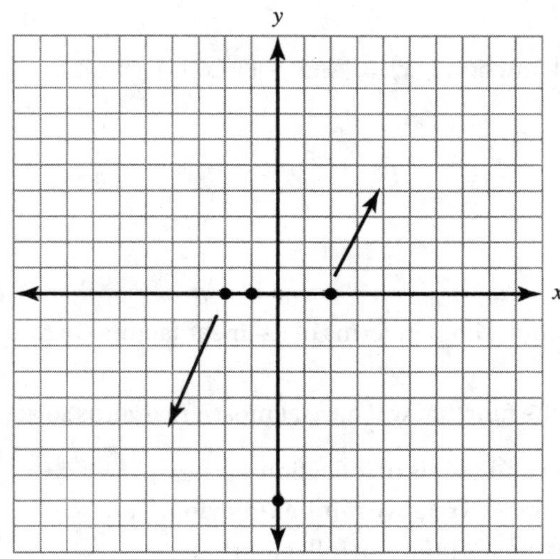

Figure 2.16

Then connect with smooth, continuous curves as shown in Figure 2.17. Check to see that there are at most 2 relative extrema and 1 point of inflection.

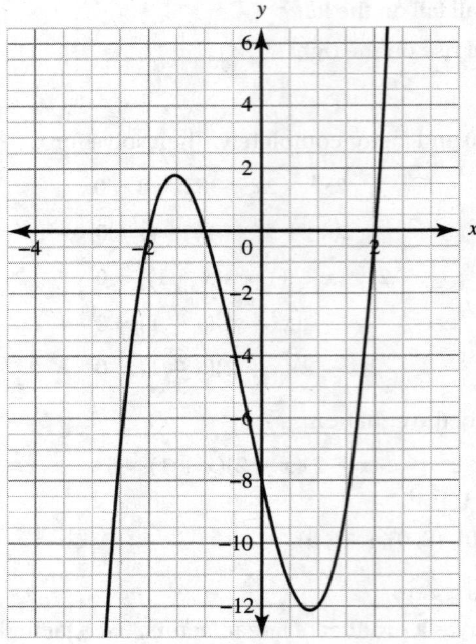

Figure 2.17

Note: Only the x-axis is scaled since the location of the relative extrema and point of inflection are being estimated. Since it is a sketch, the primary concern is the overall shape of the graph. The exact location of relative extrema and points of inflection is saved for a calculus course.

Example

Sketch the graph of the polynomial function $g(x) = -3(x-1)^2(x+1)(x-5)$.

Solution

Utilize the seven-step strategy.

STEP 1 State the degree

The function $g(x)$ is already written in terms of its linear factors. To find the degree, find the sum of the multiplicities.

Degree $= 2 + 1 + 1 = 4$, the function will have at most 4 real zeros so at most 4 x-intercepts

STEP 2 Relative minima/maxima and points of inflection

$4 - 1 = 3$, there will be at most 3 relative minima/maxima

$4 - 2 = 2$, there will be at most 2 points of inflection

STEP 3 Test for end behavior

Degree = 4, even

To find the leading coefficient, multiply all the leading coefficients in each term together.

Leading coefficient = $(-3)(1)^2(1)(1) = -3$, negative

$\lim_{x\to-\infty} g(x) = -\infty$: graph will fall on the left

$\lim_{x\to\infty} g(x) = -\infty$: graph will fall on the right

STEP 4 Find the x-intercept(s)

The function $g(x)$ is already written as a product of linear factors. Setting each factor equal to 0 shows that the zeros are $x = 1, x = -1, x = 5$. Therefore, the x-intercepts are (1, 0), (−1, 0), and (5, 0).

STEP 5 State the multiplicity of each zero

Table 2.8 shows a way to organize the information.

Table 2.8

Factor	$(x-1)^2$	$(x+1)$	$(x-5)$
Zero	$x = 1$	$x = -1$	$x = 5$
Multiplicity	2	1	1
Graph Behavior	Tangent to x-axis	Crosses x-axis	Crosses x-axis

STEP 6 Find the y-intercept

$$g(0) = -3(0-1)^2(0+1)(0-5) = -3(-1)^2(1)(-5) = 15$$

The y-intercept is at the point (0, 15).

STEP 7 Graph

First graph the information found in Steps 1–6 on a set of axes as shown in Figure 2.18.

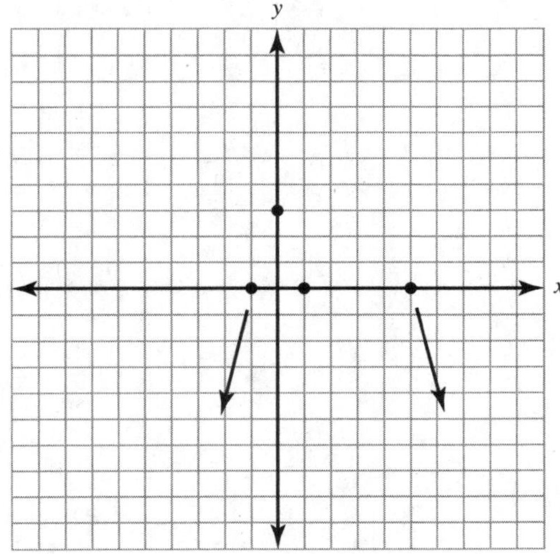

Figure 2.18

Then connect with smooth, continuous curves as shown in Figure 2.19. Check to see that there are at most 3 relative extrema and 2 points of inflection.

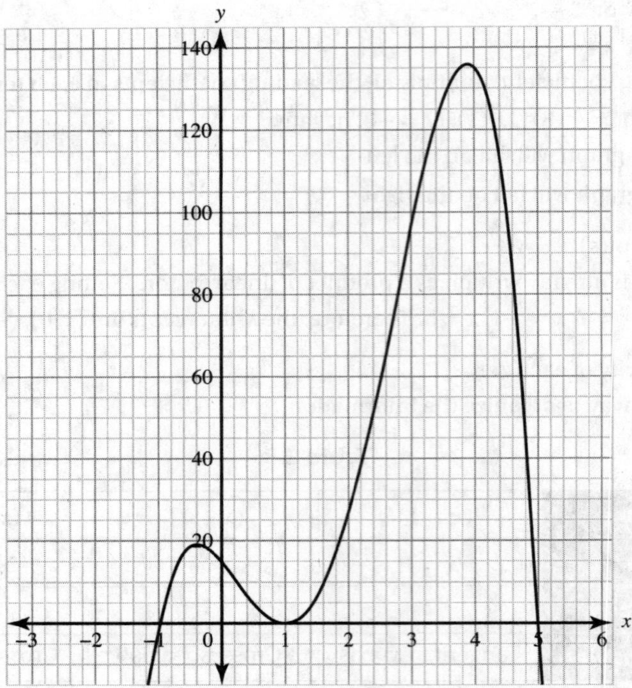

Figure 2.19

Note: Only the x-axis is scaled since the location of the relative extrema and points of inflection are being estimated. Since it is a sketch, the primary concern is the overall shape of the graph. The exact location of relative extrema and points of inflection is saved for a calculus course.

Example

Write a possible polynomial function for the graph shown in Figure 2.20.

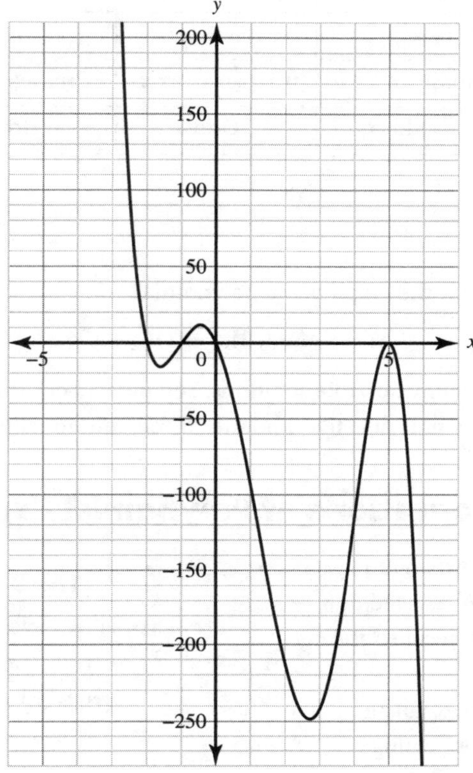

Figure 2.20

✓ Solution

Utilizing the seven-step strategy for graphing helps to identify important parts of the graph to pay attention to when writing an equation for a possible polynomial function.

STEP 1 Possible degree of the function
Since there are 4 x-intercepts, the degree is at least 4.

STEP 2 State the number of possible relative extrema and points of inflection
There are two relative minima and two relative maxima, meaning $n - 1 = 4$, determining that the degree is at least 5.
From left to right, the graph is concave up, then concave down, then concave up, and then concave down again. Therefore, there are 3 points of inflection. This means $n - 2 = 3$, proving the degree is 5.

STEP 3 Test for end behavior
The graph rises on the left and falls on the right. Therefore, the leading coefficient is negative, and the degree is odd.

STEP 4 Find the x-intercept(s)
The x-intercepts are $(-2, 0)$, $(-1, 0)$, $(0, 0)$, $(5, 0)$.

STEP 5 State the multiplicity of each zero
Table 2.9 organizes all of the information.

Table 2.9

x-intercept	$(-2, 0)$	$(-1, 0)$	$(0, 0)$	$(5, 0)$
Zero	$x = -2$	$x = -1$	$x = 0$	$x = 5$
Factor	$(x + 2)$	$(x + 1)$	$(x - 0) = x$	$(x - 5)$
Graph Behavior	Crosses x-axis	Crosses x-axis	Crosses x-axis	Tangent to x-axis
Multiplicity	Odd	Odd	Odd	Even

STEP 6 Find the y-intercept

The y-intercept is at $(0, 0)$.

STEP 7 Write a possible function for the graph of the polynomial

$$f(x) = -x(x + 2)(x + 1)(x - 5)^2$$

Note: There are infinitely many possible solutions. You must check that the sum of the multiplicities is at least 5, the leading coefficient is negative, and the correct multiplicities are given for each factor.

2.5 Equivalent Representations of Polynomial Expressions

A binomial such as $(a + b)^n$ may be expanded by repeated multiplication using $(a + b)$ as a factor n times or by using an algebraic formula called the binomial theorem.

The binomial theorem tells how to expand a binomial to a given power without performing repeated multiplication.

> **The Binomial Theorem**
>
> If n is a positive whole number, then
> $(a + b)^n = {}_nC_0a^nb^0 + {}_nC_1a^{n-1}b^1 + {}_nC_2a^{n-2}b^2 + \ldots + {}_nC_ra^0b^n$
> where ${}_nC_r = \dfrac{n!}{r!(n-r)!}$.

When the values of ${}_nC_r$ are arranged in a triangular pattern in which the rows correspond to successive values of n, starting with $n = 0$, the result is an array of numbers called Pascal's triangle, shown in Figure 2.21 below.

row 0:	${}_0C_0$	1
row 1:	${}_1C_0 \ {}_1C_1$	1 1
row 2:	${}_2C_0 \ {}_2C_1 \ {}_2C_2$	1 2 1
row 3:	${}_3C_0 \ {}_3C_1 \ {}_3C_2 \ {}_3C_3$	1 3 3 1
row 4:	${}_4C_0 \ {}_4C_1 \ {}_4C_2 \ {}_4C_3 \ {}_4C_4$	1 4 6 4 1
row 5:	${}_5C_0 \ {}_5C_1 \ {}_5C_2 \ {}_5C_3 \ {}_5C_4 \ {}_5C_5$	1 5 10 10 5 1
⋮	...	...

Figure 2.21

In Pascal's triangle, each number after 1 is the sum of the two numbers directly above it. Pascal's triangle makes it possible to find the value of ${}_nC_r$ for different values of n and r, where n is the row number and $r = 0$ corresponds to the first entry on each row.

Example

Write the expansion of $(x - 3y)^4$.

Solution

Use the binomial theorem to expand $(x - 3y)^4$, where $a = x$, $b = -3y$, and $n = 4$.

$$(x - 3y)^4 = {}_4C_0x^4(-3y)^0 + {}_4C_1x^3(-3y)^1 + {}_4C_2x^2(-3y)^2 + {}_4C_3x^1(-3y)^3 + {}_4C_4x^0(-3y)^4$$

Evaluate the coefficients in front of each term either by using the formula ${}_nC_r = \dfrac{n!}{r!(n-r)!}$ or by copying the values from row 4 of Pascal's triangle in Figure 2.21.

$$\begin{aligned}(x - 3y)^4 &= 1x^4 + 4x^3(-3y) + 6x^2(-3y)^2 + 4x(-3y)^3 + 1(-3y)^4\\ &= x^4 - 12x^3y + 54x^2y^2 - 108xy^3 + 81y^4\end{aligned}$$

Example

Write the expansion of $(x + y)^6$.

Solution

Use the binomial theorem to expand $(x + y)^6$, where $a = x$, $b = y$, and $n = 6$.

$$(x + y)^6 = {}_6C_0x^6y^0 + {}_6C_1x^5y^1 + {}_6C_2x^4y^2 + {}_6C_3x^3y^3 + {}_6C_4x^2y^4 + {}_6C_5x^1y^5 + {}_6C_6x^0y^6$$

Evaluate the coefficients in front of each term either by using the formula ${}_nC_r = \dfrac{n!}{r!(n-r)!}$ or by using the values from row 6 of Pascal's triangle. Figure 2.22 shows how to obtain row 6 from row 5 of Pascal's triangle. Write 1 as the first and the last member of row 6. Then obtain each of the remaining numbers in row 6 by adding the two numbers directly above from row 5.

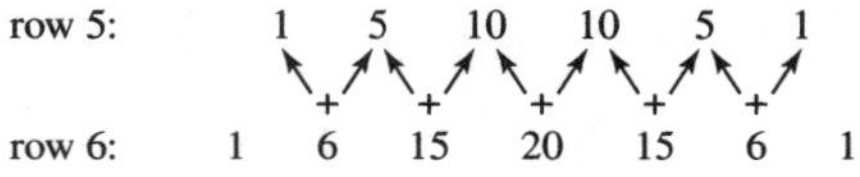

Figure 2.22

$$(x + y)^6 = 1x^6 + 6x^5y + 15x^4y^2 + 20x^3y^3 + 15x^2y^4 + 6xy^5 + 1y^6$$

2.6 Polynomial Inequalities

The key to solving polynomial inequalities comes from the geometric property that the graph of $y = f(x)$ lies above the x-axis when $f(x) > 0$ and lies below the x-axis when $f(x) < 0$.

The solutions of $f(x) > 0$ are the numbers x for which the graph of $f(x)$ lies above the x-axis. The solutions of $f(x) < 0$ are the numbers x for which the graph of $f(x)$ lies below the x-axis.

Solving Polynomial Inequalities

1. Write the inequality in one of these forms:
   $f(x) > 0$ $f(x) \geq 0$ $f(x) < 0$ $f(x) \leq 0$
2. Determine the roots of $f(x)$.
3. Use test points to determine whether the graph of $f(x)$ is above or below the x-axis on each of the intervals determined by the roots.
4. Find the solutions of the inequality.

Example

Solve: $2x^3 - 15x < x^2$

Solution

1. Write the inequality in the form $f(x) < 0$:
$$2x^3 - x^2 - 15x < 0$$
2. Determine the roots of $f(x) = 2x^3 - x^2 - 15x$.
   Factor the function: $2x^3 - x^2 - 15x = x(2x^2 - x - 15) = x(2x + 5)(x - 3)$
   The roots are $x = 0$, $x = -\frac{5}{2}$, and $x = 3$.
3. Plot the roots on a number line, as shown in Figure 2.23. Determine the sign of the polynomial on each of the intervals determined by the roots. Each root has an open circle since the inequality is strictly less than 0.

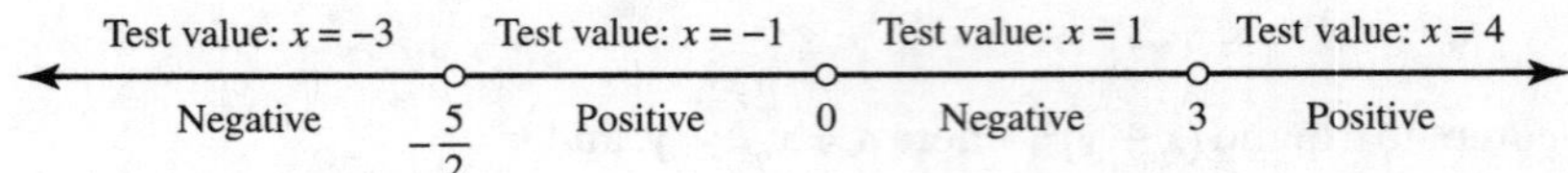

Figure 2.23

4. Since $f(x) < 0$, the solution sets lie in the intervals where $f(x)$ is negative. The solution intervals are $\left(-\infty, -\frac{5}{2}\right) \cup (0, 3)$.

Example

Solve: $(x + 15)(x - 2)^6(x - 10) \leq 0$

Solution

1. The inequality is already in the form $f(x) \leq 0$.
2. Determine the roots of $f(x) = (x + 15)(x - 2)^6(x - 10)$.
   The function is already factored.
   The roots are $x = -15$, $x = 2$, and $x = 10$.
3. Plot the roots on a number line, as shown in Figure 2.24. Determine the sign of the polynomial on each of the intervals determined by the roots. Each root has a closed circle since the inequality is less than or equal to 0.

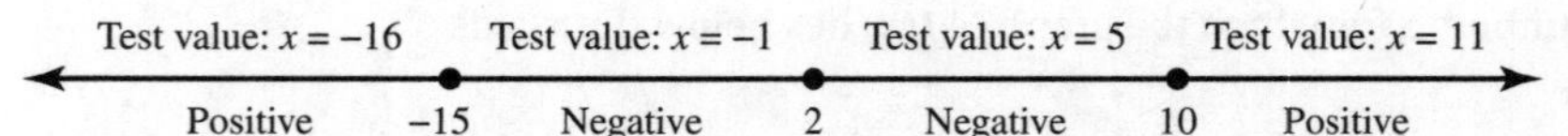

Figure 2.24

4. Since $f(x) \leq 0$, the solution sets lie in the intervals where $f(x)$ is negative. The solution intervals are $[-15, 2] \cup [2, 10]$. The reason why the interval is broken up at $x = 2$ is because at $x = 2$, the function equals 0, which is not negative.

Multiple-Choice Questions

1. Three of the roots of a polynomial equation with rational coefficients are -3, $2i$, and $-4 + 5\sqrt{6}$. What is the lowest possible degree of the equation?

   (A) 3
   (B) 4
   (C) 5
   (D) 6

2. Let $f(x)$ represent a third-degree polynomial function with real coefficients. When $y = f(x)$ is graphed, it is found that $f(3) = -1$ and $f(4) = 2$. Which statement *must* be true?

   (A) The function $f(x)$ has exactly one real zero in the interval [3, 4].
   (B) The function $f(x)$ has exactly one real zero in the interval [−1, 2].
   (C) The function $f(x)$ has exactly one or three real zeros in the interval [3, 4].
   (D) The function $f(x)$ has exactly one or two real zeros in the interval [−1, 2].

3. Let $c(x)$ represent a fourth-degree polynomial function with real coefficients. When $y = c(x)$ is graphed, it is found that $c(5) = 3$ and $c(9) = 4$. Which statement *must* be true?

   (A) The function $c(x)$ has an odd number of real zeros in the interval [5, 9].
   (B) The function $c(x)$ has either no real zeros or an even number of real zeros in the interval [5, 9].
   (C) The function $c(x)$ has at least one real zero in the interval [3, 4].
   (D) No conclusion is possible.

4. Based on the accompanying graph of the polynomial function $y = h(x)$, which statement *must* be true?

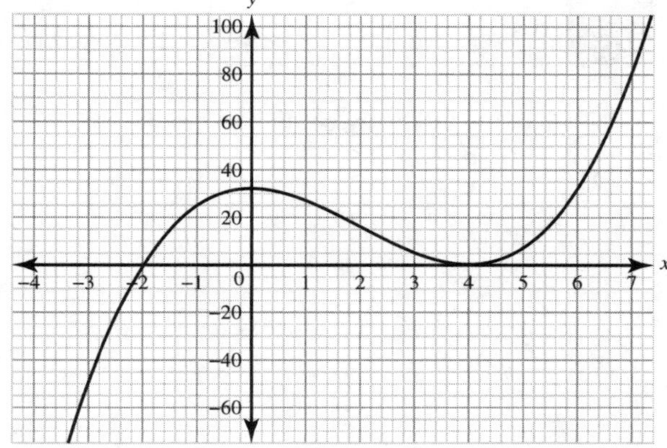

   (A) The degree of the function h is even.
   (B) The leading coefficient of the function h is negative.
   (C) $x = -2$ is a zero of multiplicity 2, and $x = 4$ is a zero of multiplicity 1.
   (D) Two of the factors of the function h are $(x + 2)$ and $(x - 4)$.

5. Given the nonconstant polynomial function $f(x)$ whose domain is $[a, e]$ where $a < b < c < d < e$, if $f(b) = f(c) = f(d) = 0$ and no other values of x in the domain are such that $f(x) = 0$, which statement *must* be true?

   (A) There is only one relative extrema in the interval $[a, e]$.
   (B) There are two relative extrema in the interval $[a, e]$.
   (C) $f(a) > 0$ and $f(e) > 0$.
   (D) The graph falls on the left and right.

6. Let f be a polynomial function that is even and continuous on the closed interval $[-3, 3]$. The function f has the properties indicated in the table below.

x	0	$0 < x < 1$	1	$1 < x < 2$	2	$2 < x < 3$
$f(x)$	1	Positive	0	Negative	-1	Negative

Which of the following could be a graph of f?

(A)

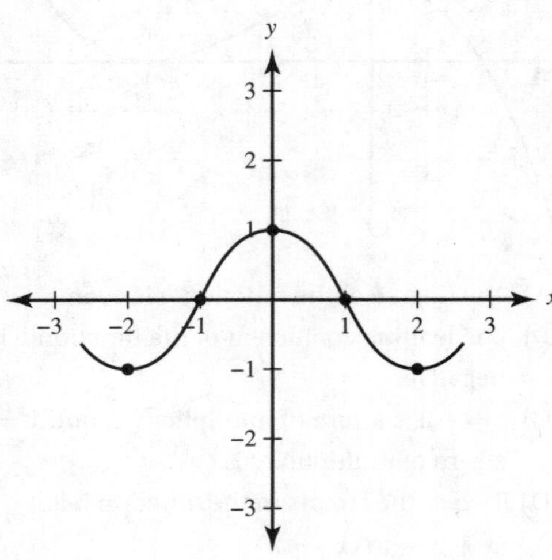

(B)

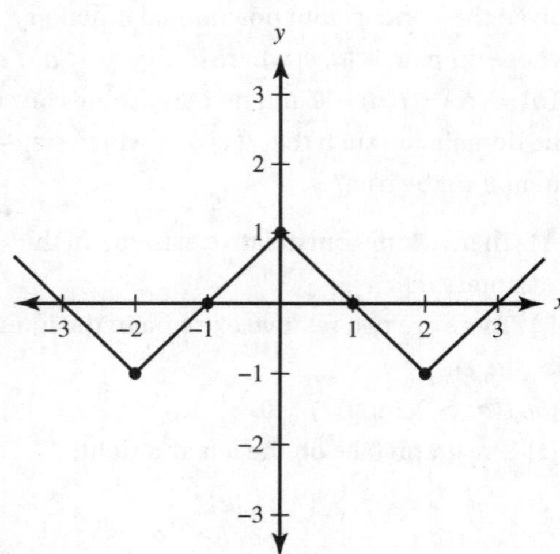

(C)

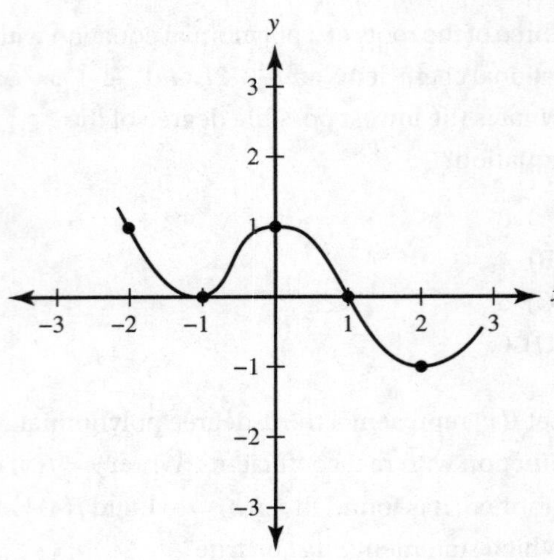

(D)

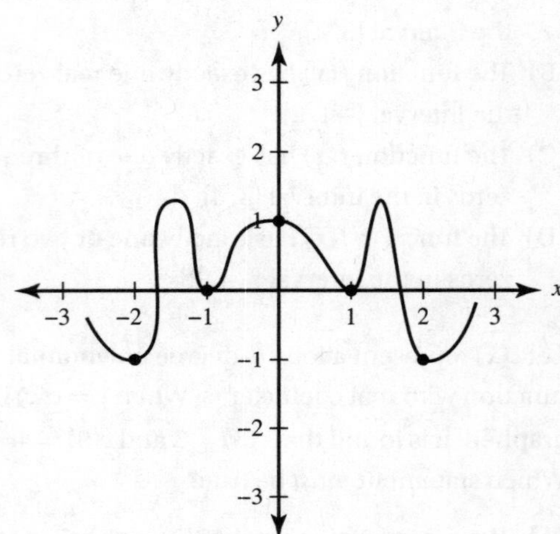

7. For what value of k is $(x + 1)$ a factor of the function $g(x) = x^4 + 6x^3 + kx^2 - 26x - 24$?

(A) -1
(B) 1
(C) 3
(D) 43

8. Which of the following is a possible polynomial function for the graph shown below?

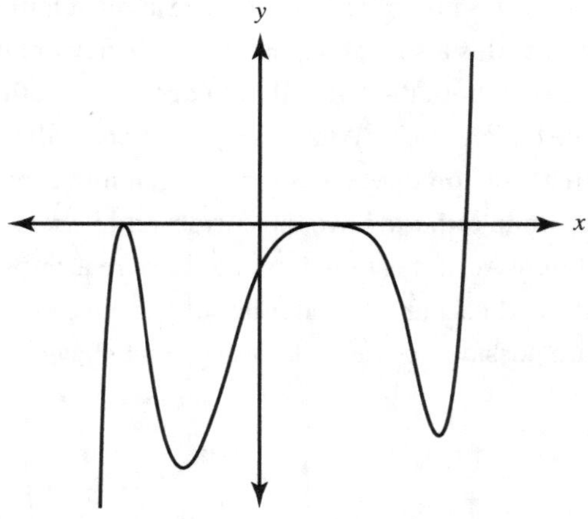

(A) $f(x) = (x - 5)^2(x + 1)^4(x + 3)$
(B) $h(x) = (x - 5)(x + 1)^3(x + 3)^2$
(C) $g(x) = (x + 5)^2(x - 1)^4(x - 3)$
(D) $m(x) = (x + 5)(x - 1)^3(x - 3)^2$

9. Which of the following statements must be true about the function that is represented by the graph shown below?

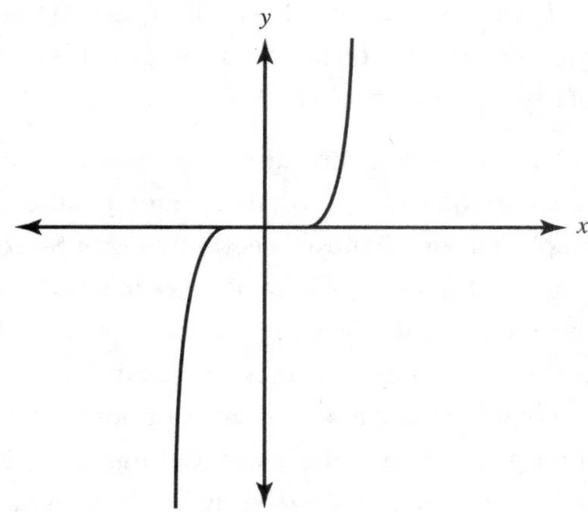

(A) $\lim_{x \to -\infty} f(x) = -\infty$ and $\lim_{x \to \infty} f(x) = -\infty$
(B) $f(-x) = f(x)$
(C) $f(-x) = -f(x)$
(D) This function has no real zeros.

10. Let g be a polynomial function over the interval $[a, e]$ with the properties listed in the table below.

x	(a, b)	(b, c)	(c, d)	(d, e)	(e, f)
Average Rate of Change	Increasing	Increasing	Decreasing	Increasing	Decreasing

Which of the following statements *must* be true?

(A) There are two relative extrema.
(B) There are three relative extrema.
(C) There are two points of inflection.
(D) There are three points of inflection.

Answer Explanations

1. **(C)** Imaginary and irrational roots come in complex conjugate pairs. So the roots are really -3, $2i$, $-2i$, $-4 + 5\sqrt{6}$, and $-4 - 5\sqrt{6}$. If each of these roots occurs only once, the lowest possible degree of the equation would be 5.

2. **(C)** Since $f(x)$ is a third-degree polynomial, the function will have at most three zeros and the graph will have at most three x-intercepts. Since the output values change from negative to positive over the interval [3, 4], the graph must cross the x-axis at least once. Since no other function values were given, you cannot assume that the graph crosses the x-axis only once. Below are two possible scenarios that would satisfy the given information.

   These graphs show how it is possible to have one or three zeros in the interval.

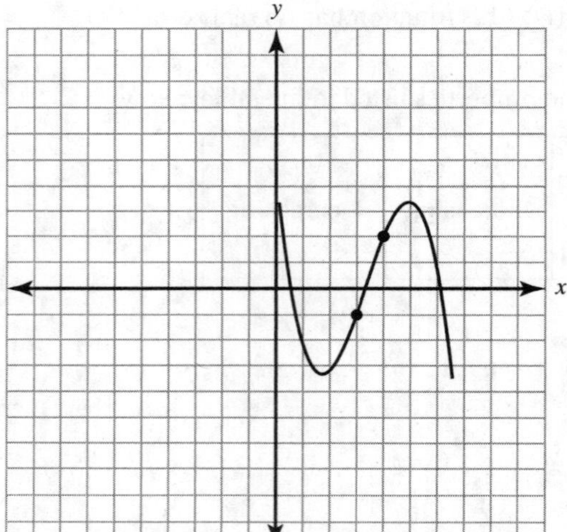

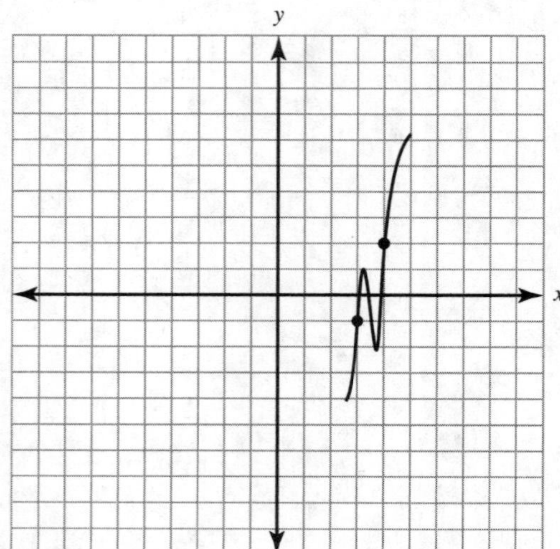

3. **(B)** It is given that $c(x)$ is a fourth-degree polynomial. This means that it can have at most four real zeros. This also means that its end behavior either rises on both the left and right or falls on both the left and right. What that guarantees is that if this function does cross the x-axis, it must cross again in order to have the correct end behavior. Therefore, if this function does have real zeros, they will occur an even number of times. Below are possible scenarios for the given behavior.

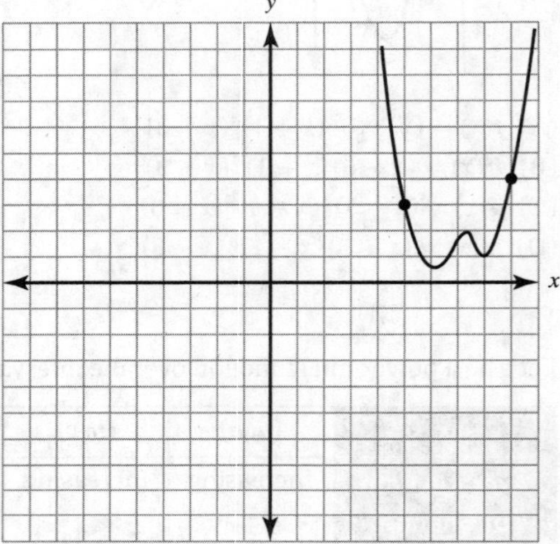

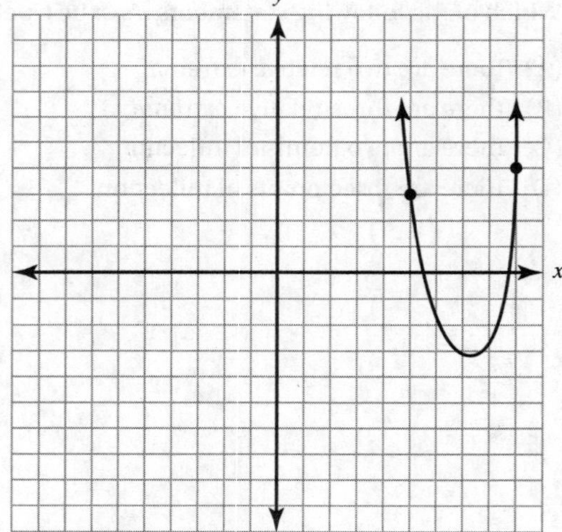

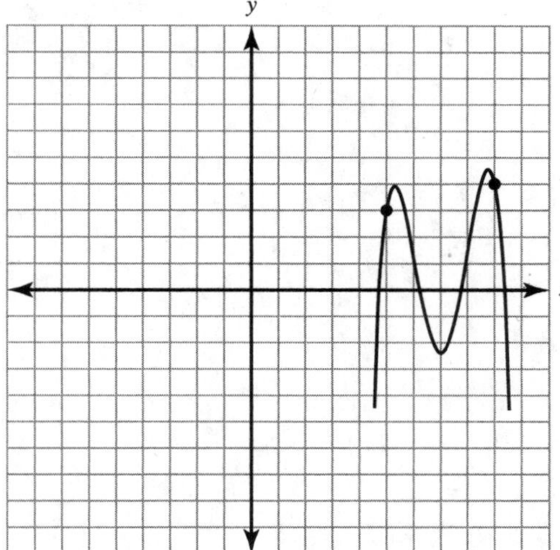

The first graph on the previous page has no real zeros, the second graph on the previous page has two real zeros, and the graph above has four real zeros. It is possible to have a graph that is tangent to the x-axis to fit the given information, but that would mean that zero has an even multiplicity and would count as two zeros.

4. **(D)** Since the end behavior of the graph falls on the left and rises on the right, the function must have an odd degree and a positive leading coefficient. Since the graph crosses the x-axis at $x = -2$, its factor $(x + 2)$ must have an odd multiplicity. Since the graph is tangent to the x-axis at $x = 4$, its factor $(x - 4)$ must have an even multiplicity.

5. **(B)** Between every two distinct real zeros of a nonconstant polynomial function there must be at least one input value corresponding to a local maximum or local minimum. Since it is given that there are exactly three and only three zeros for this function, b, c, and d, there exists a zero between $x = b$ and $x = c$ and a second zero between $x = c$ and $x = d$.

6. **(A)** Since f is a polynomial function, there must be smooth bends in the graph, which eliminates Choice (B) with its sharp corners. It was given that f is an even function. So it must be symmetric with respect to the y-axis, which eliminates Choice (C). Choice (D) is eliminated because it does not follow the behavior listed in the table since the graph has some positive output values in the interval $(1, 2)$.

7. **(C)** If $(x + 1)$ is a factor of $g(x)$, then $x = -1$ is a zero and $g(-1) = 0$. By setting the equation equal to 0 and substituting -1 for x, k can be isolated and solved for:

$$g(-1) = (-1)^4 + 6(-1)^3 + k(-1)^2 - 26(-1) - 24 = 0$$
$$1 - 6 + k + 26 - 24 = 0$$
$$k = 3$$

8. **(C)** The graph has one negative zero and two positive zeros since there is one x-intercept to the left of the origin and two x-intercepts to the right of the origin. Assuming the values of these zeros are $x = -5, 1, 3$, the factors would be $(x + 5)$, $(x - 1)$, and $(x - 3)$, respectively. This eliminates Choices (A) and (B). Since the graph is tangent to the x-axis at the first two x-intercepts, where we are assuming $x = -5$ and $x = 1$, their respective factors need to have an even multiplicity.

9. **(C)** The graph of this function is symmetric with respect to the origin, which means that it is an odd function. If a function is odd, $f(-x) = -f(x)$.

10. **(D)** The table provides information regarding the behavior of the average rates of change of the function over intervals. Points of inflection of a polynomial function occur at input values where the rate of change of the function changes from increasing to decreasing or from decreasing to increasing. Choices (A) and (B) are eliminated since this table does not yield definite conclusions about relative extrema. Since the function makes three changes, from either increasing to decreasing or decreasing to increasing, there are three points of inflection.

3

Rational Functions

Learning Objectives

In this chapter, you will learn:

- ➔ The different properties of rational functions
- ➔ How to graph rational functions
- ➔ How to divide rational functions
- ➔ How to solve rational inequalities

A rational function is represented as a quotient of two polynomial functions such as $h(x) = \frac{f(x)}{g(x)}$, where $f(x)$ and $g(x)$ are polynomial functions. Unlike polynomial functions, graphs of rational functions may be discontinuous, meaning that there are breaks in the graph. These discontinuities may be asymptotes or holes on the graph.

Since a rational function gives a measure of the relative size of the polynomial function in the numerator compared with the polynomial function in the denominator for each value in the domain, the end behavior of a rational function is found differently from that of a polynomial function. When discussing the behavior of a rational function, it is helpful to know any intercepts, asymptotes, holes, and end behavior of the graph.

3.1 Equivalent Representations of Rational Expressions: Long Division

As with polynomial functions, the factored form of a rational function provides information about zeros, x-intercepts, asymptotes, holes, domain, and range. However, the standard form of rational functions, just like with polynomial functions, is also useful in that it can reveal information about the end behavior of the function.

> Being able to represent polynomial and rational functions in different, equivalent forms is an important skill as they each help answer questions about the function.

With polynomial functions, factoring was an important technique when rewriting in equivalent forms. For rational functions, using polynomial long division is necessary. Polynomial long division is an algebraic process that, like numerical long division, involves a quotient and remainder. If $h(x) = \frac{f(x)}{g(x)}$, then $f(x)$ can be rewritten as $f(x) = g(x)q(x) + r(x)$, where q is the quotient, r is the remainder, and the degree of r is less than the degree of g.

The long division of polynomials in standard form follows a pattern like that for the long division of whole numbers.

Example

Find the quotient and remainder when x^3 is divided by $x^2 - 9$ using long division.

✓ Solution

Table 3.1 shows the steps to solve this problem.

Table 3.1

1. Divide the first term of the dividend, x^3, by the first term of the divisor, x^2. $$\frac{x^3}{x^2} = x$$ Write x above the leading term of the dividend.	$x^2 - 9\overline{)x^3}$ with x above
2. Multiply x by the divisor $x^2 - 9$. $$x(x^2 - 9) = x^3 - 9x$$ Write the result on the line below the dividend, aligning like terms.	$x^2 - 9\overline{)x^3}$ with x above; $x^3 - 9x$ below
3. Subtract and bring down the remaining terms.	$x^2 - 9\overline{)x^3}$ with x above; $\underline{-(x^3 - 9x)}$; $9x$

The process is finished here because dividing $9x$ by $x^2 - 9$ does not result in a monomial. The quotient is x, and the remainder is $9x$.

The solution can be written as:

$$\frac{x^3}{x^2 - 9} = x + \frac{9x}{x^2 - 9}$$

› Example

Using long division, write an equivalent form of the rational function $f(x) = \frac{2x^3 + 5x^2 - 3x + 7}{x^2 + 6}$.

✓ Solution

When setting up the long division problem, $x^2 + 6$ is the divisor and $2x^3 + 5x^2 - 3x + 7$ is the dividend. Then follow the steps shown in Table 3.2.

Table 3.2

1. Divide the first term of the dividend, $2x^3$, by the first term of the divisor, x^2. $$\frac{2x^3}{x^2} = 2x$$ Write $2x$ above the leading term of the dividend.	$\begin{array}{r} 2x \qquad\qquad\qquad \\ x^2+6\overline{)2x^3+5x^2-3x+7} \end{array}$
2. Multiply $2x$ by the divisor $x^2 + 6$. $$2x(x^2+6) = 2x^3 + 12x$$ Write the result on the line below the dividend, aligning like terms.	$\begin{array}{r} 2x \qquad\qquad\qquad \\ x^2+6\overline{)2x^3+5x^2-3x+7} \\ \underline{2x^3 \qquad\quad +12x} \quad \end{array}$
3. Subtract and bring down the remaining terms.	$\begin{array}{r} 2x \qquad\qquad\qquad \\ x^2+6\overline{)2x^3+5x^2-3x+7} \\ \underline{-(2x^3 \qquad\quad +12x)} \quad \\ 5x^2-15x+7 \end{array}$
4. Repeat the first three steps, using $5x^2 - 15x + 7$ as the dividend. $$\frac{5x^2}{x^2} = 5$$ $$5(x^2+6) = 5x^2 + 30$$	$\begin{array}{r} 2x \qquad\qquad\qquad \\ x^2+6\overline{)2x^3+5x^2-3x+7} \\ \underline{-(2x^3 \qquad\quad +12x)} \quad \\ 5x^2-15x+7 \\ \underline{-(5x^2 \qquad\quad +30)} \\ -15x-23 \end{array}$

The process is finished here because dividing $-15x - 23$ by $x^2 + 6$ does not result in a monomial. The quotient is $2x + 5$, and the remainder is $-15x - 23$.

The solution can be written as:

$$f(x) = \frac{2x^3 + 5x^2 - 3x + 7}{x^2 + 6} = 2x + 5 + \frac{-15x - 23}{x^2 + 6}$$

3.2 Rational Functions and Zeros

The real zeros of a rational function correspond to the real zeros of the numerator for such values in its domain. Graphically, the zeros of the function are also the x-intercepts of the graph of the rational function.

Finding Intercepts

To determine the points at which the graph of the rational function $f(x) = \frac{P(x)}{Q(x)}$ intersects the coordinate axes:

- Set $P(x) = 0$. Any solution of this equation that does not also make $Q(x) = 0$ is an x-intercept.
- Evaluate $f(0)$. The y-intercept is $(0, f(0))$ if it is defined.

Example

Find the x- and y-intercepts of the rational function $f(x) = \frac{x^2 - 64}{x + 2}$.

✓ **Solution**

To find the x-intercept(s), set the polynomial function in the numerator equal to 0 and solve for x:

$$x^2 - 64 = 0$$
$$(x + 8)(x - 8) = 0$$
$$x = \pm 8$$

Since the denominator does not equal 0 when $x = \pm 8$, the x-intercepts are $(-8, 0)$ and $(8, 0)$.

To find the y-intercept, evaluate $f(0)$:

$$f(0) = \frac{(0)^2 - 64}{(0) + 2} = \frac{-64}{2} = -32$$

The y-intercept is $(0, -32)$.

3.3 Rational Functions and Vertical Asymptotes

An asymptote is a line whose distance from a curve approaches 0 as the curve tends to infinity. The graph of a rational function has a vertical asymptote at a value of x for which the denominator evaluates to 0 but the numerator does not evaluate to 0.

Locating Vertical Asymptotes

Let $f(x) = \frac{P(x)}{Q(x)}$. The line $x = a$ is a vertical asymptote of the graph of function f when $Q(a) = 0$ and $P(a) \neq 0$.

› **Example**

Find and graph the vertical asymptotes of $f(x) = \frac{x^2}{x^2 - 5x + 6}$.

✓ **Solution**

To find the vertical asymptotes, set the denominator equal to 0 and solve for x:

$$x^2 - 5x + 6 = 0$$
$$(x - 3)(x - 2) = 0$$
$$x = 3,\ x = 2$$

Since the numerator does not equal 0 when $x = 3$ or $x = 2$, the vertical asymptotes for f are $x = 3$ and $x = 2$. The vertical asymptotes are graphed with a dashed vertical line as shown in Figure 3.1. For now, the rest of the graph is found by using a graphing calculator.

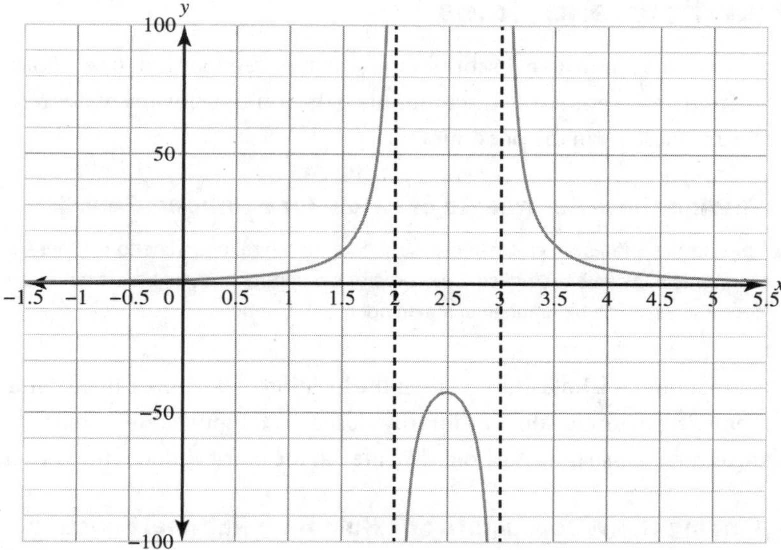

Figure 3.1

Limit notation is also useful here to show the behavior of a rational function at a vertical asymptote. Near a vertical asymptote, $x = a$, of a rational function, the values of the polynomial in the denominator are arbitrarily close to 0, so the values of the rational function increase or decrease without bound. This can be notated by $\lim_{x \to a^-} f(x) = -\infty$ or $\lim_{x \to a^-} f(x) = \infty$. This is read as "the limit as x approaches a from the left of the function $f(x)$ equals negative or positive infinity." The limit as x approaches a from the left means the input values begin less than a and then get closer to a.

Notice the graph of the function gets closer and closer to the asymptote without ever intersecting. This is what is meant by the asymptote being a line whose distance from a curve approaches 0 as the curve tends to infinity.

Similarly, coming in from the other direction can be notated by $\lim_{x \to a^+} f(x) = -\infty$ or $\lim_{x \to a^+} f(x) = \infty$. This is read as "the limit as x approaches a from the right of the function $f(x)$ equals negative or positive infinity." The limit as x approaches a from the right means the input values begin greater than a as they get closer to a.

Table 3.3 demonstrates this with a table of values from our previous example using the function $f(x) = \dfrac{x^2}{x^2 - 5x + 6}$ and its vertical asymptote $x = 2$.

Table 3.3

x	1	1.9	1.999999	2	2.00001	2.01	2.1
$f(x)$	4.5	90	10000000	undefined	−1000000	−1011	−112.2

As x is approaching 2 from the left, the output values get larger and larger, increasing without bound. As x is approaching 2 from the right, the output values are getting smaller and smaller, decreasing without bound.

Therefore, $\lim_{x \to 2^-} f(x) = \infty$ and $\lim_{x \to 2^+} f(x) = -\infty$.

3.4 Rational Functions and Holes

A second type of discontinuity is a removable discontinuity, which is also referred to as a hole because of how it is graphed. A rational function has a removable discontinuity if the multiplicity of a real zero in the numerator is greater than or equal to its multiplicity in the denominator.

Finding the *x*-Coordinate of a Hole for a Rational Function

Completely factor the polynomials in the numerator and denominator of the rational function. If there is a common factor in the numerator and denominator, it can be removed. The zeros of the common factors represent the *x*-coordinates of the hole. After finding the coordinates, it can be graphed using an open circle, or hole.

If the graph of a rational function has a hole at $x = c$, then the location of the hole can be determined by examining the output values corresponding to input values arbitrarily close to c. If input values arbitrarily close to c correspond to output values arbitrarily close to L, the hole is located at the point with coordinates (c, L).

Finding the *y*-Coordinate of a Hole for a Rational Function

After the common factor is removed from the numerator and denominator of the rational function, substitute each *x*-value of the holes into all the remaining factors of the rational function. The solutions represent the *y*-coordinate of each respective *x*-value.

Example

Find the discontinuities for the function $h(x) = \dfrac{x^2 + 9x + 8}{x + 8}$ and graph $h(x)$.

Solution

When working with rational functions, factor the polynomials in the numerator and denominator to correctly identify all removable discontinuities:

$$h(x) = \frac{x^2 + 9x + 8}{x + 8} = \frac{(x + 8)(x + 1)}{x + 8}$$

There is a common factor of $(x + 8)$ in the numerator and denominator that can be removed. Note that although this factor is removed, it doesn't disappear and still influences the graph of the function. The zero of the removable factor is the x-coordinate of the hole. The x-coordinate of the hole is $x = -8$.

After removing this discontinuity, the remaining factors make up the rational function. They are used to find the y-coordinate of the hole and contribute to the other behaviors of the rational function:

$$h(x) = x + 1$$

To find the y-coordinate, substitute $x = -8$ into the new function:

$$h(-8) = (-8) + 1 = -7$$

The coordinates of the hole are $(-8, -7)$.

With the removal of this discontinuity, there are no longer any terms left in the denominator, so there are no vertical asymptotes. Using technology, the graph is shown in Figure 3.2.

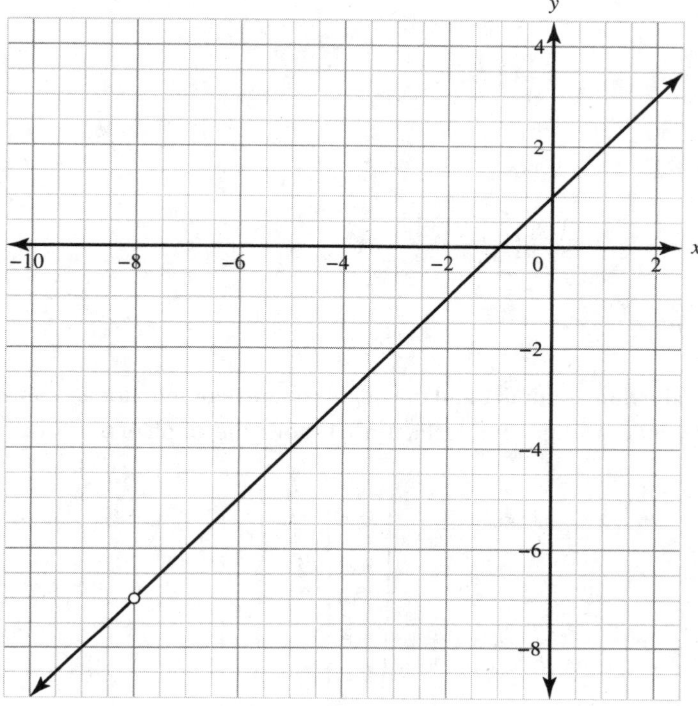

Figure 3.2

It is important to always factor first; then holes can be identified correctly and removed. Any remaining factors in the denominator can be correctly identified as vertical asymptotes.

3.5 Rational Functions and End Behavior

The end behavior of a rational function will be most affected by the polynomial with the greater degree, as its values will dominate the values of the rational function for input values of large magnitude. For input values of large magnitude, a polynomial is dominated by its leading term. Therefore, the end behavior of a rational function can be understood by examining the corresponding quotient of the leading terms.

There are two different possibilities for the end behavior of a rational function, which we call end behavior asymptotes: a horizontal asymptote or a slant asymptote.

Finding End Behavior Asymptotes

Let $f(x) = \frac{P(x)}{Q(x)} = \frac{a_n x^n + a_{n-1} x^{n-1} + \ldots + a_1 x + a_0}{b_m x^m + b_{m-1} x^{m-1} + \ldots + b_1 x + b_0}$, where $a_n \neq 0$ and $b_m \neq 0$.

- If $n < m$, the rational function has a horizontal asymptote at $y = 0$.
- If $n = m$, the rational function has a horizontal asymptote at $y = \frac{a_n}{b_m}$.
- If $n > m$, the rational function has no horizontal asymptote. The end behavior asymptote is the quotient polynomial function that is found when the polynomial $P(x)$ is divided into $Q(x)$. If that polynomial is linear, the rational function has a slant asymptote parallel to the graph of the line.

Example

Find the horizontal asymptotes of each of the rational functions. Use technology to graph each.

1. $f(x) = \dfrac{5x}{x^2 - 9}$

2. $g(x) = \dfrac{3x^2 - x - 4}{2x^2 + 5}$

3. $h(x) = \dfrac{x^3 - 3x^2 + 3x + 1}{x - 1}$

4. $j(x) = \dfrac{x^3}{x^2 - 9}$

Solution

1. For $f(x)$, the degree of the numerator is 1 and the degree of the denominator is 2. Since $1 < 2$, the rational function has the horizontal asymptote $y = 0$. Figure 3.3 shows the graph of this function.

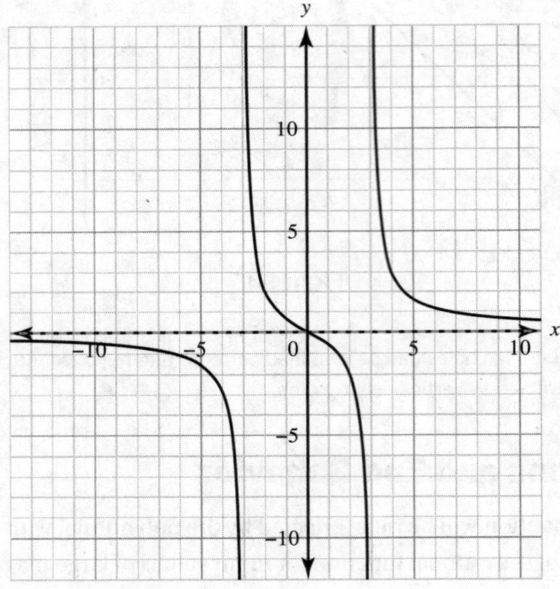

Figure 3.3

As x approaches $\pm\infty$, $f(x)$ approaches 0. This graphically shows how the end behavior of a rational function is shown through its horizontal asymptote. Note that at $x = 0$, the graph intersects the origin, showing that the middle portion of the graph can intersect the horizontal asymptote. The ends of the graph, though, approach the x-axis, or $y = 0$, and follow along the horizontal asymptote, which is graphed as a dashed horizontal line.

On the graph, it appears that there are two vertical asymptotes at $x = -3$ and $x = 3$. This is verified when the denominator of $f(x)$ is factored and set equal to 0.

2. For $g(x)$, the degree of the numerator is 2 and the degree of the denominator is 2. Since $2 = 2$, the rational function has the horizontal asymptote $y = \frac{3}{2}$. Figure 3.4 shows the graph of this function.

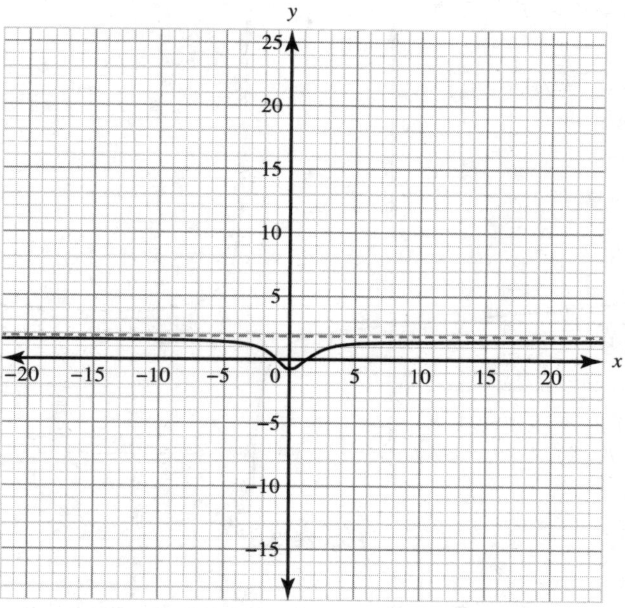

Figure 3.4

3. For $h(x)$, the degree of the numerator is 3 and the degree of the denominator is 1. Since $3 > 1$, the rational function has no horizontal asymptote. However, the end behavior asymptote is found by using polynomial long division:

$$h(x) = \frac{x^3 - 3x^2 + 3x + 1}{x - 1} = x^2 - 2x + 1 + \frac{2}{x - 1}$$

The end behavior asymptote of $h(x)$ is $y = x^2 - 2x + 1$, the quotient after completing long division.

Figure 3.5 shows the graph of $h(x)$. It can also be seen that there is a vertical asymptote at $x = 1$, which is confirmed algebraically by setting the denominator equal to 0.

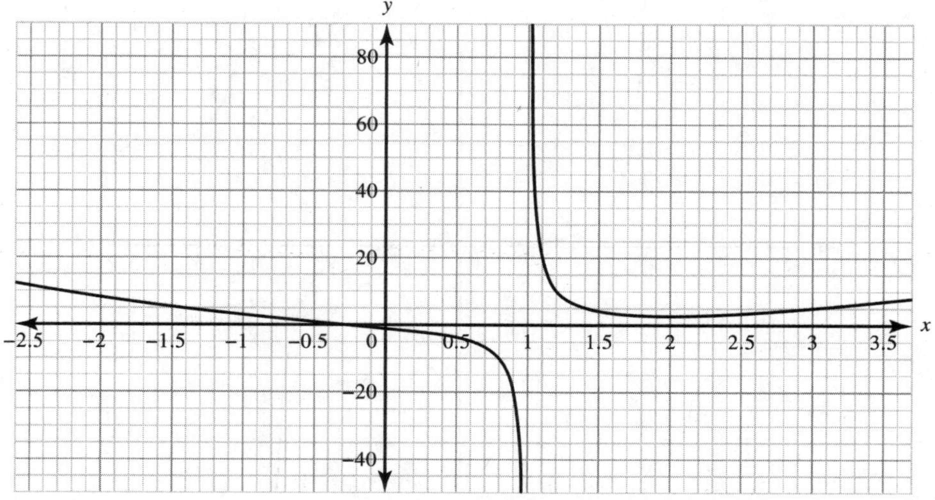

Figure 3.5

Figure 3.6 is another graph of $h(x)$ that shows its end behavior asymptote $y = x^2 - 2x + 1$ as a dashed curve. As x approaches $\pm\infty$, the graph of $h(x)$ and its asymptote appear to be identical.

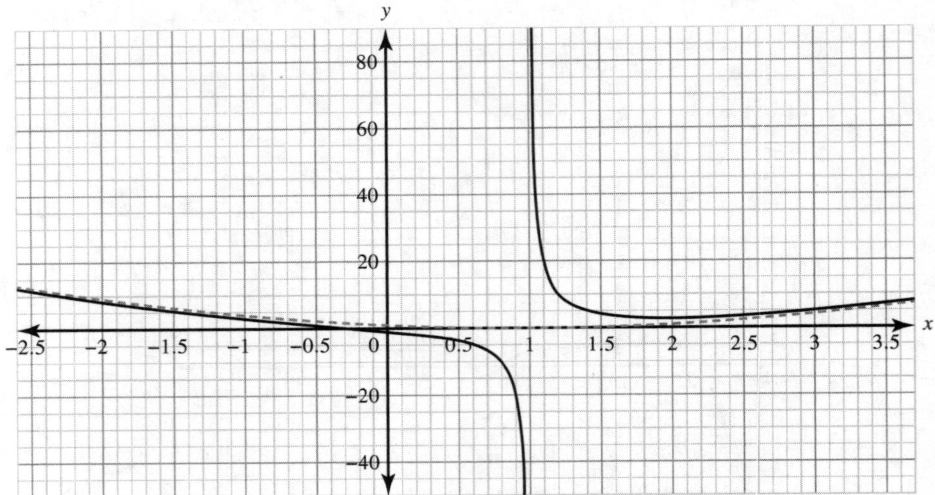

Figure 3.6

4. For $j(x)$, the degree of the numerator is 3 and the degree of the denominator is 2. Since $3 > 2$, the rational function has no horizontal asymptote. However, the end behavior asymptote is found by using polynomial long division:

$$j(x) = \frac{x^3}{x^2 - 9} = x + \frac{9x}{x^2 - 9}$$

The end behavior asymptote of $j(x)$ is $y = x$, the quotient after completing long division. Since this is a linear function, this is known as a slant asymptote.

Figure 3.7 shows the graph of $j(x)$. Vertical asymptotes can be seen at $x = -3$ and $x = 3$, which is confirmed algebraically by setting the denominator equal to 0.

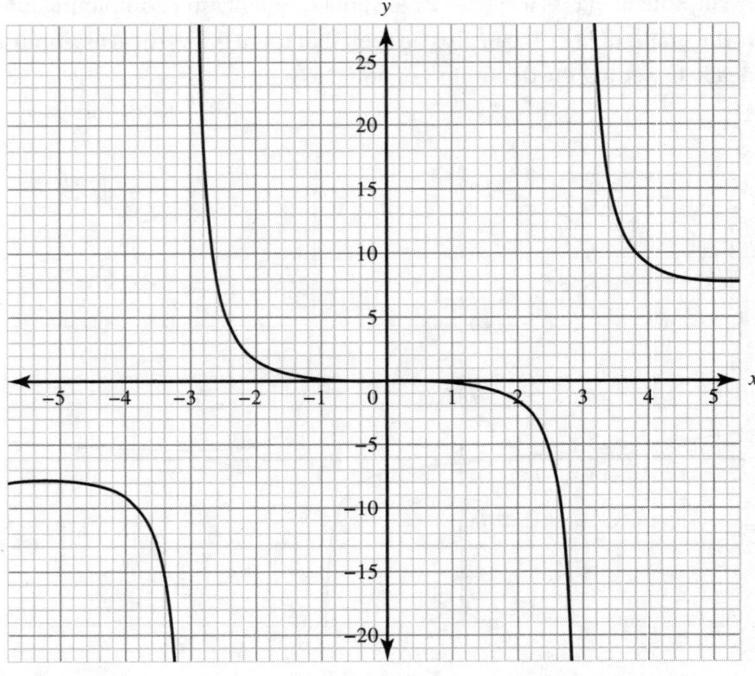

Figure 3.7

Figure 3.8 is another graph of $j(x)$ that shows its end behavior asymptote $y = x$. As x approaches $\pm\infty$, the graph of $j(x)$ and its asymptote appear to be identical as the graph of $j(x)$ runs parallel to the graph of $y = x$.

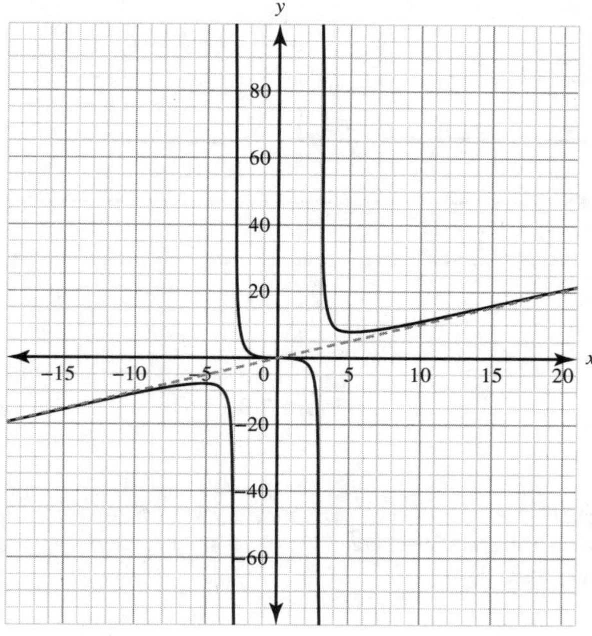

Figure 3.8

> It is possible for the graph of a rational function to intersect a horizontal asymptote and a slant asymptote. The graph of a rational function will never intersect a vertical asymptote.

Example

If $f(x) = \dfrac{10x^4 - 2x^2 + 3x + 7}{5x^4 + x^3 - 2}$, evaluate each of the following limits:

(a) $\lim\limits_{x \to -\infty} f(x)$

(b) $\lim\limits_{x \to \infty} f(x)$

Solution

Limits approaching infinity are really asking about the end behavior of a function. Since this is a rational function, the end behavior is found by examining the degrees. The degree of the numerator is 4, and the degree of the denominator is 4. Since the degrees are equal, the graph has a horizontal asymptote at $y = \frac{10}{5} = 2$.

(a) As x gets more negative, the y-values will get closer to 2. Therefore:

$$\lim_{x \to -\infty} f(x) = 2$$

(b) As x gets more positive, the y-values will get closer to 2. Therefore:

$$\lim_{x \to \infty} f(x) = 2$$

These limits can both be verified with the graph of $f(x)$ shown in Figure 3.9.

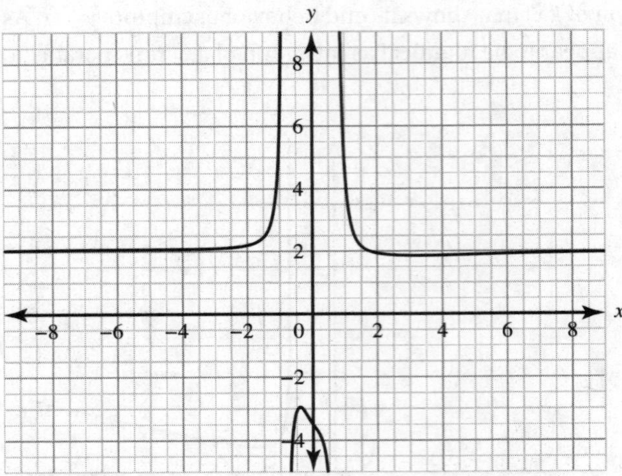

Figure 3.9

When the graph of a rational function, $f(x)$, has a horizontal asymptote $y = b$, where b is a constant, the output values of the rational function get arbitrarily close to b and stay arbitrarily close to b as input values increase or decrease without bound. In other words, $\lim_{x \to -\infty} f(x) = b$ and $\lim_{x \to \infty} f(x) = b$.

3.6 Graphs of Rational Functions

As with polynomial functions, there is a suggested order to follow to graph rational functions.

Seven-Step Strategy for Graphing a Rational Function

1. Factor the numerator and denominator of the rational function to find any holes. Remove any discontinuities.
2. Find all possible x- and y-intercepts.
3. Determine the degree of the numerator and denominator. Determine the end behavior asymptote.
4. Find all vertical asymptotes with any remaining factors in the denominator.
5. Find any symmetry the graph displays by evaluating $f(-x)$.
6. The vertical asymptotes will divide the graph into regions. In each region, graph at least one point that will determine whether the graph will be above or below any horizontal asymptotes.
7. Sketch a graph using the above information.

Example

Find the key features of the graph of $f(x) = \dfrac{x^2 - 64}{x^2 - 16}$ and then sketch the graph.

✓ Solution

To find the key features of the graph, utilize the seven-step strategy.

STEP 1 Find any holes

Factor the numerator and denominator, and look for any common factors:

$$f(x) = \frac{x^2 - 64}{x^2 - 16} = \frac{(x-8)(x+8)}{(x-4)(x+4)}$$

Since there are no common factors, there are no removable discontinuities. The graph of $f(x)$ does not have any holes.

STEP 2 Find all possible x- and y-intercepts

To find the x-intercepts, set the numerator equal to 0 and solve for x:

$$x^2 - 64 = 0, x = \pm 8$$

The x-intercepts are $(-8, 0)$ and $(8, 0)$.

To find the y-intercept, evaluate $f(0)$:

$$f(0) = \frac{(0)^2 - 64}{(0)^2 - 16} = \frac{-64}{-16} = 4$$

The y-intercept is $(0, 4)$.

STEP 3 Find the end behavior asymptote

The degree of the numerator and denominator is 2. Since the degrees are equal, there is a horizontal asymptote. Its equation is $y = \frac{1}{1} = 1$.

STEP 4 Find all vertical asymptotes

Set the denominator equal to 0 and solve for x.

$$(x-4)(x+4) = 0, x = \pm 4$$

Since the numerator does not equal 0 at these x-values, the vertical asymptotes are $x = -4$ and $x = 4$.

STEP 5 Evaluate $f(-x)$

$$f(-x) = \frac{(-x)^2 - 64}{(-x)^2 - 16} = \frac{x^2 - 64}{x^2 - 16} = f(x)$$

$f(x)$ is an even function that has symmetry with respect to the y-axis.

STEP 6 Test points

As shown in Table 3.4, choose x-values within each interval broken up by the vertical asymptotes. Then find their corresponding y-values.

Table 3.4

x	-5	1	5
$f(x)$	$\frac{(-5)^2 - 64}{(-5)^2 - 16} = \frac{-39}{9} = -4.333$	$\frac{(1)^2 - 64}{(1)^2 - 16} = \frac{-63}{-15} = 4.2$	$\frac{(5)^2 - 64}{(5)^2 - 16} = \frac{-39}{9} = -4.333$

Since the function is even, the point $(1, 4.2)$ will also lie on the graph.

STEP 7 Graph

Plot and graph everything found in Steps 1–6, as shown in Figure 3.10.

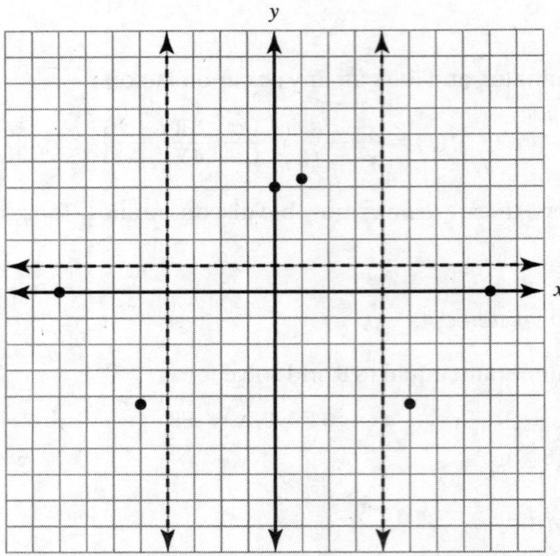

Figure 3.10

To create the graph, as shown in Figure 3.11, follow along the horizontal and vertical asymptotes while passing through the intercepts. Remember to observe the symmetry with respect to the *y*-axis.

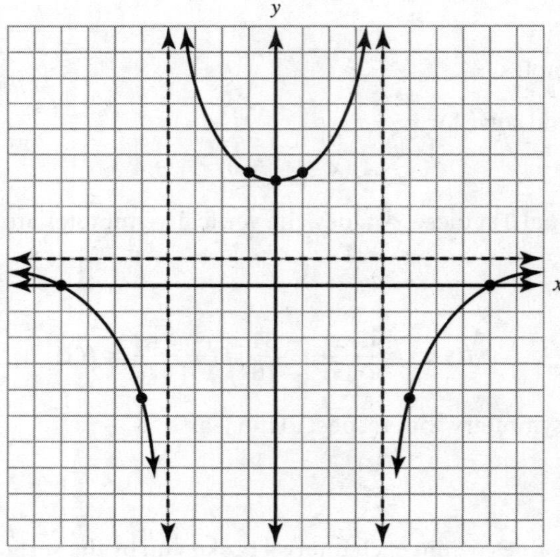

Figure 3.11

Example

Find the key features of the graph of $g(x) = \dfrac{(x+1)(3x^2 - 12x + 2)}{(x+1)(x-4)}$ and then sketch the graph.

✓ Solution

To find the key features of the graph, utilize the seven-step strategy.

STEP 1 Find any holes

Since the given rational function is already factored, look for any common factors:

$$g(x) = \frac{(x+1)(3x^2 - 12x + 2)}{(x+1)(x-4)}$$

There is a common factor, $(x + 1)$. Remove this discontinuity to produce the following function:

$$g(x) = \frac{3x^2 - 12x + 2}{x - 4}$$

The x-coordinate of the hole is $x = -1$. To find its corresponding y-value, evaluate $g(-1)$:

$$g(-1) = \frac{3(-1)^2 - 12(-1) + 2}{(-1) - 4} = \frac{17}{-5} = -3.4$$

The coordinates of the hole are $(-1, -3.4)$.

The remaining steps will now be determined using the following function: $g(x) = \frac{3x^2 - 12x + 2}{x - 4}$

STEP 2 Find all possible x- and y-intercepts

To find the x-intercepts, set the numerator equal to 0 and solve for x:

$$3x^2 - 12x + 2 = 0$$

To solve for x, use the quadratic formula:

$$x = \frac{-(-12) \pm \sqrt{(-12)^2 - 4(3)(2)}}{2(3)} = \frac{12 \pm \sqrt{120}}{6} = \frac{12 \pm \sqrt{4}\sqrt{30}}{6} = \frac{12 \pm 2\sqrt{30}}{6} = \frac{6 \pm \sqrt{30}}{3}$$

The x-intercepts are $\left(\frac{6 - \sqrt{30}}{3}, 0\right) \approx (0.174258, 0)$ and $\left(\frac{6 + \sqrt{30}}{3}, 0\right) \approx (3.825741, 0)$.

To find the y-intercept, evaluate $g(0)$:

$$g(0) = \frac{3(0)^2 - 12(0) + 2}{(0) - 4} = -\frac{2}{4} = -\frac{1}{2}$$

The y-intercept is $\left(0, -\frac{1}{2}\right)$.

STEP 3 Find the end behavior asymptote

The degree of the numerator is 2, and the degree of the denominator is 1. Since $2 > 1$, there is no horizontal asymptote.

Use polynomial long division to find the end behavior asymptote:

$$\begin{array}{r} 3x \\ x - 4 \overline{\smash{)}\, 3x^2 - 12x + 2} \\ \underline{-(3x^2 - 12x)} \\ 2 \end{array}$$

$$g(x) = \frac{3x^2 - 12x + 2}{x - 4} = 3x + \frac{2}{x - 4}$$

The end behavior asymptote is $y = 3x$. Since it is linear, it is a slant asymptote.

STEP 4 Find all vertical asymptotes

Set the denominator equal to 0 and solve for x:

$$x - 4 = 0, x = 4$$

Since the numerator does not equal 0 at this x-value, the vertical asymptote is $x = 4$.

STEP 5 Evaluate $g(-x)$

$$g(-x) = \frac{3(-x)^2 - 12(-x) + 2}{(-x) - 4} = \frac{3x^2 + 12x + 2}{-x - 4}$$

This does not equal either $g(x)$ or $-g(x)$. This means that $g(x)$ is not an even or an odd function. So there will not be any symmetry with respect to the y-axis or the origin.

STEP 6 Test points

As shown in Table 3.5, choose x-values within each interval broken up by the vertical asymptotes. Then find their corresponding y-values.

Table 3.5

x	-2	5
$g(x)$	$\frac{3(-2)^2 - 12(-2) + 2}{(-2) - 4} = \frac{38}{-6} = -6.333$	$\frac{3(5)^2 - 12(5) + 2}{(5) - 4} = \frac{17}{1} = 17$

STEP 7 Graph

Plot and graph everything found in Steps 1–6, as shown in Figure 3.12.

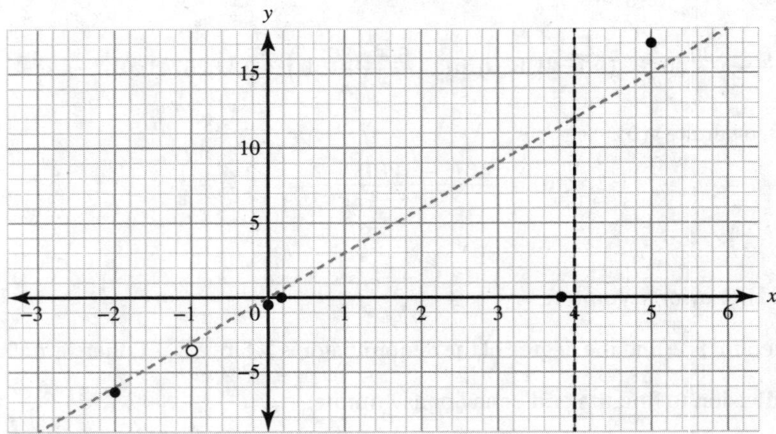

Figure 3.12

To create the graph, as shown in Figure 3.13, follow along the horizontal and vertical asymptotes while passing through the intercepts.

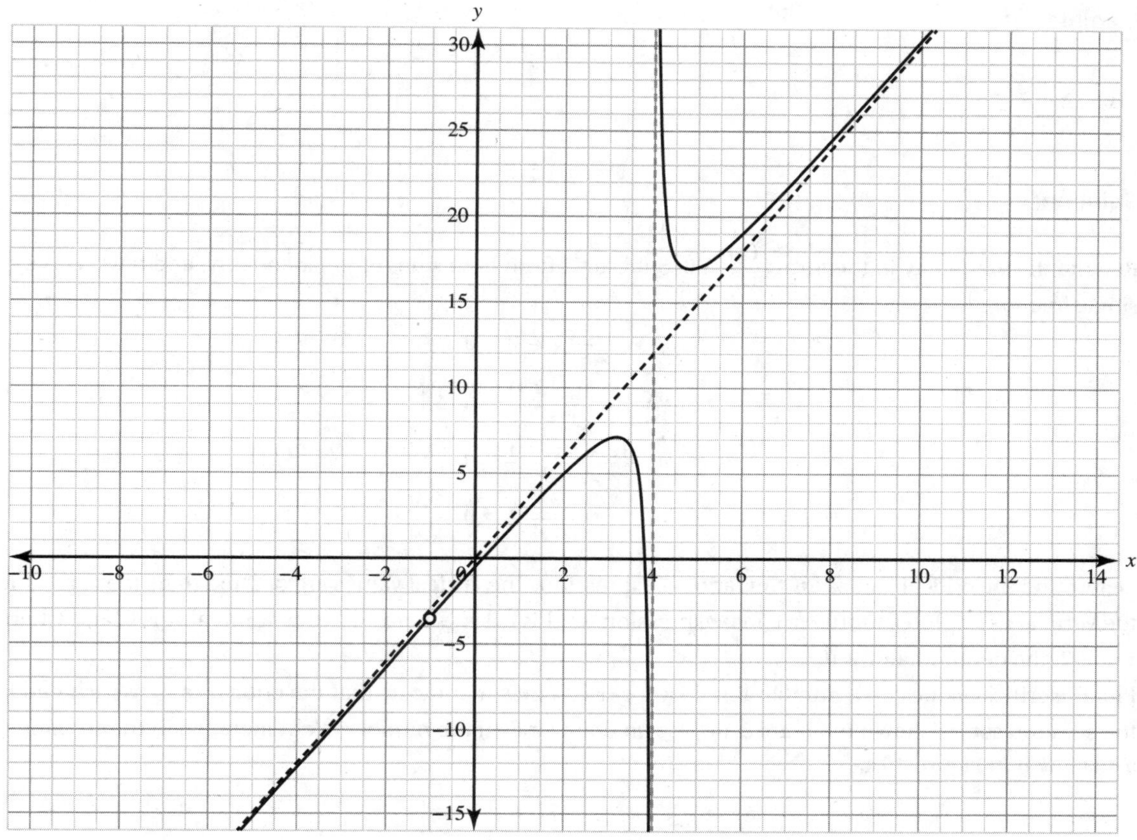

Figure 3.13

3.7 Rational Inequalities

The key to solving rational inequalities comes from the geometric property that the graph of $y = f(x)$ lies above the x-axis when $f(x) > 0$ and below the x-axis when $f(x) < 0$.

Solving Rational Inequalities

1. Write the inequality in one of these forms:
   $f(x) > 0$ $f(x) \geq 0$ $f(x) < 0$ $f(x) \leq 0$
2. Determine the critical values of $f(x)$. These are the x-intercepts and vertical asymptotes of the rational expression.
3. Plot the critical values on a number line.
4. Use test points to determine whether the graph of $f(x)$ is above or below the x-axis on each of the intervals determined by the critical values.
5. Find the solutions of the inequality.

Example

Solve: $\frac{3x}{x+2} \geq 2$

Solution

1. Rewrite the inequality in the form $\frac{f(x)}{g(x)} \geq 0$. Subtract 2 from both sides of the inequality, and rewrite as a single fraction:

$$\frac{3x}{x+2} - 2 \geq 0$$

$$\frac{3x}{x+2} - \frac{2(x+2)}{(x+2)} \geq 0$$

$$\frac{3x - 2x - 4}{x+2} \geq 0$$

$$\frac{x-4}{x+2} \geq 0$$

2. Determine the critical values for the rational expression. The critical values are defined to be the zeros and the vertical asymptotes for the rational expression. In this case, when $x = 4$, the numerator is 0, and when $x = -2$, the denominator is 0.
3. Plot the critical values on a number line. Use a closed circle to indicate that the critical value is included in the solution set, as shown in Figure 3.14. Use an open circle if the number is not included in the solution set, also shown in Figure 3.14.

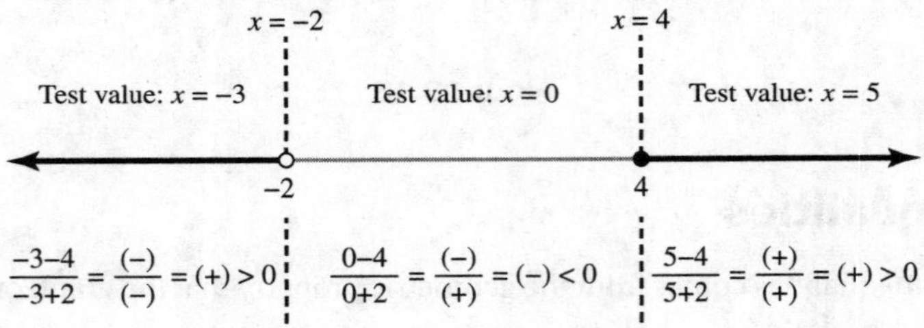

Figure 3.14

4. Choose a test value for x in each of the intervals determined by the critical values. Substitute that test value into the rational expression. Use the substitution to determine the signs of the numerator and denominator as well as their quotient, as shown in Figure 3.14. For example, in the left interval, let $x = -3$ represent the test value. Substitute into the rational expression:

$$\frac{(-3) - 4}{(-3) + 2} = \frac{-7}{-1} = +7 \geq 0$$

5. The solution intervals are $(-\infty, -2) \cup [4, \infty)$.

Example

Solve for x: $\frac{5}{x-3} \leq \frac{2}{x}$

Solution

1. Rewrite the given inequality in the form $\frac{f(x)}{g(x)} \leq 0$, and rewrite as a single fraction:

$$\frac{5}{x-3} - \frac{2}{x} \leq 0$$

$$\frac{5x - 2(x-3)}{x(x-3)} \leq 0$$

$$\frac{3x+6}{x(x-3)} \leq 0$$

2. To solve for the critical values, set the numerator and denominator each equal to 0 and solve for x:

$$3x + 6 = 0 \rightarrow x = -2$$

$$x(x-3) = 0 \rightarrow x = 0, x = 3$$

3. Plot the critical values on a number line. Place a closed circle around -2 since the rational expression can be equal to 0. Use open circles around 0 and 3 since, for these values of x, the rational expression is undefined. This is shown in Figure 3.15.

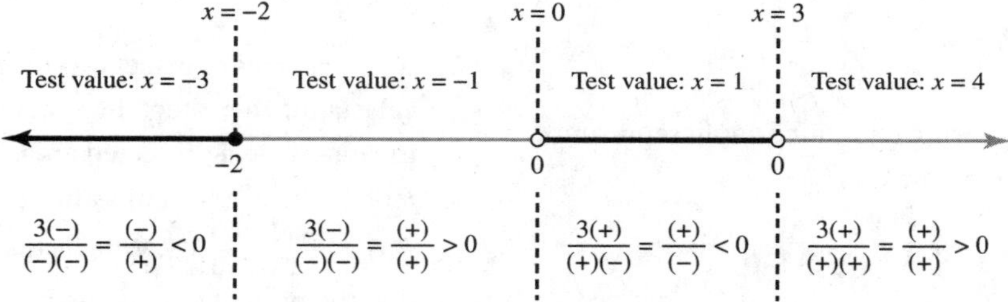

Figure 3.15

4. Determine the sign of the rational expression on each of the four intervals determined by the three critical values, as shown in Figure 3.15.
5. The solution intervals are $(-\infty, -2] \cup (0, 3)$.

Multiple-Choice Questions

1. Which of the following is equivalent to $\lim_{x\to\infty} \frac{4x^2}{x^2 + 5{,}000x}$?

(A) 0
(B) $\frac{1}{1{,}250}$
(C) 1
(D) 4

2. The graph of which of the following equations has $y = 1$ as an asymptote?

(A) $y = \frac{x^2 - 1}{x - 1}$
(B) $y = \frac{x}{x + 1}$
(C) $y = \frac{x^2}{x - 1}$
(D) none of these

3. Which of the following is the end behavior asymptote for the function $f(x) = \frac{x^3 - 8}{x^2 - 3x - 4}$?

(A) $y = 13x + 4$
(B) $y = x - 3$
(C) $y = x + 3$
(D) $y = x - 11$

4. Which of the following equations could represent the graph shown below?

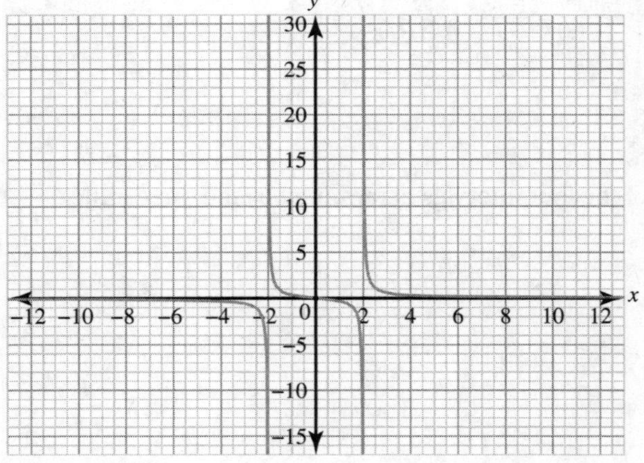

(A) $y = \frac{x}{x^2 - 4}$
(B) $y = \frac{x^2}{x^2 - 4}$
(C) $y = \frac{x^3}{x^2 - 4}$
(D) $y = \frac{x}{x^2 - 2}$

5. Which of the following functions could represent $f(x)$ if $\lim_{x\to 0^-} f(x) = -\infty$ and $\lim_{x\to 0^+} f(x) = \infty$?

(A) $f(x) = \frac{1}{x^2}$
(B) $f(x) = \frac{1}{x(x - 3)}$
(C) $f(x) = \frac{1}{x - 1}$
(D) $f(x) = \frac{1}{x}$

6. A rare species of insect was discovered on a tropical island. To protect the species, scientists moved the insects to a protected area. The population P of the insect t months after being moved is $P(t) = \frac{50(1 + 0.5t)}{(2 + 0.01t)}$. How many insects were initially discovered?

(A) 5
(B) 25
(C) 37
(D) 50

7. A rare species of insect was discovered on a tropical island. To protect the species, scientists moved the insects to a protected area. The population P of the insect t months after being moved is $P(t) = \frac{50(1 + 0.5t)}{(2 + 0.01t)}$. What is the largest population that the protected area can sustain?

(A) 5
(B) 50
(C) 2,500
(D) none of these

8. $\lim_{n\to\infty} \frac{-5n + 3n^3}{n^3 - 2n^2 + 1} =$

(A) −5
(B) 1
(C) 3
(D) nonexistent

9. If the graph of $y = \frac{ax + b}{x + c}$ has a horizontal asymptote $y = 2$ and a vertical asymptote $x = -3$, what is the sum of a and c?

    (A) -5
    (B) -1
    (C) 1
    (D) 5

10. Given the graph of the rational function, $c(x)$, which of the following could be its equation?

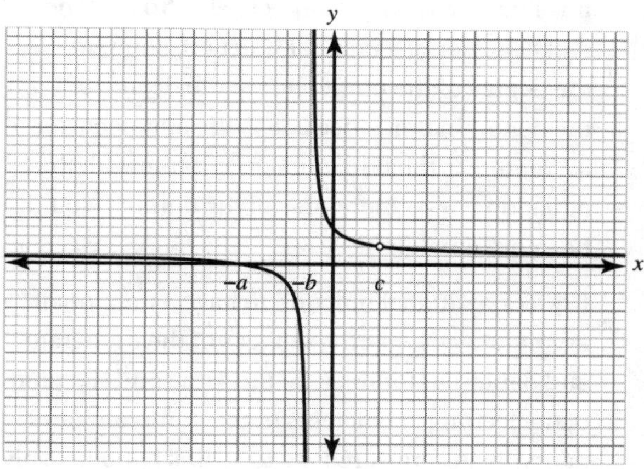

(A) $c(x) = \frac{(x-c)(x+a)}{(x-c)(x+b)}$

(B) $c(x) = \frac{(x+c)(x-a)}{(x+c)(x-b)}$

(C) $c(x) = \frac{(x+b)(x+a)}{(x+b)(x-c)}$

(D) $c(x) = \frac{(x-b)(x+a)}{(x-b)(x+c)}$

Answer Explanations

1. **(D)** Since the limit has x approaching infinity, this limit is assessing end behavior. To find the end behavior of a rational function, compare the degree of the numerator to the degree of the denominator. Since $2 = 2$, the end behavior asymptote is the horizontal asymptote with the equation $y = 4$. Therefore, all the outputs would be approximately 4.

2. **(B)** For a rational function to have a horizontal asymptote at $y = 1$, the degrees of the numerator and denominator must be the same and the ratio of the leading coefficients of the numerator and denominator must reduce to 1. This happens only with the equation $y = \frac{x}{x+1}$.

3. **(C)** To find the end behavior asymptote, compare the degree of the numerator to the degree of the denominator. The degree of the numerator is greater than the degree of the denominator, which means there is no horizontal asymptote. Perform polynomial long division to find the end behavior asymptote:

$$\begin{array}{r} x+3 \\ x^2-3x-4\overline{)\,x^3+8} \\ \underline{-(x^3-3x^2-4x)} \\ 3x^2+4x-8 \\ \underline{-(3x^2-9x-12)} \\ 13x+4 \end{array}$$

$$f(x) = \frac{x^3-8}{x^2-3x-4} = x+3+\frac{13x+4}{x^2-3x-4}$$

The end behavior asymptote is a slant asymptote with the equation $y = x + 3$.

4. **(A)** The graph has vertical asymptotes at $x = -2$ and $x = 2$. The denominator must have zeros that are $x = -2$ and $x = 2$, eliminating Choice (D). The graph has a horizontal asymptote of $y = 0$. The degree of the numerator must be less than the degree of the denominator.

5. **(D)** The unbounded behavior of the given limits occurs only when there is a vertical asymptote at the value x is approaching. The function must have a vertical asymptote at $x = 0$, which eliminates Choice (C). Choices (A) and (B) are eliminated since as x approaches 0 from the left, the function approaches positive infinity. The graph of $f(x) = \frac{1}{x}$ that follows satisfies the given limits.

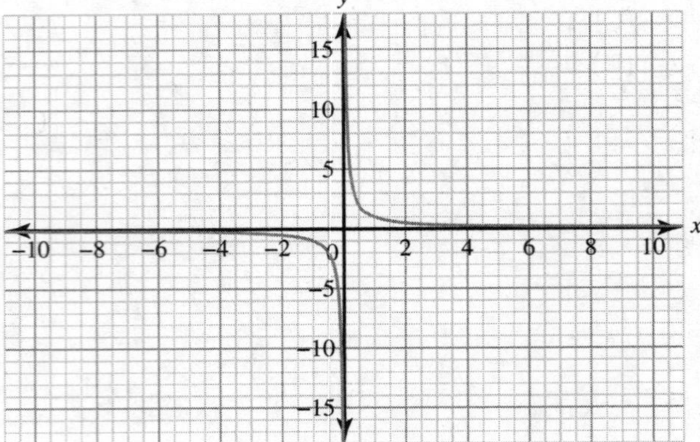

6. **(B)** The initial population can be determined by evaluating $P(0)$:

$$P(0) = \frac{50(1+0.5(0))}{(2+0.01(0))} = \frac{50}{2} = 25$$

7. **(C)** Determining the largest population that the area can sustain refers to the end behavior of the rational function. To determine end behavior, compare the degree of the numerator to the degree of the denominator. Since the degrees of the numerator and denominator both equal 1, the end behavior is a horizontal asymptote that is the ratio of the leading coefficients. To determine the leading coefficients, multiply the numerator by the factored-out constant:

$$P(t) = \frac{50(1+0.5t)}{(2+0.01t)} = \frac{50+25t}{2+0.01t}$$

The horizontal asymptote is $y = \frac{25}{0.01} = 2{,}500$.

8. **(C)** Evaluating infinite limits of rational functions can be determined by their end behavior. To find the end behavior, compare the degree of the numerator to the degree of the denominator. The degree of the numerator and denominator both equal 3. Therefore, the end behavior follows the horizontal asymptote. Its equation is determined by taking the ratio of the leading coefficients:

$$y = \frac{3}{1} = 3$$

9. **(D)** Since the rational function has a vertical asymptote at $x = -3$, its denominator must have a zero of $x = -3$. Therefore, the denominator must be the factor $x + 3$, so $c = 3$. Since the rational function has a horizontal asymptote of $y = 2$, the ratio of the leading coefficients must also be 2. Therefore, $y = 2 = \frac{a}{1}$ and $a = 2$. The sum of a and c is $2 + 3 = 5$.

10. **(A)** The graph has an x-intercept at $-a$, meaning there is a zero at $x = -a$. So there must be a factor of $(x - (-a)) = (x + a)$ in the numerator of the function. There is a vertical asymptote at $x = -b$, so there must be a factor of $(x - (-b)) = (x + b)$ in the denominator. There is a hole at $x = c$, so there must be a common factor in the numerator and denominator of $(x - c)$.

4

Parent Functions

Learning Objectives

In this chapter, you will learn:

- ➔ The different properties and graphs of parent functions
- ➔ How to transform the graphs of parent functions
- ➔ How to model scenarios using functions
- ➔ How to compose functions
- ➔ How to find the inverse of a function

In previous chapters, we have seen that linear functions are represented graphically by lines and quadratic functions are represented by parabolas. Each function in its most basic form has a corresponding graph that can be changed based upon transformations to its basic function, also called a parent function.

4.1 Graphs and Properties of Basic Functions

Table 4.1 discusses the graphs and properties of basic parent functions. Other special functions will be discussed in their respective units.

Table 4.1

Parent Function	Equation	Graph	Properties
Linear	$f(x) = x$	y; x; −5 −4 −3 −2 −1 1 2 3 4 5; 4 3 2 1 −1 −2 −3 −4	▪ Domain and range is the set of real numbers ▪ Odd function; has origin symmetry ▪ Includes point (0, 0) ▪ Increases throughout its domain

(*Continued*)

Parent Function	Equation	Graph	Properties
Quadratic	$f(x) = x^2$	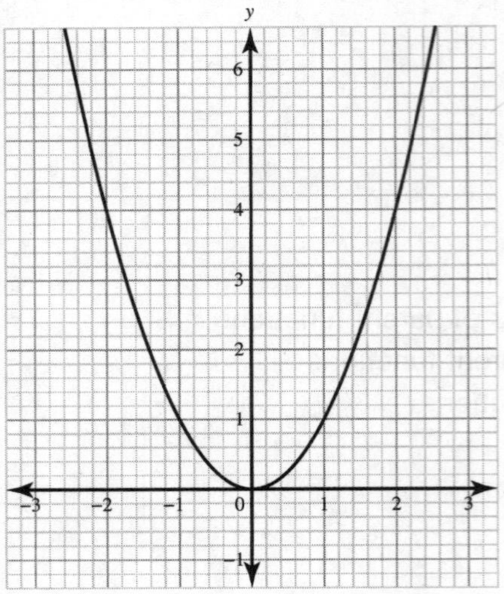	▪ Domain and range is the set of real numbers ▪ Even function; has y-axis symmetry ▪ Includes point (0, 0) ▪ Decreases then increases ▪ Has a relative extremum
Cubic	$f(x) = x^3$	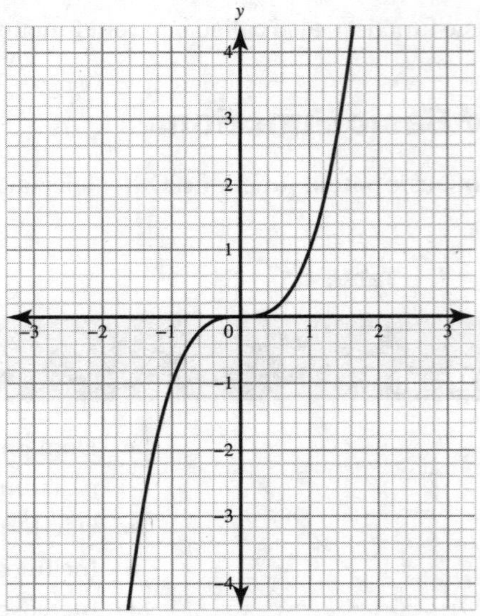	▪ Domain and range is the set of real numbers ▪ Odd function; has origin symmetry ▪ Includes point (0, 0) ▪ Increases throughout its domain ▪ Has a point of inflection
Square root	$f(x) = \sqrt{x}$	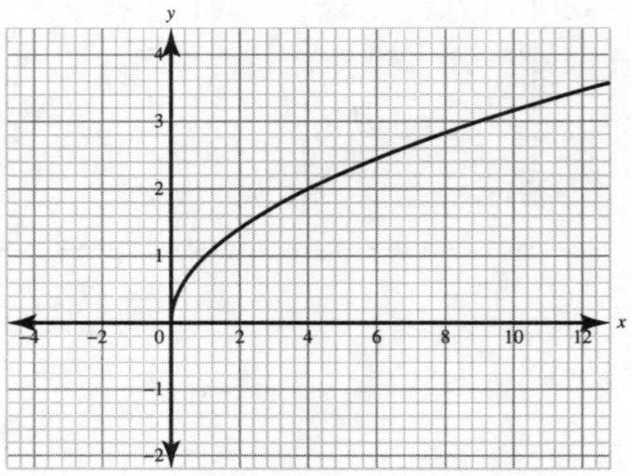	▪ Domain and range is the set of all nonnegative real numbers ▪ Neither even nor odd function ▪ Includes point (0, 0) ▪ Increases throughout its domain

(*Continued*)

Parent Function	Equation	Graph	Properties
Reciprocal	$f(x) = \frac{1}{x}$	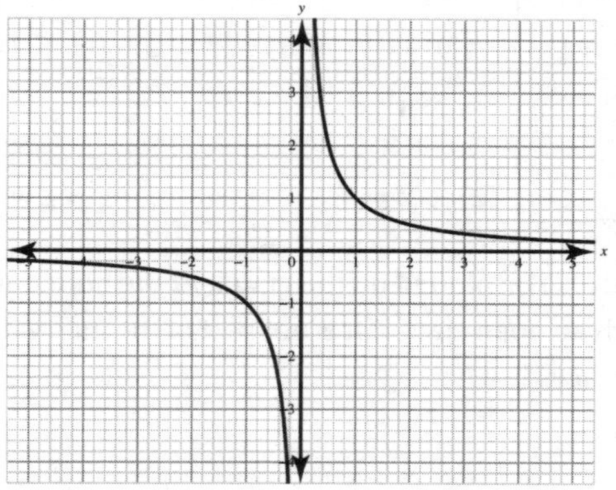	▪ Domain and range is the set of all real numbers except 0 ▪ Odd function; symmetric with respect to the y-axis ▪ x- and y-axes are asymptotes ▪ Decreases on the intervals $(-\infty, 0)$ and $(0, \infty)$
Absolute value	$f(x) = \lvert x \rvert$	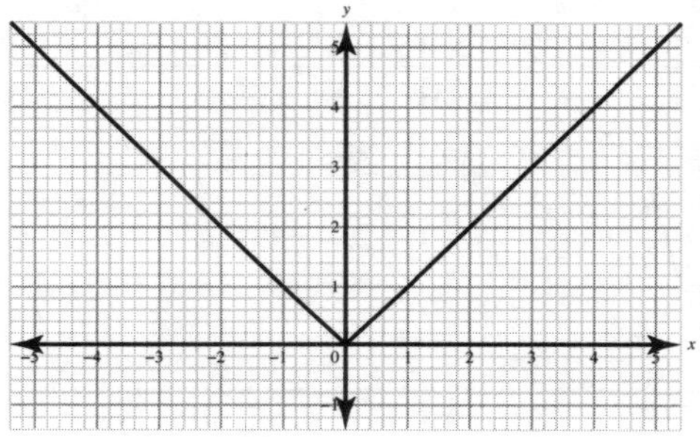	▪ Domain and range is the set of real numbers ▪ Odd function; has origin symmetry ▪ Includes point (0, 0) ▪ Decreases then increases

Piecewise-Defined Functions

A function may change its definition over its domain. The absolute value function, $f(x) = |x|$, is an example of a piecewise-defined function. When $x < 0, f(x) = -x$, and when $x \geq 0, f(x) = x$. This can also be written as:

$$f(x) = |x| = \begin{cases} -x, x < 0 \\ x, x \geq 0 \end{cases}$$

To graph a piecewise function, it is important to check the inequalities associated with each function and where the boundaries change. Making a table of values can be helpful.

Example

Graph the piecewise function: $f(x) = \begin{cases} x^2, x < 2 \\ x + 2, 2 \leq x < 4 \\ \sqrt{x}, x \geq 4 \end{cases}$

Solution

Three different functions make up $f(x)$. To help organize, create a table of values, as shown in Table 4.2. Pay attention to the boundaries of the inequality. At each boundary point, the two different functions should be evaluated. The values in bold are points that lie on the function.

Table 4.2

x	0	1	2	3	4	9
$f(x)$	$\mathbf{x^2}$ $\mathbf{(0)^2 = 0}$	$\mathbf{x^2}$ $\mathbf{(1)^2 = 1}$	x^2 $(2)^2 = 4$ $\mathbf{x + 2}$ $\mathbf{(2) + 2 = 4}$	$\mathbf{x + 2}$ $\mathbf{(3) + 2 = 5}$	$x + 2$ $(4) + 2 = 6$ $\mathbf{\sqrt{x}}$ $\mathbf{\sqrt{4} = 2}$	$\mathbf{\sqrt{x}}$ $\mathbf{\sqrt{9} = 3}$

When graphing, all the points in bold should also be plotted on the graph. At the boundary points, graph up to that point. However, plot the boundary point with an open circle if it is not part of the interval. An open circle shows that the graph goes up to that point but does not include it.

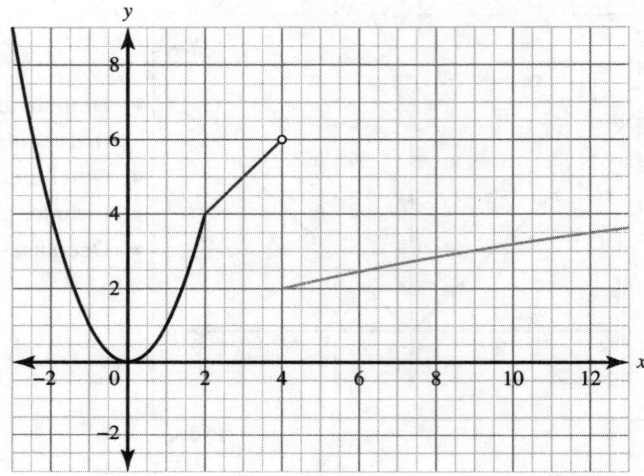

Figure 4.1

The graph of $f(x)$ is shown in Figure 4.1. At $x = 2$, the output values for both functions are the same. So while the graph of the quadratic would have an open circle at (2, 4), the graph of the line has a closed circle at (2, 4), closing the open circle. Also note that at $x = 4$, the two functions have different output values. The graph of the line goes up to but does not include the point (4, 6), which is why it is graphed with an open circle. The graph drops down to (4, 2) for the square root and is closed there. It is important to note that often piecewise functions are not continuous. At $x = 4$ there is a discontinuity, specifically called a jump discontinuity since you must lift your pencil to jump from one graph to the other. These discontinuities often occur at the boundaries of the different intervals of each piece.

4.2 Transformations of Functions

A transformation of a function moves each point of the graph of the function according to a given rule. Flipping a graph over a line is a reflection over that line. Shifting a graph vertically, horizontally, or both vertically and horizontally is a translation. Reflections and translations do not affect the size or shape of the graph. Stretching or shrinking a graph is known as a dilation and does affect the size or shape of the graph. Each of these transformations has a corresponding additive and/or multiplicative rule that can be performed on the function. If two functions are related to each other by a simple transformation, knowing the graph of one function allows you to determine the graph of the transformed function.

Vertical Translations

Adding a number k to a function shifts the graph up if k is positive and shifts it down if k is negative.

Vertical Translation of Function $f(x)$

Given the function $f(x)$:

The function $g(x) = f(x) + k$ is an additive transformation of the function $f(x)$ that results in a vertical translation of the graph of f by k units.

Example

Given $f(x) = x^2$, describe the transformations of $g(x)$ and $h(x)$ as they relate to $f(x)$. Graph each to confirm your answer.

1. $g(x) = x^2 + 3$
2. $h(x) = x^2 - 4$

Solution

The parent function for each is $f(x) = x^2$, which graphs a parabola that contains the point (0, 0).

1. The function $g(x)$ is based upon $f(x)$ but with an addition of 3. By adding 3 to the function, the graph of $f(x)$ shifts up 3 units. This can easily be seen in Figure 4.2 as the point (0, 0) moves to (0, 3). All the points in the graph move up 3 units.

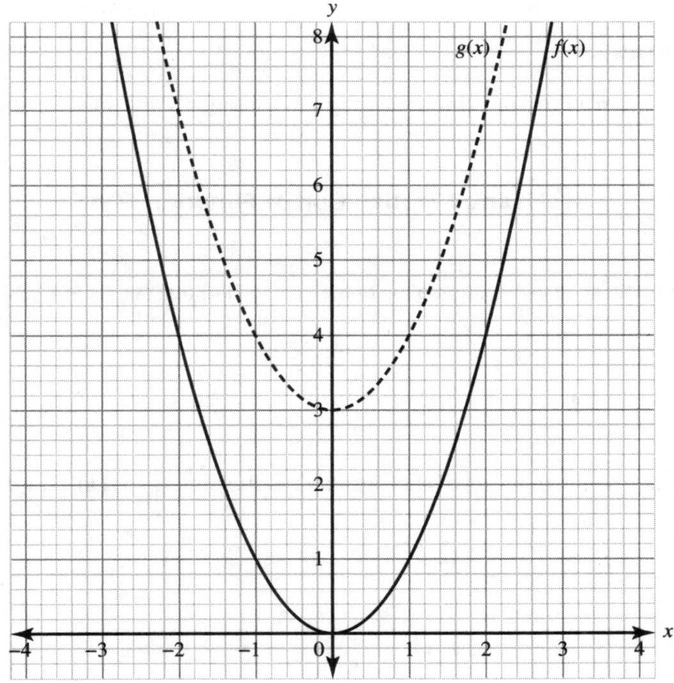

Figure 4.2

2. The function $h(x)$ is based upon $f(x)$ but with a subtraction of 4. By subtracting the function by 4, the graph of $f(x)$ shifts 4 units down. This can easily be seen in Figure 4.3 as the point (0, 0) moves to (0, −4). All the points in the graph move down 4 units.

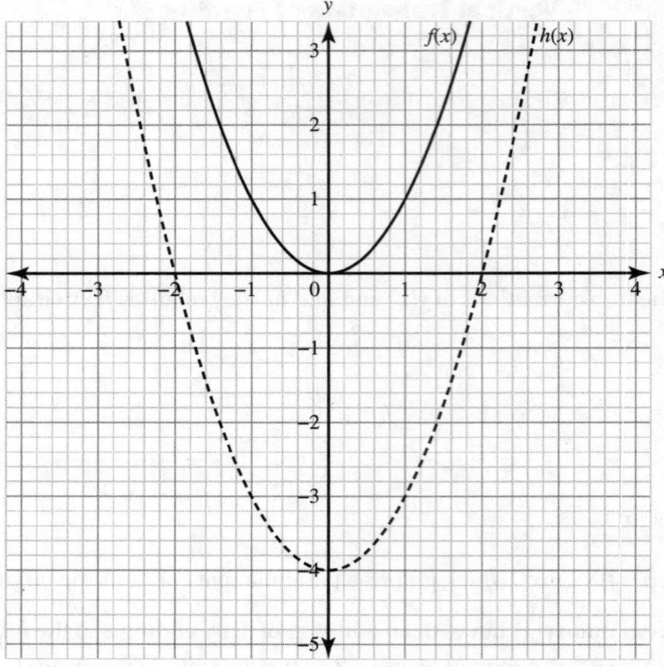

Figure 4.3

Horizontal Translations

Subtracting a number h from x in a function shifts the graph to the right if h is positive and shifts it to the left if h is negative.

Horizontal Translation of Function *f* (*x*)

Given the function $f(x)$:

The function $g(x) = f(x + h)$ is an additive transformation of the function $f(x)$ that results in a horizontal translation of the graph of f by $-h$ units.

Example

Given $f(x) = |x|$, describe the transformations of $g(x)$ and $h(x)$ as they relate to $f(x)$. Graph each to confirm your answer.

1. $g(x) = |x + 3|$
2. $h(x) = |x - 4|$

Solution

The parent function for each is $f(x) = |x|$, which graphs the piecewise function made up of $y = -x$ and $y = x$ and contains the point (0, 0).

1. In $g(x)$, adding 3 to x shifts the graph of $f(x)$ 3 units left. This can easily be seen in Figure 4.4 as the point (0, 0) moves to (−3, 0). All the points in the graph move left 3 units.

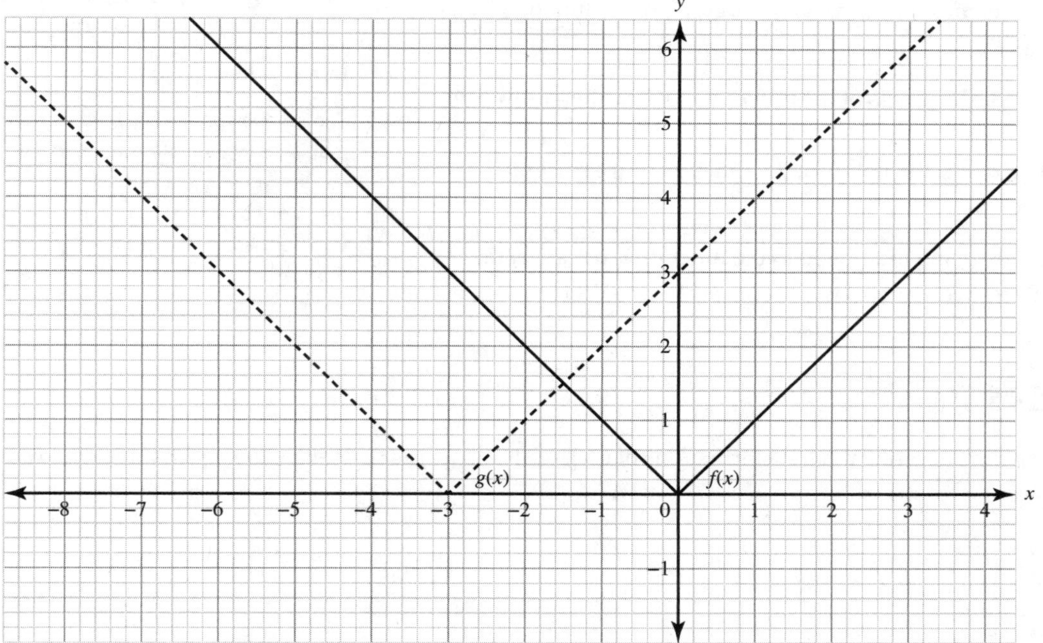

Figure 4.4

2. In $h(x)$, subtracting 4 from x shifts the graph of $f(x)$ 4 units right. This can easily be seen in Figure 4.5 as the point (0, 0) moves to (4, 0). All the points in the graph move right 4 units.

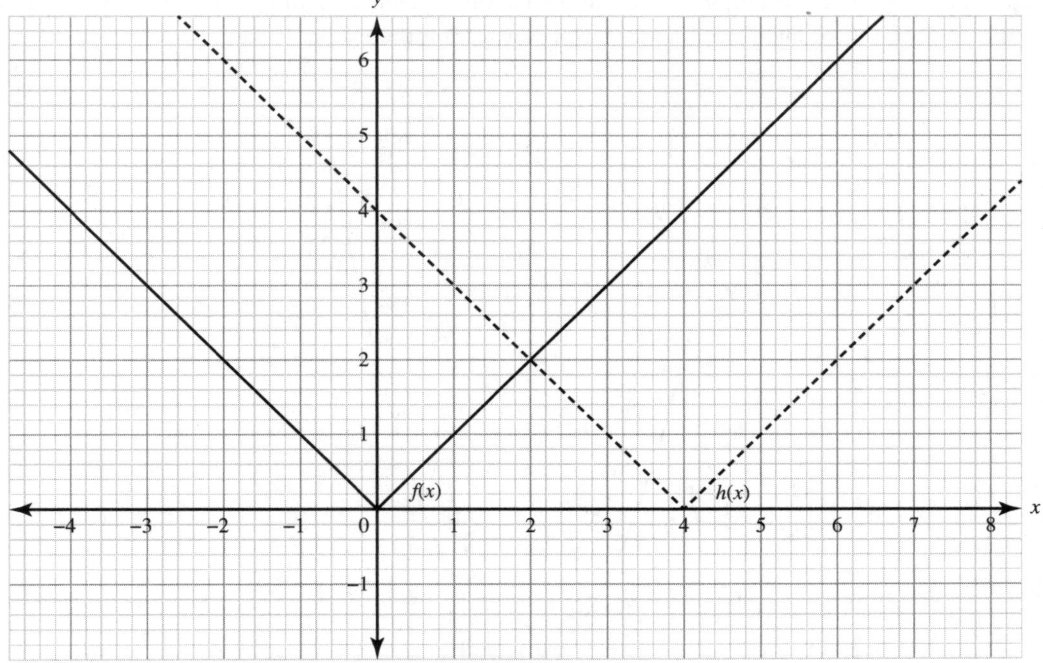

Figure 4.5

Vertical Dilation and Reflection over *x*-axis

Multiplying a positive number, a, by a function, where $a \neq 0$, vertically stretches or compresses the graph of the function by a factor of a.

- If $a > 1$, the points of the function move farther away from the x-axis.
- If $0 < a < 1$, the points of the function move closer to the x-axis.

If the number being multiplied to the function is negative, the graph reflects over the x-axis.

Vertical Dilation and Reflection of Function *f*(*x*)

Given the function $f(x)$:

The function $g(x) = af(x)$, where $a \neq 0$, is a multiplicative transformation of the function $f(x)$ that results in a vertical dilation of the graph of f by a factor of $|a|$. If $a < 0$, the transformation involves a reflection over the x-axis.

Example

Given $f(x) = x^2$, describe the transformations of $g(x)$, $h(x)$, and $i(x)$ as they relate to $f(x)$. Graph each to confirm your answer.

1. $g(x) = 2x^2$
2. $h(x) = \frac{1}{2}x^2$
3. $i(x) = -3x^2$

Solution

The parent function for each is $f(x) = x^2$, which graphs a parabola that contains the point (1, 1).

1. By multiplying the function by 2, the graph of $f(x)$ vertically dilates. All the x-values remain fixed, and their corresponding y-values are multiplied by 2. This can easily be seen in Figure 4.6 as the point (1, 1) moves to (1, 2). The points on the dilated graph move farther away from the x-axis.

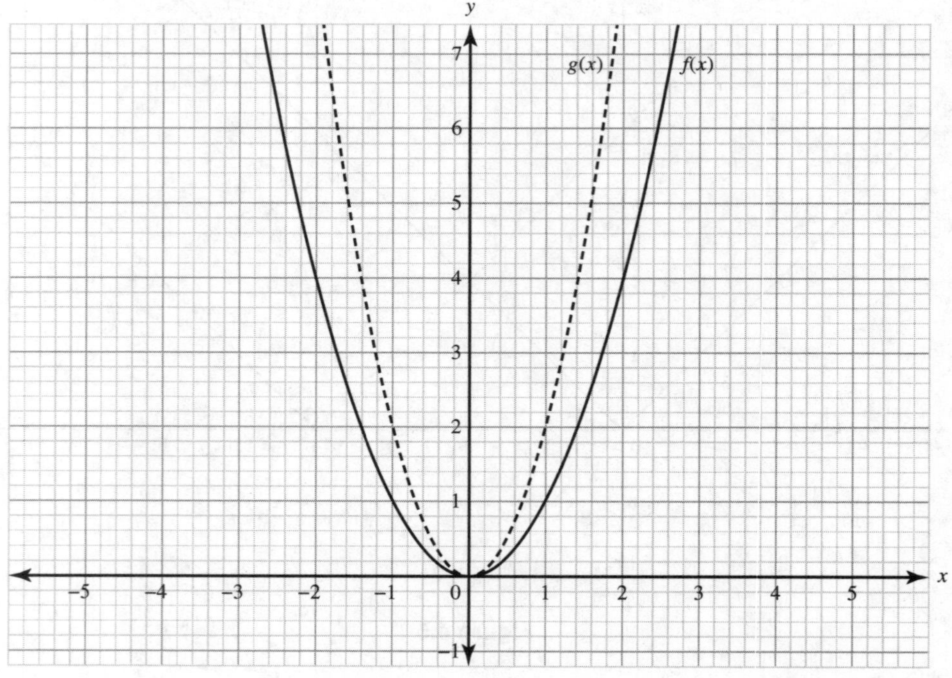

Figure 4.6

2. By multiplying the function by $\frac{1}{2}$, the graph of $f(x)$ vertically dilates. All the x-values remain fixed, and their corresponding y-values are multiplied by $\frac{1}{2}$. This can easily be seen in Figure 4.7 as the point (1, 1) moves to $(1, \frac{1}{2})$. The points on the dilated graph move closer to the x-axis.

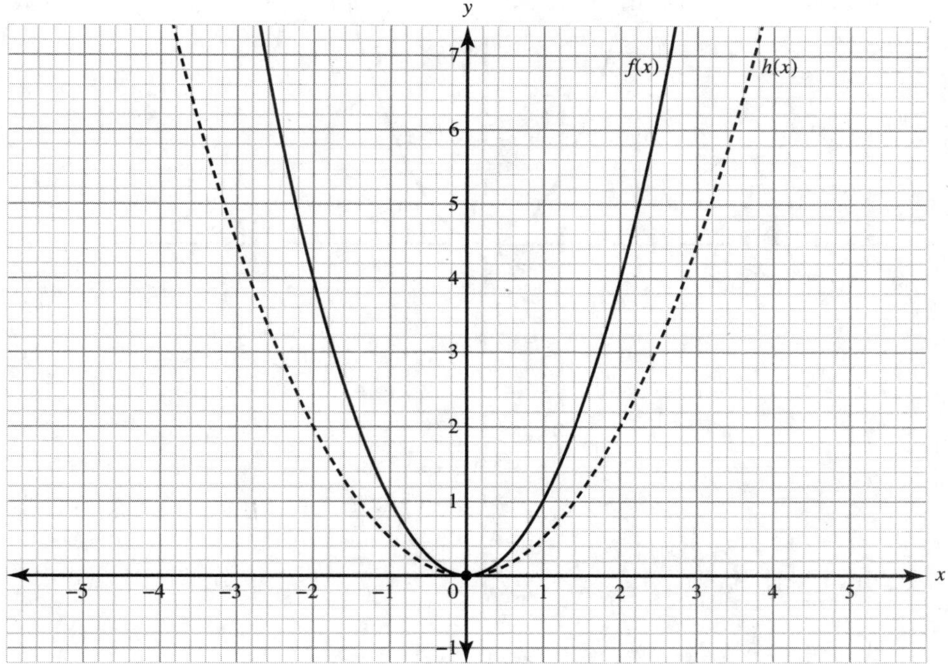

Figure 4.7

3. By multiplying the function by −3, the graph of $f(x)$ vertically dilates by a factor of $|-3|$ and is reflected over the x-axis. All the x-values remain fixed, and their corresponding y-values are multiplied by −3. This can easily be seen in Figure 4.8 as the point (1, 1) moves to (1, −3). The points on the dilated graph move farther away from the x-axis and are also reflected over the x-axis.

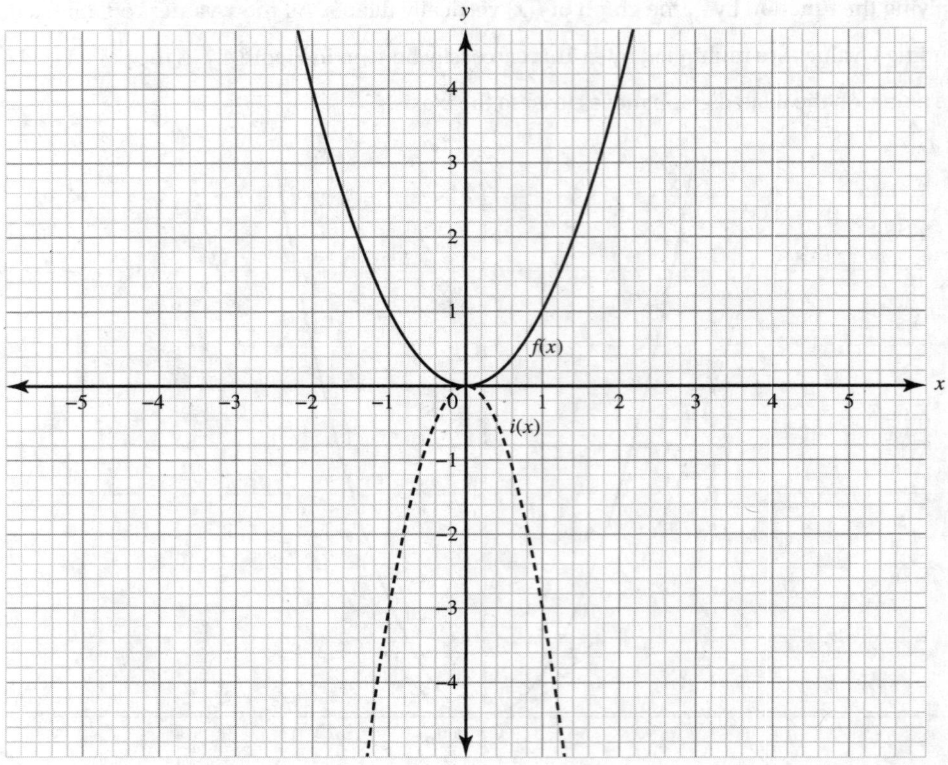

Figure 4.8

Horizontal Dilation and Reflection over *y*-axis

Multiplying a positive number, b, by x in a function, where $b \neq 0$, horizontally stretches or compresses the graph of the function by a factor of $\frac{1}{b}$.

- If $b > 1$, the points of the function move closer to the y-axis.
- If $0 < b < 1$, the points of the function move farther away from the y-axis.

If the number being multiplied by x in the function is negative, the graph reflects over the y-axis.

Horizontal Dilation and Reflection of Function $f(x)$

Given the function $f(x)$:

The function $g(x) = f(bx)$, where $b \neq 0$, is a multiplicative transformation of the function $f(x)$ that results in a horizontal dilation of the graph of f by a factor of $\left|\frac{1}{b}\right|$. If $b < 0$, the transformation involves a reflection over the y-axis.

Example

Given $f(x) = x^3$, describe the transformations of $g(x)$, $h(x)$, and $i(x)$ as they relate to $f(x)$. Graph each to confirm your answer.

1. $g(x) = (2x)^3$
2. $h(x) = \left(\frac{1}{2}x\right)^3$
3. $i(x) = (-2x)^3$

✓ Solution

The parent function for each is $f(x) = x^3$, which graphs a cubic polynomial that contains the point (1, 1).

1. By multiplying the x of the function by 2, the graph of $f(x)$ horizontally dilates. This isn't the same as a vertical dilation where the output values are scaled; here the input values are scaled. The x in the original $f(x)$ becomes $8x$ in $g(x)$, meaning $g(x)$ reaches a given input value 8 times faster than $f(x)$. This can easily be seen in Figure 4.9 as the point (1, 1) moves to (1, 8).

   Another way to find the points is by using the factor $\left|\frac{1}{2}\right|$. All the y-values remain fixed on the parent function, and their corresponding x-values are multiplied by $\left|\frac{1}{2}\right|$. The graph of $f(x) = x^3$ contains the point (2, 8). Multiplying the input value 2 by the scale factor $\left|\frac{1}{2}\right|$ simplifies to 1. So the point on the graph of $g(x) = (2x)^3$ will be (1, 8). The point (4, 64) lies on the graph of $f(x) = x^3$, so the point $\left(4 \cdot \frac{1}{2}, 64\right) = (2, 64)$ will lie on the graph of $g(x) = (2x)^3$. As shown in Figure 4.9, the points on the dilated graph move closer to the y-axis.

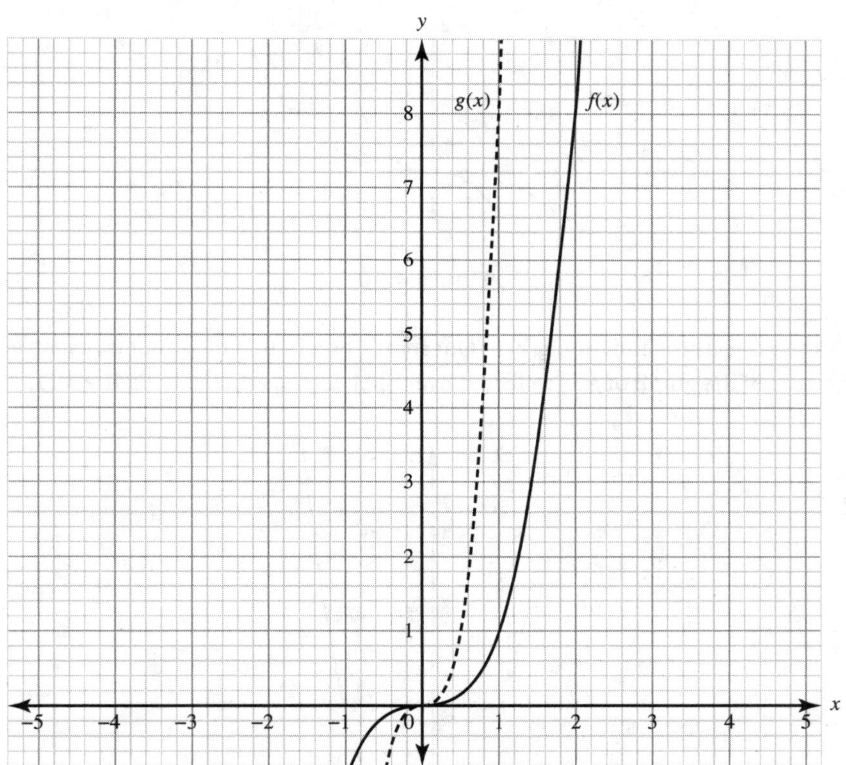

Figure 4.9

2. By multiplying the x of the function by $\frac{1}{2}$, the graph of $f(x)$ horizontally dilates. This isn't the same as a vertical dilation where the output values are scaled; here the input values are scaled. The x in the original $f(x)$ becomes $\frac{1}{8}x$ in $h(x)$. This can easily be seen in Figure 4.10 as the point (1, 1) moves to $\left(1, \frac{1}{8}\right)$.

   Another way to find the points is by using the factor $\left|\frac{1}{\frac{1}{2}}\right| = |2|$. All the y-values remain fixed on the parent function, and their corresponding x-values are multiplied by $|2|$. The graph of $f(x) = x^3$ contains the point (2, 8). Multiplying the input value 2 by the scale factor $|2|$, simplifies to 4. So the point on the graph of $h(x) = \left(\frac{1}{2}x\right)^3$ will be (4, 8). The point (4, 64) lies on the graph of $f(x) = x^3$, so the point $(4 \cdot 2, 64) = (8, 64)$ lies on the graph of $h(x) = \left(\frac{1}{2}x\right)^3$. As shown in Figure 4.10, the points on the dilated graph move farther away from the y-axis.

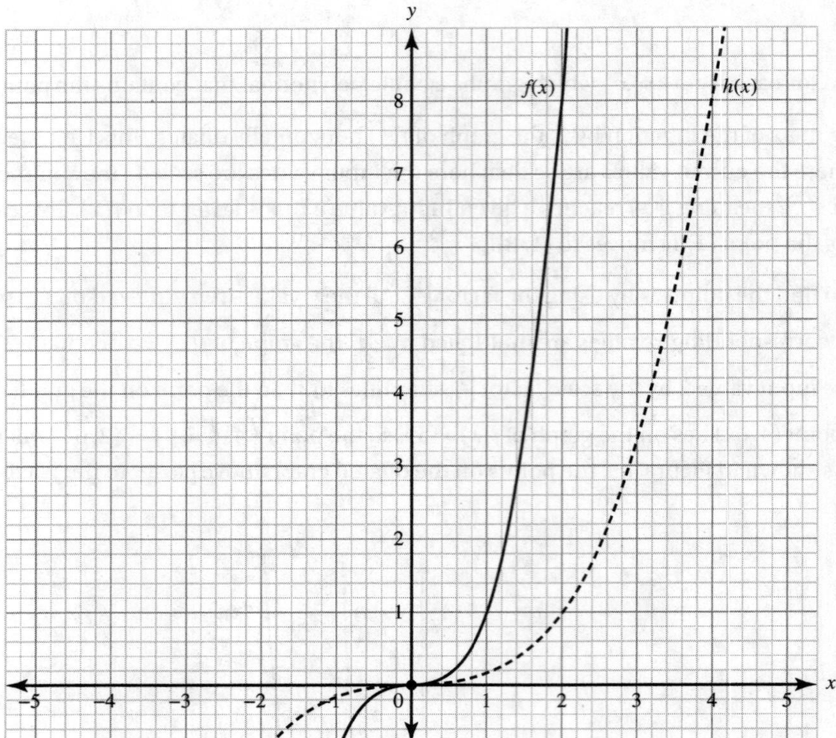

Figure 4.10

3. Function $i(x)$ is the same as the function $g(x)$ in Question 1 except that the input is negative. The graph of $i(x)$ will be the reflection of the graph of $g(x)$ reflected over the y-axis, as shown in Figure 4.11.

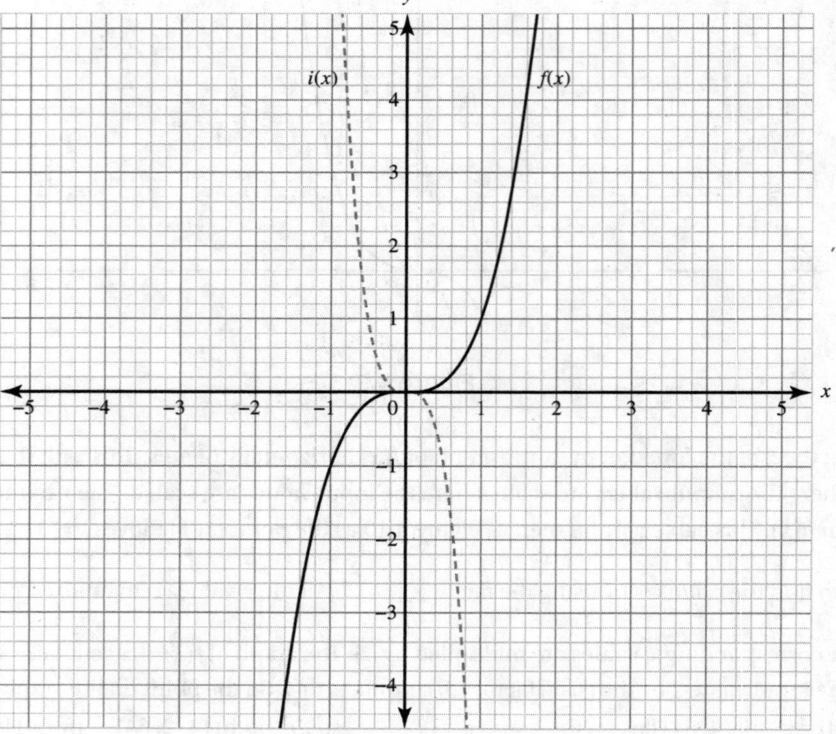

Figure 4.11

Combinations of Transformations

Additive and multiplicative transformations can be combined, resulting in combinations of horizontal and vertical translations, dilations, and reflections. It is often necessary to perform transformations in a certain order to guarantee the arrival at the correct graph. The following steps will help to ensure the correct order of operations.

Steps for a Sequence of Transformations

Apply the following steps when graphing a function containing more than one transformation.

1. Start with the parentheses. Look for a possible horizontal shift.
2. Deal with horizontal or vertical dilations.
3. Deal with reflections over the x- and/or y-axes.
4. Deal with vertical shifts.

This is very similar to the algebraic order of operations known as PEMDAS, which stands for Parentheses, Exponents, Multiplication, Division, Addition, and Subtraction.

Given the function $y = f(x)$, a combination of transformations can be expressed algebraically as:

$$y = af(b(x + c)) + d$$

where
$a =$ vertical stretch or compression
$b =$ horizontal stretch or compression
$c =$ horizontal shift
$d =$ vertical shift

A function that is made up of a combination of transformations is also referred to as a composite function.

Example

The graph of $y = f(x)$ is shown in Figure 4.12 below. Determine the graph of the composite function $y = 2f(x + 1) - 3$ by showing the effect of a sequence of transformations on the graph of $y = f(x)$.

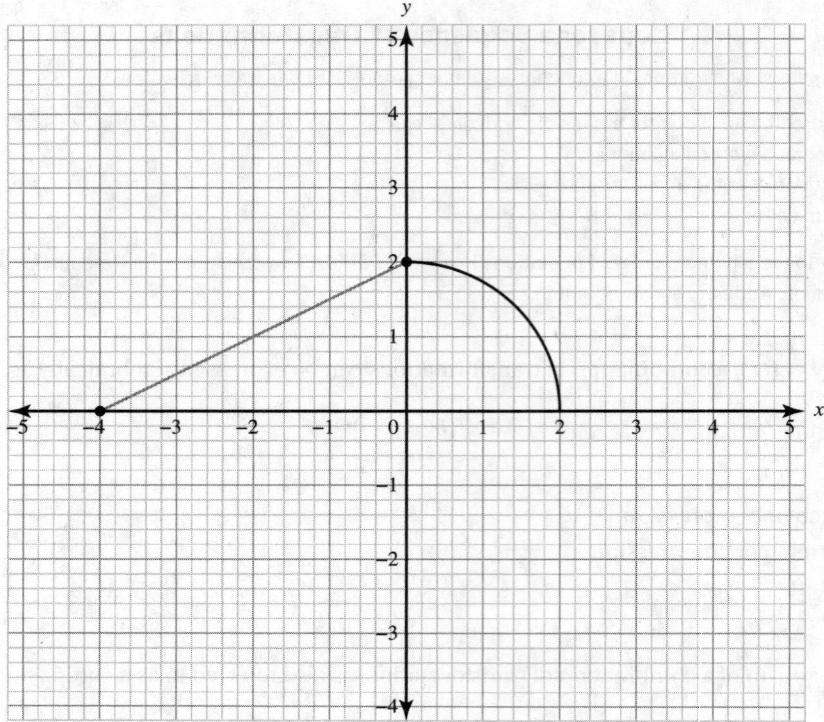

Figure 4.12

Solution

1. Starting with the parentheses, there is a horizontal translation 1 unit to the left, as shown in Figure 4.13.

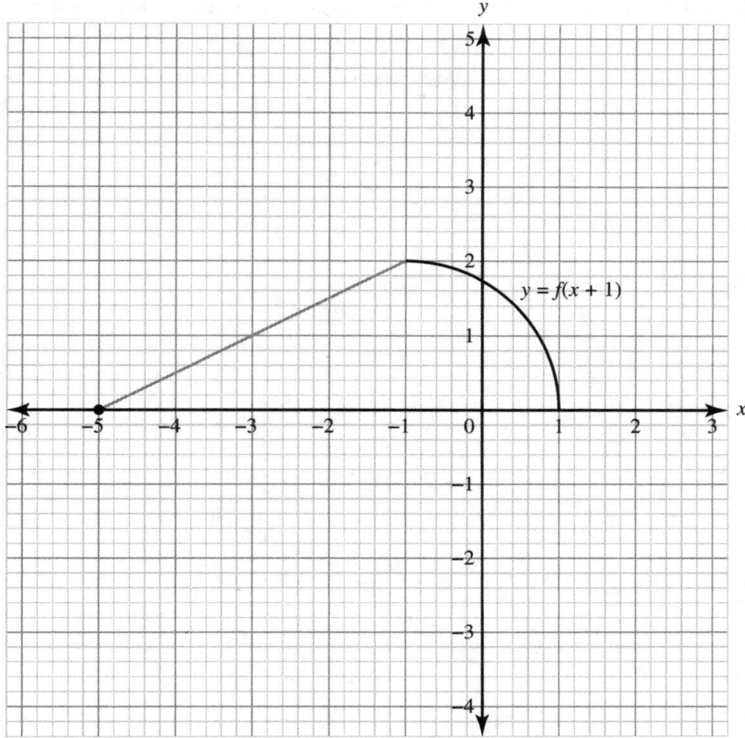

Figure 4.13

2. Next deal with the multiplication. Figure 4.14 shows a vertical dilation by a factor of 2.

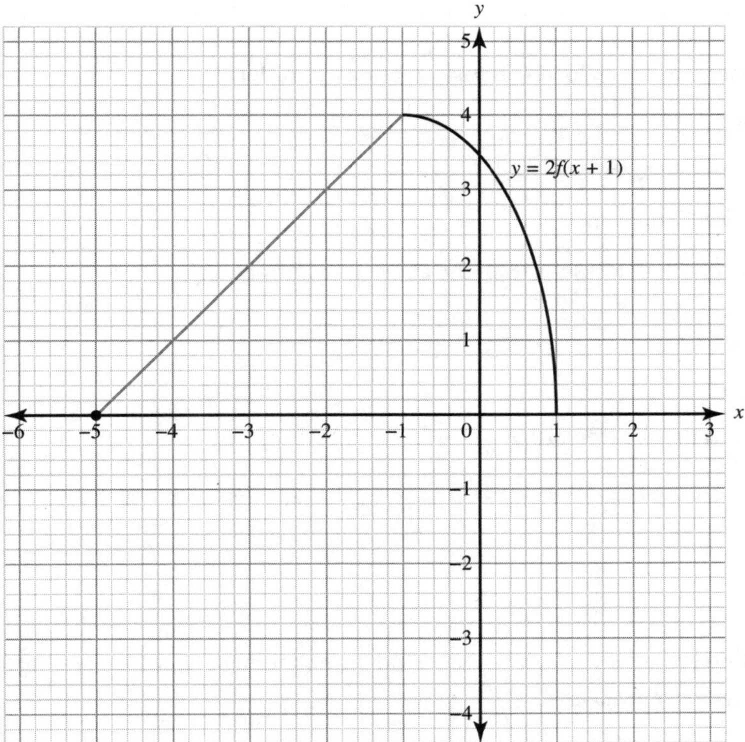

Figure 4.14

3. Finally, deal with the addition/subtraction. There is a vertical translation down 3, as shown in Figure 4.15.

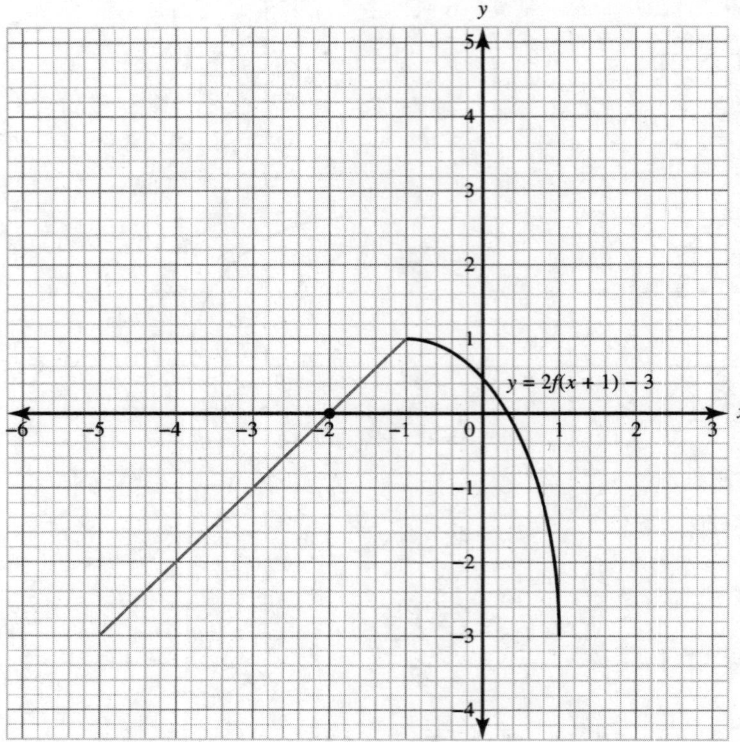

Figure 4.15

Example

The graph of $y = |x|$ undergoes the following transformations, in order:

- A horizontal shift 2 units to the right
- A vertical stretch by a factor of $\frac{1}{3}$
- A vertical translation 7 units up

Find the equation of the graph that results from the above transformations.

Solution

Apply the transformations in the given order:

$$y = |x| \rightarrow y = |x - 2| \rightarrow y = \frac{1}{3}|x - 2| \rightarrow y = \frac{1}{3}|x - 2| + 7$$

Example

Let C_1 be the curve defined by $y_1 = f(x) = x^3 - 16x$. Find the equation of C_2, which is a horizontal shrink of C_1 by a factor of $\frac{1}{2}$ and is reflected over the x-axis.

Solution

Let the equation for C_2 be denoted by y_2. Then:

$$y_2 = -f\left(\frac{x}{\frac{1}{2}}\right) = -f(2x) = -((2x)^3 - 16(2x)) = -(8x^3 - 32x) = -8x^3 + 32x$$

All these different types of transformations alter the graph in certain ways, either by changing their position or their size. Therefore, it is common to find that the domain and range of a function that is a transformation of a parent function may be different from those of the parent function.

Example

Given each of the following, describe a basic graph and a sequence of transformations that can be used to produce a graph of the given function. State the domain and range of the given function and its parent function.

1. $y = (3x)^2 - 4$
2. $y = -3\sqrt{x+1}$

Solution

1. The parent function is $y = x^2$, as shown in Figure 4.16. The domain of $y = x^2$ is all real numbers, and the range is $y \geq 0$.

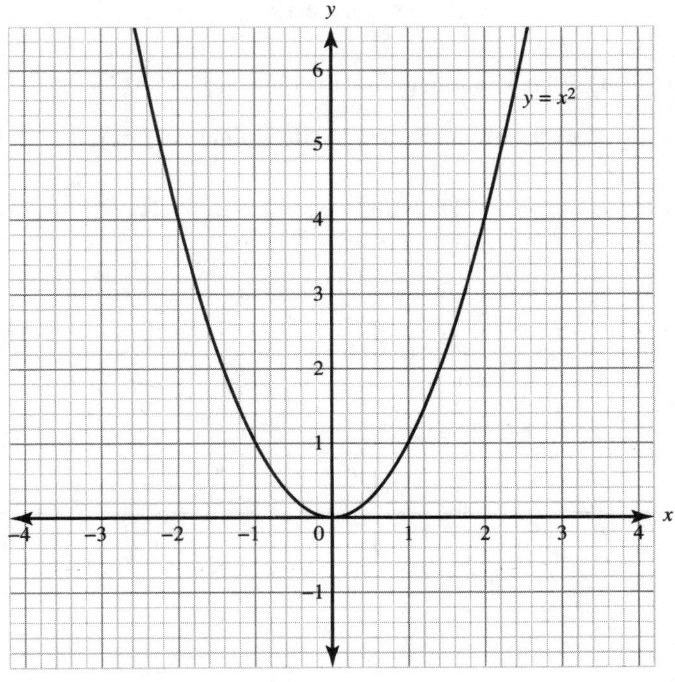

Figure 4.16

Starting with the parentheses, there is a horizontal dilation with a factor of $\frac{1}{3}$. Then there is a vertical shift down 4 units. The result is graphed in Figure 4.17.

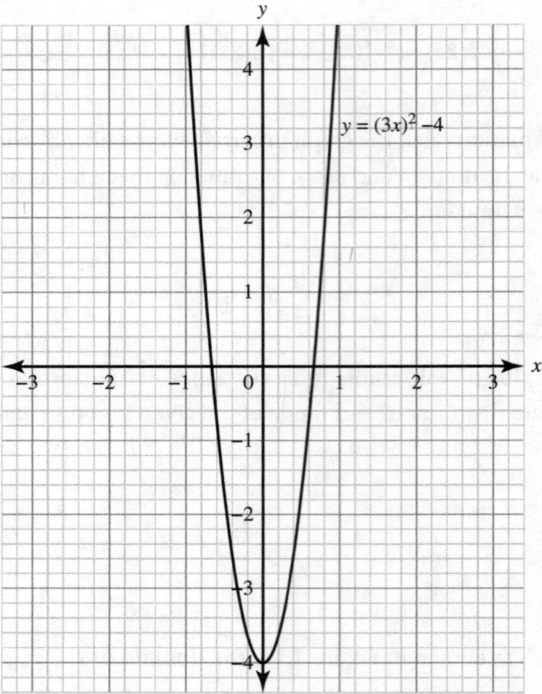

Figure 4.17

The domain of $y = (3x)^2 - 4$ is all real numbers, and the range is $y \geq -4$.

2. The parent function is $y = \sqrt{x}$, as shown in Figure 4.18. The domain of $y = \sqrt{x}$ is $x \geq 0$, and the range is $y \geq 0$.

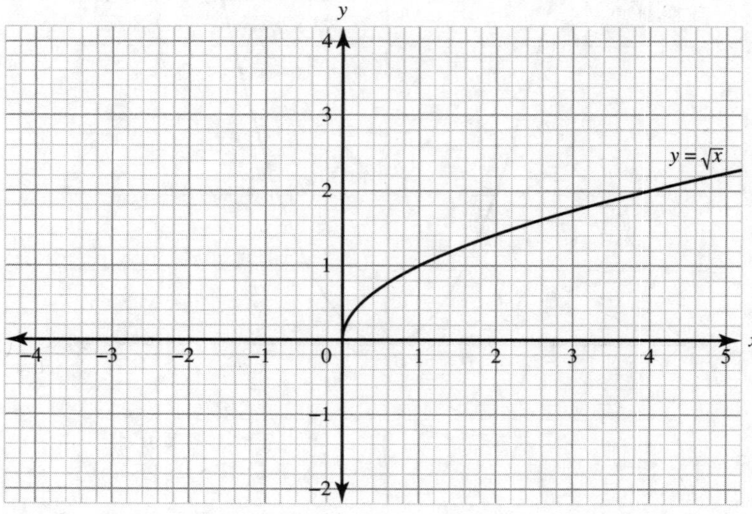

Figure 4.18

Starting with the parentheses, which in this case are under the radical, there is a horizontal shift to the left 1 unit. Then there is a vertical dilation with a scale factor of 3 and finally a reflection over the x-axis. The result is graphed in Figure 4.19.

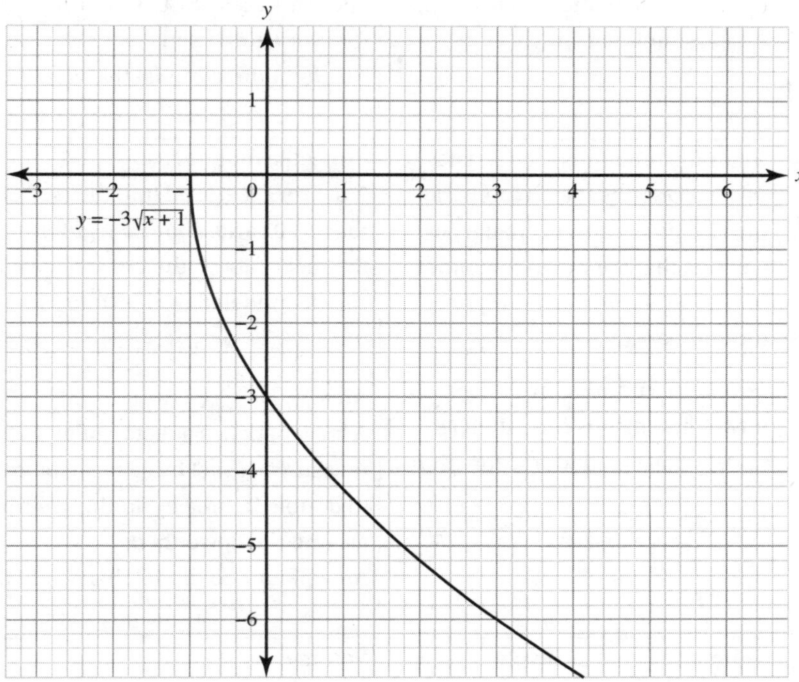

Figure 4.19

The domain of $y = -3\sqrt{x+1}$ is $x \leq -1$, and the range is $y \leq 0$.

4.3 Function Model Selection, Construction, and Application

Mathematical equations are a useful tool to model real-life scenarios that can help make predictions. It is important that an appropriate function type is used to construct a function model for a given scenario.

Table 4.3 describes the different functions discussed so far and what type of scenarios they best demonstrate.

Table 4.3

Function Type	Model Description	Example
Linear functions	Linear functions model data sets or scenarios that demonstrate roughly constant rates of change.	Renting a car: There is a fixed cost of \$150 and then a cost of \$5.50 for every hour used. An equation to model this would be $y = 5.5x + 150$
Quadratic functions	Quadratic functions model data sets or scenarios that demonstrate roughly linear rates of change, or data sets that are roughly symmetric with a unique maximum or minimum value. Geometric contexts involving area or two dimensions can often be modeled using quadratic functions.	Projectile motion: A toy rocket is fired into the air from the top of a barn that is 20 yards above the ground. The height, h, of the rocket above the ground in yards after t seconds can be modeled by this function $h(t) = -5t^2 + 10t + 20$ Area: A local park has a rectangular flower bed that measures 10 feet by 15 feet. The park gardener wants to double its area by adding a strip of uniform width around the flower bed. An equation to model this would be $4x^2 + 50x + 150 = 300$
Cubic functions	Geometric contexts involving volume or three dimensions can often be modeled by cubic functions.	Volume: A box has a square base where the length of a side of the square base is x. The box is twice as tall as it is wide. An equation to model the volume of the box would be $V = (x)(x)(2x) = 2x^3$
Polynomial functions	Polynomial functions model data sets or scenarios with multiple real zeros or multiple maxima or minima.	Temperature: A house thermostat is set to 72°F. The actual temperature in the home changes between maximum and minimum temperatures.
Piecewise functions	Piecewise functions model a data set or scenario that demonstrates different characteristics over different intervals.	Income tax: How much people earn determines their income tax rate.
Rational functions	Data sets and scenarios involving quantities that are inversely proportional can often be modeled by rational functions.	Forces: The magnitudes of both gravitational force and electromagnetic force between objects are inversely proportional to the objects' squared distance.

Linear Models

Linear functions model data sets or scenarios that demonstrate roughly constant rates of change.

Example

The data in Table 4.4 represent the price, p, and quantity demanded per day, q, of a digital camera.

Table 4.4

Price, *p* (in dollars)	Quantity Demanded, *q*
150	120
200	100
250	80
300	60

(a) Show that the quantity demanded, q, is a linear function of the price, p.
(b) State and interpret the slope.
(c) Determine the linear function that describes the relationship between p and q.
(d) State and interpret the value of the intercepts.
(e) What is the implied domain of the linear function?

Solution

(a) Since each input corresponds to a single output, the quantity demanded is a function of price. Since the average rate of change is a constant—$0.40 per camera—the function is linear.

(b) The slope is -0.4, meaning that if the price increases by $1, the quantity demanded of digital cameras decreases by 0.4 camera.

(c) Since the slope is known, the rest of the equation can be found by substituting a known ordered pair from the table into the point-slope formula:

$$y - y_1 = m(x - x_1)$$

Each ordered pair (x, y) is represented by (p, q). Substitute the ordered pair (150, 120) into the linear equation:

$$q - 120 = -0.4(p - 150)$$

Solve for q:

$$q = -0.4p + 180$$

(d) Using the given slope-intercept form from part (c), the q-intercept is 180. This means when the price is $0, 180 digital cameras will be demanded. To find the p-intercept, substitute 0 for q and solve for p:

$$0 = -0.4p + 180$$

$$p = \frac{-180}{-0.4} = 450$$

The p-intercept is 450, meaning that there will be 0 digital cameras demanded when the price is $450.

(e) The implied domain is $0 \leq p \leq 450$.

Models can be built from linear descriptions but also from given data as shown in the previous example. When a lot of data points are given, there isn't always a constant rate of change. However, when the data points are graphed, which is called a scatterplot, it can be obvious what model best fits most data points. This is known as a regression model.

Regression models can be found using graphing utilities, such as a graphing calculator. The steps listed below show how to find regression models using the TI-84.

Steps for Finding Regression Models

1. The data must be entered into lists in the calculator. Press STAT and ENTER for 1:Edit.
2. Enter the data for the independent variable into list 1, L1. Enter data for the dependent variable into list 2, L2. When necessary, to delete data from a list, highlight the name of the list at the top, and press the CLEAR button.
3. After entering in the data, press STAT, move to the right to select CALC, and choose the appropriate regression model for your data. Typical models would be a line of best fit, or linear regression, which is option 4. There is also a quadratic regression (option 5).
4. After selecting the appropriate regression model, make sure the Xlist has L1 and the Ylist has L2. The FreqList and Store RegEQ can be left blank. Select Calculate at the bottom, and press ENTER.
5. The general regression model will be listed along with the values of each coefficient.

Example

A pediatrician wanted to find a model that related a child's height, h, to their head circumference, c. She randomly selects nine children from her practice, measures their height and head circumference, and obtains the data shown in Table 4.5. She plots the data on a graph to create a scatterplot, shown in Figure 4.20, letting h represent the independent variable and c represent the dependent variable.

Table 4.5

Height, *h* (inches)	Head Circumference, *c* (inches)
25.25	16.4
25.75	16.9
25	16.9
27.75	17.6
26.5	17.3
27	17.5
26.75	17.3
26.75	17.5
27.5	17.5

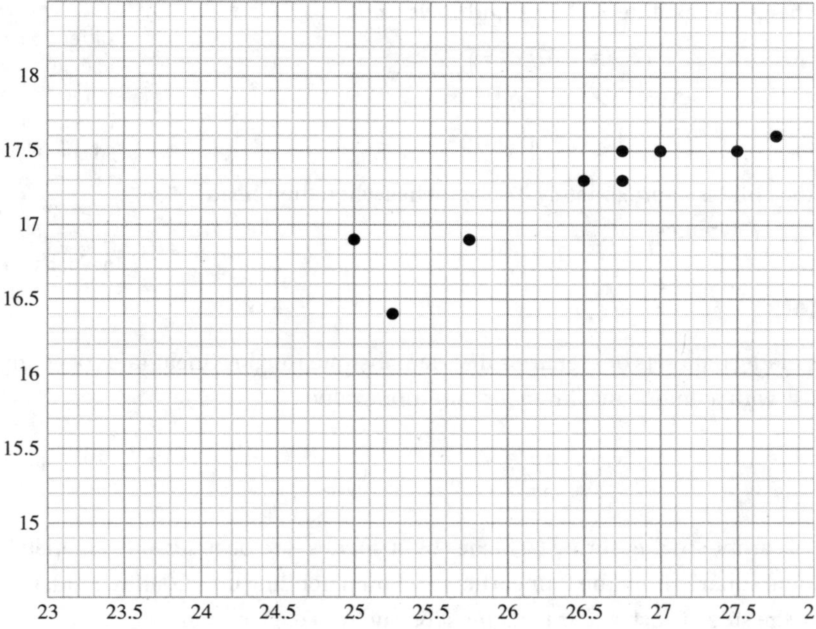

Figure 4.20

(a) Use a graphing calculator to find the line of best fit that models the relation between height and head circumference. Round all values to the nearest thousandth.

(b) Interpret the slope.

(c) Predict the head circumference of a child who is 26 inches tall.

(d) What is the height of a child whose head circumference is 17.4 inches?

✓ Solution

(a) To find the line of best fit, enter the data into L1 and L2 by pressing STAT and ENTER. After the data have been entered, press STAT and select CALC on the right. Type 4 for the linear regression model, which will provide the line of best fit. Check that L1 and L2 are listed before selecting Calculate. The calculator will display the following:

$$y = ax + b$$

$$a = .3733840304$$

$$b = 7.326806084$$

The line of best fit can be modeled by the equation $y = 0.373x + 7.327$. By substituting the variables given in the problem, this model will change to $c = 0.373h + 7.327$.

(b) The slope of the line is approximately 0.373. This means that for every inch in height, the head circumference will increase by 0.373 inches.

(c) To predict the head circumference, substitute 26 for h and solve for c:

$$c = 0.373(26) + 7.327 = 17.025$$

The head circumference is approximately 17.025 inches when the child is 26 inches tall.

(d) To predict the height, substitute 17.4 for c and solve for h.

$$17.4 = 0.373h + 7.327$$
$$10.073 = 0.373h$$
$$h = 27.00536193$$

The height of the child is approximately 27.005 inches when their head circumference is 17.4 inches.

Different strategies can be used to select the function to model a situation. One is to create a diagram. This will help identify variables, restrict domain and range if necessary, and identify the most appropriate function needed to model.

Quadratic Models

Quadratic functions model data sets or scenarios that demonstrate roughly linear rates of change, or data sets that are roughly symmetric with a unique maximum or minimum value.

Example

A rectangular piece of cardboard measuring 50 inches by 30 inches is to be made into an open box with a base of 900 in.^{2} by cutting equal squares from the four corners and then bending up the sides. Find, to the nearest tenth of an inch, the length of the side of each square that must be cut from each corner.

Solution

First create a diagram that models the given scenario. Start with a rectangle measuring 50 inches by 30 inches, as shown in Figure 4.21.

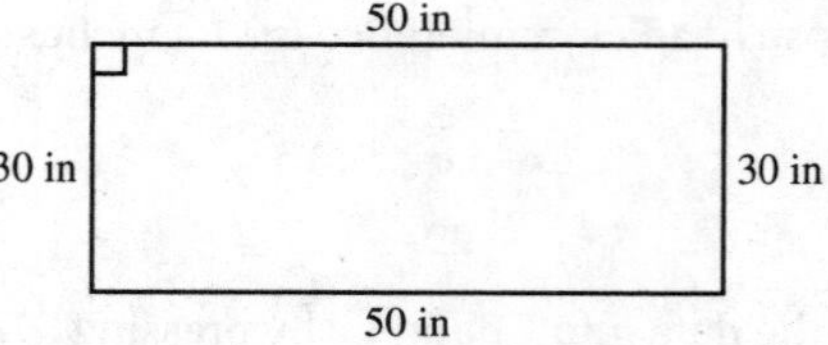

Figure 4.21

Cut out the four equal squares from each corner. Since the length of the squares is unknown, let x represent the length of the side of a square.

Since the cardboard has set dimensions and it is given that the squares are cut out so the cardboard can be bent upward, the set of numbers that x can represent needs to be restricted. See Figure 4.22. This is what is known as restricting the domain. The domain for this example is $0 < x < 15$.

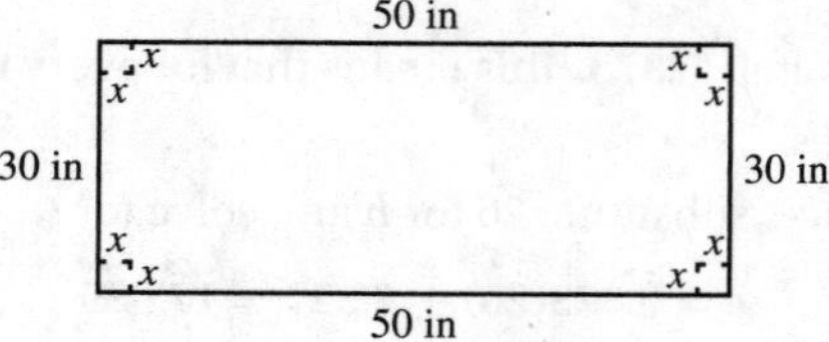

Figure 4.22

At each corner of the cardboard, the squares are being cut out. This affects each side of the cardboard since its length will be reduced by x on both ends. As shown in Figure 4.23, this changes the length and width of the cardboard.

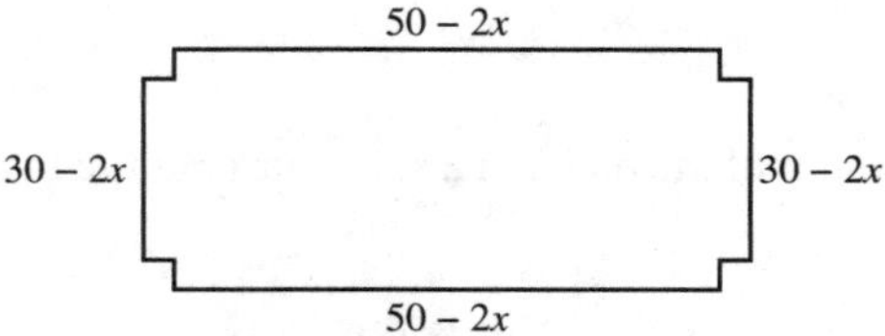

Figure 4.23

When the sides are bent up, as shown in Figure 4.24, the three-dimensional box is created and the height of the box is x.

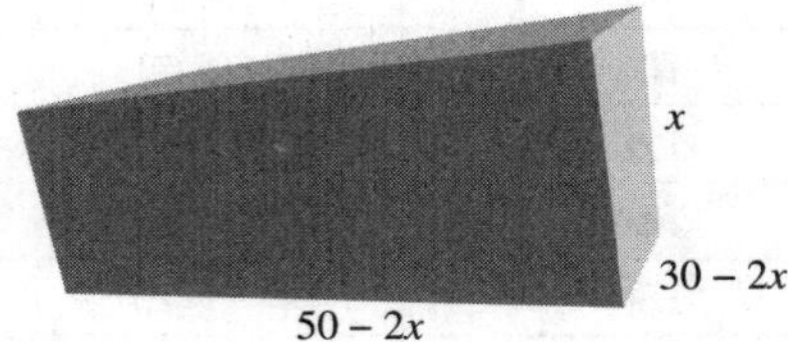

Figure 4.24

The length $= 50 - 2x$, the width $= 30 - 2x$, and the height $= x$.

The area of the rectangle is the product of its length and width. The area of the rectangle was given to be 900 in.^{2}

To solve the problem, set up the required equation:

$$900 = (50 - 2x)(30 - 2x)$$

Multiply the terms and simplify:

$$900 = 1500 - 160x + 4x^2$$

The equation can be identified as a quadratic equation. To solve for x, the equation must be set equal to 0. To solve for x, use the quadratic formula, factor, or graph to find the x-intercepts:

$$4x^2 - 160x + 600 = 0$$

Divide the equation by 4:

$$x^2 - 40x + 150 = 0$$

Use the quadratic formula:

$$x = \frac{-(-40) \pm \sqrt{(-40)^2 - 4(1)(150)}}{2(1)}$$

$$x = \frac{40 \pm \sqrt{1600 - 600}}{2}$$

$$x = \frac{40 \pm \sqrt{1000}}{2}$$

$$x \approx 35.8113883,\ x \approx 4.188611699$$

Since the restricted domain is $0 < x < 15$, the only acceptable answer is $x \approx 4.188611699$.

The length of the side of each square is 4.2 inches.

Example

The data in Table 4.6 represent the height of the ball h of a shot-putter at the instant that it has traveled x feet horizontally.

Table 4.6

Distance, x	Height, h
20	25
40	40
60	55
80	65
100	71
120	77
140	77
160	75
180	71
200	64

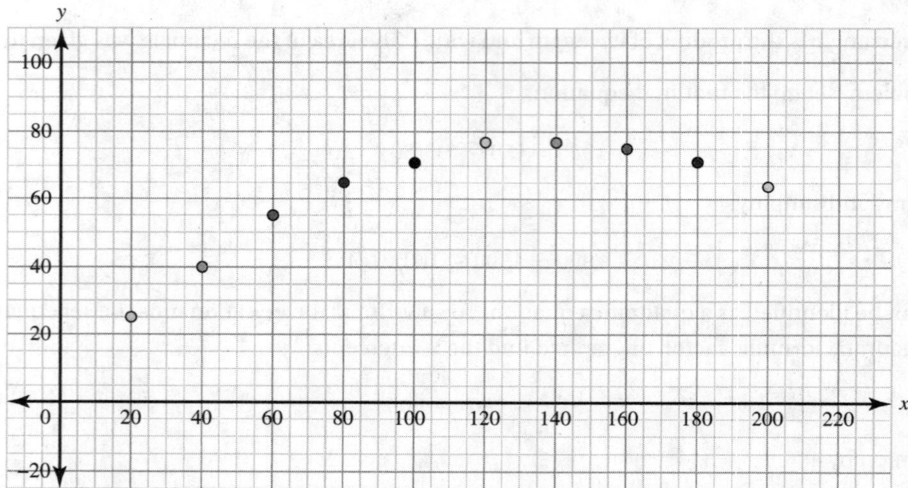

Figure 4.25

(a) Using the scatterplot in Figure 4.25, describe the type of relation that may exist between the two variables.

(b) Using your answer from part (a), find the function of best fit that models the relation between distance and height. Round all coefficients to the nearest thousandth.

(c) Use the function of best fit found in part (b) to determine how far the ball will travel before it reaches its maximum height.

(d) Use the function found in part (b) to find the maximum height of the ball.

Solution

(a) Based on the scatterplot, it appears that a quadratic regression model would best fit the data.

(b) To find the quadratic regression model, enter the data into L1 and L2 by pressing STAT and ENTER. After the data have been entered, press STAT and select CALC on the right. Type 5 for the quadratic regression model. Check that L1 and L2 are listed before selecting Calculate. The calculator will display the following:

$$y = ax^2 + bx + c$$

$$a = -0.0037121212$$

$$b = 1.031818182$$

$$c = 5.666666667$$

The quadratic regression can be modeled by the equation $y = -0.004x^2 + 1.032x + 5.667$. By substituting the variables given in the problem, this model will change to $h = -0.004x^2 + 1.032x + 5.667$.

(c) A graphing calculator can be used to find the coordinates of the maximum of the quadratic equation. Type the quadratic regression model into Y=. To see the graph better, change the window settings to the following:

Xmin = 0
Xmax = 300
Xscl = 10
Ymin = −5
Ymax = 80
Yscl = 10

Press the GRAPH button. Press 2ND TRACE and select 4:maximum. Move the cursor to the left of the maximum point and press ENTER, move the cursor to the right and press ENTER, and move the cursor close to the maximum and press ENTER for guess.

To the nearest integers, the coordinates of the maximum are (129, 72).

Horizontally, the ball will travel 129 feet just as it reaches its maximum height.

(d) Using the solution above, the maximum height of the ball is 72 feet.

Note that the answers in parts (c) and (d) differ from the maximum height that is provided in the table of values. This is because the regression equation was used to calculate the maximum height. The regression equation passes through most data points, not all points. Also, the equation used had rounded coefficients, which also contributes to the values being different.

When provided with a model, it is important to closely read the problem to make sense of the context of the problem. This will help identify variables, restrict domain and range if necessary, and identify the most appropriate function needed to model.

Polynomial Models

Polynomial functions model data sets or scenarios with multiple real zeros or multiple maxima or minima.

Example

An engineer working for an automobile company is trying to help design a crash test barrier whose ideal characteristics are shown in Figure 4.26. The time after impact, t, is measured in milliseconds. The distance that the barrier has been depressed after impact, d, is measured in millimeters.

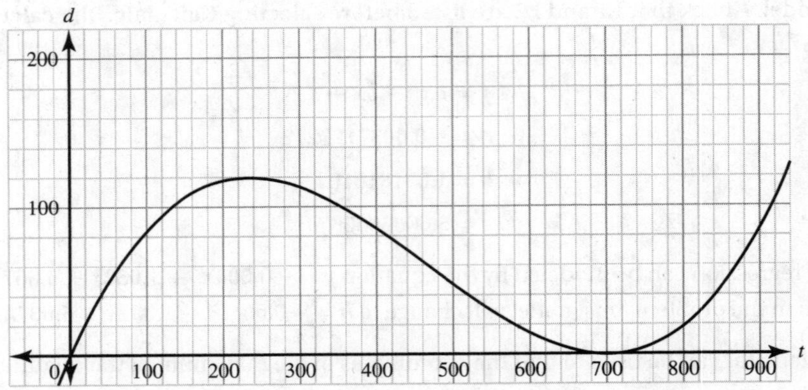

Figure 4.26

(a) If the equation for this graph is of the form $d = kt(t - a)^2$, what is the value of a? What does a represent?

(b) If the ideal crash barrier is depressed by 100 millimeters after 130 milliseconds, find the equation of the graph given.

(c) What is the maximum amount of depression, and when does it occur? Round each to the nearest integer.

✓ Solution

(a) The given equation is a polynomial equation written in factored form. The x-intercepts will give the zeros of the graph, which yield the factors. The x-intercepts are $x = 0$ and $x = 700$. The factors of these zeros are $(x - 0) = x$ and $(x - 700)$. Since the graph is tangent at $x = 700$, the multiplicity of that zero is even, which means that $a = 700$.

When $a = 700$, this is the time that the barrier has returned to its original position.

(b) The provided information gives the ordered pair $(t, d) = (130, 100)$. Substitute these values into the equation along with $a = 700$:

$$100 = k(130)(130 - 700)^2$$
$$100 = 42237000k$$
$$k = \frac{100}{42237000} \approx 0.00000236759$$

The equation is $d = 0.00000236759t(t - 700)^2$.

(c) After the barrier returns to its initial position at $t = 700$, it won't continue depressing. It is important to note that this affects the range of the given graph; it won't continue increasing when $t > 700$. Therefore, the range is restricted to $0 \leq y \leq 120$. To find the maximum, a graphing calculator will be used to find the coordinates of the relative maximum shown in the graph.

Using a TI-84, enter in the equation by pressing the Y= button.

Type the equation into Y1. Note that d will be represented by Y1 and t will be represented by x.

To see the graph, the window will need to be changed. This can be done by pressing the WINDOW button and setting the following:

Xmin = 0
Xmax = 900
Xscl = 200
Ymin = 0
Ymax = 120
Yscl = 20

Press the GRAPH button. Press the 2ND TRACE button and select 4:maximum. Move the cursor to the left of the relative maximum, and press ENTER. Move the cursor to the right of the relative maximum, and press ENTER. Move the cursor close to the relative maximum for GUESS, and press ENTER.

The coordinates are approximately (233.333332, 119.92296).

The maximum amount of depression is 120 millimeters at 233 milliseconds.

Piecewise Models

Often scenarios don't always follow the same behavior and might change based upon different constraints such as time or money. When this occurs, the situation needs to be modeled by a piecewise function. This allows for a better fit of the situation over the different intervals.

› Example

An electric company supplies electricity to residences for a monthly customer charge of \$7.58 plus 8.275 cents per kilowatt-hour (kWhr) for the first 400 kWhr supplied in the month. It then charges 6.280 cents per kWhr for all usage over 400kWhr in the month.

(a) What is the charge for using 250 kWhr in a month?

(b) What is the charge for using 900 kWhr in a month?

(c) If C is the monthly charge for x kWhr, write a model relating the monthly charge and kilowatt-hours used.

✓ Solution

(a) To figure out the charge for using 250 kWhr, the information given for the first 400 kWhr must be used. There is the fixed monthly charge of \$7.58 and then 8.275 cents for every kWhr.

Therefore, the total charge = \$7.58 + \$0.08275(250) = \$28.27.

(b) To figure out the charge for using 900 kWhr, the cost needs to be broken into two parts: the first 400 kWhr and the usage over 400 kWhr. There is the fixed monthly charge of \$7.58, then 8.275 cents for every kWhr for the first 400 kWhr, and then the charge changes to 6.280 cents for the remaining 500 kWhr.

Therefore, the total charge = \$7.58 + \$0.08275(400) + \$0.06280(500) = \$72.08.

(c) The steps to finding the solution in part (b) can help write a general equation that models the given scenario. It is important to first identify the independent variable. Let x = the number of kilowatt-hours used. If $0 \leq x \leq 400$, the monthly charge, C, in dollars can be found by adding the monthly charge \$7.58 to the product of x and \$0.08275. So if $0 \leq x \leq 400$, $C(x) = \$7.58 + \$0.08275x$.

If $x > 400$, the increased kWhr charge of \$0.06280 is multiplied by the amount of kWhr that is more than 400. This is represented by $x - 400$. So if $x > 400$, $C(x) = \$7.58 + \$0.08275x + \$0.06280(x - 400)$.

Written as a piecewise function:

$$C(x) = \begin{cases} \$7.58 + \$0.08275x, & 0 \leq x \leq 400 \\ \$7.58 + \$0.08275x + \$0.06280(x - 400), & x > 400 \end{cases}$$

Rational Models

Data sets and aspects of contextual scenarios involving quantities that are inversely proportional can often be modeled by a rational function.

Inverse variation is the relationship between variables that are represented in the form $y = \frac{k}{x}$, where x and y are two variables and k is the constant value or constant of proportionality. It states that if the value of one quantity increases, the value of the other quantity decreases. Inverse variation means that a variable is inversely varying with respect to another variable. It represents the inverse relationship between two quantities. Hence, a variable is inversely proportional to another variable.

There are many real-life examples of inverse variation. One example is if the distance traveled by a train at a constant speed increases, the time remaining for the ride decreases. A second example is if the number of people added to a job increases, the time taken to accomplish the job decreases.

Example

If x varies inversely with y and if $x = 10$ and $y = 4$, then what is the value of the constant of variation?

Solution

Let k represent the constant of variation. Since x varies inversely with y, $x = \frac{k}{y}$ or $xy = k$. Substitute the given values for x and y:

$$10 = \frac{k}{4}$$

$$k = 40$$

The constant of variation is 40.

Example

If 24 workers can build a house in 40 days, how many workers will be required to build the same house in 20 days?

Solution

Let x_2 represent the number of workers employed to build the house in 20 days. The time taken to build the house is inversely proportional to the number of workers required. So the following equation can be used:

$$x_1y_1 = x_2y_2$$

Substitute the known information:

$$(24)(40) = (x_2)(20)$$

$$x_2 = 48$$

It will take 48 workers to build the house in 20 days.

Rational functions can also appear in geometric models as well.

Example

A rectangular area adjacent to a river is to be fenced in; no fence is needed on the river side. The enclosed area is to be 1,000 square feet. Fencing for the side parallel to the river is \$5 per foot, and fencing for the other two sides is \$8 per foot. There are four corner posts that are \$25 each. Let x be the length of one of the sides perpendicular to the river.

(a) Write a function $C(x)$ that describes the cost of the project.
(b) State the domain of $C(x)$.
(c) Find the dimensions of the least expensive enclosure.

Solution

First create a diagram that models the given scenario. Start with a rectangle with an area of 1,000 square feet, as shown in Figure 4.27. Label one side as the river and the two perpendicular sides to the river as x.

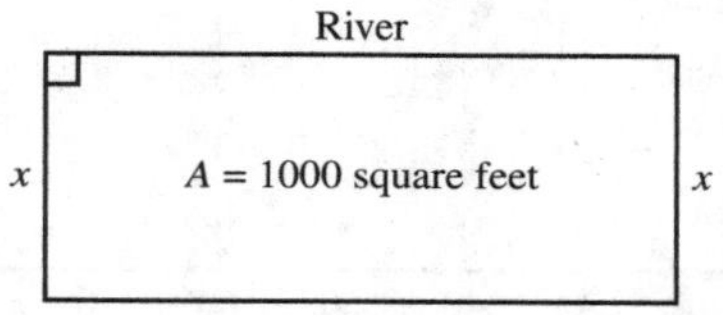

Figure 4.27

The length of the side opposite the river is also unknown. Instead of assigning it a new variable, use the given area information to represent its length in terms of x:

$$\text{Area of a rectangle} = (\text{length})(\text{width})$$
$$1000 = (\text{length})(x)$$
$$\text{length} = \frac{1000}{x}$$

Figure 4.28 shows the diagram with all four sides labeled.

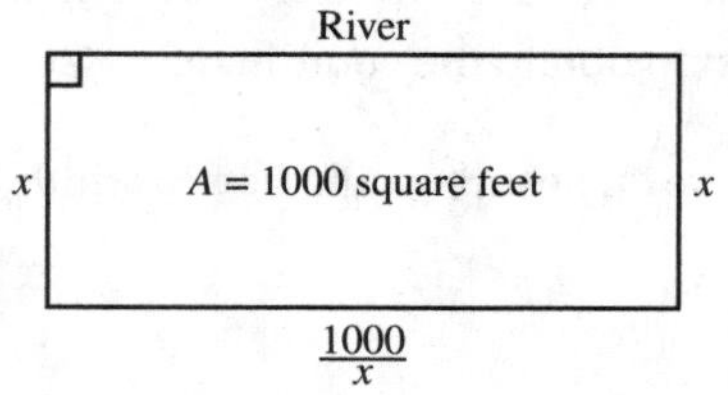

Figure 4.28

Since the rectangle has a set area, the set of numbers that x can represent needs to be restricted. The domain for this example is $x > 0$.

Now that the sides of the model are labeled, the cost of the project can be determined by multiplying each side by its respective cost per foot and adding on any additional charges.

(a) $C(x) = 5\left(\frac{1000}{x}\right) + 8(x) + 8(x) + 25(4) = 100 + 16x + \frac{5000}{x}$.

(b) The domain is $x > 0$.

(c) To find the dimensions of the least expensive enclosure, graph the function as shown in Figure 4.29 and use the calculator to find the coordinates of the minimum point.

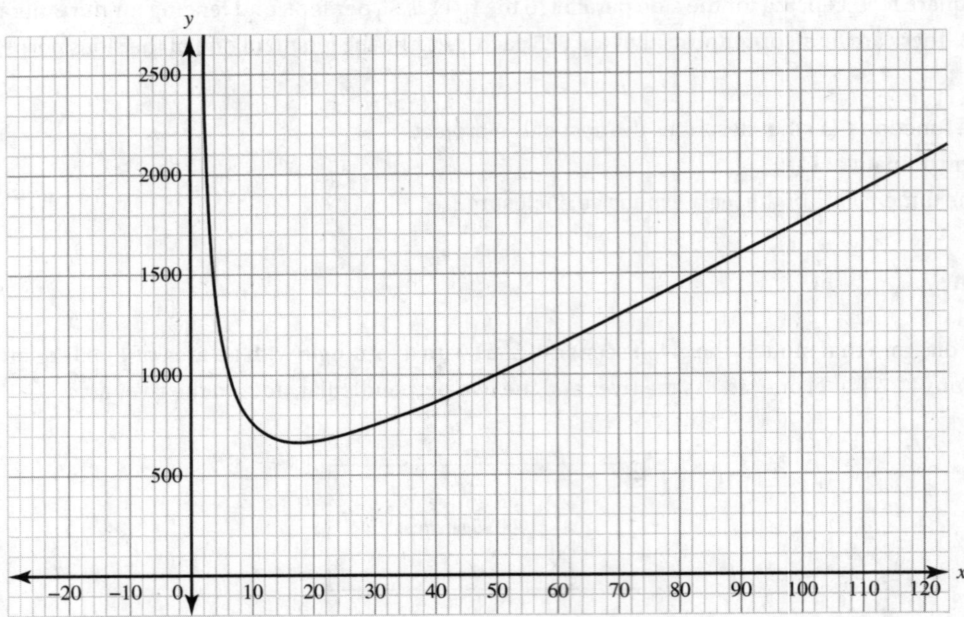

Figure 4.29

A good viewing window would be:

Xmin = 0
Xmax = 120
Xscl = 10
Ymin = −5
Ymax = 2500
Yscl = 100

Press the GRAPH button. To find the minimum point, press 2ND TRACE and select 3:minimum. Move the cursor to the left and right of the minimum point to obtain the coordinates (17.67761, 665.68542).

The dimensions to yield the least expensive enclosure would have a width of approximately 17.7 feet and a length of approximately $\frac{1000}{17.67761} = 56.6$ feet.

4.4 Composition of Functions

Two functions may be linked to form a new function by using the output of one function as the input to the other function. The new function that is created is called a composite function.

Given two functions f and g, the composite function, denoted by $f \circ g(x)$, which is read as "f composed with g" or "f of g of x," is defined by:

$$f \circ g(x) = f(g(x))$$

where the domain of $f \circ g$ is the set of all numbers x in the domain of g such that $g(x)$ is in the domain of f.

In other words, if f and g are functions, the composite function $f(g(x))$ maps a set of input values to a set of output values such that the output values of g are used as input values of f.

There are many properties associated with composite functions. To help derive these properties, the following composite functions will be evaluated.

Example

Let $f(x) = x - 2$ and $g(x) = 10x$. Find each of the following.

1. $f \circ g(1)$
2. $g \circ f(1)$
3. $f \circ f(-2)$
4. $g \circ g(-1)$

Solution

For functions f and g, if the output of function g is used as the input to function f, the composite function is denoted as $f \circ g(x) = f(g(x))$. This composition allows you to consider g the "inside" function and shows where the problem should begin to be evaluated. When evaluating composite functions, first evaluate the inside function and work your way to the outside functions.

Order of Composition

The composition of functions is not commutative. The values of $f(g(x))$ and $g(f(x))$ are usually not equal for the same x. The order in which functions are composed matters.

1. $f \circ g(1) = f(g(1))$. First evaluate $g(1) = 10(1) = 10$. Then $f(g(1)) = f(10) = (10) - 2 = 8$.
2. $g \circ f(1) = g(f(1))$. First evaluate $f(1) = (1) - 2 = -1$. Then $g(f(1)) = g(-1) = 10(-1) = -10$.
3. $f \circ f(-2) = f(f(-2))$. First evaluate $f(-2) = (-2) - 2 = -4$. Then $f(f(-2)) = f(-4) = (-4) - 2 = -6$.
4. $g \circ g(-1) = g(g(-1))$. First evaluate $g(-1) = 10(-1) = -10$. Then $g \circ g(-1) = g(-10) = 10(-10) = -100$.

When considering the solutions from (1) and (2), it can be observed that $f \circ g(1) \neq g \circ f(1)$. Therefore, the composition of functions is not commutative.

Example

Let $f(x) = 2x - 1$, $g(x) = \sqrt{x + 7}$, and $h(x) = x$. Find each of the following.

1. $f(g(x))$
2. $g(f(x))$
3. $f(h(x))$
4. $h(f(x))$
5. $f(g(x + 2))$

Solution

1. The inside function is $g(x)$, which equals $\sqrt{x + 7}$. That becomes the input for the outside function, f. Substitute $\sqrt{x + 7}$ for every x-term in $f(x)$:

$$f(g(x)) = f(\sqrt{x + 7}) = 2(\sqrt{x + 7}) - 1 = 2\sqrt{x + 7} - 1$$

2. The inside function is $f(x)$, which equals $2x - 1$. That becomes the input for the outside function, g. Substitute $2x - 1$ for every x-term in $g(x)$:

$$g(f(x)) = g(2x - 1) = \sqrt{(2x - 1) + 7} = \sqrt{2x - 1 + 7} = \sqrt{2x + 6}$$

3. The inside function is $h(x)$, which equals x. That becomes the input for the outside function, f. Substitute x for every x-term in $f(x)$:

$$f(h(x)) = f(x) = 2(x) - 1 = 2x - 1$$

4. The inside function is $f(x)$, which equals $2x - 1$. That becomes the input for the outside function, h. Substitute $2x - 1$ for every x-term in $h(x)$:

$$h(f(x)) = h(2x - 1) = (2x - 1) = 2x - 1$$

5. The inside function is $g(x)$. However, the input for $g(x)$ is something other than x. First evaluate $g(x + 2)$ by substituting $x + 2$ in for every x in $g(x)$:

$$g(x + 2) = \sqrt{(x + 2) + 7} = \sqrt{x + 2 + 7} = \sqrt{x + 9}$$

That output becomes the input for the outside function, f. Substitute $\sqrt{x + 9}$ for every x-term in $f(x)$:

$$f(g(x + 2)) = f(\sqrt{x + 9}) = 2(\sqrt{x + 9}) - 1 = 2\sqrt{x + 9} - 1$$

Like in the previous example, when considering the solution from (1) and (2), it can be observed that $f(g(x)) \neq g(f(x))$. It has been proved in general that the composition of functions is not commutative; that is, $f(g(x))$ and $g(f(x))$ are typically different functions.

When considering the solutions from (3) and (4), it can be observed that $f(h(x)) = f(x)$ and $h(f(x)) = f(x)$. This might appear to contradict the noncommutative property of composite functions. However, this is a special case.

A Special Case for Compositions

If the function $f(x) = x$ is composed with any function g, the resulting composite function is the same as g; that is, $g(f(x)) = f(g(x)) = g(x)$.

The function $f(x) = x$ is called the identity function.

Values for the composite function $f(g(x))$ can be calculated or estimated for the algebraic, graphical, numerical, or verbal representations of f and g by using output values from g as input values for f.

Example

Evaluate each expression using the values given in Table 4.7.

Table 4.7

x	-3	-2	-1	0	1	2	3
$f(x)$	-7	-5	-3	-1	3	5	5
$g(x)$	8	3	0	-1	0	3	8

1. $f(g(2))$
2. $g(f(0))$
3. $f \circ g(0)$
4. $g \circ g(-2)$

Solution

1. To evaluate $f(g(2))$, first evaluate $g(2)$ by locating $x = 2$ on the table and sliding down to the row of $g(x)$: $g(2) = 3$. Then evaluate $f(3)$ by locating $x = 3$ and sliding down to the row of $f(x)$:
$$f(g(2)) = f(3) = 5$$
2. To evaluate $g(f(0))$, first evaluate $f(0)$ by locating $x = 0$ on the table and sliding down to the row of $f(x)$: $f(0) = -1$. Then evaluate $g(-1)$ by locating $x = -1$ and sliding down to the row of $g(x)$:
$$g(f(0)) = g(-1) = 0$$
3. To evaluate $f \circ g(0)$, first evaluate $g(0)$ by locating $x = 0$ on the table and sliding down to the row of $g(x)$: $g(0) = -1$. Then evaluate $f(-1)$ by locating $x = -1$ and sliding down to the row of $f(x)$:
$$f \circ g(0) = f(g(0)) = f(-1) = -3$$
4. To evaluate $g \circ g(-2)$, first evaluate $g(-2)$ by locating $x = -2$ on the table and sliding down to the row of $g(x)$: $g(-2) = 3$. Then evaluate $g(3)$ by locating $x = 3$ and sliding down to the row of $g(x)$:
$$g \circ g(-2) = g(g(-2)) = g(3) = 8$$

Example

Evaluate each expression using the values given in the graphs of $f(x)$ and $g(x)$ shown in Figure 4.30.

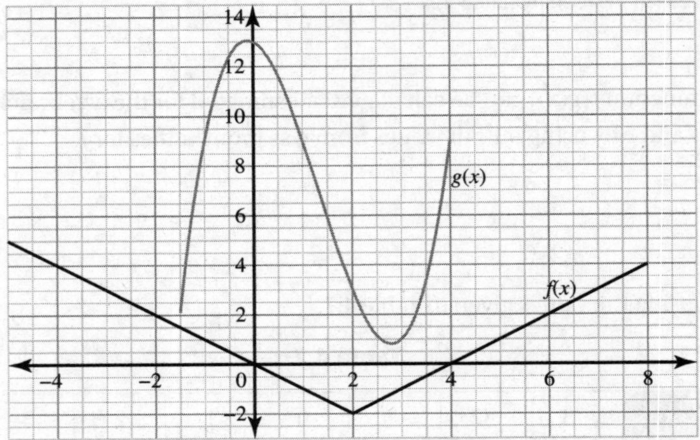

Figure 4.30

1. $f(g(2))$
2. $g(f(0))$
3. $f(f(6))$
4. $g(g(-1.5))$

Solution

1. To evaluate $f(g(2))$, first evaluate $g(2)$ by locating $x = 2$ on the x-axis and sliding up to the graph of $g(x)$ to find the y-coordinate of the point on the graph: $g(2) = 3$. Then evaluate $f(3)$ by locating $x = 3$ on the x-axis and sliding down to the graph of $f(x)$ to find the y-coordinate of the point on the graph:

$$f(g(2)) = f(3) = -1$$

2. To evaluate $g(f(0))$, first evaluate $f(0)$ by locating $x = 0$ on the x-axis and locating the point on the graph of $f(x)$ to find the y-coordinate: $f(0) = 0$. Then evaluate $g(0)$ by locating $x = 0$ on the x-axis and sliding up to the graph of $g(x)$ to find the y-coordinate of the point on the graph:

$$g(f(0)) = g(0) = 13$$

3. To evaluate $f(f(6))$, first evaluate $f(6)$ by locating $x = 6$ on the x-axis and sliding up to the graph of $f(x)$ to find the y-coordinate of the point on the graph: $f(6) = 2$. Then evaluate $f(2)$ by locating $x = 2$ on the x-axis and sliding down to the graph of $f(x)$ to find the y-coordinate of the point on the graph:

$$f(f(6)) = f(2) = -2$$

4. To evaluate $g(g(-1.5))$, first evaluate $g(-1.5)$ by locating $x = -1.5$ on the x-axis and sliding up to the graph of $g(x)$ to find the y-coordinate of the point on the graph: $g(-1.5) = 2$. Then evaluate $g(2)$ by locating $x = 2$ on the x-axis and sliding up to the graph of $g(x)$ to find the y-coordinate of the point on the graph:

$$g(g(-1.5)) = g(2) = 3$$

Domain of a Composite Function

The domain of the composite function $f(g(x))$ is restricted to those input values of g for which the corresponding output values is in the domain of f. The composite function $f(g(x))$ uniquely maps input values of g to output values of f, dependent on the domain restrictions of f and g.

Example

Let $f(x) = x^2 - 3x + 2$ and $g(x) = -4x + 7$. Find:

(a) $f \circ g(x)$

(b) The domain of $f \circ g(x)$

Solution

(a) The inside function is $g(x)$, which equals $-4x + 7$. That becomes the input for the outside function, f. Substitute $-4x + 7$ for every x-term in $f(x)$:

$$f(g(x)) = f(-4x + 7) = (-4x + 7)^2 - 3(-4x + 7) + 2$$
$$= 16x^2 - 28x - 28x + 49 + 12x - 21 + 2$$
$$f(g(x)) = 16x^2 - 44x + 30$$

(b) Since the domains of both f and g are the set of real numbers, the domain of $f \circ g(x)$ is the set of real numbers.

Example

Find the domain of $f \circ g(x)$ if $f(x) = \frac{1}{x-3}$ and $g(x) = \frac{-3}{x+7}$.

Solution

Since $f \circ g(x) = f(g(x))$, first consider the domain of $g(x)$. The domain of g is all real numbers, $x \neq -7$. So $x = -7$ is excluded from the domain of $f \circ g(x)$. The domain of f is all real numbers, $x \neq 3$. So $g(x)$ cannot equal 3. It needs to be determined what x-values are the solution to $g(x) = 3$ so those x-values can be excluded from the domain:

$$g(x) = 3$$
$$\frac{-3}{x+7} = 3$$
$$-3 = 3(x + 7)$$
$$-1 = x + 7$$
$$x = -8$$

The domain of $f \circ g(x)$ is all real numbers, $x \neq -8$, $x \neq -7$.

Some techniques in calculus make it necessary to be able to determine the different functions of a composite function. For example, the function $c(x) = |x^2 - 3x|$ is the composition of the functions f and g, where $f(x) = |x|$ and $g(x) = x^2 - 3x$, because $c(x) = f(g(x)) = f(x^2 - 3x) = |x^2 - 3x|$.

Example

Find functions $f(x)$ and $g(x)$ such that $f(g(x)) = c(x)$ and $c(x) = \sqrt{3x - 7}$.

Solution

The outer function is the square root function, and the inner function is the radicand $3x - 7$. Let $g(x) = 3x - 7$ and $f(x) = \sqrt{x}$. Then $f(g(x)) = f(3x - 7) = \sqrt{3x - 7} = c(x)$.

Example

Find functions $f(x)$ and $g(x)$ such that $f(g(x)) = c(x)$ and $c(x) = 2 \cdot \frac{1}{x+5} - 3$.

Solution

The inner function is the rational function $\frac{1}{x+5}$. If that expression is replaced by an x, it reveals that the outer function is $2x - 3$. Let $g(x) = \frac{1}{x+5}$ and $f(x) = 2x - 3$. Then $f(g(x)) = f\left(\frac{1}{x+5}\right) = 2 \cdot \frac{1}{x+5} - 3 = c(x)$.

4.5 Inverse Functions

For a relation to be a function, every element in its domain must have one and only one corresponding element in its range. On a specified domain, a function, f, has an inverse function or is invertible if each output value of f is mapped from a unique input value. The domain of a function may be restricted to make the function invertible.

An inverse function can be thought of as a reverse mapping of the function. An inverse function, f^{-1}, maps the output values of a function, f, on its invertible domain to their corresponding input values.

Inverse Functions and Input-Output Pairs

Given function $f(x)$, if $f(a) = b$, then $f^{-1}(b) = a$.

If a function consists of input-output pairs (a, b), the inverse function consists of input-output pairs (b, a).

Example

(a) Find the inverse of the following function:

$$\{(-3, -27), (-2, -8), (-1, -1), (0, 0), (1, 1), (2, 8), (3, 27)\}$$

(b) State the domain and range of the function and its inverse.

(c) Determine if the inverse is also a function.

✓ Solution

(a) The inverse of the given function is found by interchanging the entries in each ordered pair:

$$\{(-27, -3), (-8, -2), (-1, -1), (0, 0), (1, 1), (8, 2), (27, 3)\}$$

(b) The domain of the given function is

$$\{-3, -2, -1, 0, 1, 2, 3\}$$

The range of the given function is

$$\{-27, -8, -1, 0, 1, 8, 27\}$$

The domain of the inverse is

$$\{-27, -8, -1, 0, 1, 8, 27\}$$

The range of the inverse is

$$\{-3, -2, -1, 0, 1, 2, 3\}$$

(c) The inverse is a function since every element in the domain maps to one and only one element in the range.

› Example

(a) Find the inverse of the following function:

$$\{(-3, 9), (-2, 4), (-1, 1), (0, 0), (1, 1), (2, 4), (3, 9)\}$$

(b) State the domain and range of the function and its inverse.

(c) Determine if the inverse is also a function.

✓ Solution

(a) The inverse of the given function is found by interchanging the entries in each ordered pair:

$$\{(9, -3), (4, -2), (1, -1), (0, 0), (1, 1), (4, 2), (9, 3)\}$$

(b) The domain of the given function is

$$\{-3, -2, -1, 0, 1, 2, 3\}$$

The range of the given function is

$$\{0, 1, 4, 9\}$$

The domain of the inverse is

$$\{0, 1, 4, 9\}$$

The range of the inverse is

$$\{-3, -2, -1, 0, 1, 2, 3\}$$

(c) The inverse is not a function since $x = 1$, $x = 4$, and $x = 9$ map to more than one element in the range.

A one-to-one function, f, has an inverse, f^{-1}, that is also a function.

A function is a one-to-one function if any two different inputs in the domain correspond to two different outputs in the range.

If x_1 and x_2 are two different inputs of a function f, then f is one-to-one if $f(x_1) \neq f(x_2)$.

A function is not one-to-one if two different inputs correspond to the same output.

Example

Determine whether the following functions are one-to-one.

1. $\{(-3, 9), (-2, 4), (-1, 1), (0, 0), (1, 1), (2, 4), (3, 9)\}$
2. $\{(-2, 6), (-1, 3), (0, 2), (1, 5), (2, 8)\}$

Solution

1. The function is not one-to-one because there are three different pairs of inputs that correspond to the same output: -3 and 3 correspond to 9, -2 and 2 correspond to 4, and -1 and 1 correspond to 1.
2. The function is one-to-one because there are no two distinct inputs that correspond to the same output.

   For functions defined by an equation $y = f(x)$ and for which the graph is known, the horizontal line test can be performed to determine whether a function is one-to-one.

The Horizontal Line Test

If every horizontal line intersects the graph of a function f in at most one point, f is one-to-one.

Example

For each function, use its graph to determine whether the function is one-to-one.

1. $f(x) = x^4$
2. $g(x) = x^5$

Solution

1. As shown in Figure 4.31, a horizontal line intersects the graph twice. So $f(x)$ is not one-to-one.

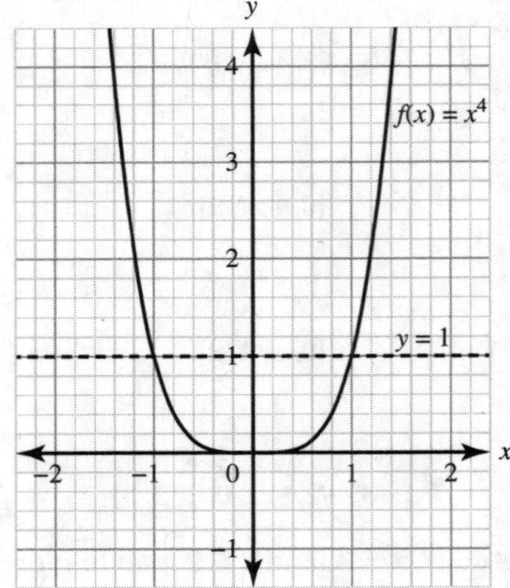

Figure 4.31

2. As shown in Figure 4.32, every horizontal line intersects the graph exactly once. So $g(x)$ is one-to-one.

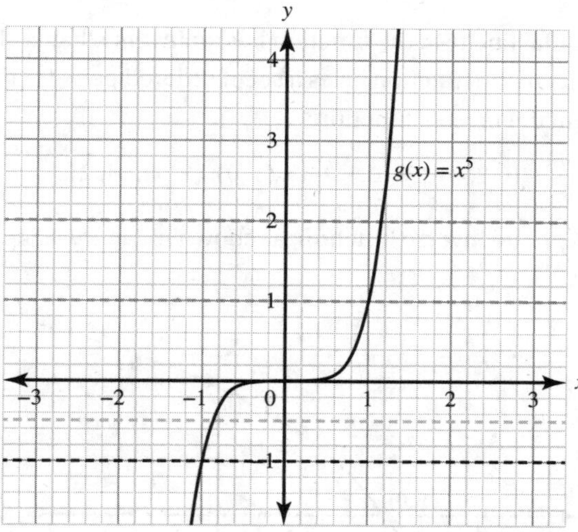

Figure 4.32

Another interesting graphical relationship between a function and its inverse deals with symmetry. The inverse of the graph of the function $y = f(x)$ can be found by reversing the roles of the x- and y-axes, that is, by reflecting the graph of the function over the line $y = x$.

Example

Given the graph of the function $f(x)$ in Figure 4.33, use the set of ordered pairs to graph its inverse function, $f^{-1}(x)$, on the same set of axes.

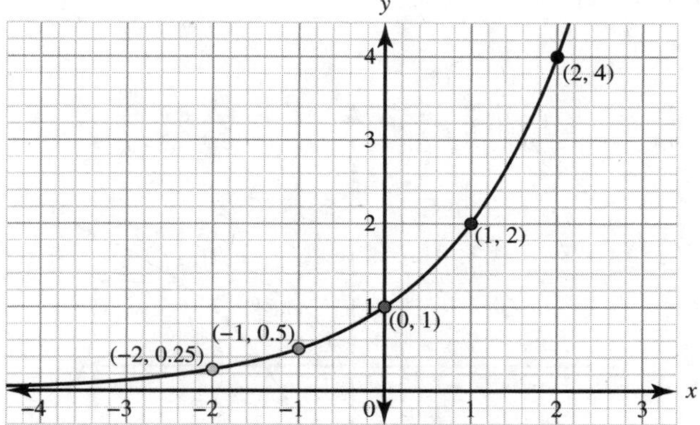

Figure 4.33

Solution

Since this graph passes the horizontal line test, the function has an inverse that is also a function.

Reversing the input-output pairs of the function gives a set of ordered pairs for the inverse function. The following is a set of points on the graph of the inverse function:

$$\{(0.25, -2), (0.5, -1), (1, 0), (2, 1), (4, 2)\}$$

Use this set of ordered pairs to graph the inverse function, as shown in Figure 4.34:

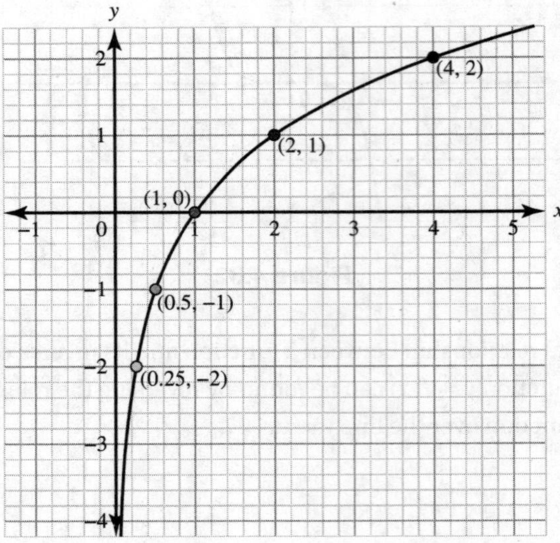

Figure 4.34

As a check, graph the original function and its inverse along with the line $y = x$ to check for symmetry. This is shown in Figure 4.35.

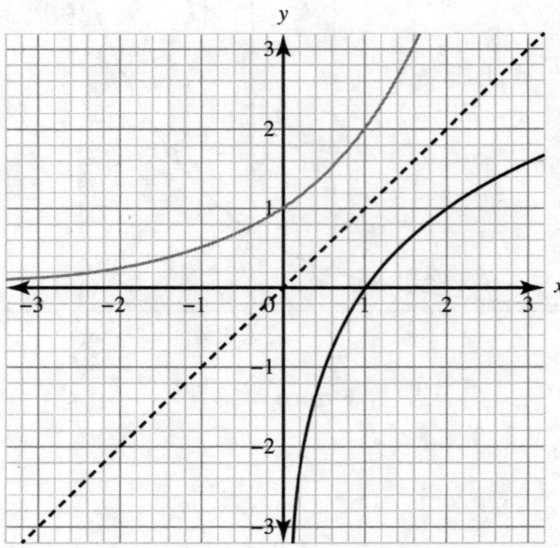

Figure 4.35

There are also algebraic techniques to determine the inverse of a function.

The Identity Function

The composition of a function, f, and its inverse function, f^{-1}, is the identity function. In other words, $f(f^{-1}(x)) = f^{-1}(f(x)) = x$.

Example

State whether the following functions are inverses of each other.

1. $f(x) = 3x - 4$ and $g(x) = \frac{x}{3} + \frac{4}{3}$
2. $h(x) = x^2 - 4$ and $i(x) = \sqrt{x^2 + 4}$

Solution

One way to verify if two equations are inverses of each other is to take their composition in both directions and show that both are equal to x.

1.

$$f(g(x)) = f\left(\frac{x}{3} + \frac{4}{3}\right) = 3\left(\frac{x}{3} + \frac{4}{3}\right) - 4 = x + 4 - 4 = x$$

$$g(f(x)) = g(3x - 4) = \frac{3x - 4}{3} + \frac{4}{3} = \frac{3x - 4 + 4}{3} = \frac{3x}{3} = x$$

Since $f(g(x)) = g(f(x)) = x$, the two functions are inverses of each other.

2.

$$h(i(x)) = h\left(\sqrt{x^2 + 4}\right) = \left(\sqrt{x^2 + 4}\right)^2 - 4 = x^2 + 4 - 4 = x^2$$

$$i(h(x)) = i(x^2 - 4) = \sqrt{(x^2 - 4)^2 + 4} = \sqrt{x^4 - 8x^2 + 16 + 4} = \sqrt{x^4 - 8x^2 + 20}$$

Since $h(i(x)) \neq i(h(x))$ and they both did not equal x, they are not inverses of each other.

It is possible to find the equation of a function's inverse algebraically. The inverse of the function can be found by determining the inverse operations to reverse the mapping.

Algebraically Finding the Inverse

One method for finding the inverse of the function f is reversing the roles of x and y in the equation $y = f(x)$ and then solving for $y = f^{-1}(x)$.

Example

Find the inverse equation of $f(x) = 3x + 7$. Verify that the original equation and its inverse equation are both functions.

Solution

The given equation graphs a line with a y-intercept at (0, 7) and has a slope of 3. This graph would pass the horizontal line test, and so both the given equation and its inverse will be a function.

To find the inverse, reverse the x and the y:

$$x = 3y + 7$$

That equation represents the inverse, but it is more common to solve for y:

$$x - 7 = 3y$$

$$y = \frac{x - 7}{3}$$

The inverse function is $f^{-1}(x) = \frac{x - 7}{3}$.

Example

Given the function $f(x) = \frac{2x - 1}{x + 1}, x \neq -1$:

(a) State whether the function is one-to-one.

(b) Find the equation of its inverse.

Solution

(a) To find whether the function is one-to-one, graph $f(x)$, as shown in Figure 4.36, to see if it passes the horizontal line test.

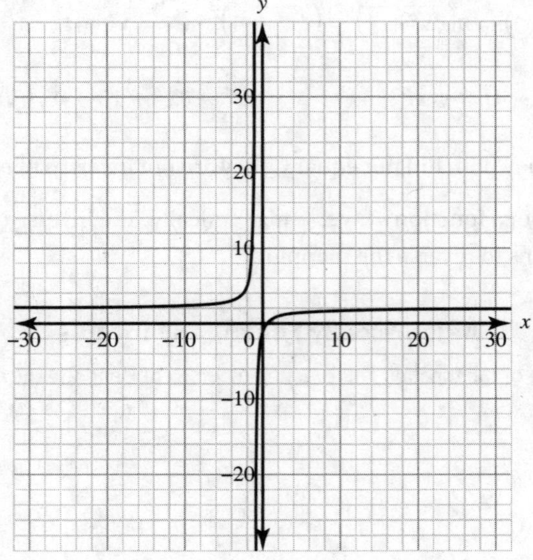

Figure 4.36

Any horizontal line drawn will intersect the graph once and only once. Therefore, the function is one-to-one and its inverse will also be a function.

(b) To find the inverse function, reverse the x and the y:

$$x = \frac{2y - 1}{y + 1}$$

That equation represents the inverse, but it is more common to solve for y:

$$x(y + 1) = 2y - 1$$
$$xy + x = 2y - 1$$
$$xy - 2y = -1 - x$$
$$y(x - 2) = -1 - x$$
$$y = \frac{-1 - x}{x - 2}$$

The inverse function is $f^{-1}(x) = \frac{-1 - x}{x - 2}$.

Not all functions are one-to-one. So to have an inverse that is also a function, it is necessary to restrict the domain of a function. Here the horizontal line test can be a useful visual aid to see where to limit the domain of a function so that the horizontal lines will intersect the function only once.

Example

Find the inverse function of $f(x) = x^2$.

Solution

While $f(x) = x^2$ is a function, it is not a one-to-one function as it fails the horizontal line test. This is shown in Figure 4.37.

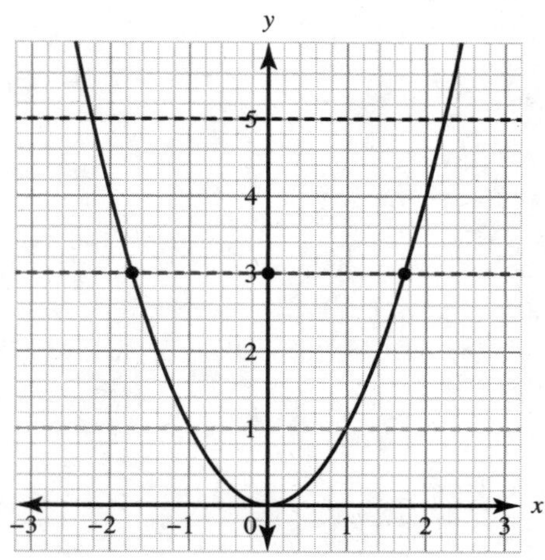

Figure 4.37

Using this visual, if the domain is restricted to $x \geq 0$, the horizontal lines intersect the graph only once, as shown in Figure 4.38.

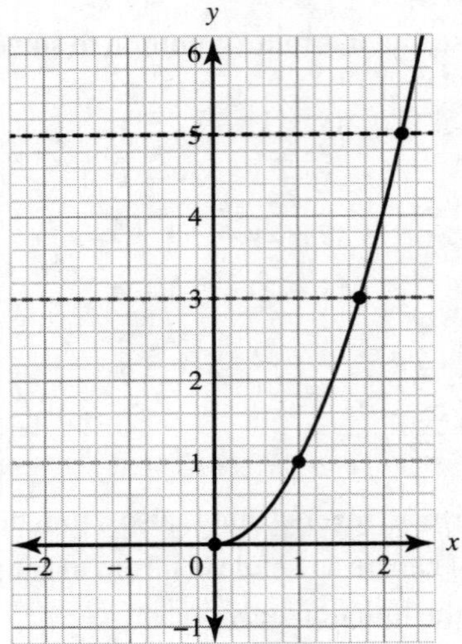

Figure 4.38

The function $f(x) = x^2, x \geq 0$ will have an inverse function.

To find the equation of the inverse function, reverse the x and the y:

$$x = y^2, y \geq 0$$

Solve for y by extracting the square root:

$$y = \pm\sqrt{x}$$

Since $y \geq 0$, there is only one solution:

$$y = \sqrt{x}$$

Therefore, the inverse function is $f^{-1}(x) = \sqrt{x}$.

Multiple-Choice Questions

1. If $f(x) = x^2 + 1$, then $f(a - 1) =$

   (A) $a^2 + 1$
   (B) a^2
   (C) $a^2 - 1$
   (D) $a^2 - 2a + 2$

2. Which of the accompanying diagrams describes a one-to-one function?

   (A)

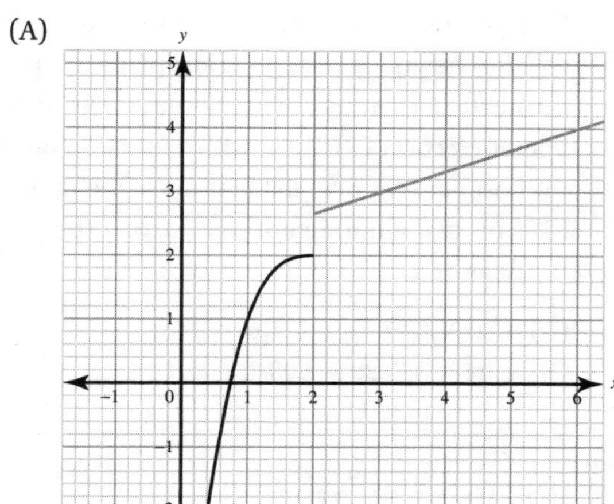

   (B)

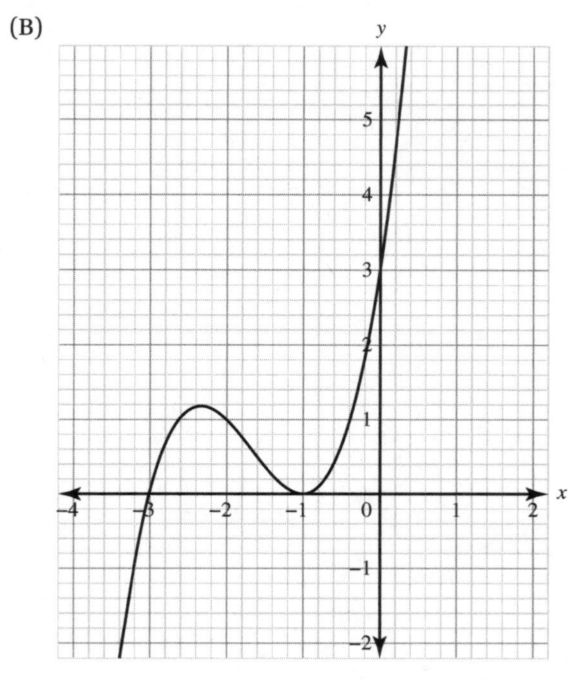

   (C)

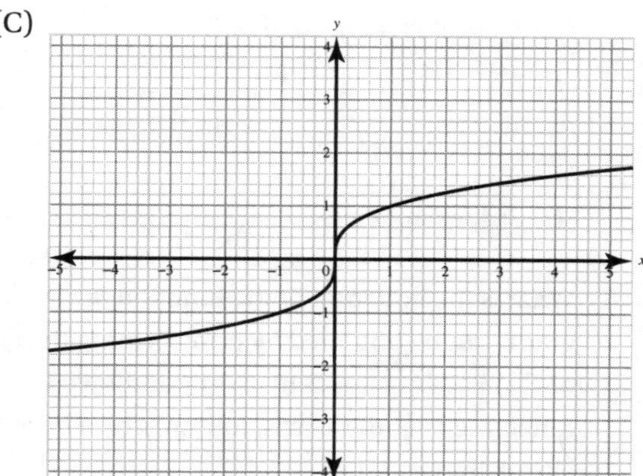

   (D)

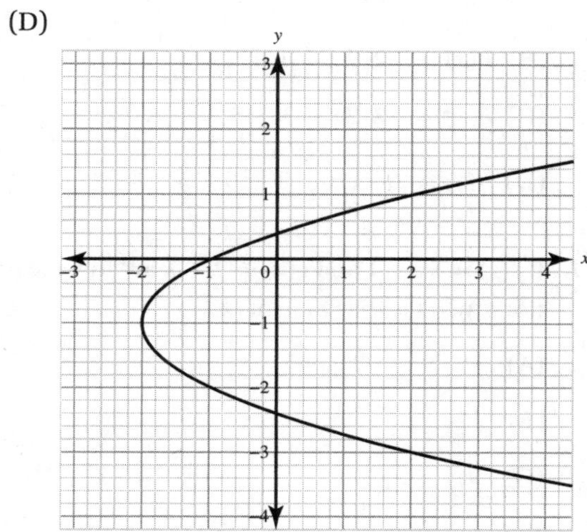

3. If $f(x) = x^2 + 1$, then $\frac{f(x+h) - f(x)}{h} =$

   (A) $2x + h$
   (B) $2x + h + 2$
   (C) h
   (D) $\frac{h^2 + 2}{h}$

4. A ball is thrown into the air such that its height, h, in feet at any time, t, in seconds is given by the function $h(t) = -16t^2 + 80t + 10$. What is the maximum height attained by the ball?

   (A) 140
   (B) 110
   (C) 85
   (D) 10

5. If the function f is defined by $f(x) = 3x + k$ and the function g is defined by $g(x) = \frac{x-1}{3}$, for what value of k is $f(g(x)) = g(f(x))$?

(A) -1
(B) 0
(C) 1
(D) k does not exist

6. What is the equation of the function $h(x)$ whose graph can be obtained by transforming the graph of $f(x) = 2\sqrt{x-1} + 3$ up 2 units, followed by reflecting over the x-axis, followed by a horizontal shrink by a factor of $\frac{1}{3}$?

(A) $h(x) = -2\sqrt{3x-1} - 5$
(B) $h(x) = -2\sqrt{\frac{1}{3}x - 1} + 5$
(C) $h(x) = -2\sqrt{3x-3} - 5$
(D) $h(x) = 2\sqrt{-\frac{1}{3}x - 3} + 5$

7. If $f(x) = 5x + 1$ and $g(x) = \frac{x}{5} + k$ are inverse functions, what is the value of k?

(A) -1
(B) $-\frac{1}{5}$
(C) $\frac{1}{5}$
(D) 5

8. If $f(x) = \frac{5x-1}{x+2}$, find the equation of its inverse, $f^{-1}(x)$.

(A) $f^{-1}(x) = \frac{x+2}{5x-1}$
(B) $f^{-1}(x) = \frac{2x+1}{x-5}$
(C) $f^{-1}(x) = \frac{2x+1}{5-x}$
(D) $f^{-1}(x) = \frac{2x-1}{x-5}$

9. Cailey wants to buy a new outdoor heater. The cost for the unit is \$349.99. If she plans to run the heater 5 months out of the year for an annual operating cost of \$121.67, which function models the cost per year over the lifetime of the unit, $C(t)$, in terms of the number of years, t, that she owns the heater?

(A) $C(t) = 349.99 + 121.67t$
(B) $C(t) = 349.99 + 608.35t$
(C) $C(t) = \frac{349.99 + 121.67t}{t}$
(D) $C(t) = \frac{349.99 + 608.35t}{t}$

10. The accompanying graph represents the equation of the polynomial function $y = f(x)$.

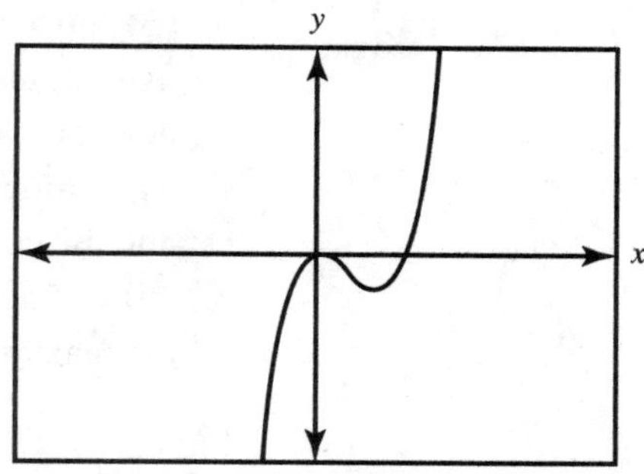

Which graph represents $g(x)$ if $g(x) = -f(x) + 3$?

(A)

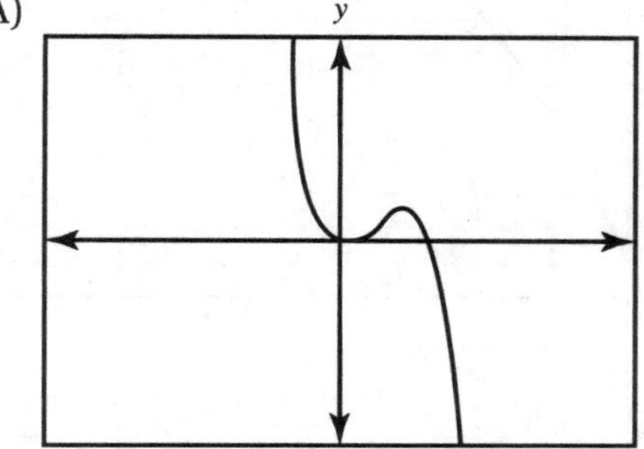

(B)

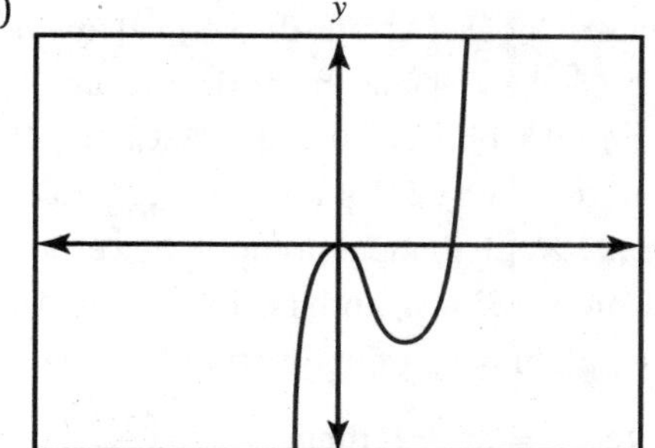

(C)

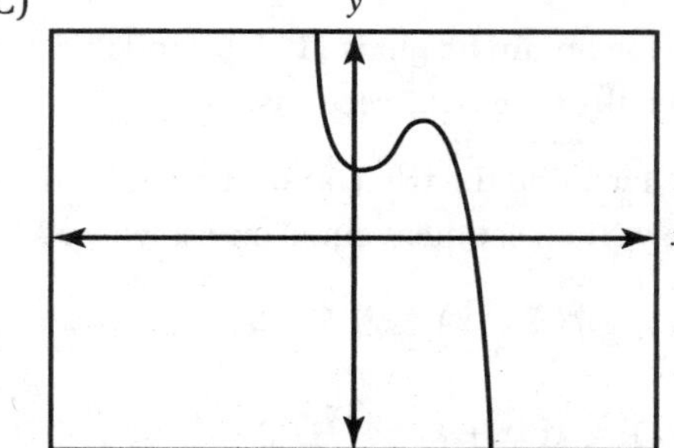

(D)

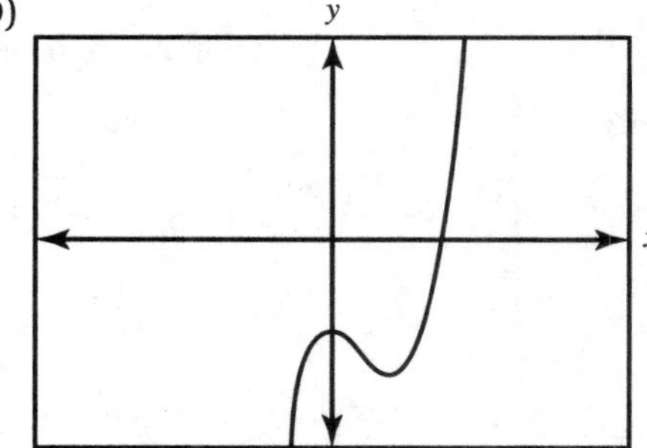

Answer Explanations

1. **(D)** To evaluate $f(a-1)$, let the input, $a-1$, be substituted into the function for x:

$$f(a-1) = (a-1)^2 + 1$$

Expand and combine like terms:

$$\begin{aligned} f(a-1) &= (a-1)^2 + 1 = (a-1)(a-1) + 1 \\ &= a^2 - 2a + 1 + 1 \\ f(a-1) &= a^2 - 2a + 2 \end{aligned}$$

2. **(C)** The graph of a one-to-one function must pass both the vertical line test and the horizontal line test. Choice (A) fails the vertical line test at $x = 2$. Choice (B) fails the horizontal line test for any y-values between 0 and 1.25. Choice (D) fails the vertical line test for any $x > -2$. Choice (C) is the only graph that passes both the vertical and horizontal line test, and it can be recognized as the graph of the parent function $f(x) = \sqrt[3]{x}$.

3. **(A)** If $f(x) = x^2 + 1$, then:

$$\begin{aligned} f(x+h) &= (x+h)^2 + 1 = (x+h)(x+h) + 1 \\ &= x^2 + 2xh + h^2 + 1 \end{aligned}$$

Then:

$$\begin{aligned} \frac{f(x+h) - f(x)}{h} &= \frac{x^2 + 2xh + h^2 + 1 - (x^2 + 1)}{h} \\ &= \frac{x^2 + 2xh + h^2 + 1 - x^2 - 1}{h} \\ &= \frac{2xh + h^2}{h} = 2x + h \end{aligned}$$

4. **(B)** Since the function is a quadratic with a negative leading coefficient, the parabola is concave down, meaning there is a maximum point. Using a graphing calculator, the function can be graphed and the coordinates of the maximum point can be found. The y-coordinate of the point represents the maximum height.

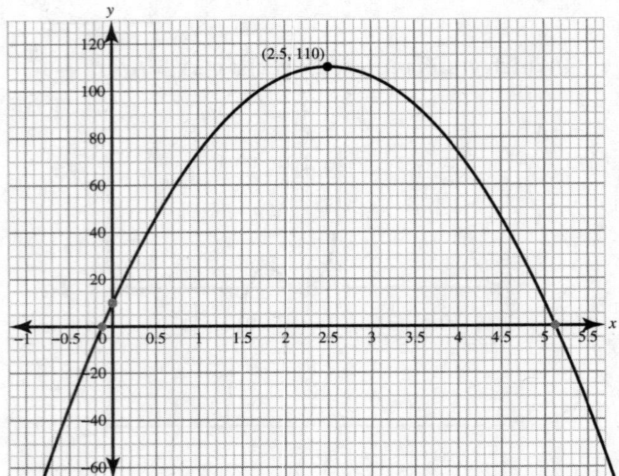

The maximum height is 110 feet, and this occurs when the time is 2.5 seconds.

5. **(C)** First find the two composite functions separately. Then set them equal and solve for k:

$$f(g(x)) = f\left(\frac{x-1}{3}\right) = 3\left(\frac{x-1}{3}\right) + k = x - 1 + k$$

$$g(f(x)) = g(3x + k) = \frac{3x + k - 1}{3}$$

Set the two expressions equal and solve for k:

$$\begin{aligned} x - 1 + k &= \frac{3x + k - 1}{3} \\ 3(x - 1 + k) &= 3x + k - 1 \\ 3x - 3 + 3k &= 3x + k - 1 \\ 2k &= 2 \\ k &= 1 \end{aligned}$$

6. **(A)** Each transformation affects the graph in a different way. A translation up 2 units is an addition of 2 to the function, a reflection over the x-axis negates the entire function, and a horizontal shrink by a factor of $\frac{1}{3}$ is a substitution of $3x$ in for x in the original function.

7. **(B)** If two functions are inverses of each other, their composition in either direction must equal x. For this problem, $f(g(x)) = g(f(x)) = x$. Find each composition, set them each equal to x, and solve for k:

$$f(g(x)) = f\left(\frac{x}{5} + k\right) = 5\left(\frac{x}{5} + k\right) + 1 = x + 5k + 1$$

$$x + 5k + 1 = x$$

$$5k + 1 = 0$$

$$k = -\frac{1}{5}$$

$$g(f(x)) = g(5x + 1) = \frac{5x + 1}{5} + k = x + \frac{1}{5} + k$$

$$x + \frac{1}{5} + k = x$$

$$\frac{1}{5} + k = 0$$

$$k = -\frac{1}{5}$$

8. **(C)** To find the equation of an inverse, reverse the x and the y and solve for y. The result will represent the inverse:

$$x = \frac{5y - 1}{y + 2}$$

$$x(y + 2) = 5y - 1$$

$$xy + 2x = 5y - 1$$

$$xy - 5y = -1 - 2x$$

$$y(x - 5) = -1 - 2x$$

$$y = f^{-1}(x) = \frac{-1 - 2x}{x - 5} = -\frac{1 + 2x}{x - 5} = \frac{1 + 2x}{-x + 5} = \frac{2x + 1}{5 - x}$$

9. **(C)** To determine the cost function, there is a fixed cost of \$349.99 that happens once and then an additional operating cost of \$121.67 that is paid every year. The operating cost is not dependent on the number of months Cailey runs the heater. To find the cost per year over the lifetime of the unit, take $349.99 + 121.67t$ and divide that by the number of years she owns the heater.

10. **(C)** The given function is $g(x) = -f(x) + 3$. Verbally, these changes to the function mean the graph is reflected over the x-axis and is shifted up 3 units. Only one graph has both transformations.

UNIT 2
Exponential and Logarithmic Functions

5

Exponential Functions

Learning Objectives

In this chapter, you will learn:

- ➔ The different properties of exponential functions
- ➔ How to graph exponential functions
- ➔ How to transform graphs of exponential functions
- ➔ How to model real-world scenarios using exponential functions

The previous chapters have dealt with algebraic functions, a function expressed as an equation in which real numbers and the variable are connected only by the operations of addition, subtraction, multiplication, division, and exponents. This chapter will discuss an important type of nonalgebraic function that can be used to model certain types of growth processes.

5.1 Exponential Functions

The following equations all have a 2 and an x: $y = 2x$, $y = x^2$, and $y = 2^x$. However, they are each very different functions. The first equation is a linear function, and its graph is a straight line. The second equation is a quadratic function, and its graph is a parabola. The third equation is called an exponential function. The exponential function has an exponent that is a variable, and its base is a number. An exponential function is a nonalgebraic function in which the independent variable is in the exponent.

General Form

The general form of an exponential function is $f(x) = ab^x$, with the initial value a, where $a \neq 0$, and the base b, where $b > 0$ and $b \neq 1$.

Exponential Graphs

The functions $f(x) = 2^x$ and $g(x) = \left(\frac{1}{2}\right)^x$ are both exponential functions. The graph of $f(x)$ is shown in Figure 5.1, and the graph of $g(x)$ is shown in Figure 5.2.

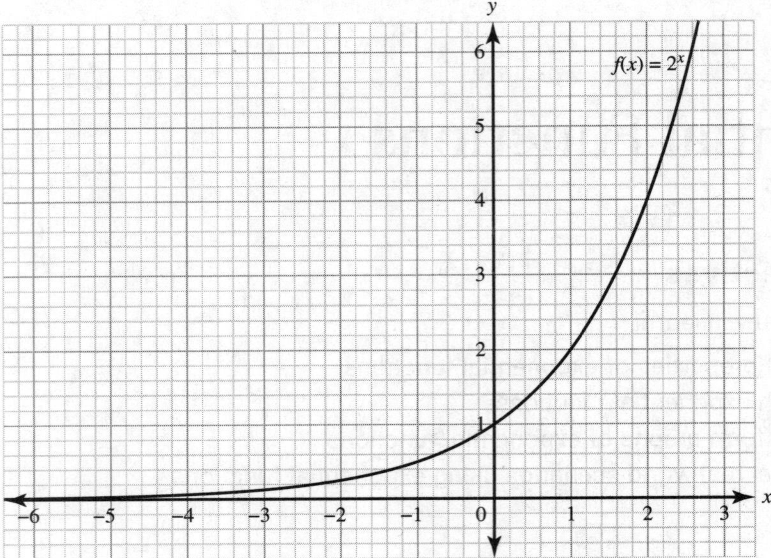

Figure 5.1

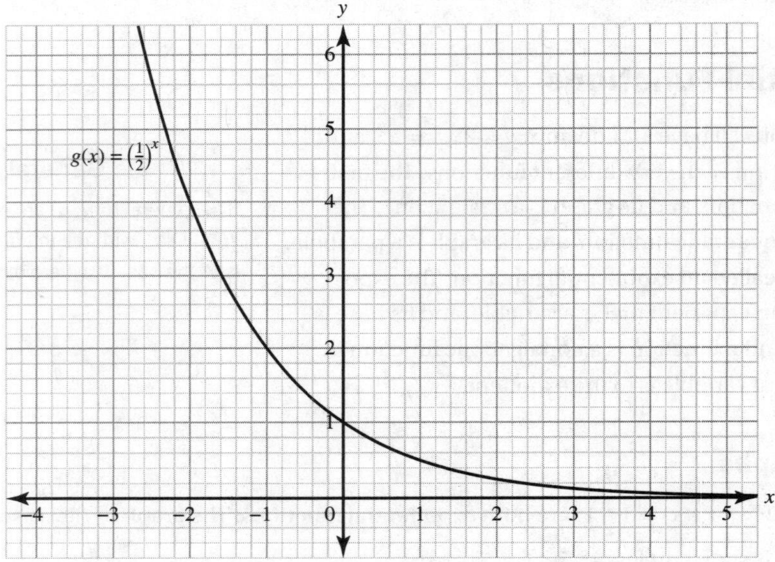

Figure 5.2

There are many similarities between the two graphs, which lead to generalizations for the graphs of exponential functions, $y = ab^x$:

1. Domain is all real numbers.
2. Range is $(0, \infty)$.
3. y-intercept is $(0, 1)$.
4. Horizontal asymptote at $y = 0$.
5. When $a > 0$ and $b > 1$, the function always increases, or demonstrates exponential growth.
6. When $a > 0$ and $0 < b < 1$, the function always decreases, or demonstrates exponential decay.
7. Exponential functions do not have extrema unless on a closed interval.
8. When $a > 0$ and $b > 0$, the graph is always concave up.

9. When $a > 0$ and $0 < b < 1$, the graph is always concave up.
10. The graphs of exponential functions do not have inflection points.

Example

Use the graphs of $f(x) = 2^x$ and $g(x) = \left(\frac{1}{2}\right)^x$, and the graph of $h(x) = -3^x$ shown in Figure 5.3, to evaluate the following limits.

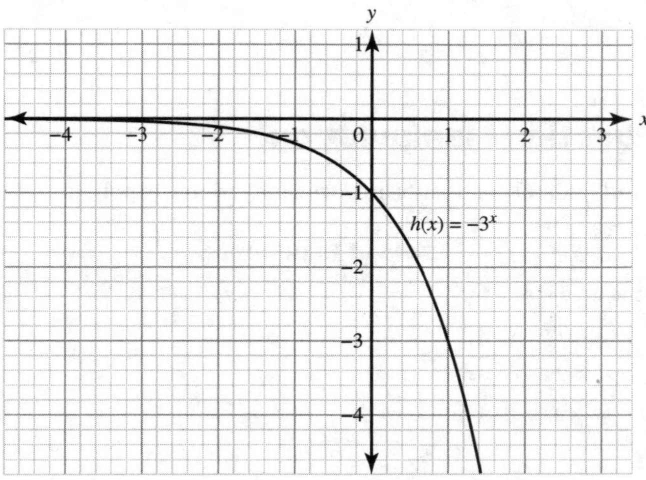

Figure 5.3

1. $\lim_{x \to -\infty} f(x)$
2. $\lim_{x \to \infty} f(x)$
3. $\lim_{x \to -\infty} g(x)$
4. $\lim_{x \to \infty} g(x)$
5. $\lim_{x \to -\infty} h(x)$
6. $\lim_{x \to \infty} h(x)$

Solution

Since all these limits have x approaching infinity, to evaluate them, consider the end behavior of each function in either the negative or positive direction.

1. $\lim_{x \to -\infty} f(x) = 0$
2. $\lim_{x \to \infty} f(x) = \infty$
3. $\lim_{x \to -\infty} g(x) = \infty$
4. $\lim_{x \to \infty} g(x) = 0$
5. $\lim_{x \to -\infty} h(x) = 0$
6. $\lim_{x \to \infty} h(x) = -\infty$

End Behavior of Exponential Functions

For an exponential function in general form, as the input values increase or decrease without bound, the output values will increase or decrease without bound or will get arbitrarily close to 0.

This means that the end behavior of an exponential function can be one of three limits:

$$\lim_{x\to\pm\infty} a\cdot b^x = \infty$$

$$\lim_{x\to\pm\infty} a\cdot b^x = -\infty$$

$$\lim_{x\to\pm\infty} a\cdot b^x = 0$$

5.2 Exponential Function Manipulation

The familiar laws of integer and rational exponents hold true for variable exponents as well.

Laws of Exponents

- Multiplication Law of Exponents: $a^x \cdot a^y = a^{x+y}$
- Quotient Law of Exponents: $\frac{a^x}{a^y} = a^{x-y}$
- Power of a Power Law of Exponents: $(a^x)^y = a^{xy}$
- Negative Exponent Law: $a^{-x} = \frac{1}{a^x}$ and $\frac{1}{a^{-x}} = a^x$
- Rational Exponent Law: $a^{\frac{x}{y}} = \sqrt[y]{a^x} = \left(\sqrt[y]{a}\right)^x$
- Zero Exponent Law: $a^0 = 1, a \neq 0$

The laws of exponents lead to very interesting graphical transformations in exponential functions.

Example

Graph $f(x) = 2^x$ and $g(x) = 2^{x+3}$ on the same set of axes. State the transformation of the graph from $f(x)$ to $g(x)$.

Solution

The graphs of the two functions are shown in Figure 5.4 on the same set of axes.

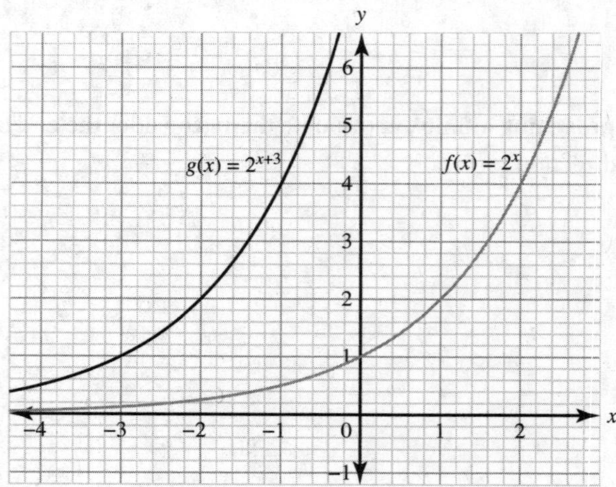

Figure 5.4

Since $g(x) = 2^{x+3} = f(x + 3)$, the $g(x)$ graph is the $f(x)$ graph shifted 3 units left.

This can also be looked at in a different way using laws of exponents. Since $g(x) = 2^{x+3} = 2^x \cdot 2^3 = 2^3 \cdot 2^x = 8 \cdot 2^x$, this can also be considered a vertical stretch of the graph of $f(x)$ by a factor of 8. Figure 5.5 is the graph of $f(x) = 2^x$ and $h(x) = 8 \cdot 2^x$.

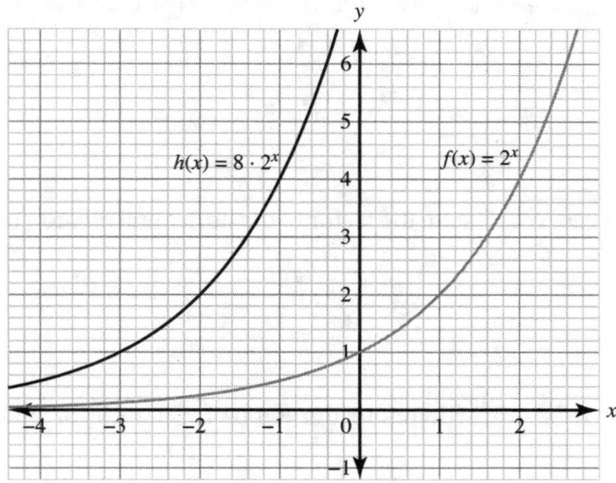

Figure 5.5

The graphs of $g(x)$ and $h(x)$ are identical.

Product Property and Vertical Dilation

The product property for exponents states that $b^m b^n = b^{(m+n)}$. Graphically, this property implies that every horizontal translation of an exponential function, $f(x) = b^{(x+k)}$, is equivalent to a vertical dilation, $f(x) = b^{(x+k)} = b^x b^k = ab^x$, where $a = b^k$.

Example

Given the exponential function $g(x) = 5^{x-1}$:

(a) State its parent function, $f(x)$.
(b) State the transformation that exists between the given function, $g(x)$, and its parent function, $f(x)$.
(c) Write a unique function, $h(x)$, that is equivalent to the function $g(x)$.
(d) State the transformation that exists between $h(x)$ and its parent function, $f(x)$.

Solution

(a) The parent function is $f(x) = 5^x$.
(b) The graph of $g(x)$ is a horizontal shift 1 unit to the right of the graph of $f(x)$.
(c) Since $g(x) = 5^{x-1} = 5^x \cdot 5^{-1} = 5^{-1} \cdot 5^x = \frac{1}{5} \cdot 5^x$, $h(x) = \frac{1}{5} \cdot 5^x$.
(d) The graph of $h(x)$ is a vertical shrink of the graph of $f(x)$ by a factor of $\frac{1}{5}$.

The power of a power law of exponents also leads to an interesting graphical transformation.

Power Property and Change of Base

The power property for exponents states that $(b^m)^n = b^{(mn)}$. Graphically, this property implies that every horizontal dilation of an exponential function, $f(x) = b^{(cx)}$, is equivalent to a change of the base of an exponential function, $f(x) = (b^c)^x$, where b^c is a constant and $c \neq 0$.

Example

Given the exponential function $g(x) = 81^x$, state the factor of the horizontal dilation of $g(x)$ if its parent function is $f(x) = 3^x$.

Solution

Since $81 = 3^4$, then $g(x) = 81^x = (3^4)^x = 3^{4x}$. Therefore, the graph of $g(x)$ is a horizontal shrink by a factor of $\frac{1}{4}$ of the graph of $f(x)$.

Example

Show that the graph of $g(x) = \frac{1}{4^x}$ is a reflection over the y-axis of the graph of $f(x) = 4^x$.

Solution

Using the negative exponent law, $g(x) = \frac{1}{4^x} = 4^{-x} = f(-x)$. $f(-x)$ reflects the graph of the original function, $f(x)$, over the y-axis.

Example

Rewrite the function $h(x) = \sqrt[5]{\frac{1}{3^x}}$ in the form $f(x) = ab^x$.

Solution

Using the exponent laws, $h(x) = \sqrt[5]{\frac{1}{3^x}} = \sqrt[5]{3^{-x}} = (3^{-x})^{\frac{1}{5}} = 3^{\frac{-x}{5}} = 1 \cdot 3^{\frac{-x}{5}}$.

5.3 Exponential Function Context and Data Modeling

Previous chapters discussed how to recognize if tables of data represented linear or quadratic functions. Exponential functions model growth patterns where successive output values over equal-length input value intervals are proportional. When the input values are whole numbers, exponential functions model situations of repeated multiplication of a constant to an initial value.

Example

Determine whether the functions in the tables below are linear, exponential, or neither. For those that are linear, find a linear function that models the data. For those that are exponential, find an exponential function that models the data.

1.

Table 5.1

x	$f(x)$
−1	5
0	2
1	−1
2	−4
3	−7

2.

Table 5.2

x	$g(x)$
−1	2
0	4
1	7
2	11
3	15

3.

Table 5.3

x	$h(x)$
−1	32
0	16
1	8
2	4
3	2

Solution

For each table, the input values are of equal length, so calculate the average rate of change. If the average rate of change is constant, the function is linear. Also compute the ratio of consecutive outputs. If the ratio is constant, the function is exponential.

1. For Table 5.1, the average rate of change for every 1 unit increase in x is -3. Since this is a constant rate of change, the function is linear. To write the equation of the linear function, $y = mx + b$, find the slope, m, and the y-intercept, b. The average rate of change is the slope, so $m = -3$. The y-intercept, b, is the value of the function at $x = 0$, so $b = 2$. The linear function that models this data is $f(x) = -3x + 2$.

2. For Table 5.2, the average rate of change from -1 to 0 is 2, and the average rate of change from 0 to 1 is 3. Because the average rate of change is not constant, the function is not linear. The ratio of consecutive outputs from -1 to 0 is 2, and the ratio of consecutive outputs from 0 to 1 is $\frac{7}{4}$. Because the ratio of consecutive outputs is not constant, the function is not an exponential function.
3. For Table 5.3, the average rate of change from -1 to 0 is -16, and the average rate of change from 0 to 1 is -8. Because the average rate of change is not constant, the function is not linear. The ratio of consecutive outputs from -1 to 0 is $\frac{1}{2}$, and the ratio of consecutive outputs from 0 to 1 is $\frac{1}{2}$. The ratio of all consecutive outputs for every 1-unit increase is $\frac{1}{2}$. Since the ratio of consecutive outputs is constant, the function is an exponential function with a growth factor of $b = \frac{1}{2}$. The initial value of the exponential function, a, can be found when $x = 0$, so $a = 16$. Therefore, the exponential function that models the data is $h(x) = a \cdot b^x = 16 \cdot \left(\frac{1}{2}\right)^x$.

An exponential function model can be constructed from an appropriate ratio and initial value as shown in the previous example. An exponential function model can also be constructed from two input-output pairs. The initial value and the base can be found by solving a system of equations resulting from two input-output pairs.

Example

Determine an equation for the exponential function whose graph is shown in Figure 5.6.

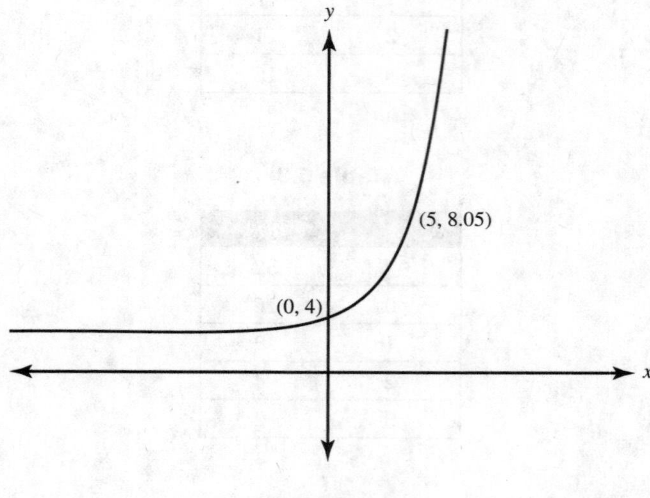

Figure 5.6

Solution

Since it was given that this models an exponential function, use the two sets of ordered pairs of the function to create a system of equations to find the a- and b-values in the general exponential equation form, $f(x) = a \cdot b^x$. Substituting (0, 4) yields the equation $4 = a \cdot b^0$. Since $b \neq 0$, then $b^0 = 1$ and $a = 4$.

Substitute (5, 8.05) and $a = 4$ into the equation $f(x) = a \cdot b^x$:

$$8.05 = 4 \cdot b^5$$

Solve for b:

$$b^5 = \frac{8.05}{4}$$

$$b = \sqrt[5]{\frac{8.05}{4}} \approx 1.15$$

An equation to model the exponential function is $f(x) = 4 \cdot (1.15)^x$.

Modeling Growth and Decay

The growth patterns of many types of biological and physical quantities closely approximate exponential curves whose equations have the general form:

$$f(x) = a \cdot b^x$$

where a represents the initial amount, b represents the constant growth factor, and x represents the time variable.

When a quantity grows linearly over time, it increases by the same amount over equal intervals of time. When a quantity grows exponentially, it increases by a fixed percent of its previous value over equal intervals of time.

If a quantity is increasing at a constant annual rate of r of its previous value, after t years, an initial amount, a, of that quantity has grown to amount $f(t)$.

Example

Suppose that $450 is deposited in a bank account that earns 8.2% interest compounded annually. State the balance in the account after each of the following time periods.

(a) 1 year
(b) 2 years
(c) t years

Solution

(a) After 1 year, the balance in the account is

$$450 + (450)(0.082) = 450[1 + 0.082] = 450(1.082)$$

(b) After 2 years, the balance in the account is

$$450(1.082) + (450(1.082)) \cdot (0.082) = 450(1.082)[1 + 0.082] = 450(1.082)(1.082) = 450(1.082)^2$$

(c) Based on (a) and (b), a pattern is emerging that leads to a general formula for t years:

$$f(t) = 450(1.082)^t$$

The function in part (c), $f(t) = 450(1.082)^t$, is called an exponential growth model. The initial amount $a = f(0) = 450$, and the annual growth factor is $1 + r = 1 + 0.082 = 1.082$. If a quantity is decreasing at a constant annual rate of r% of its previous value, then $r < 0$, and the exponential function that describes the process is referred to as an exponential decay model.

Example

Determine the exponential function that has an initial value of 15 and increases at a rate of 12% each year.

Solution

The given information provides the initial value, $a = 15$, and the rate of increase, $r = 12\% = 0.12$. Therefore, the exponential function is $f(t) = 15(1.12)^t$, where t represents time in years.

Example

Suppose a culture of 100 bacteria in a Petri dish doubles every hour. Write a model that fits this bacterial growth, where $P(t)$ represents the total bacteria after t hours. Use the model to predict when the number of bacteria will be 5,000.

Solution

To recognize the type of model that is being described with the given information, create a table of values until a pattern is recognized. This has been done in Table 5.4.

Table 5.4

Total bacteria after 1 hour	$100 \cdot 2 = 200$
Total bacteria after 2 hours	$200 \cdot 2 = 400$ or $100 \cdot 2 \cdot 2 = 400$
Total bacteria after 3 hours	$400 \cdot 2 = 800$ or $100 \cdot 2 \cdot 2 \cdot 2 = 800$

An equation that models this behavior is $P(t) = 100 \cdot 2^t$.

To predict when the number of bacteria will be 5,000, use technology to graph the exponential model and the line $y = 5{,}000$ and find their point of intersection. This is shown in Figure 5.7.

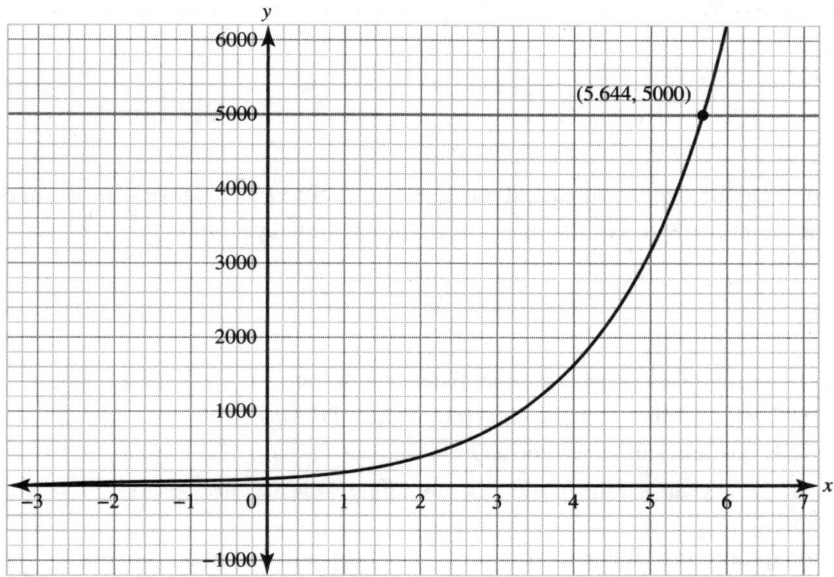

Figure 5.7

The point of intersection is (5.644, 5000). The number of bacteria will be 5,000 after 5.644 hours.

Example

Suppose the half-life of a certain radioactive substance is 20 days and there are 5 grams present initially. Write a function that models this situation, and use the model to predict the time when there will be 1 gram of the substance remaining.

Solution

If t is the time in days, the number of half-lives is $\frac{t}{20}$. The function $f(t) = 5\left(\frac{1}{2}\right)^{t/20}$ models the mass in grams of the radioactive substance at time t.

To find when there will be 1 gram of radioactive substance left, use technology to graph the exponential model and the line $y = 1$ and find their point of intersection. This is shown in Figure 5.8.

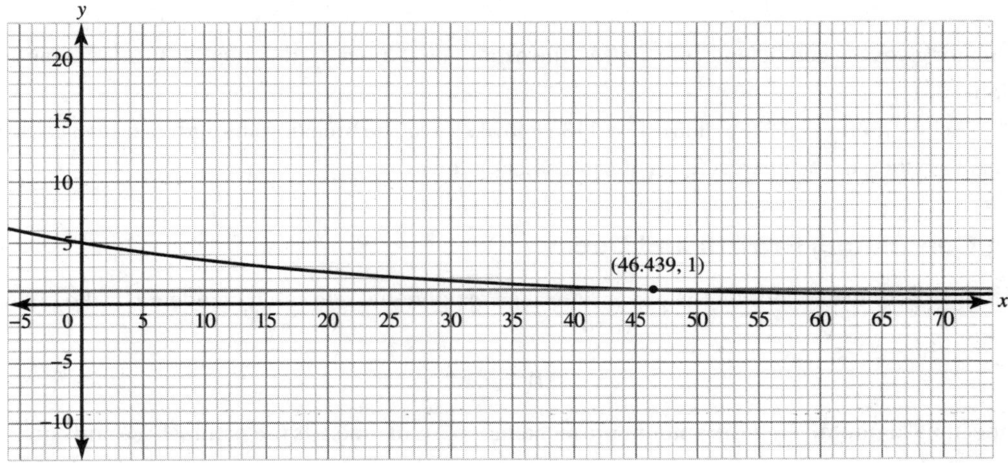

Figure 5.8

The point of intersection is (46.439, 1). There will be 1 gram of the radioactive substance left after approximately 46.439 days.

Example

Christopher buys a new car for $32,750. The car depreciates by about 11.75% per year. Write an equation to model this situation. Use the equation to determine the value of the car after 4 years.

Solution

The given situation can be modeled by an exponential decay function, $f(t) = a \cdot (1 + r)^t$, where $r = -11.75\% = -0.1175$, $a = 32{,}750$, $t =$ years, and $f(t) =$ value of the car.

An equation to model the data can be written as $f(t) = 32{,}750 \cdot (1 - 0.1175)^t$.

The value of the car after 4 years can be found by evaluating $f(4) = 32{,}750\,(1 - 0.1175)^4 = \$19{,}864.16$.

Like with linear and quadratic functions, exponential function models can be constructed for a data set with technology using exponential regressions.

Example

The breaking strength in tons, y, of a steel cable with diameter x, in inches, is given in Table 5.5.

Table 5.5

x (inches)	0.5	0.75	1	1.25	1.5
y (tons)	9.85	14.8	31.3	48.2	61.4

Using technology, fit an exponential regression equation to the data, expressing the coefficients to the nearest thousandth. Using this regression model, estimate, to the nearest tenth of a ton, the breaking strength of a steel cable with a diameter of 1.75 inches.

Solution

Enter the data into L1 and L2 by pressing STAT EDIT. After the data have been entered, press STAT, CALC, and 0:ExpReg.

The calculator gives the coefficients for a and b. The exponential function that models the data is approximated by $y = 3.848(6.933)^x$.

If $x = 1.75$ inches, then $y = 3.848(6.933)^{1.75} \approx 113.9848583$ tons. The breaking strength of a steel cable with a diameter of 1.75 inches, to the nearest tenth of a ton, is 114.0 tons.

Modeling Continuous Growth and Decay

Quantities that grow or decay continuously over time at an exponential rate can be modeled by the following function:

$$f(t) = a \cdot e^{kt}$$

where a is the initial amount, $f(t)$ is the amount present after t units of time, e is the natural base and equals approximately 2.718, and k is some constant specific to the particular growth or decay process that the function is modeling.

Example

If $12,000 is invested for 7 years and interest is compounded continuously at an annual rate of 5%, what is the balance at the end of 7 years?

Solution

Since the interest is compounded continuously, the function $f(t) = a \cdot e^{kt}$ can be used to model the data. It is also given that the initial amount, a, is 12,000, the annual rate, $r = k$, is 5% = 0.05, and the time, t, is 7. Substitute the values into the equation:

$$f(7) = (12{,}000) \cdot e^{(0.05)(7)} \approx \$17{,}028.81$$

5.4 Competing Function Model Validation

Depending on the patterns of a set of data of an event, it is possible to construct linear, quadratic, and exponential models that can help make predictions of the event. Two variables in a data set that demonstrate a slightly changing rate of change can be modeled by linear, exponential, and quadratic function models.

Models can be compared based on contextual clues, by considering scatterplots of data, or by examining the relationships with consecutive terms or rates of change to determine which model is most appropriate.

A model can be deemed appropriate by finding its coefficient of determination, which is notated as r^2. The r^2-value is a measure of the goodness of fit of a model. In regression, the r^2 coefficient of determination is a statistical measure of how well the regression predictions approximate the real data points. An r^2 of 1 indicates that the regression predictions perfectly fit the data. If using a TI calculator to find this value, press 2ND 0, and scroll down to DiagnosticOn. Press ENTER twice so that the word Done appears on the screen. Once this is complete, anytime a regression equation is found, the r^2-value will appear at the bottom of the regression model.

Example

Three different tables of data are given below. For each, determine whether a linear, quadratic, or exponential model is most appropriate for the data. Justify your answer. Using technology, write a regression equation that is a best fit model for each set of data.

1. Table 5.6 shows fuel efficiencies of a vehicle at different speeds.

Table 5.6

Miles per hour, x	20	24	30	36	40	45	50	56	60	70
Miles per gallon, y	14.5	17.5	21.2	23.7	25.2	25.8	25.8	25.1	24	19.5

2. Table 5.7 shows the humerus lengths (in centimeters) and heights (in centimeters) of several females.

Table 5.7

Humerus length (cm), x	22	25	26	27	28	30	32	33
Height (cm), y	130	142	141	145	152	154	159	166

3. Table 5.8 shows the U.S. population in millions between the years 1900–2011.

Table 5.8

Year, x	1900	1910	1920	1930	1940	1950	1960	1970	1980	1990	2000	2011
Population (in millions), y	76.2	92.2	106	123.2	132.2	151.3	179.3	203.3	226.5	248.7	281.4	311.6

✓ Solution

Since all the tables have x-values that are not equally spaced, the differences between the outputs cannot be analyzed. Instead, use technology to create a scatterplot of each set of data to find a function that models the data.

1. Figure 5.9 is the scatterplot for the data found in Table 5.6.

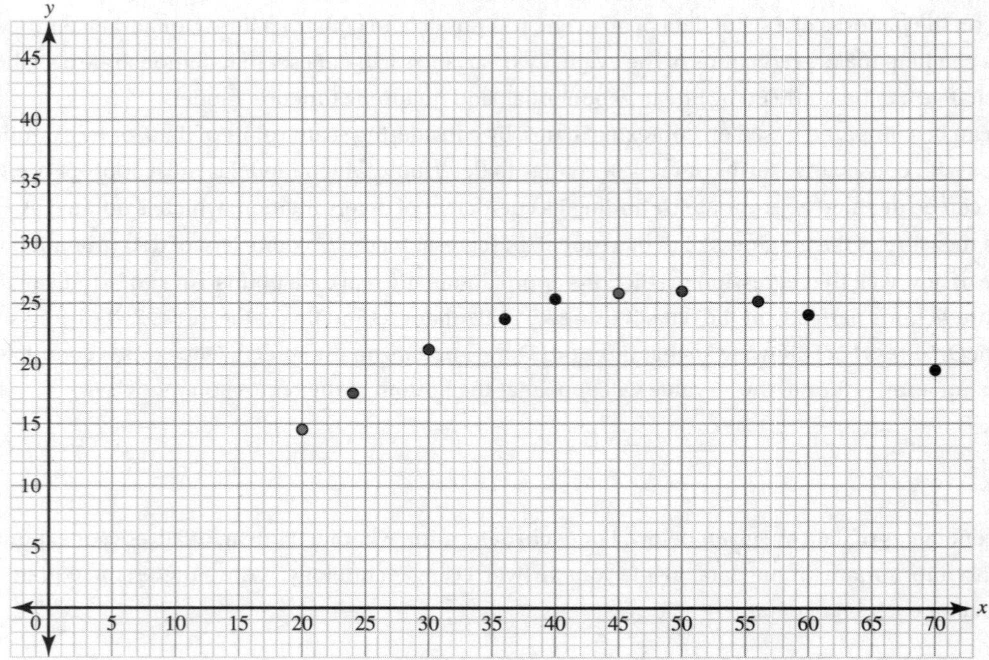

Figure 5.9

Based on the scatterplot, the data can best be modeled by a quadratic function.

Use a graphing calculator to find the best-fit equation by calculating the quadratic regression. The quadratic regression equation is $f(x) = -0.014x^2 + 1.366x - 7.144$ with $r^2 = 0.999$. Since the r^2-value is very close to 1, this indicates an appropriate model for the data set.

2. Figure 5.10 is the scatterplot for the data found in Table 5.7.

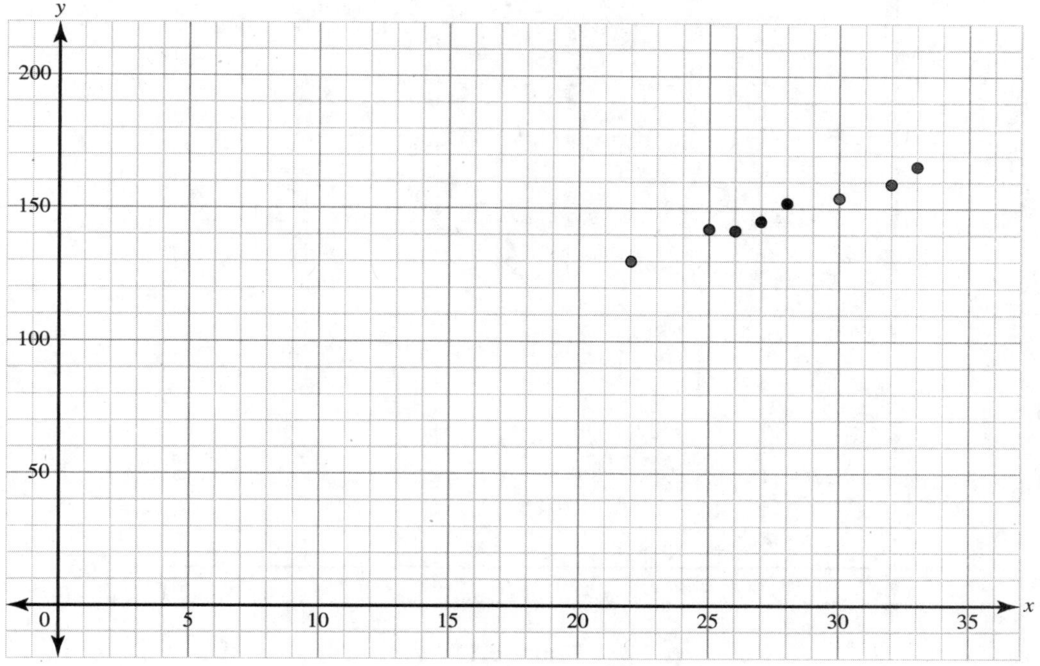

Figure 5.10

Based on the scatterplot, the data can best be modeled by a linear function.

Use a graphing calculator to find the best-fit equation by calculating the linear regression. The linear regression equation is $f(x) = 3.0527x + 63.531$ with $r^2 = 0.9847$. Since the r^2-value is very close to 1, this indicates an appropriate model for the data set.

3. When working with years, often it is easier to scale the years to have a better-fitting scatterplot. For the data in Table 5.8, let x represent the number of years after 1900. So in Figure 5.11, 0 represents 1900, 10 represents 1910, and so on.

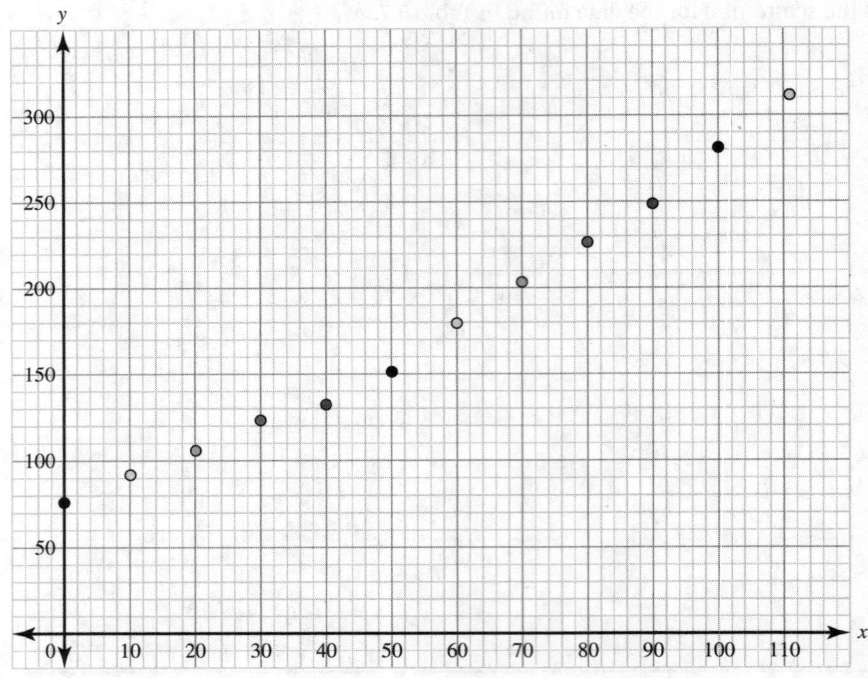

Figure 5.11

Based on the scatterplot, the data can best be modeled by an exponential function.

Use a graphing calculator to find the best-fit equation by calculating the exponential regression. Enter the data in the calculator using the scaled x-values. The exponential regression equation is $f(x) = 81.27455 \cdot 1.0126^x$ with $r^2 = 0.9946$. Since the r^2-value is very close to 1, this indicates an appropriate model for the data set.

Example

Table 5.9 shows the U.S. population in millions between the years 1900–2000. Using technology, find an exponential regression model that represents the values shown in the table. Use that regression equation to predict the U.S. population for the year 2011. Compare the result with the actual 2011 population that was 311.6 million.

Table 5.9

Year, x	1900	1910	1920	1930	1940	1950	1960	1970	1980	1990	2000
Population (in millions), y	76.2	92.2	106	123.2	132.2	151.3	179.3	203.3	226.5	248.7	281.4

Solution

Enter the data into a calculator. The exponential regression equation is $f(x) = 80.55136 \cdot 1.01289^x$ with $r^2 = 0.995$, which indicates that this is a close fit.

To find the predicted population in the year 2011, evaluate

$$f(111) = 80.55136 \cdot 1.01289^{(111)} \approx 333.8$$

Compare this predicted value to the actual value in Table 5.8; the predicted value is an overestimation of the population by about 22 million persons. The difference between the predicted and the actual values is known as the error in the model.

This error can also be shown visually by graphing the regression equation with the scatterplot as shown in Figure 5.12.

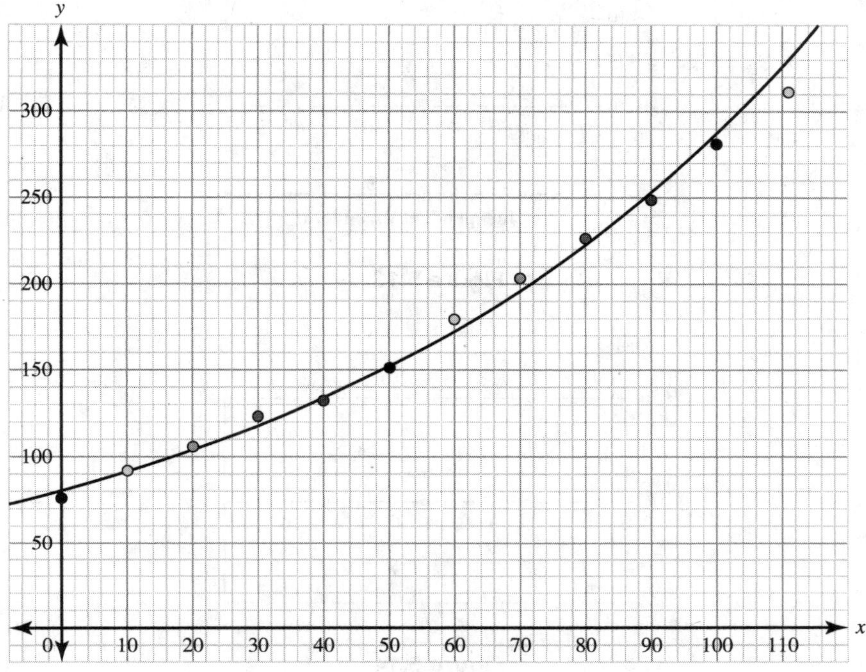

Figure 5.12

The graph of the regression equation is well above the data point (111, 311.6). The distance between the regression equation and the other data points is not as great as the distance between the equation and the last data point, showing that the greatest error occurs with the 2011 estimation.

Residuals

The difference between an observed value of data and the predicted value from a regression equation is called a residual. Each data point has one residual.

A residual plot is a graph that shows the residuals on the vertical axis and the independent variable on the horizontal axis. If the points in a residual plot are randomly dispersed around the horizontal axis, a linear regression model is appropriate for the data; otherwise, a nonlinear model is more appropriate.

Calculating the Residual

Residual = Observed value − Predicted value

Example

Below are four residual plots. Using each graph, comment on the most appropriate model for each set of data.

1.

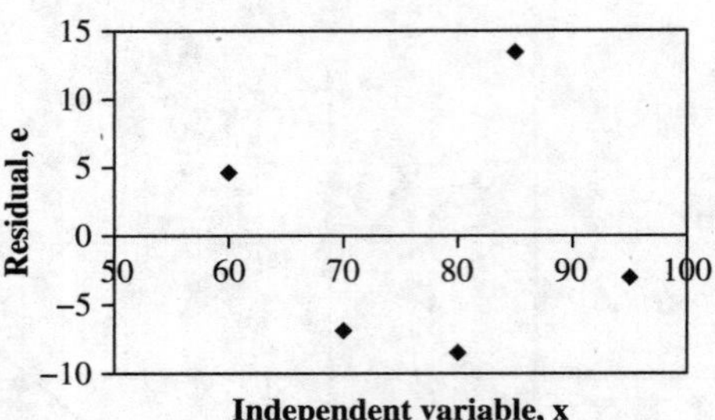

Figure 5.13

2.

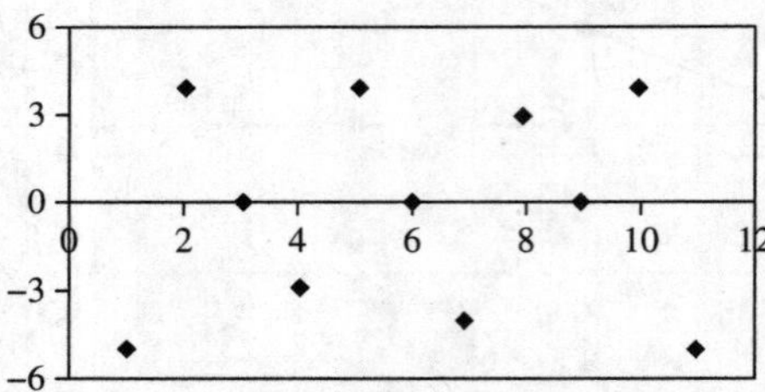

Figure 5.14

3.

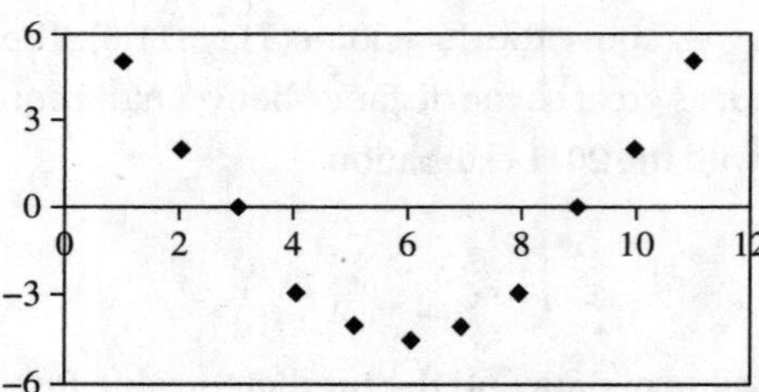

Figure 5.15

4.

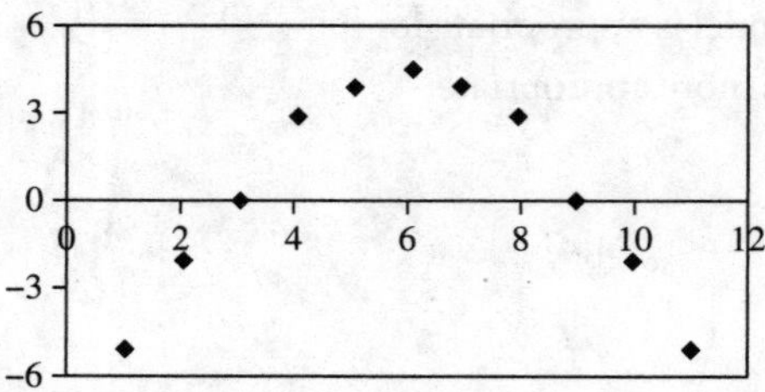

Figure 5.16

Solution

1. The plot in Figure 5.13 shows a random pattern. This random pattern indicates that a linear model provides an appropriate fit for the data.
2. The plot in Figure 5.14 shows a random pattern. This random pattern indicates that a linear model provides an appropriate fit for the data.
3. The plot in Figure 5.15 is U-shaped and not random. This nonrandom pattern indicates that a nonlinear model provides an appropriate fit for the data.
4. The plot in Figure 5.16 is U-shaped and not random. This nonrandom pattern indicates that a nonlinear model provides an appropriate fit for the data.

Example

Table 5.10 gives the years since 1890 and the population in millions of the people in California.

Table 5.10

Years since 1890, x	0	20	40	60	80	100
Population in millions, y	1.21	2.38	5.68	10.59	19.97	39.76

(a) Create a scatterplot using this data.
(b) Find the equation for the curve of best fit for the data.
(c) Create a residual plot, and comment on the appropriateness of your model from part (b).

Solution

(a) Figure 5.17 is a scatterplot using this data.

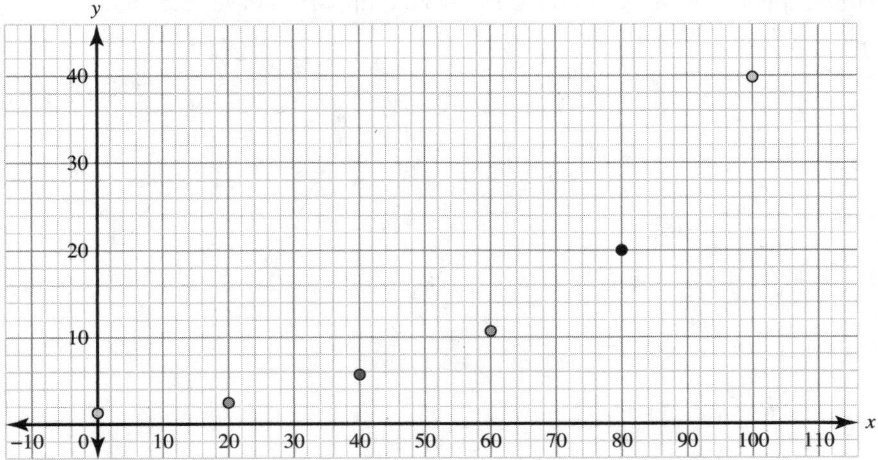

Figure 5.17

(b) Based on the scatterplot, it looks like an exponential regression equation would best fit the data. Using technology, the exponential regression equation is $y = 1.251579 \cdot (1.0355688)^x$ with $r^2 = 0.997558$.
(c) First create a table of values to help graph the residual plot, as shown in Table 5.11.

Table 5.11

x	y	Predicted y-value from Regression Equation	Residual = Observed value − Predicted value
0	1.21	1.2516	−0.0416
20	2.38	2.5179	−0.1379
40	5.68	5.0654	0.6146
60	10.59	10.191	0.399
80	19.97	20.501	−0.531
100	39.76	41.243	−1.483

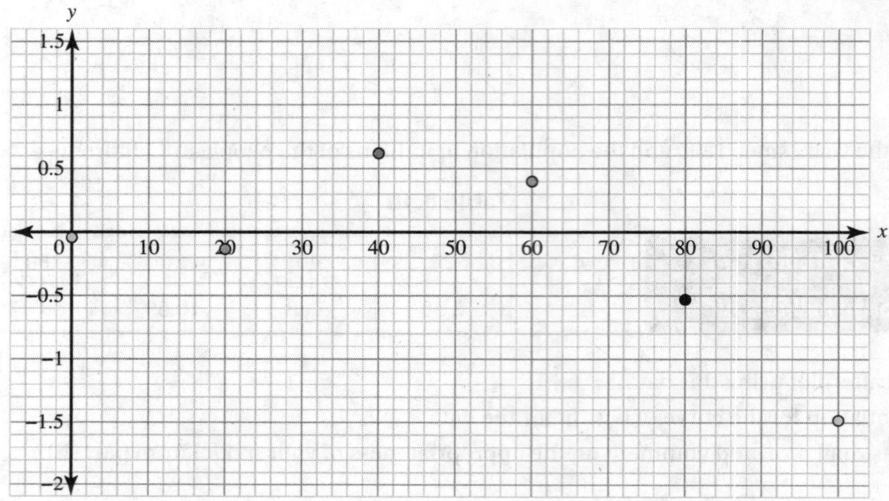

Figure 5.18

Overall, the residual plot shown in Figure 5.18 looks U-shaped. This nonrandom pattern indicates that a nonlinear model provides a more appropriate fit for the data. This would support using an exponential regression equation to model the data.

Multiple-Choice Questions

1. A scientist studying bacteria growth notices interesting behavior with the replication of cells over the course of 24 hours. She records her description of the behavior in hopes of writing a function that can be used to make predictions for future bacteria growth. She observed that for the first 8 hours, the number of bacteria, y, increases at a constant rate of 10 cells per minute. After the initial 8 hours, the number of bacteria varies inversely with time, x, with a proportional constant of 3 for the next 10 hours. Between hour 18 through hour 20, the bacteria decrease at a constant rate of 5 cells per minute. In the last 4 hours, the bacteria double in growth every minute. Which of the following models best fits the scientist's observations?

(A) $y = \begin{cases} 10x,\ 0 \le x \le 8 \\ \frac{3}{x},\ 8 < x \le 18 \\ -5x,\ 18 < x \le 20 \\ 2^x,\ 20 < x \le 24 \end{cases}$

(B) $y = \begin{cases} 10x,\ 0 \le x \le 8 \\ \frac{3}{x},\ 8 < x \le 18 \\ 5x,\ 18 < x \le 20 \\ 2x,\ 20 < x \le 24 \end{cases}$

(C) $y = \begin{cases} 10x,\ 0 \le x \le 8 \\ \frac{3}{x},\ 8 < x \le 18 \\ 5x,\ 18 < x \le 20 \\ 2^x,\ 20 < x \le 24 \end{cases}$

(D) $y = \begin{cases} 10x,\ 0 \le x \le 8 \\ \frac{3}{x},\ 8 < x \le 18 \\ -5x,\ 18 < x \le 20 \\ 2x,\ 20 < x \le 24 \end{cases}$

2. A radioactive substance has an initial mass of 250 grams, and its mass halves every 5 years. Which equation shows the number of grams, $M(t)$, remaining after t years?

(A) $M(t) = 250(5)^{t/4}$

(B) $M(t) = 250\left(\frac{1}{2}\right)^{t/5}$

(C) $M(t) = 250(5)^{-2t}$

(D) $M(t) = 250\left(\frac{1}{2}\right)^{5t}$

3. A culture of 3,000 bacteria triples every 40 minutes. If $P(t)$ represents the size of the bacteria population after t minutes, which equation can be used to model the exponential growth of these bacteria?

(A) $P(t) = 3{,}000(40)^{3t}$

(B) $P(t) = 3{,}000(3)^{t/40}$

(C) $P(t) = 3{,}000(3)^{40t}$

(D) $P(t) = 3{,}000(40)^{t/3}$

4. When exposed to sunlight, the number of bacteria in a culture decreases exponentially at the rate of 10% per hour. What is the best approximation for the number of hours required for the initial number of bacteria to decrease by 50%?

(A) 6.6

(B) 6.0

(C) 5.4

(D) 4.6

5. Which of the following functions given below are equivalent?

I. $f(x) = (2)^{3-x}$

II. $g(x) = 8 \cdot 2^{-x}$

III. $h(x) = \frac{8}{2^x}$

(A) I and II only

(B) I and III only

(C) II and III only

(D) I, II, and III

6. Which of the following functions given below are equivalent?

I. $f(x) = (9)^{x/2}$

II. $g(x) = 3^x$

III. $h(x) = 4.5^x$

(A) I and II only

(B) I and III only

(C) II and III only

(D) I, II, and III

7. Michael buys a new computer at full price. He learns that each year the value of the computer depreciates at a rate of 7%. Which of the following residual plots would best fit the model of depreciation?

(A)

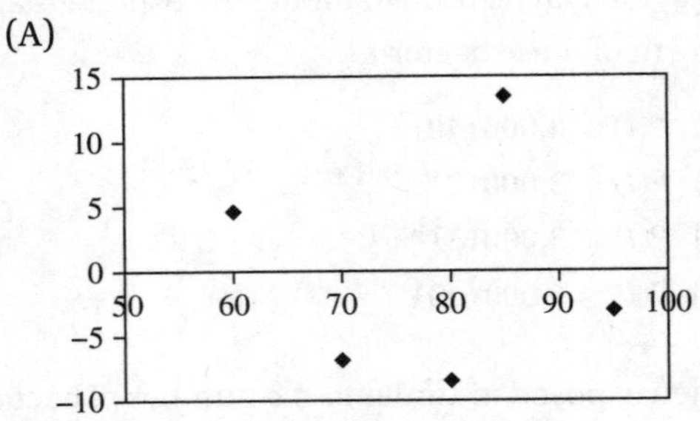

(B)

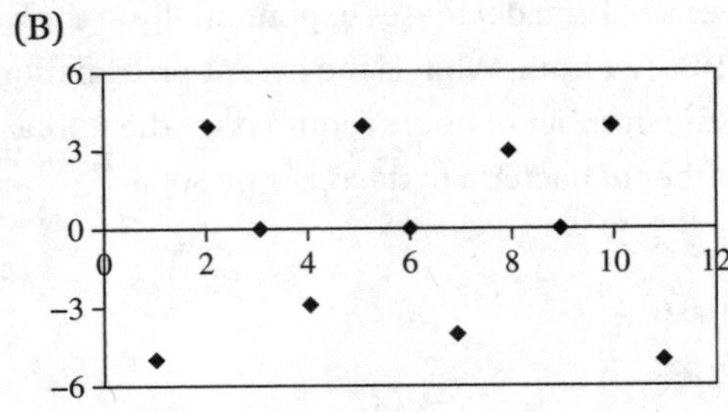

(C)

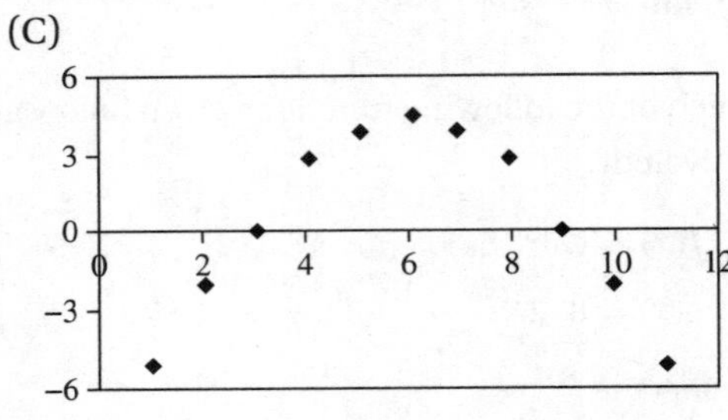

(D)

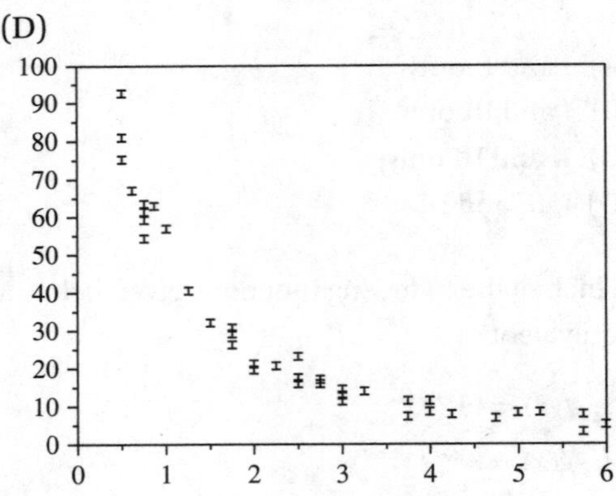

8. Which of the following tables of data best fits an exponential regression model?

(A)

x	−2	−1	0	1
y	.08	.4	2	10

(B)

x	−2	−1	0	1	2
y	2	4	4.8	5.5	6

(C)

x	−3	−2	−1	0	1
y	−7	−5	−3	−1	1

(D)

x	−2	−1	0	1	2
y	8	2	0	2	8

9. If $a > 0$, $b > 1$, and $c < 0$, evaluate $\lim_{x\to\infty} a \cdot b^{cx}$.

(A) $-\infty$
(B) 0
(C) 1
(D) ∞

10. Given $f(t) = -5 \cdot e^{2t}$, find $\lim_{t\to\infty} f(t)$.

(A) $-\infty$
(B) 0
(C) 1
(D) ∞

Answer Explanations

1. **(A)** Look at the different time intervals. In the first 8 hours, the bacteria grow at a constant rate of change, which implies that it is a linear model with a slope of 10. The first equation should be $y = 10x$. In the second interval, between hours 8 and 18, the growth is described as an inverse variation, which takes on the model $y = \frac{k}{x}$. Since the constant of proportionality is given as 3, the equation is $y = \frac{3}{x}$. Between hours 18 and 20, the bacteria decrease at a constant rate of change, implying that a linear model with a negative slope best fits; $y = -5x$. Finally in the last 4 hours, the cells are doubling, which means they are growing in such a way that it would be 2 then 4 then 8 then 16. This pattern follows the equation $y = 2^x$.

2. **(B)** If t is the time in years, the number of half-lives is $\frac{t}{5}$. The function $M(t) = 250\left(\frac{1}{2}\right)^{t/5}$ models the mass in grams of the radioactive substance at time t.

3. **(B)** If t is the time in minutes, the number of times the bacteria triples is $\frac{t}{40}$. The function $P(t) = 3{,}000\,(3)^{t/40}$ models the size of the bacteria population after t minutes.

4. **(A)** Since the model is exponentially decreasing, an equation to fit this data is $f(t) = a(1 - r)^t$. Let $r = 10\% = 0.10$, $a = 1$, and $f(t) = \frac{1}{2}$:

$$\frac{1}{2} = 1\,(1 - 0.10)^t$$

Using technology, graph both $y = \frac{1}{2}$ and $y = 1\,(1 - 0.10)^t$ on the same set of axes and then find their point of intersection.

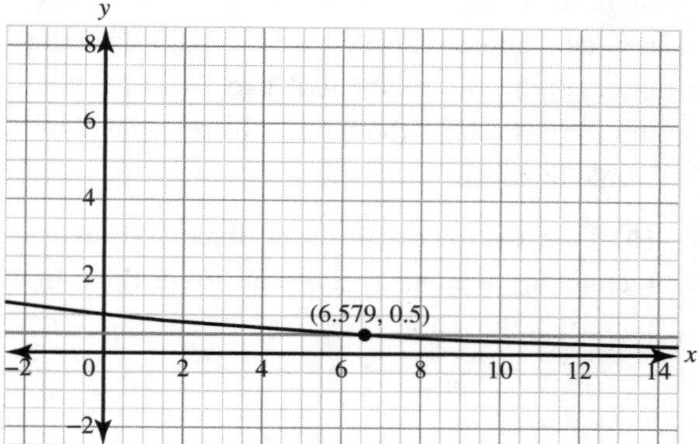

The point of intersection is (6.579, 0.5). The time the bacteria will be half of its initial value is approximately 6.579 hours.

5. **(D)** Beginning with the first equation, $f(x) = (2)^{3-x}$ and using the laws of exponents, $(2)^{3-x} = 2^3 \cdot 2^{-x} = 8 \cdot 2^{-x} = g(x)$. So I and II are equivalent. Using the law of negative exponents, $8 \cdot 2^{-x} = \frac{8}{2^x} = h(x)$. So I, II, and III are all equivalent.

6. **(A)** Since $3 = 9^{1/2}$, then $3^x = (9^{1/2})^x = 9^{x/2}$. So I and II are equivalent. There is no connection between base numbers 4.5 and 3 and 9, so III is not equivalent to the other two.

7. **(C)** The description of the situation fits an exponential model. Choices (A) and (B) show a random pattern. This random pattern indicates that a linear model provides an appropriate fit. The residual plot in Choice (C) is U-shaped and not random. This nonrandom pattern indicates that a nonlinear model provides a more appropriate fit for the data. The plot in Choice (D) is an exponential scatterplot, which is most likely what the graph of the data would look like. However, the question is asking about what the residual plot would look like.

8. **(A)** One way to choose a best-fit model is to use technology to create scatterplots to see the overall pattern of the data.

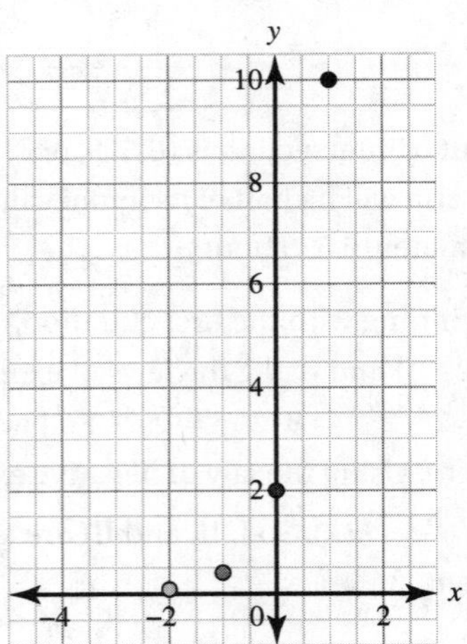

The scatterplot for Choice (A) follows an exponential model.

9. **(B)** To evaluate $\lim_{x \to \infty} a \cdot b^{cx}$, think of the graph of the function $y = a \cdot b^{cx}$. Since a and b are both positive and $b > 1$, and c is negative, it will follow the behavior of an exponential graph that is reflected over the y-axis, as shown below.

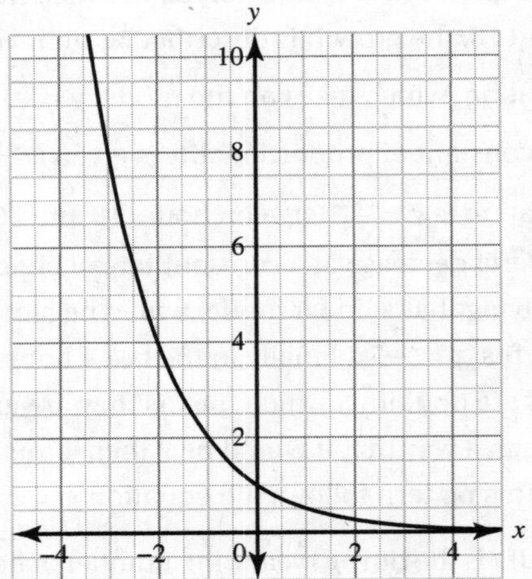

Therefore, $\lim_{x \to \infty} a \cdot b^{cx} = 0$.

10. **(A)** Graph the function $f(t) = -5 \cdot e^{2t}$.

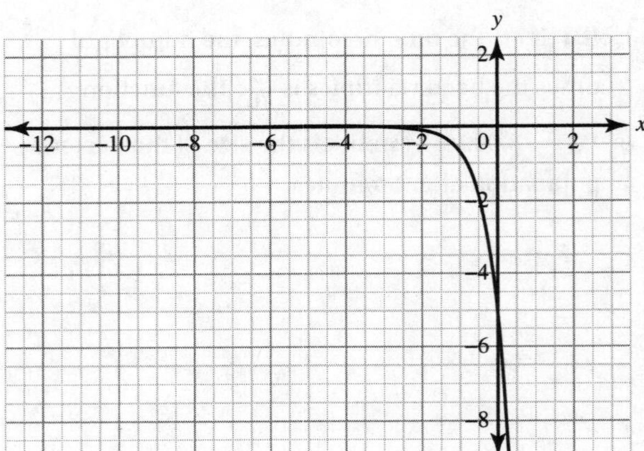

Graphically, $\lim_{t \to \infty} f(t)$ approaches $-\infty$.

6

Sequences

Learning Objectives

In this chapter, you will learn:

- ➔ The difference between arithmetic and geometric sequences
- ➔ How to derive formulas to find terms for arithmetic and geometric sequences
- ➔ How to compare and contrast sequences to linear and exponential functions

In math, a sequence is a list of numbers written in a specific order that often has some type of pattern or rule that connects one term to another.

The following sets of terms are examples of sequences:

1. $\{1, 3, 5, 7, 9, \ldots\}$
2. $\{2, 4, 8, 16, 32\}$
3. $\{0.5, -1, -2.5, -4, -5.5\}$
4. $\{1, 1, 2, 3, 5, 8, 13, \ldots\}$

The first and last examples show infinite sequences as indicated by..., which means that the sequence continues indefinitely following the same pattern.

The second and third examples are finite sequences because there is a clear first term and last term in each sequence.

Each example has a type of rule, or function, that connects each term. Writing out a sequence as a function makes it easier to figure out a specific term in the sequence without having to write out all the previous terms. Depending on the rule, these functions can often appear linear or exponential.

In Example 1, each term is found by adding 2 to the previous term. This implies that a possible rule or function is $y = x + 2$. When working with sequences, we typically let n represent the specific term and a_n represent the value of that term. So the first term is $a_1 = 1$; the second term is $a_2 = 3$; the third term is $a_3 = 5$; the fourth term is $a_4 = 7$; the fifth term is $a_5 = 9$; and the general term is $a_n = 1 + 2(n - 1)$.

In Example 2, each term is found by multiplying 2 by the previous term. The first term is $a_1 = 2$, and every other term after is multiplied by 2. Therefore, the general term for this sequence is $a_n = 2 \cdot 2^{n-1}$.

In Example 3, each term is found by subtracting 1.5 from the previous term. The first term is $a_1 = 0.5$. Therefore, the general term for this sequence is $a_n = 0.5 - 1.5(n - 1)$.

In Example 4, starting with $a_1 = 1$ and $a_2 = 1$, each term is found by adding together the previous two terms. This is more difficult to write as a rule since it is based upon other terms in the sequence and is known as a recursive sequence. This can be studied more in a calculus course.

All four examples have a common domain. The inputs, n, are all whole numbers. The outputs, a_n, are all real numbers. This is true of all sequences.

Sequence Domain and Range

A sequence is a function whose domain is the set of whole numbers that corresponds to a range that is the set of real numbers.

6.1 Arithmetic Sequences

Arithmetic sequences describe sequences where each pair of consecutive terms have a common difference, *d*, or a constant rate of change.

Examples 1 and 3 are both arithmetic sequences. In Example 1, the common difference is 2 since $a_2 - a_1 = a_3 - a_2 = a_4 - a_3 = a_5 - a_4 = 2$. In Example 3, the common difference is -1.5 since $a_2 - a_1 = a_3 - a_2 = a_4 - a_3 = a_5 - a_4 = -1.5$.

Consider a finite arithmetic sequence with *n* terms in which a_1 is the first term, a_n is the *n*th term, and *d* is the common difference. The second term of this sequence is $a_1 + 1d$, the third term is $a_1 + 2d$, the fourth term is $a_1 + 3d$, and so on. Based on this pattern, a generalized equation can be written to find the *n*th term.

In the sequence $\{1, 3, 5, 7, 9, \ldots\}$, the next term, which is the 6th term, is 11 since $9 + 2 = 11$. Using the general equation would produce the same results since it is known that the first term is 1 and the common difference is 2:

$$a_6 = 1 + 2(6 - 1) = 1 + (2)(5) = 1 + 10 = 11$$

Sometimes the first term of an arithmetic sequence is not known. If this is the case, there is another generalized equation that can be used to find the *n*th term of an arithmetic sequence.

General Equations for Arithmetic Sequences

The *n*th term of an arithmetic sequence can be found using the following equation:

$$a_n = a_1 + d(n - 1)$$

where a_n is the *n*th term, a_1 is the first term, and *d* is the common difference.

If the first term is correlated with $n = 0$, the *n*th term of an arithmetic sequence can be written as:

$$a_n = a_0 + d(n)$$

Another Equation for Arithmetic Sequences

If a_1 is not known, the *n*th term of an arithmetic sequence can be found using the following equation:

$$a_n = a_k + d(n - k)$$

where a_k is the *k*th term of the sequence.

Example

An arithmetic sequence with a common difference of -3.5 has an 11th term equal to 7. What is the value of the 50th term?

Solution

This question gives the arithmetic sequence where $d = -3.5$ and $a_{11} = 7$. Finding the 50th term means that $k = 50$. Substitute the given information into the *n*th term of an arithmetic sequence formula:

$$\begin{aligned} a_{11} &= a_{50} + (-3.5)(11 - 50) \\ 7 &= a_{50} + (-3.5)(-39) \\ 7 &= a_{50} + 136.5 \\ a_{50} &= -129.5 \end{aligned}$$

The 50th term of the arithmetic sequence is -129.5.

Example

An arithmetic sequence whose terms increase in value includes -2 and 13. If -2 and 13 are separated by 4 terms of this sequence, what are the values of those terms?

Solution

Since the placement of each of these terms is not given, the n-value is unknown. Think of this problem instead as finding 4 numbers that make the sequence of 6 numbers from -2 to 13 an arithmetic sequence. This can be shown visually.

$$-2, \_\_\_, \_\_\_, \_\_\_, \_\_\_, 13$$

Since it is given that this is an arithmetic sequence, to get from term to term, the common difference, d, needs to be added each time. It is important to find the value of d.

Let $a_1 = -2$, $a_6 = 13$, and $n = 6$. Use the nth term of an arithmetic sequence formula:

$$\begin{aligned} a_6 &= a_1 + d(6 - 1) \\ 13 &= -2 + d(6 - 1) \\ 13 &= -2 + 5d \\ 15 &= 5d \\ d &= 3 \end{aligned}$$

Using the common difference, d, the remaining terms can be found:

$$\begin{aligned} a_2 &= -2 + 3 = 1 \\ a_3 &= 1 + 3 = 4 \\ a_4 &= 4 + 3 = 7 \\ a_5 &= 7 + 3 = 10 \end{aligned}$$

Note that the nth term of an arithmetic sequence formula is a function that has a domain different from the domain of the actual sequence.

6.2 Geometric Sequences

Geometric sequences describe sequences where each term after the first is obtained by multiplying the preceding term by the same number.

The finite sequence {2, 4, 8, 16, 32} is called a geometric sequence because each term after the first was multiplied by 2 to get to the next term.

In a geometric sequence, dividing any term after the first by the term that precedes it always results in the same nonzero number. This number is called the common ratio, or constant proportional change, and is denoted as r. The common ratio for the geometric sequence 6, 18, 54, 162, ... is 3 since:

$$\frac{18}{6} = \frac{54}{18} = \frac{162}{54} = \ldots = 3$$

Like with the arithmetic sequence, there are two formulas to determine the nth term of a geometric sequence.

General Equations for Geometric Sequences

The nth term of a geometric sequence can be found using either of the following equations:

$$g_n = g_1 r^{n-1}$$
$$g_n = g_k r^{n-k}$$

where g_n is the nth term, g_1 is the first term, g_k is the kth term, and r is the common ratio ($r \neq 0, 1$).

If the first term is correlated with $n = 0$, the nth term of a geometric sequence can be written as:

$$g_n = g_0 r^n$$

Example

If the first term of a geometric sequence is 36 and the fifth term is $\frac{64}{9}$, what could be the 6th term of the sequence?

✓ Solution

Using the nth term of a geometric sequence formula, $g_n = g_1 r^{n-1}$, substitute in the given values $g_1 = 36$, $g_5 = \frac{64}{9}$, and $n = 5$ to solve for the common ratio, r:

$$\frac{64}{9} = 36r^{5-1}$$

$$\frac{64}{9} = 36r^4$$

$$r^4 = \frac{16}{81}$$

$$r = \pm\sqrt[4]{\frac{16}{81}} = \pm\frac{2}{3}$$

To find the 6th term, substitute the given values into the nth term of a geometric sequence formula.

If $r = -\frac{2}{3}$, then:

$$g_6 = (36)\left(-\frac{2}{3}\right)^{6-1}$$

$$g_6 = -\frac{128}{27}$$

If $r = \frac{2}{3}$, then:

$$g_6 = (36)\left(\frac{2}{3}\right)^{6-1}$$

$$g_6 = \frac{128}{27}$$

The 6th term of the geometric sequence is either $-\frac{128}{27}$ or $\frac{128}{27}$.

› Example

Given the arithmetic sequence $-1, 4, 9, 14, 19, 24, 29, \ldots$ and the geometric sequence $1, 5, 25, 125, 625, \ldots$, find their respective common difference and common ratio. Find the average rate of change for each sequence between the first and second terms and the average rate of change between the second and third terms.

✓ Solution

For the arithmetic sequence, the common difference is $d = 4 - (-1) = 9 - 4 = 5$. For the geometric sequence, the common ratio is $r = \frac{5}{1} = \frac{25}{5} = 5$.

To calculate the average rates of change, write the specified terms as ordered pairs, (n, a_n) or (n, g_n), and calculate $\frac{\Delta y}{\Delta x}$.

For the arithmetic sequence, the first, second, and third terms can be written as the ordered pairs $(1, -1)$, $(2, 4)$, and $(3, 9)$.

The average rate of change between the first and second terms is

$$\frac{4-(-1)}{2-1} = \frac{5}{1} = 5$$

The average rate of change between the second and third terms is

$$\frac{9-4}{3-2} = \frac{5}{1} = 5$$

For the geometric sequence, the first, second, and third terms can be written as the ordered pairs $(1, 1)$, $(2, 5)$, and $(3, 25)$.

The average rate of change between the first and second terms is

$$\frac{5-1}{2-1}=\frac{4}{1}=4$$

The average rate of change between the second and third terms is

$$\frac{25-5}{3-2}=\frac{20}{1}=20$$

Notice that for the arithmetic sequence, the average rates of change are constant, showing that the terms of this increasing arithmetic sequence increase equally with each step. The average rates of change of the geometric sequence are not constant and are increasing. This shows that the terms of this geometric sequence increase by a larger amount with each step.

Increasing arithmetic sequences increase equally with each step, whereas increasing geometric sequences increase by a larger amount with each successive step.

6.3 Change in Linear and Exponential Functions

Linear functions of the form $y = mx + b$ are like arithmetic sequences of the form $a_n = a_1 + d(n - 1)$, which can also be written as $a_n = a_0 + dn$.

Both can be expressed as an initial value:

- For the linear function, the initial value is the y-intercept, b.
- For the arithmetic sequence, the initial value is the initial term a_0.

Both have repeated addition as a constant rate of change:

- For the linear function, the rate of change is the slope, m.
- For the arithmetic sequence, the rate of change is the common difference, d.

Both linear functions and arithmetic sequences can be written in different forms as well. Similar to arithmetic sequences of the form $a_n = a_k + d(n - k)$, which is based on the common difference, d, and the kth term, linear functions can be expressed in point-slope form $y = y_1 + m(x - x_1)$, which is based on a known slope, m, and a point, (x_1, y_1).

Exponential functions of the form $f(x) = ab^x$ are like geometric sequences of the form $g_n = g_1 r^{n-1}$ or $g_n = g_0 r^n$.

Both can be expressed as an initial value:

- For the exponential function, the initial value is the constant, a.
- For the geometric sequence, the initial value is the initial term, g_0.

Both have repeated multiplication of a constant proportion:

- For the exponential function, the constant proportion is the b-value.
- For the geometric sequence, the constant proportion is the common ratio, r.

Both exponential functions and geometric sequences can be written in different forms as well. Similar to geometric sequences of the form $g_n = g_k r^{n-k}$, which is based on the common ratio, r, and the kth term, exponential functions can be expressed in the form $y = y_1 r^{(x-x_1)}$, which is based on a known ratio, r, and a point, (x_1, y_1).

Example

Given the following equations, describe a sequence that best fits each equation.

1. $f(n) = -3 + 4n$
2. $g(n) = \left(\frac{1}{2}\right)^n$
3. $h(n) = 4(-3.5)^{n-1}$
4. $j(n) = -3 + 6(n - 4)$

Solution

1. This is a linear function that best describes an arithmetic sequence where $a_0 = -3$ and $d = 4$.
2. This is an exponential function that best describes a geometric sequence where $g_0 = 1$ and $r = \frac{1}{2}$.
3. This is an exponential function that best describes a geometric sequence whose first term is 4 and common ratio is -3.5.
4. This is a linear function that best describes an arithmetic sequence whose fourth term is -3 and common difference is 6.

Sequences are sets of ordered pairs. Since the domain is comprised of only whole numbers, when graphing the ordered pairs of the term and its value, (n, a_n) or (n, g_n), the graph of a sequence consists of discrete points instead of a curve.

The equations that determine the terms of a sequence are functions, linear and exponential. Their domain is the set of all real numbers, and their graphs are lines or curves.

This means that sequences and their corresponding functions may have different domains.

Example

Given the sequence $\{-4, -0.5, 3, 6.5, 10, 13.5, 17, 20.5\}$, do the following.

(a) State the type of sequence it is.
(b) Write an equation to determine the *n*th term of the sequence.
(c) State the domain and range of the sequence.
(d) State the domain and range of the equation found in part (b).

Solution

(a) The sequence is arithmetic since there is a common difference, $d = 3.5$.
(b) An equation to determine the *n*th term of the sequence is $a_n = -4 + 3.5(n - 1)$.
(c) The domain of the sequence is $\{1, 2, 3, 4, 5, 6, 7, 8\}$ since there are 8 terms in the sequence. The range of the sequence is $\{-4, -0.5, 3, 6.5, 10, 13.5, 17, 20.5\}$ since these are the values of the 8 terms in the sequence.
(d) Both the domain and range of the equation found in part (b) are all real numbers since this is a linear function.

Example

Given the sequence $\left\{15, 5, \frac{5}{3}, \frac{5}{9}, \frac{5}{27}\right\}$, do the following.

(a) State the type of sequence it is.
(b) Write an equation to determine the nth term of the sequence.
(c) State the domain and range of the sequence.
(d) State the domain and range of the equation found in part (b).

Solution

(a) The sequence is geometric since there is a common ratio, $r = \frac{1}{3}$.
(b) An equation to determine the nth term of the sequence is $g_n = 15\left(\frac{1}{3}\right)^{n-1}$.
(c) The domain of the sequence is {1, 2, 3, 4, 5} since there are 5 terms in the sequence. The range of the sequence is $\left\{15, 5, \frac{5}{3}, \frac{5}{9}, \frac{5}{27}\right\}$ since these are the values of the 5 terms in the sequence.
(d) The domain of the equation found in part (b) is all real numbers and the range is $(0, \infty)$ since this is an exponential function.

There are similarities and differences between linear and exponential functions, which are outlined in Table 6.1.

Table 6.1

Similarities of Linear and Exponential Functions	Differences of Linear and Exponential Functions
▪ Linear functions of the form $f(x) = mx + b$ and exponential functions of the form $f(x) = ab^x$ can both be expressed in terms of an initial value and a constant involved with change.	▪ Over equal-length input value intervals, if the output values of a function change at a constant rate, the function is linear. If the output values of a function change proportionally, the function is exponential.
▪ Arithmetic sequences, linear functions, geometric sequences, and exponential functions can all be determined by two distinct sequence or function values.	▪ Linear functions are based on addition, while exponential functions are based on multiplication.

Multiple-Choice Questions

1. What is the first term of an arithmetic sequence whose 19th term is 123 and 2nd term is 4?

   (A) -3
   (B) 1
   (C) 2
   (D) 7

2. The last term of a geometric sequence is $\frac{8}{125}$. If the first term is $\frac{5}{2}$ and the sequence has 5 terms, what is the value of the second term?

   (A) $\frac{2}{5}$
   (B) $\frac{4}{5}$
   (C) 1
   (D) 2

3. In the sequence shown below, $g_0 = \frac{1}{2}$ and g_n represents the nth term.

$$\frac{1}{2}, -\frac{1}{4}, \frac{1}{8}, -\frac{1}{16}, \ldots, g_n$$

   Which equation expresses g_n in terms of n?

   (A) $g_n = (-1)^n \frac{1}{2^n}$
   (B) $g_n = (-1)^{n+1} \frac{1}{2^n}$
   (C) $g_n = (-1)^n \frac{1}{2^{n+1}}$
   (D) $g_n = \left(-\frac{1}{2}\right)^n$

4. Given the two arithmetic sequences whose general term can be found using the following equations:

$$a_n = k - 2(n - 1)$$
$$a_n = 4 - 3(n + 1)$$

   What would be the value of k for the two sequences to have the same third term?

   (A) -8
   (B) -4
   (C) -2
   (D) 1

5. Which of the following equations would produce sequences with the same terms?

   I. $g_n = 4 \cdot 2^n$
   II. $g_n = 2^{2+n}$
   III. $g_n = 2^{4n}$

   (A) None of them
   (B) All of them
   (C) I and II only
   (D) I and III only

6. In the sequence shown below, $a_0 = -7.5$ and a_n represents the nth term.

$$-7.5, -4.3, -1.1, 2.1, 5.3, 8.5, \ldots, a_n$$

   Which equation expresses a_n in terms of n?

   (A) $a_n = -7.5 + 3.2(n - 1)$
   (B) $a_n = -7.5 + 3.2n$
   (C) $a_n = -7.5 - 3.2(n - 1)$
   (D) $a_n = -4.3 + 3.2n$

7. Which of the following represents the graph of the sequence whose terms can be found using the equation $a_n = -5 + 3(n - 1)$?

(A)

(B)

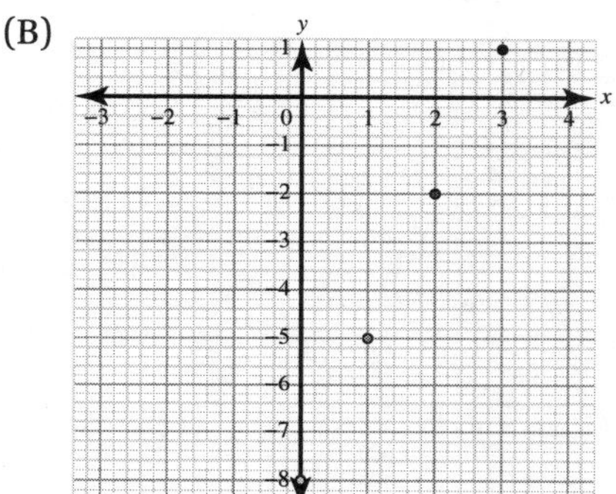

(C)

(D)

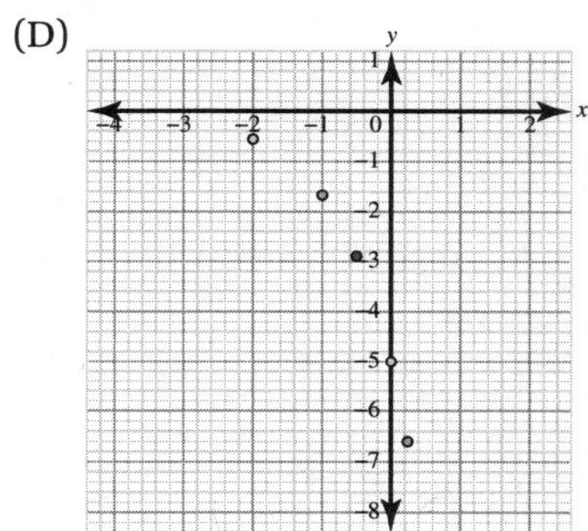

8. Which of the following represents the graph of the sequence whose terms can be found using the equation $g_n = -5 \cdot 3^n$?

(A)

(B)

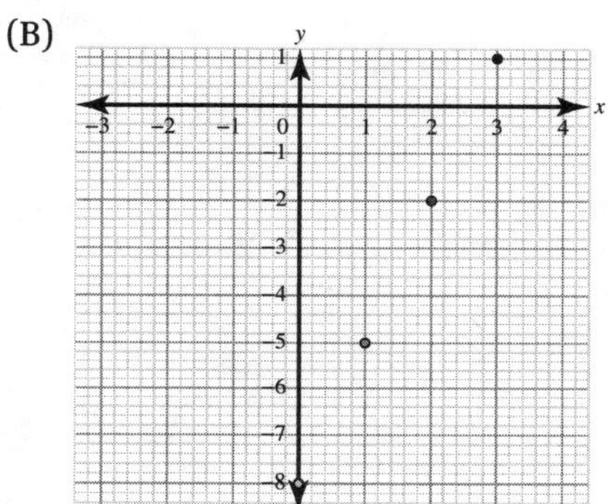

(C)

(D)

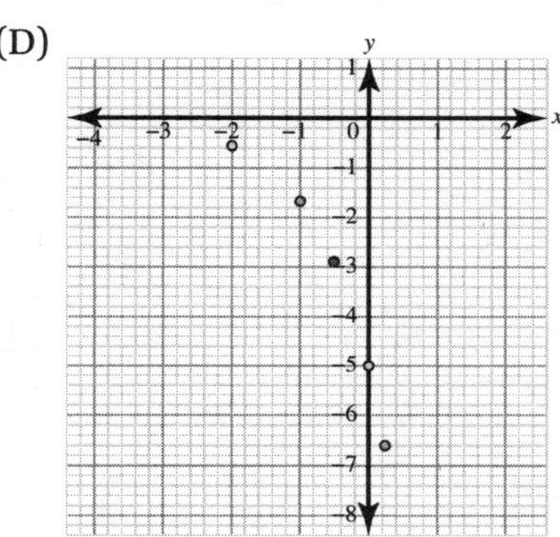

9. Given the geometric sequence whose second term is ae^2 and fifth term is ae^5, which of the following equations can represent the kth term of this sequence?

(A) $g_k = a(e^k)$
(B) $g_k = a(e)^{k-1}$
(C) $g_n = a(e^n)$
(D) $g_n = a(e)^{n-1}$

10. Which of the following sequences have the same common ratio or same common difference?

I. $-1, 3, -9, 27, \ldots$
II. $-1, 2, 5, 8, \ldots$
III. $a_n = -3(n-1) + 1$
IV. $g_n = (-1)^n (3)^{n+2}$

(A) I and II only
(B) II and IV only
(C) I, III, and IV
(D) All of them

Answer Explanations

1. **(A)** The given arithmetic sequence has $a_2 = 4$ and $a_{19} = 123$. Substitute these values into the arithmetic term formula:

$$4 = a_1 + d(2 - 1)$$
$$123 = a_1 + d(19 - 1)$$

Solve the system of equations using the elimination method:

$$\begin{array}{r} 4 = a_1 + d \\ \underline{-(123 = a_1 + 18d)} \\ -119 = -17d \end{array}$$

Solving for d, the common difference is $d = 7$.

If the second term of the arithmetic sequence is 4, subtracting the common difference yields the first term

$$4 - 7 = -3$$

The first term of the sequence is -3.

2. **(C)** The given geometric sequence has $g_5 = \frac{8}{125}$ and $g_1 = \frac{5}{2}$. Substitute these values into the geometric term formula:

$$\frac{8}{125} = \frac{5}{2} \cdot r^{5-1}$$

Solve for r:

$$r^4 = 0.0256$$
$$r = (0.0256)^{1/4}$$
$$r = 0.4 = \frac{4}{10} = \frac{2}{5}$$

Since the first term of the sequence is $\frac{5}{2}$, multiplying this term by the common ratio yields the second term:

$$\frac{5}{2} \cdot \frac{2}{5} = 1$$

The second term of the sequence is 1.

3. **(C)** The given geometric sequence has $g_0 = \frac{1}{2}$. Taking the ratio of the first and second terms of the sequence determines the common ratio:

$$r = \frac{-\frac{1}{4}}{\frac{1}{2}} = -\frac{1}{2}$$

Substitute these values into the nth term formula for a geometric sequence, $g_n = g_0 r^n$:

$$g_n = \left(\frac{1}{2}\right)\left(-\frac{1}{2}\right)^n = \left(\frac{1}{2}\right)(-1)^n\left(\frac{1}{2}\right)^n$$
$$= (-1)^n\left(\frac{1^{n+1}}{2^{n+1}}\right) = (-1)^n\left(\frac{1}{2^{n+1}}\right)$$

4. **(B)** Arithmetic sequences can be written as linear functions. The two sequences are written in a form like point-slope:

$$a_n = k - 2(n - 1) \rightarrow y = k - 2(x - 1)$$
$$a_n = 4 - 3(n + 1) \rightarrow y = 4 - 3(x + 1)$$

It was given that the third term should have the same value. Therefore, $n = 3$ and so $x = 3$.

To find the value of k, substitute $x = 3$ and look for a point of intersection. This can be done either graphically or algebraically:

$$y = k - 2(3 - 1) \rightarrow y = k - 2(2) \rightarrow y = k - 4$$
$$y = 4 - 3(3 + 1) \rightarrow y = 4 - 3(4) \rightarrow y = 4 - 12 = -8$$

Set the first equation equal to -8, and solve for k:

$$-8 = k - 4$$
$$k = -4$$

5. **(C)** Geometric sequences can be written as exponential functions. All the equations are examples of exponential functions. Using the laws of exponents, two of the equations are equivalent to each other. Start with the first equation:

$$g_n = 4 \cdot 2^n = 2^2 \cdot 2^n = 2^{2+n}$$

This is the equation in II. Equation III would have same value terms but not in the same order.

6. **(B)** Since there is a constant rate of change between each consecutive term, this is an arithmetic sequence with $d = 3.2$. Given that $a_0 = -7.5$, use the nth term formula for arithmetic sequences. An equation that represents this sequences. is $a_n = -7.5 + 3.2n$.

7. **(B)** Since the question asks for the graph of the sequence, the graph should consist of a set of ordered pairs. This eliminates Choices (A) and (C). The equation given is a linear function, which represents an arithmetic sequence. The graph in Choice (B) has sets of ordered pairs that have a constant rate of change.

8. **(D)** Since the question asks for the graph of the sequence, the graph should consist of a set of ordered pairs. This eliminate Choices (A) and (C). The equation given is an exponential function, which represents a geometric sequence. The graph in Choice (D) has sets of ordered pairs that have a proportional rate of change.

9. **(A)** It is given that $g_2 = ae^2$ and $g_5 = ae^5$. Following this pattern, an equation to represent the *k*th term would be represented by $g_k = a(e^k)$. Choices (C) and (D) use the letter *n*, which would represent the *n*th term.

10. **(C)** Sequence I is a geometric sequence with $r = -3$. Sequence II is an arithmetic sequence with $d = 3$. Sequence III is an arithmetic sequence with $d = -3$. Sequence IV is a geometric sequence with $r = -3$. The sequences that have the same common ratio or same common difference are I, III, and IV.

7

Logarithmic Functions

Learning Objectives

In this chapter, you will learn:

- ➔ The different properties of logarithmic functions
- ➔ How to graph logarithmic functions
- ➔ How to transform the graphs of logarithmic functions
- ➔ How to model real-world scenarios using logarithmic functions
- ➔ How best to model data and functions using semi-log plots

7.1 Logarithmic Expressions

A logarithmic expression, which is notated as $\log_b c$ and read as "log base b of c," represents the value that the base b must be exponentially raised to in order to obtain the value of c.

For example, $\log_3 81 = 4$ since the base 3 when raised to the 4th power is 81 ($3^4 = 81$).

Converting Logarithms and Exponents

Given that a and c are constants such that $b > 0$ and $b \neq 1$:

If $\log_b c = a$, then $b^a = c$
If $b^a = c$, then $\log_b c = a$

Example

Evaluate each of the following logarithmic expressions.

1. $\log_2 32$
2. $\log_5 \sqrt{5}$
3. $\log_3 \frac{1}{27}$
4. $\log_{10} 1$
5. $\log_{10} 10$
6. $\log_{10} 100$

Solution

1. $\log_2 32 = 5$ since $2^5 = 32$.
2. $\log_5 \sqrt{5} = \frac{1}{2}$ since $5^{1/2} = \sqrt{5}$.
3. $\log_3 \frac{1}{27} = -3$ since $3^{-3} = \frac{1}{3^3} = \frac{1}{27}$.
4. $\log_{10} 1 = 0$ since $10^0 = 1$.
5. $\log_{10} 10 = 1$ since $10^1 = 10$.
6. $\log_{10} 100 = 2$ since $10^2 = 100$.

The last three problems in the previous example appear to show a pattern. The number of zeros in the *c*-value represents the solution to the logarithmic expression. This is true only because in each of these examples, the base of the logarithm is 10. When a logarithm is base 10, this is known as a common logarithm. Common logarithms are indicated when the base is omitted from the logarithm expression. Thus, $\log x$ is understood to mean $\log_{10} x$.

The common logarithm of a power of 10 is the exponent of 10. For example, $\log 1{,}000 = 3$ because $1{,}000 = 10^3$. Therefore $\log\left(\frac{1}{10}\right) = -1$ since $\frac{1}{10} = 10^{-1}$.

The values of the previous examples were determined using basic arithmetic. However, often logarithmic expressions can be evaluated only using technology.

For example, to find the value of a common logarithm such as log 35, using a calculator, press the LOG button and enter 35. The calculator should display 1.544068044. Therefore $\log 35 \approx 1.544068044$, meaning that $10^{1.544068044} \approx 35$.

Evaluating Logarithms

If the logarithmic expression being evaluated is not a common logarithm, there are two ways that expression can be evaluated.

1. Change of Base Formula

A logarithm with base *b* can be rewritten as an equivalent common logarithm by using the following formula:

$$\log_b x = \frac{\log x}{\log b}$$

Note: The change of base formula can be used for bases other than 10:

$$\log_b x = \frac{\log_a x}{\log_a b}$$

2. Technology

Many calculators are now updated so that the logarithmic expression can be typed in. For example, if using a TI graphing calculator, press MATH and the UP arrow to get to A:logBASE (on the selection screen). Press ENTER. Then enter the base and the input of the logarithmic expression in the empty fields, and press ENTER again.

Example

Evaluate each of the following logarithmic expressions, rounding all answers to the nearest hundredth.

1. $\log 45$
2. $\log \frac{2}{3}$
3. $\log_5 67$
4. $\log_4 \frac{8}{11}$

✓ Solution

Technology is needed to evaluate each of these logarithmic expressions. Rewriting these expressions using the change of base formula would still require the use of technology, so each should be typed into the calculator just as written.

1. $\log 45 \approx 1.65$
2. $\log \frac{2}{3} \approx -0.18$
3. $\log_5 67 \approx 2.61$
4. $\log_4 \frac{8}{11} \approx -0.23$

7.2 Inverses of Exponential Functions

The general form of a logarithmic function is $f(x) = a\log_b x$, with base b, where $b > 0$, $b \neq 1$, and $a \neq 0$.

The previous section demonstrated that a logarithm is an exponent and that logarithms can be rewritten in exponential form.

A logarithmic equation is defined so that $\log_b x = y$ means $b^y = x$. Since $b^y = x$ is the inverse of $y = b^x$, then $\log_b x = y$ is the inverse of $y = b^x$.

Definition of the Logarithmic Function

The logarithmic function with base b, denoted as $y = \log_b x$, is the inverse of the exponential function $y = b^x$ with base b, where $x > 0$ and $b > 0$ and $b \neq 1$. Thus:

$$y = \log_b x \text{ means } b^y = x$$

Since logarithmic functions are inverse functions, all the properties about inverse functions that were discussed in Section 4.5 apply.

Composition of Inverse Functions

If $f(x) = \log_b x$ and $g(x) = b^x$, where $b > 0$ and $b \neq 1$, are inverse functions, then $g(f(x)) = f(g(x)) = x$.

The composition rule of inverse functions will come in handy when working with logarithmic equations. It helps to develop the following simplifying rule.

Simplifying Rule

When given the inverse functions $f(x) = \log_2 x$ and $g(x) = 2^x$, composing one with the other gives:

$$f(g(x)) = f(2^x) = \log_2 2^x = x$$

and

$$g(f(x)) = g(\log_2 x) = 2^{\log_2 x} = x$$

Exponential and Logarithmic Identities

The following is a generalization about logarithms and exponential expressions with the same base:

$$\log_b b^x = x \text{ and } b^{\log_b x} = x$$

Graphs of Logarithmic Functions

The following two properties of inverse functions will help when figuring out the graph of the parent function $f(x) = \log_b x$.

1. The graph of the logarithmic function $f(x) = \log_b x$, where $b > 0$ and $b \neq 1$, is a reflection of the graph of the exponential function $g(x) = b^x$, where $b > 0$ and $b \neq 1$, over the line $y = x$.
2. If (s, t) is an ordered pair of the exponential function $g(x) = b^x$, where $b > 0$ and $b \neq 1$, then (t, s) is an ordered pair of the logarithmic function $f(x) = \log_b x$, where $b > 0$ and $b \neq 1$.

Example

Using the graph of $g(x) = 2^x$, graph the logarithmic function $f(x) = \log_2(x)$ and state its domain and range.

Solution

Table 7.1 is a table of values for $g(x) = 2^x$. Figure 7.1 is a graph of the function.

Table 7.1

x	$g(x)$
−3	$\frac{1}{8}$
−2	$\frac{1}{4}$
−1	$\frac{1}{2}$
0	1
1	2
2	4
3	8

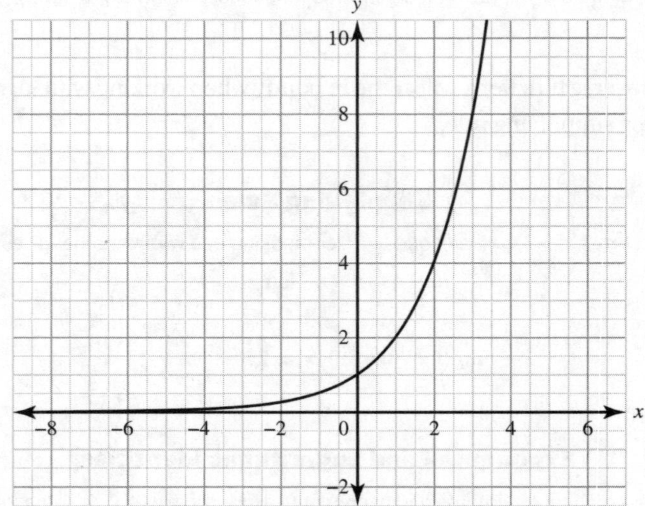

Figure 7.1

For $g(x)$, the domain is $(-\infty, \infty)$ and the range is $(0, \infty)$.

To create the table of values for the graph of $f(x) = \log_2(x)$, switch the input and output values. This is shown in Table 7.2. Then plot the ordered pairs to reveal the graph of $f(x)$, as shown in Figure 7.2.

Table 7.2

x	f(x)
$\frac{1}{8}$	−3
$\frac{1}{4}$	−2
$\frac{1}{2}$	−1
1	0
2	1
4	2
8	3

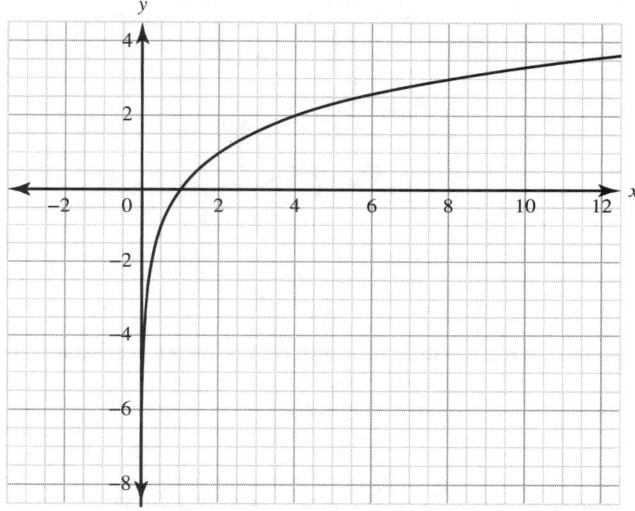

Figure 7.2

For $f(x)$, the domain is $(0, \infty)$ and the range is $(-\infty, \infty)$.

When graphed on the same set of axes, as shown in Figure 7.3, the graphs of the exponential function and its inverse are symmetric to the line $y = x$.

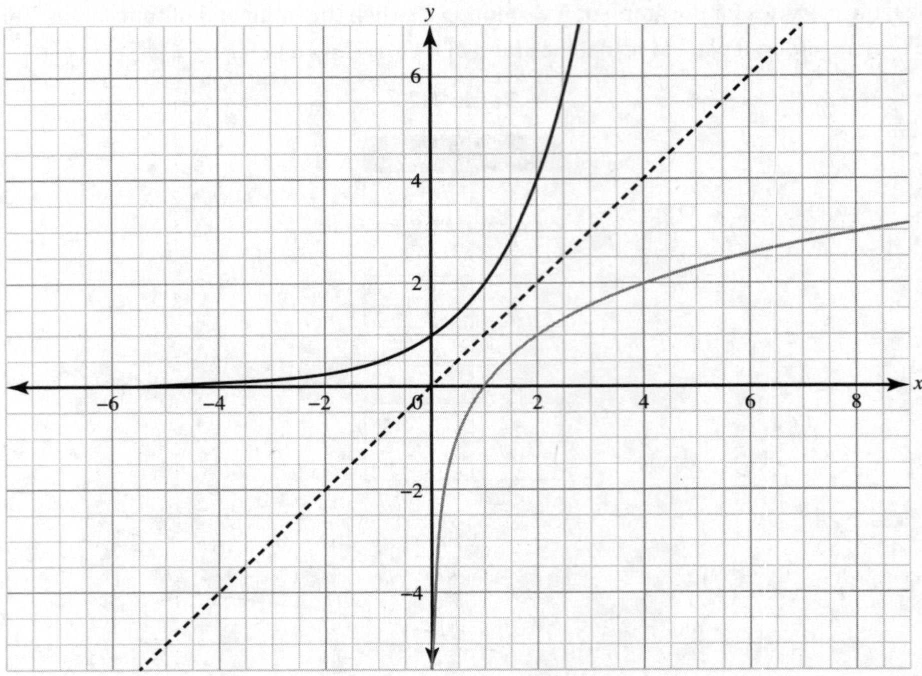

Figure 7.3

Example

Using the graph of $g(x) = \frac{1}{2}^{x}$, graph the logarithmic function $f(x) = \log_{\frac{1}{2}}(x)$ and state its domain and range.

Solution

Table 7.3 is a table of values for $g(x) = \frac{1}{2}^{x}$. Figure 7.4 is a graph of the function.

Table 7.3

x	$g(x)$
−3	8
−2	4
−1	2
0	1
1	$\frac{1}{2}$
2	$\frac{1}{4}$
3	$\frac{1}{8}$

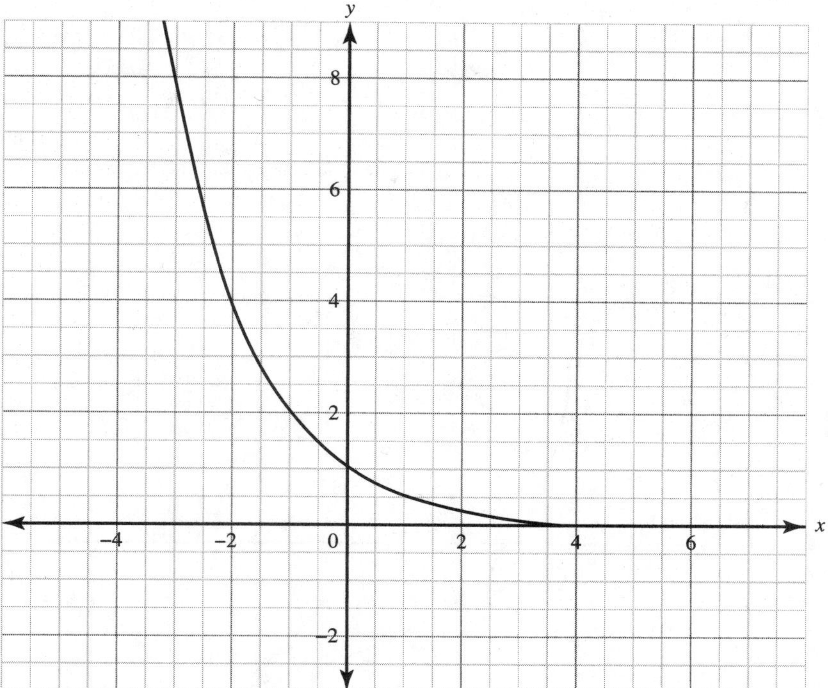

Figure 7.4

For $g(x)$, the domain is $(-\infty, \infty)$ and the range is $(0, \infty)$.

To create the table of values for the graph of $f(x) = \log_{\frac{1}{2}}(x)$, switch the input and output values and plot the ordered pairs to reveal the graph of $f(x)$. This is shown in Table 7.4. Figure 7.5 is a plot of the ordered pairs.

Table 7.4

x	$f(x)$
8	−3
4	−2
2	−1
1	0
$\frac{1}{2}$	1
$\frac{1}{4}$	2
$\frac{1}{8}$	3

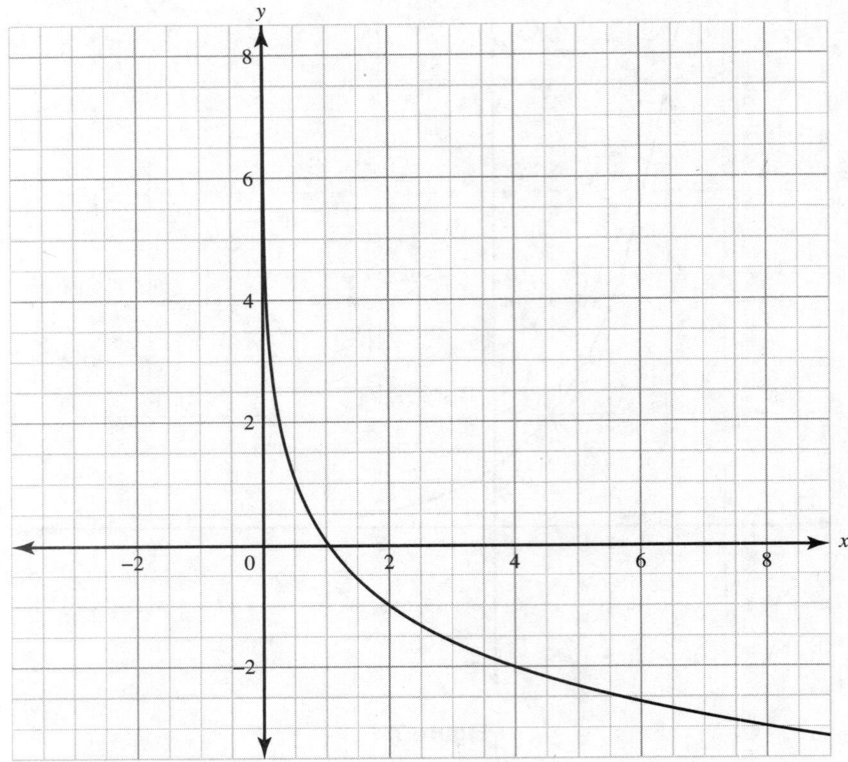

Figure 7.5

For $f(x)$, the domain is $(0, \infty)$ and the range is $(-\infty, \infty)$.

When graphed on the same set of axes, as shown in Figure 7.6, the graphs of the exponential function and its inverse are symmetric to the line $y = x$.

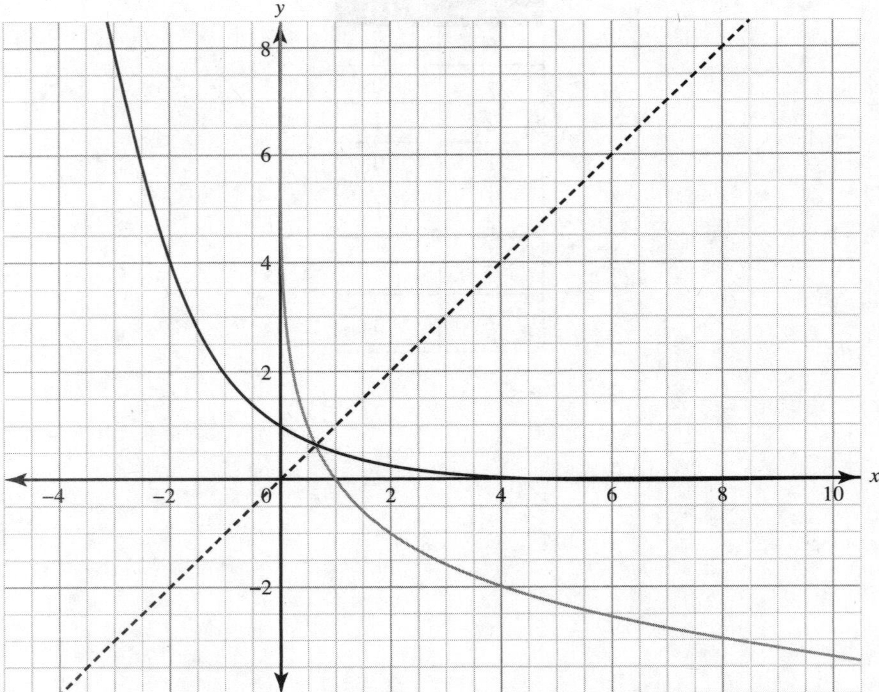

Figure 7.6

Based on these graphs, the following generalizations can be made about the graphs of logarithmic functions:

- The domain is limited to the set of positive real numbers, while the range is the set of all real numbers.
- The x-intercept of the graph is $(1, 0)$.
- There is no y-intercept of the graph.
- There is a vertical asymptote at $x = 0$.
- If the base is greater than 1, the graph rises as x increases. As x decreases, the graph is asymptotic to the negative y-axis.
- If the base is between 0 and 1, the graph falls as x increases. As x decreases, the graph is asymptotic to the positive y-axis.

7.3 Logarithmic Functions

In the previous section, the graphs of logarithmic functions were discussed and with them many properties were illustrated. These properties all follow from the fact that exponential and logarithmic functions are inverse functions. Table 7.5 provides a comparison of the graphs of the two parent functions.

Table 7.5 Comparison of the Graphs of Exponential and Logarithmic Functions

Property	Graph of $y = b^x$	Graph of $y = \log_b x$
Quadrants	I and II	I and IV
Domain	$(-\infty, \infty)$	$(0, \infty)$
Range	$(0, \infty)$	$(-\infty, \infty)$
x-intercept	None	$(1, 0)$
y-intercept	$(0, 1)$	None
Asymptote	x-axis	y-axis
Rises as x increases	$b > 1$	$b > 1$
Falls as x increases	$0 < b < 1$	$0 < b < 1$

As shown by their graphs, logarithmic functions are always increasing or decreasing, and their graphs are either always concave up or concave down. As a result, logarithmic functions do not have extrema except on a closed interval, and their graphs do not have points of inflection.

By identifying these key characteristics of logarithmic functions, the limits of these functions can be evaluated.

Example

If $f(x) = \log_4 x$ and $g(x) = \log_{\frac{1}{5}} x$, evaluate the following limits.

1. $\lim_{x \to \infty} f(x)$
2. $\lim_{x \to 0^+} f(x)$
3. $\lim_{x \to 1} f(x)$
4. $\lim_{x \to \infty} g(x)$
5. $\lim_{x \to 0^+} g(x)$
6. $\lim_{x \to 1} g(x)$

Solution

To help evaluate the limits, use Table 7.5 or consider the graph of $f(x)$ in Figure 7.7 and the graph of $g(x)$ in Figure 7.8.

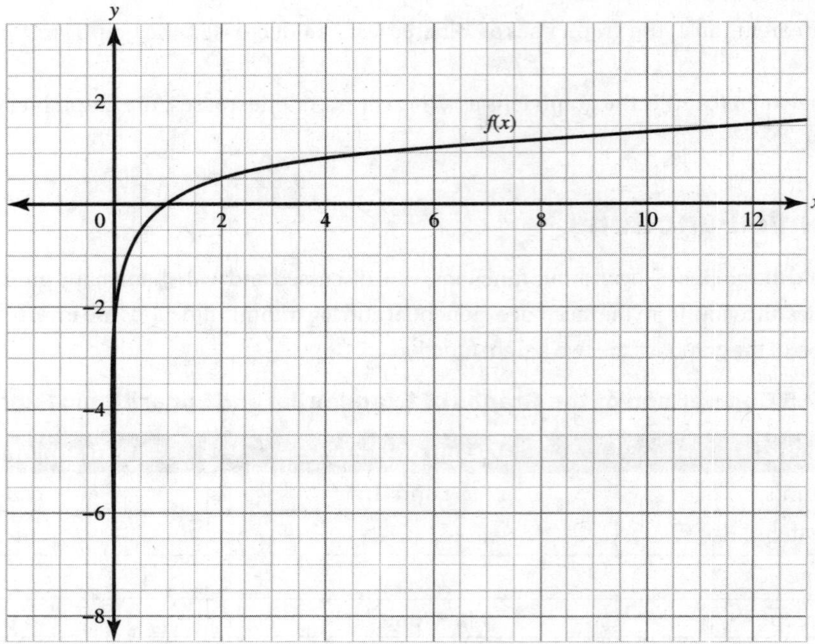

Figure 7.7

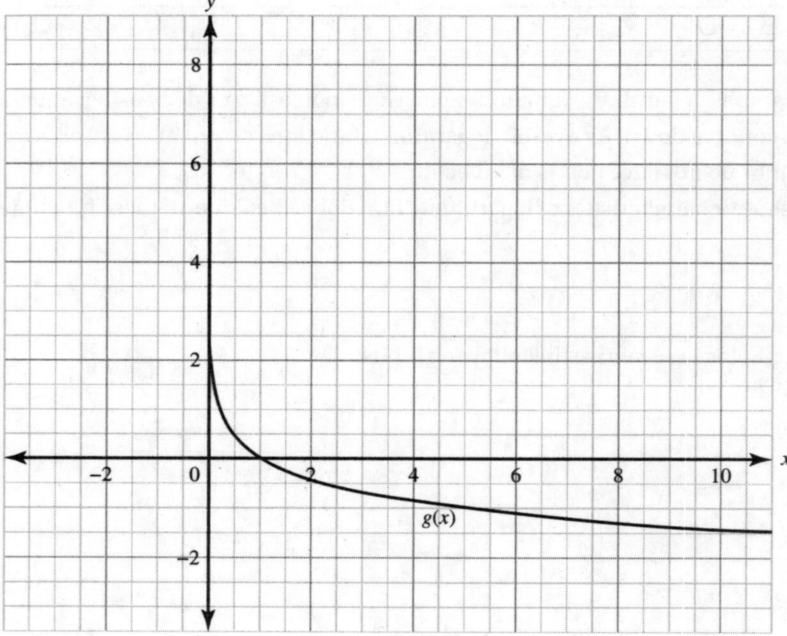

Figure 7.8

1. $\lim_{x\to\infty} f(x) = \infty$
2. $\lim_{x\to 0^+} f(x) = -\infty$
3. $\lim_{x\to 1} f(x) = 0$
4. $\lim_{x\to\infty} g(x) = -\infty$
5. $\lim_{x\to 0^+} g(x) = \infty$
6. $\lim_{x\to 1} g(x) = 0$

With their limited domain, logarithmic functions in general form are vertically asymptotic to $x = 0$, with an end behavior that is unbounded. That is, for a logarithmic function in general form,

$$\lim_{x\to 0^+} a\log_b x = \pm\infty$$
and $$\lim_{x\to\infty} a\log_b x = \pm\infty$$

7.4 Logarithmic Function Manipulation

Since a logarithm is an exponent, the laws for evaluating logarithms follow the laws of exponents listed in Section 5.2. Logarithms of products, quotients, powers, and roots can be broken down into their component parts by using one or more of the logarithm laws outlined in Table 7.6.

Table 7.6 Logarithm Laws

Law	General Rule	Examples
Product	$\log_b(xy) = \log_b x + \log_b y$	▪ $\log 8 = \log(4\cdot 2) = \log 4 + \log 2$ ▪ $\log(100a) = \log 100 + \log a = 2 + \log a$
Quotient	$\log_b\left(\frac{x}{y}\right) = \log_b x - \log_b y$	▪ $\log\left(\frac{ab}{6}\right) = \log a + \log b - \log 6$ ▪ $\log\left(\frac{c}{10}\right) = \log c - \log 10 = \log c - 1$
Power	$\log_b(x^n) = n\cdot\log_b x$	▪ $\log 49 = \log 7^2 = 2\log 7$ ▪ $\log\sqrt{x} = \log x^{1/2} = \frac{1}{2}\log x$
Change of base	$\log_b x = \frac{\log_a x}{\log_a b}$, $a > 0$ and $a \neq 1$	▪ $\log_x 10 = \frac{\log 10}{\log x} = \frac{1}{\log x}$ ▪ $\log_9 5 = \frac{\log_3 5}{\log_3 9} = \frac{\log_3 5}{2} = \frac{1}{2}\log_3 5$

Example

Express $\log_5\sqrt[3]{\frac{x}{y}}$ in terms of $\log_5 x$ and $\log_5 y$.

Solution

First use the power law:

$$\log_5\sqrt[3]{\frac{x}{y}} = \log_5\left(\frac{x}{y}\right)^{1/3} = \frac{1}{3}\log_5\left(\frac{x}{y}\right)$$

Then use the quotient law:

$$\frac{1}{3}\log_5\left(\frac{x}{y}\right) = \frac{1}{3}[\log_5 x - \log_5 y]$$

Therefore:

$$\log_5 \sqrt[3]{\frac{x}{y}} = \frac{1}{3}[\log_5 x - \log_5 y]$$

Example

If $\log 2 = x$ and $\log 3 = y$, express each of the following in terms of x and y.

1. $\log \sqrt{6}$
2. $\log 24$
3. $\log 1.5$

Solution

For each of the examples, rewrite the numbers as factors of 2 and 3 and then use the logarithm laws.

1. $\log \sqrt{6} = \log \sqrt{2 \cdot 3} = \log (2 \cdot 3)^{\frac{1}{2}} = \frac{1}{2}[\log(2 \cdot 3)] = \frac{1}{2}[\log(2) + \log(3)] = \frac{1}{2}(x + y)$
2. $\log 24 = \log(2^3 \cdot 3) = \log(2^3) + \log(3) = 3(\log 2) + \log(3) = 3x + y$
3. $\log 1.5 = \log \frac{3}{2} = \log 3 - \log 2 = y - x$

Each of these logarithm laws has a different graphical implication. These graphic properties are outlined in Table 7.7.

Table 7.7 Logarithm Graphical Transformations

Law	General Rule	Transformation
Product	$f(x) = \log_b(kx)$ $= \log_b k + \log_b x$ $= a + \log_b x$	Every horizontal dilation of a logarithmic function is equivalent to a vertical translation.
Power	$f(x) = \log_b x^k$ $= k \log_b x$	Raising the input of a logarithmic function to a power results in a vertical dilation.
Change of base	$\log_b x = \dfrac{\log_a x}{\log_a b} = \dfrac{1}{k}\log_a x$, $a > 0$ and $a \neq 1$	All logarithmic functions are vertical dilations of each other.

Example

Describe the transformation of each of the following functions from its parent function.

1. $f(x) = \log_2 2x$
2. $g(x) = \log_3 x^2$
3. $h(x) = \log_5 x$
4. $i(x) = \log_{\frac{1}{2}} \sqrt{4x}$

Solution

For each of the functions, use the logarithm laws to rewrite each function as a form of the parent function $y = a\log_b x$. Then use the differences to describe the transformations.

1. $f(x) = \log_2 2x = \log_2 2 + \log_2 x = 1 + \log_2 x$
   When written in this form, $f(x)$ is a vertical translation 1 unit up of the graph of the function $y = \log_2 x$.
2. $g(x) = \log_3 x^2 = 2\log_3 x$
   When written in this form, $g(x)$ is a vertical dilation of the function $y = \log_3 x$ by a factor of 2.
3. $h(x) = \log_{25} x = \dfrac{\log_5 x}{\log_5 25} = \dfrac{\log_5 x}{2} = \dfrac{1}{2}\log_5 x$
   When written in this form, $h(x)$ is a vertical dilation of the function $y = \log_5 x$ by a factor of $\frac{1}{2}$. Depending on the base chosen for the change of base, there are infinitely many different vertical dilations of this function that would result in an equivalent graph.
4. $i(x) = \log_5 \sqrt{125x} = \log_5 (125x)^{1/2} = \frac{1}{2}[\log_5 125 + \log_5 x] = \frac{1}{2}[3 + \log_5 x]$
   When written in this form, $i(x)$ is a vertical translation 3 units up of the graph of the function $y = \log_5 x$ followed by a vertical dilation by a factor of $\frac{1}{2}$.

The Natural Logarithmic Function

The natural logarithmic function, denoted as $\ln x$, is the logarithmic function whose base is the irrational number, e, called Euler's constant. The base of $\log x$ is understood to be 10 with $\log 10 = 1$. Similarly, the base of $\ln x$ is understood to be e with $\ln e = 1$.

To evaluate natural log expressions, a graphing calculator has an LN key with its inverse, e^x, above it.

A function $f(x) = \ln x$ is a logarithmic function with the natural base e; that is, $\ln x = \log_e x$.

Example

Evaluate the following expressions, rounding all answers to the nearest ten-thousandth.

1. $\ln 5$
2. $2\ln 3$
3. $\ln(3)^2$
4. $\ln e$
5. $\ln e^{-4}$
6. $e^{\ln 10}$

Solution

Each of the expressions should be evaluated with the use of a calculator by using the LN key.

1. $\ln 5 = 1.6094$
2. $2\ln 3 = 2.1972$
3. $\ln(3)^2 = 2.1972$
4. $\ln e = 1$
5. $\ln e^{-4} = -4$
6. $e^{\ln 10} = 10$

The same laws and properties that common logarithms have also apply to natural logarithms as outlined in Table 7.8.

Table 7.8 Natural Logarithm Laws

Law	General Rule	Example
Product	$\ln(xy) = \ln x + \ln y$	$\ln(6x) = \ln 6 + \ln x$
Quotient	$\ln\left(\frac{x}{y}\right) = \ln x - \ln y$	$\ln\frac{x}{7} = \ln x - \ln 7$
Power	$\ln(x)^n = n \cdot \ln x$	$\ln\sqrt[3]{x} = \ln x^{1/3} = \frac{1}{3}\ln x$
Composition of inverse functions	$\ln e^a = a$ $e^{\ln a} = a$	$\ln e^5 = 5$ $e^{\ln 5} = 5$

7.5 Exponential and Logarithmic Equations and Inequalities

Properties of exponents, properties of logarithms, and the inverse relationship between exponential and logarithmic functions can be used to solve equations and inequalities involving exponents and logarithms. This section will go through the different types of equations and inequalities containing exponents and logarithms and discuss how to find their solutions.

Solving a Logarithmic Equation

When each term of an equation is a logarithm with the same base, rewrite each side of the equation as a single logarithm. Then write an equation without logarithms, using the one-to-one function property of logarithms that states if $\log_b A = \log_b B$, then $A = B$.

Example

If $\log_5 N = 3\log_5 x + \log_5 y$, solve for N in terms of x and y.

Solution

Every term in this equation contains a logarithm of base 5. Rewrite each side of the equation as a single logarithm to solve for N:

$$\log_5 N = 3\log_5 x + \log_5 y$$

$$\log_5 N = \log_5 x^3 + \log_5 y$$

$$\log_5 N = \log_5(x^3 y)$$

$$N = x^3 y$$

Example

Solve for x: $\ln x - \frac{1}{3}\ln 8 = \ln 10$

Solution

Every term in this equation contains a natural log. Isolate $\ln x$, and rewrite each side of the equation as a single logarithm to solve for x:

$$\ln x - \frac{1}{3}\ln 8 = \ln 10$$

$$\ln x = \frac{1}{3}\ln 8 + \ln 10$$

$$\ln x = \ln 8^{\frac{1}{3}} + \ln 10$$

$$\ln x = \ln 2 + \ln 10$$

$$\ln x = \ln (2 \cdot 10)$$

$$\ln x = \ln 20$$

Therefore, $x = 20$.

Solving a Logarithmic Equation with a Constant

If every term in an equation does not contain a logarithm and, instead, some terms are a constant, follow the steps outlined below.

Solving a Log Equation Containing a Constant Term

1. Bring all the logarithms with the same base to the same side of the equation.
2. Write the equation in logarithmic form: $\log_b c = a$.
3. Rewrite the equation in exponential form and solve the equation.
4. Check each solution in the original equation. Reject any solution that leads to taking the logarithm of a negative number.

Example

Solve for x: $\log_3 80 + \log_3 x = 2 + \log_3(x^2 - 1)$

Solution

Since this equation contains a constant term, consolidate the logarithmic terms and rewrite in exponential form:

$$\log_3 80 + \log_3 x = 2 + \log_3(x^2 - 1)$$

$$\log_3(80x) = 2 + \log_3(x^2 - 1)$$

$$\log_3(80x) - \log_3(x^2 - 1) = 2$$

$$\log_3\left(\frac{80x}{x^2 - 1}\right) = 2$$

$$3^2 = \frac{80x}{x^2 - 1}$$

Solve the equation by multiplying both sides by $x^2 - 1$ to eliminate the fraction:

$$9(x^2 - 1) = 80x$$

Distribute the 9. Then set the quadratic equation equal to 0 and factor:

$$9x^2 - 80x - 9 = 0$$
$$(9x + 1)(x - 9) = 0$$
$$x = -\frac{1}{9}, x = 9$$

Check both solutions by substituting into the original equation.

First try $x = -\frac{1}{9}$:

$$\log_3 80 + \log_3\left(-\tfrac{1}{9}\right) = 2 + \log_3\left(\left(-\tfrac{1}{9}\right)^2 - 1\right)$$

This solution is rejected because the domain of $\log_3 x$ is all positive real numbers. There cannot be an input of $-\frac{1}{9}$.

Next try $x = 9$:

$$\log_3 80 + \log_3(9) = 2 + \log_3((9)^2 - 1)$$
$$\log_3 80 + 2 = 2 + \log_3(81 - 1)$$
$$\log_3 80 + 2 = 2 + \log_3 80$$

The two sides of the equation are equal and there are no issues with the domain of the logarithm. Therefore, $x = 9$ is the only solution.

Solving an Exponential Equation Using Logarithms

Using logarithms is a useful way to solve exponential equations algebraically, especially when it may not be possible to write each side of an exponential equation as a power of the same base.

Example

Solve for x: $2^{3x} = 6$

Solution

Since 6 cannot be rewritten as a base of 2 to a power, to eliminate the exponent, take the logarithm of each side of the equation. Note that any logarithm to any base can be used:

$$\log(2^{3x}) = \log 6$$
$$3x \cdot \log 2 = \log 6$$
$$x = \frac{\log 6}{3 \cdot \log 2} \approx 0.8616541669$$

Example

Solve for x: $3^x = 21$

✓ Solution

Since 21 cannot be rewritten as a base of 3 to a power, to eliminate the exponent, take the logarithm of each side of the equation. Note that any logarithm to any base can be used:

$$\ln(3^x) = \ln 21$$

$$x \cdot \ln 3 = \ln 21$$

$$x = \frac{\ln 21}{\ln 3} \approx 2.771243749$$

> Example

The amount of a 20-milligram dose of a medicinal drug remaining in the bloodstream falls continuously at an exponential rate. After 2 hours, 17 milligrams remain in the bloodstream. Find, to the nearest tenth, how much time is required for at least half of the original drug dose to leave the bloodstream.

✓ Solution

Using the exponential growth model $f(t) = a\,e^{kt}$ from Section 5.3, first find k when $f(t) = 17$, $a = 20$, and $t = 2$:

$$17 = 20\,e^{2k}$$

Isolate the exponential expression:

$$e^{2k} = \frac{17}{20}$$

Take the log of both sides:

$$\ln(e^{2k}) = \ln\left(\frac{17}{20}\right)$$

Solve for k:

$$2k = \ln\left(\frac{17}{20}\right)$$

$$k = \frac{\ln\left(\frac{17}{20}\right)}{2} \approx -0.08126$$

Use the exponential model $f(t) = 20\,e^{-0.08126t}$ to find t when $f(t) = 10$:

$$10 = 20\,e^{-0.08126t}$$

$$e^{-0.08126t} = \frac{1}{2}$$

$$\ln(e^{-0.08126t}) = \ln\left(\frac{1}{2}\right)$$

$$-0.08126t = \ln\left(\frac{1}{2}\right)$$

$$t = \frac{\ln\left(\frac{1}{2}\right)}{-0.08126} \approx 8.5$$

The techniques used for solving equations can still be applied to solving inequalities involving exponential and logarithmic functions.

Solving Exponential Inequalities

Example

Find the values of x such that $2^x > 35$.

Solution

An algebraic approach to solving is to change the inequality sign to an equal sign and use logarithms to solve for x:

$$2^x = 35$$
$$\log(2^x) = \log 35$$
$$x \cdot \log 2 = \log 35$$
$$x = \frac{\log 35}{\log 2} \approx 5.129283017$$

For that inequality to be true, $x > \frac{\log 35}{\log 2} \approx 5.129283017$.

Solving Logarithmic Inequalities

Example

Find all values of x such that $\log x < 2$.

Solution

Replace the inequality sign with an equal sign and solve for x:

$$\log x = 2$$
$$10^2 = x$$
$$x = 100$$

Since the input values of a logarithm must be greater than 0, the solution to the inequality is $0 < x < 100$.

Example

What values of x satisfy the inequality $\log_2(2x + 3) > \log_2(3x)$?

Solution

Since both logs are base 2, this implies that $2x + 3 > 3x$. Solving the inequality for x gives $x < 3$. Since the input values of a logarithm must be greater than 0, the solution to the inequality is $0 < x < 3$.

Example

What values of x satisfy the following inequality: $\log_2(x + 1) > \log_4(x^2)$?

✓ Solution

The bases of the logarithms are different. Since $4 = 2^2$, the change of base formula can be used in the following way:

$$\log_2(x+1) = \frac{\log_4(x+1)}{\log_4 2} = \frac{\log_4(x+1)}{\frac{1}{2}} = 2\log_4(x+1) = \log_4((x+1)^2)$$

Therefore, the inequality can be rewritten as base 4 in the following way:

$$\log_4((x+1)^2) > \log_4(x^2)$$

Now that the bases are equal, the following inequality can be solved:

$$(x+1)^2 > x^2$$

Change the inequality sign to an equal sign, and solve for x:

$$(x+1)^2 = x^2$$

$$x^2 + 2x + 1 = x^2$$

$$2x + 1 = 0$$

$$x = -\frac{1}{2}$$

To determine if $x < -\frac{1}{2}$ or $x > -\frac{1}{2}$, choose test points in those intervals. Then substitute the values into the original inequality to see if they are true or false.

Test $x = -\frac{3}{4}$:

$$\log_2\left(-\frac{3}{4}+1\right) > \log_4\left(\left(-\frac{3}{4}\right)^2\right)$$

$$\log_2\frac{1}{4} > \log_4\frac{9}{16}$$

$$-2 > -0.415037$$

This is not a true statement, so x cannot be less than $-\frac{1}{2}$.

Test $x = 1$:

$$\log_2(1+1) > \log_4((1)^2)$$

$$\log_2(2) > \log_4(1)$$

$$1 > 0$$

This is a true statement, so $x > -\frac{1}{2}$.

For the final solution, it is important to consider the domain of a logarithmic function, which is all input values greater than 0. Since $0 > -\frac{1}{2}$ and cannot be substituted into the original inequality without causing an issue, the final answer must exclude the case where $x = 0$. Therefore, the final solution is $x > -\frac{1}{2}, x \neq 0$.

When solving exponential and logarithmic equations found through analytical or graphical methods, the results should be considered for extraneous solutions. Extraneous solutions are solutions that cannot be a part of the domain of the functions or solutions because they would cause the original equation or inequality to not exist.

Example

Find the values of x that satisfy the following inequality: $\log_7(x+5) > \log_5(x+5)$.

Solution

Since both log expressions are of different bases that do not have a relationship to each other, use the change of base formula to rewrite each of the logarithmic expressions:

$$\log_7(x+5) > \log_5(x+5)$$

$$\frac{\log(x+5)}{\log 7} > \frac{\log(x+5)}{\log 5}$$

Since $\log 7 > \log 5$, for this inequality to be true, both fractions must be negative, meaning that $\log(x+5)$ must be negative.

This leads to the following inequality:

$$\log(x+5) < 0$$

Change the inequality to an equal sign and rewrite in exponential form:

$$10^0 = x + 5$$

$$x + 5 = 1$$

$$x = -4$$

To determine if $x < -4$ or $x > -4$, choose an x-value in those intervals. Then test those values in the original inequality.

Test $x = -4.5$:

$$\log_7((-4.5)+5) > \log_5((-4.5)+5)$$

$$\log_7(0.5) > \log_5(0.5)$$

$$-0.3562071871 > -0.4306765581$$

This is a true statement, so x is less than -4.

Test $x = 0$:

$$\log_7((0)+5) > \log_5((0)+5)$$

$$\log_7(5) > \log_5(5)$$

$$0.8270874753 > 1$$

This is not a true statement, so x cannot be greater than -4.

For the final solution, it is important to consider the domain of a logarithmic function, which is all input values greater than 0. Since $x + 5 > 0$, then $x > -5$. Therefore, the final solution is $-5 < x < -4$.

Transformations of Graphs of Exponential and Logarithmic Functions

In previous sections, the graphs of the basic parent exponential and logarithmic functions were shown along with their properties.

Table 7.9 shows that shifting the graphs of exponential and logarithmic functions vertically and horizontally involves a combination of additive transformations of the general forms of their equations.

Table 7.9

Exponential Function	Logarithmic Function
$f(x) = ab^{(x-h)} + k$	$f(x) = a\log_b(x - h) + k$

Example

For each of the following functions, describe the transformation of the graph based on the parent function. Include a sketch of the graph.

1. $f(x) = e^{x+7} - 3$
2. $g(x) = \log_2(x - 1) + 4$

Solution

1. The parent function for $f(x)$ is the exponential function $y = e^x$. The graph of $f(x)$ is shifted to the left 7 units and down 3 units. The graph of $f(x)$ and its parent function is shown in Figure 7.9. The easiest way to see the transformation is to follow the point (0, 1), which lies on the graph $y = e^x$. Move the point left 7 units and down 3 units to the point $(-7, -2)$, which lies on the graph of $f(x) = e^{x+7} - 3$.

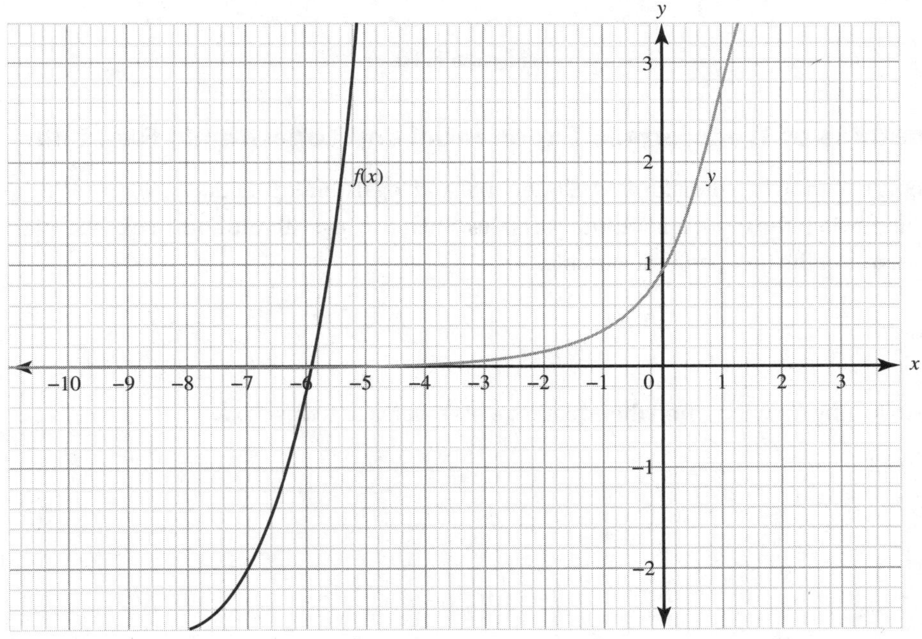

Figure 7.9

2. The parent function for $g(x)$ is the logarithmic function $y = \log_2 x$. The graph of $g(x)$ is shifted to the right 1 unit and up 4 units. A graph of $g(x)$ and its parent function is shown in Figure 7.10. The easiest way to see the transformation is to follow the point (1, 0), which lies on the graph $y = \log_2 x$. Move the point right 1 unit and up 4 units to the point (2, 4), which lies on the graph of $g(x) = \log_2(x - 1) + 4$.

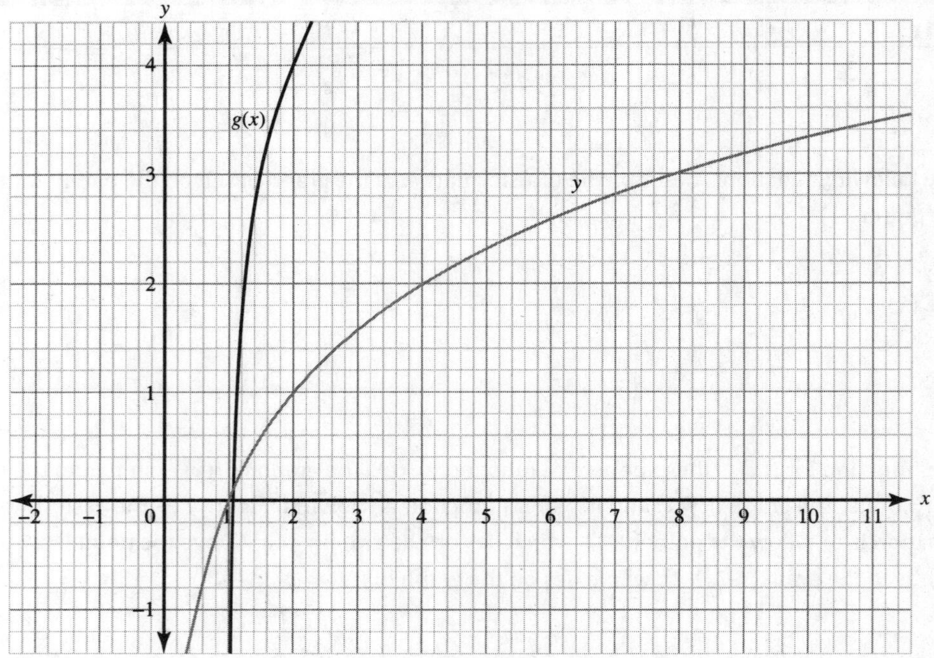

Figure 7.10

Inverse Equations of Transformed Exponential and Logarithmic Functions

To find the inverse equation of a transformed exponential and logarithmic function, use the technique discussed in Section 4.5. First switch the input and output variables. Then use logarithmic and/or exponential laws to rewrite the equation in terms of the independent variable.

Example

For each of the following functions, find the equation of its inverse in terms of x.

1. $f(x) = e^{x+7} - 3$
2. $g(x) = \log_2(x - 1) + 4$

Solution

1. Switch the input and output variables:

$$x = e^{y+7} - 3$$

Isolate the exponential expression:

$$x + 3 = e^{y+7}$$

Rewrite the exponential equation as a logarithmic equation:

$$y + 7 = \ln(x + 3)$$

Solve for y:

$$y = \ln(x + 3) - 7$$

$$f^{-1}(x) = \ln(x + 3) - 7$$

Figure 7.11 shows $f(x)$ and $f^{-1}(x)$ as reflections of each other over the line $y = x$, which confirms that the equations are inverses of one another.

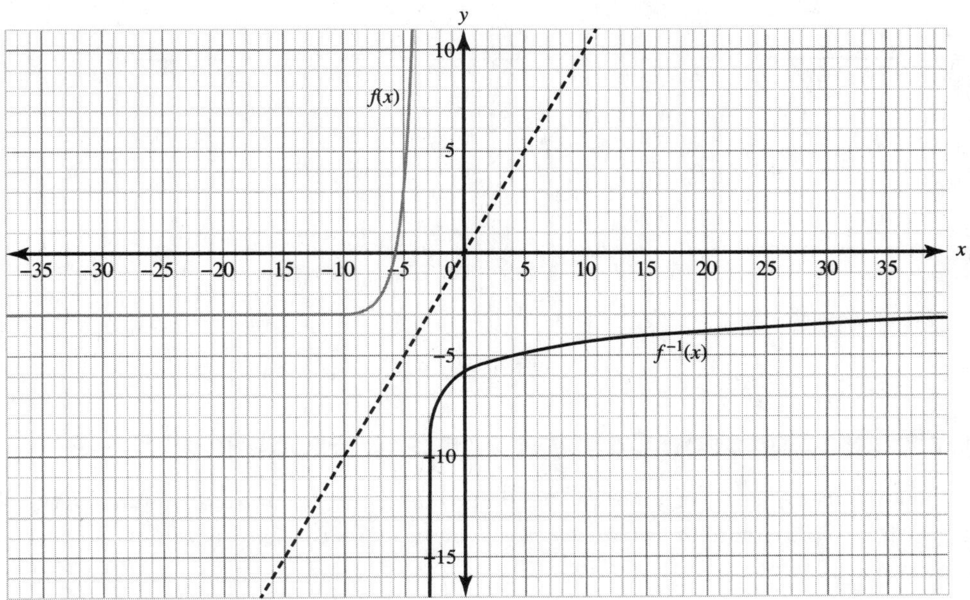

Figure 7.11

2. Switch the input and output variables:

$$x = \log_2(y - 1) + 4$$

Isolate the logarithmic expression:

$$x - 4 = \log_2(y - 1)$$

Rewrite the logarithmic equation as an exponential equation:

$$y - 1 = 2^{x-4}$$

Solve for y:

$$y = 2^{x-4} + 1$$

$$g^{-1}(x) = 2^{x-4} + 1$$

Figure 7.12 shows $g(x)$ and $g^{-1}(x)$ as reflections of each other over the line $y = x$, which confirms that the equations are inverses of one another.

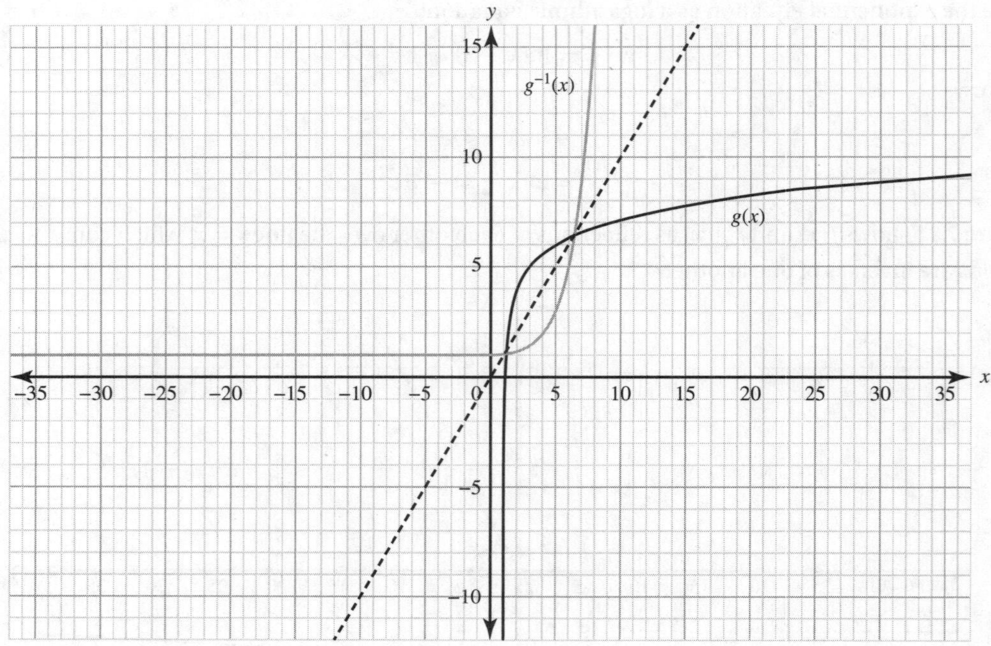

Figure 7.12

7.6 Logarithmic Function Context and Data Modeling

Logarithmic functions are inverses of exponential functions. They can be used to model situations involving proportional growth, or repeated multiplication, where the input values change proportionally over equal-length output value intervals.

Logarithmic regressions are mostly used for phenomena that grow quickly at first and then slow down over time, but the growth continues to increase without bound. There are a few different ways to find the equation of a logarithmic best-fit model. Each is discussed in this section.

A logarithmic function model can be constructed from an appropriate proportion and a real zero or from two input-output pairs.

Example

The Richter magnitude scale is a device used to compare the intensity of earthquakes. Table 7.10 shows the relationship between earthquake magnitude and the intensity of an earthquake as measured by ground motion.

Table 7.10

Magnitude (x)	Ground Motion (y)
1	10
2	100
3	1,000
4	10,000
5	100,000
6	1,000,000

(a) Model the data with an exponential function.

(b) Model the data with a logarithmic function.

✓ Solution

(a) To create an exponential function to model the data, use $f(x) = a \cdot b^x$ and ordered pairs from the table to solve for a and b.

Using (1, 10) and (2, 100), the following system of equations can be solved:

$$10 = a \cdot b^1$$

$$100 = a \cdot b^2$$

Dividing the two equations yields:

$$10 = b$$

Substituting this value into the first system of equations yields:

$$10 = a \cdot (10)^1$$

$$1 = a$$

An exponential model is $f(x) = 1 \cdot 10^x$ or $y = 10^x$.

(b) To create a logarithmic function to model the data, use the equation found in (a) and write it in logarithmic form. A logarithmic model is $x = \log y$.

Logarithmic function models can be constructed by applying transformations to $f(x) = a\log_b x$ or $f(x) = a\ln x$ based on the context or data set. Often the natural logarithmic function is used when modeling natural phenomena.

› Example

Given the data in Table 7.11, construct a scatterplot of the data, state which function is the best model for the data, and write its equation.

Table 7.11

x	1	2	3	4
y	3	4.4	5.2	5.8

✓ Solution

The scatterplot of the data is shown in Figure 7.13.

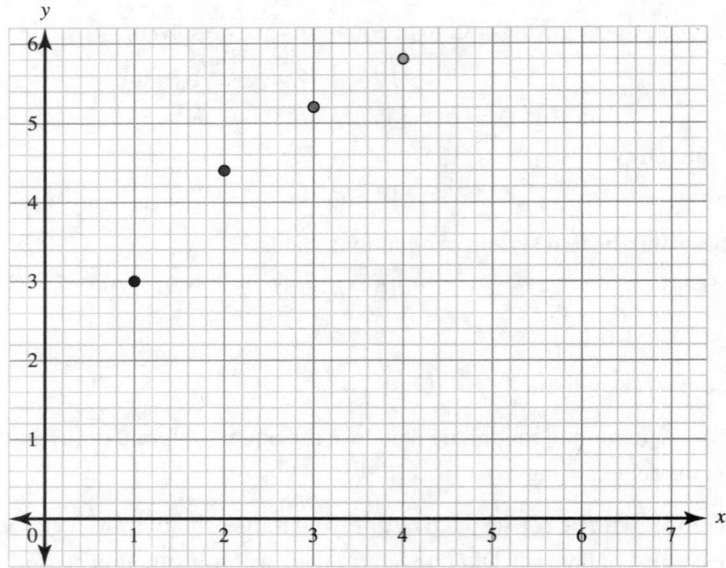

Figure 7.13

The scatterplot looks most like a logarithmic function with its quick growth in the beginning and gradual increase toward the end. Figure 7.14 is the graph of the basic parent function $y = \ln(x)$ graphed with the scatterplot.

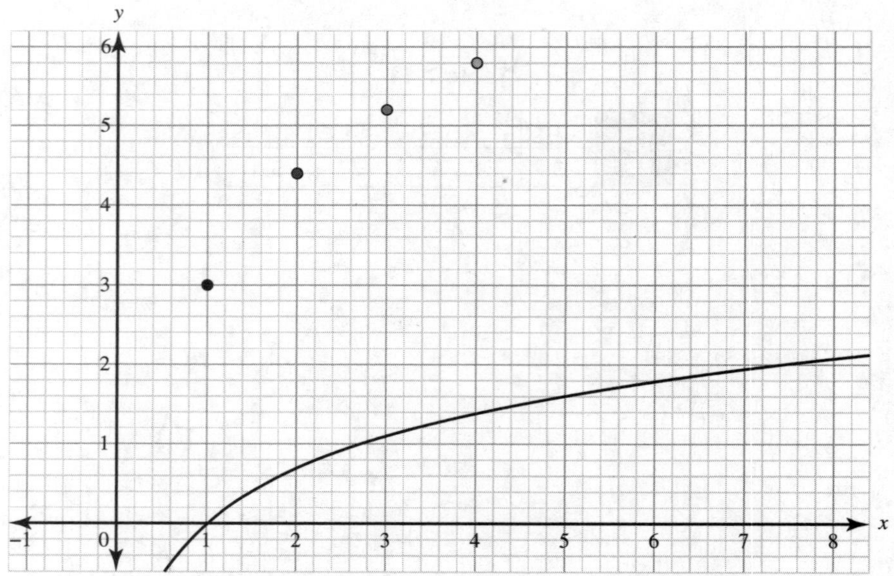

Figure 7.14

If the graph of $y = \ln x$ is shifted up 3 units, the point (1, 0) will correspond to the point (1, 3) on the scatterplot, as shown in Figure 7.15.

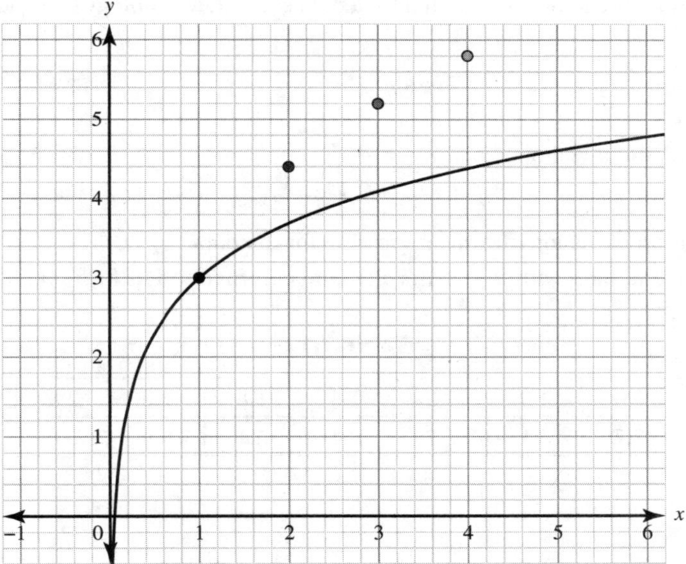

Figure 7.15

Choosing another point on the scatterplot can help to find the factor of any vertical stretching. Substitute the point (2, 4.4) into the general equation $y = a \ln x + 3$:

$$4.4 = a \ln(2) + 3$$
$$1.4 = a \ln(2)$$
$$a = \frac{1.4}{\ln(2)} \approx 2$$

The graph of the equation $y = 2 \ln x + 3$ along with the scatterplot is shown in Figure 7.16.

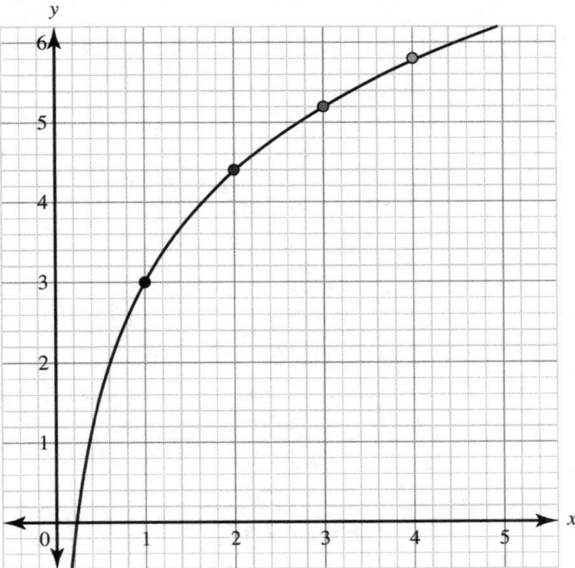

Figure 7.16

This equation appears to be a best-fit model for the given data.

Logarithmic function models can be constructed for a data set with technology using logarithmic regressions.

Example

Given the data in Table 7.12, do the following.

(a) Create a scatterplot.
(b) Determine the equation of the logarithmic regression for the data.
(c) Graph the regression equation with the scatterplot, and comment on its fit.

Table 7.12

x	0.5	0.7	0.9	1.0	1.2	1.4	1.8	2.0	2.3	2.7	3.2	3.8
y	0.5	1.6	2.7	3.1	3.7	4.4	5.1	5.8	6.4	7.0	7.7	8.3

Solution

(a) The scatterplot of the data is shown in Figure 7.17.

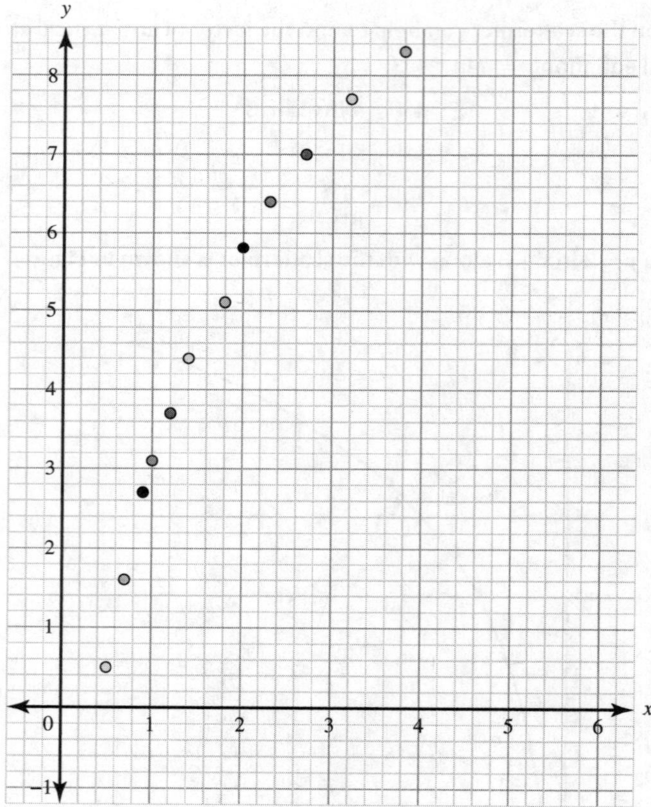

Figure 7.17

The scatterplot looks most like a logarithmic function with its quick growth in the beginning and gradual increase toward the end.

(b) Using a graphing calculator, enter the data into lists L1 and L2, press STAT, select CALC, select 9:LnReg, and calculate the logarithmic regression model.

The screen will display:

$$y = a + b \ln x$$
$$a = 3.075114904$$
$$b = 3.904040334$$

Therefore, the logarithmic regression model is $y = 3.075 + 3.904 \ln x$.

(c) The graph of the regression model and the scatterplot is shown in Figure 7.18.

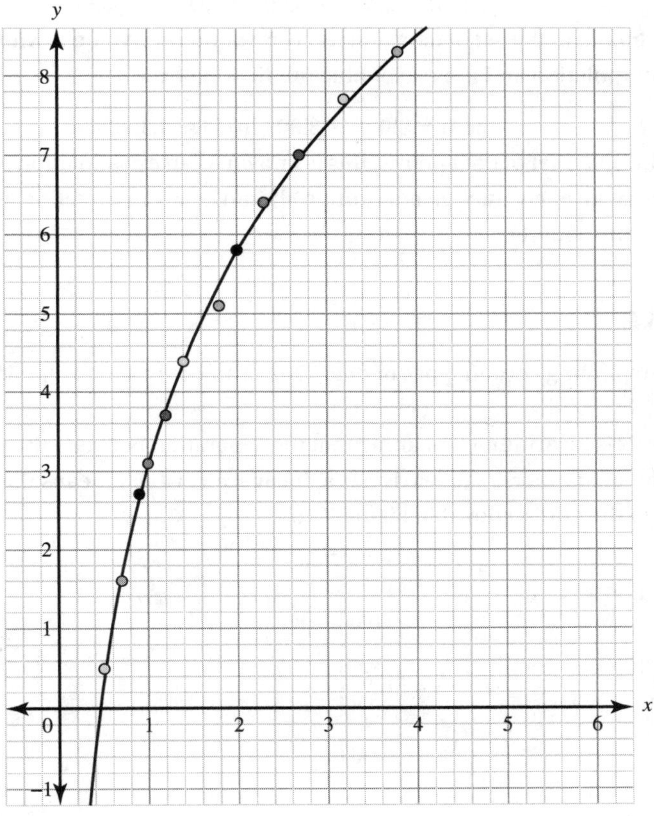

Figure 7.18

The regression equation seems to be a good fit for the scatterplot because it passes through many of the data points.

Logarithmic function models can be used to predict values for the dependent variable.

Example

Use the logarithmic regression model found in the previous example to do the following.

(a) Evaluate $y(3)$ and comment on the reasonableness of its value.
(b) Evaluate $y(5)$ and comment on the reasonableness of its value.

Solution

(a)
$$y(3) = 3.075 + 3.904 \ln(3) \approx 7.36398$$
Since 3 is within the domain of the set of data and its corresponding y-value is within the range of the data, this is known as an interpolation and is a reasonable prediction of $x = 3$.

(b)
$$y(5) = 3.075 + 3.904 \ln(5) \approx 9.3582456$$
Since 5 is outside of the given set of data points, this is known as an extrapolation. While its value seems to follow the trend of the range of y-values given, caution should be given to stating its reasonableness without knowing the context of the data set.

7.7 Semi-log Plots

Logarithms are helpful when plotting certain types of data. One example of a log graph is called a semi-log graph or semi-log plot.

In a semi-log plot the y-axis is logarithmic, meaning that the separation between the ticks on the graph is proportional to the logarithm of numbers. The x-axis has a linear scale, which means that the ticks on the graph are evenly spaced. An example of the grid is shown in Figure 7.19.

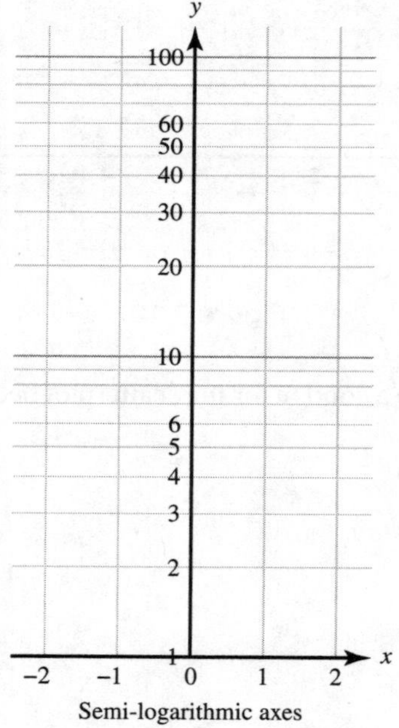

Figure 7.19

A semi-log plot is useful when graphing exponential functions where one variable covers a large range of values. When the exponential function $y = a \cdot b^x$ is graphed on a semi-log grid, this function produces a straight line with slope $= \log(b)$ and y-intercept $(0, a)$.

Example

Graph the function $y = 5^x$ on a Cartesian coordinate plane and on a semi-log axis.

✓ Solution

To graph this function, first create a table of values as shown in Table 7.13. Then plot the points.

Table 7.13

x	0	1	2	3	4	5	6
y	1	5	25	125	625	3125	15625

It can be seen from Table 7.13 that as the x-values increase so do the y-values. When $x = 3$, the function's values become so large that it is hard to graph the points on a regular coordinate plane. On a semi-log axis, the y-values are spaced differently. The range increases but at a slower rate, allowing the points to fit on the graph without having such a large scale.

Graphs of the points from Table 7.13 on the different axes are shown in Figure 7.20.

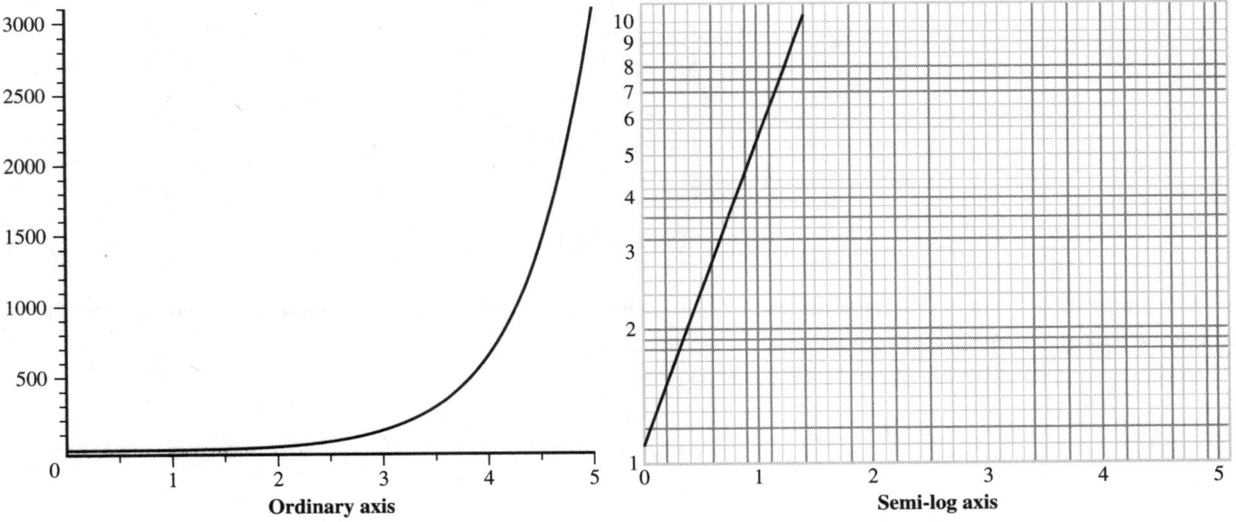

Figure 7.20

On the ordinary xy-axes, the graph looks exponential with a greater and greater increase toward the right. On the semi-log graph, the exponential function appears to be linear.

Example

Given the data in Table 7.14 about the growth of a microbial population at different temperatures, write an equation that models the data. Graph the data on a Cartesian coordinate plane and a semi-log plot. Comment on the two graphs.

Table 7.14

T (°C) Temperature	−2	−1	0	1	2	3	4
P Population	0.020	0.143	1	7	49	343	2401

Solution

Since the population increases 7 times for each 1 degree rise in temperature, the data can be modeled using the function $P = 7^T$.

In Figure 7.21, the data are plotted on the Cartesian coordinate plane.

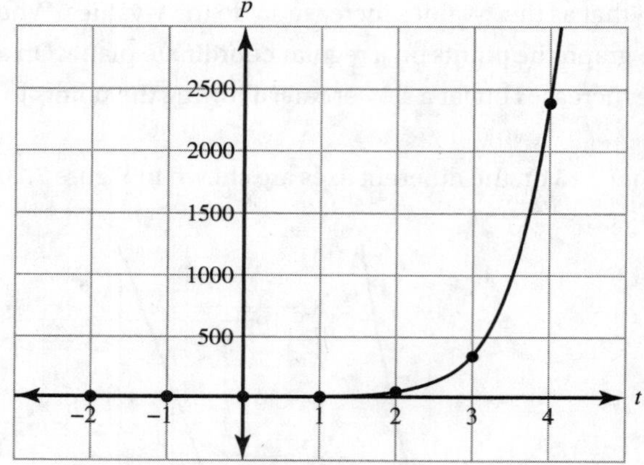

Figure 7.21

This graph shows a lot of detail for values of $T > 1$. However, for values of $T < 1$, the points are too close on the T-axis even to make out their P-values.

In Figure 7.22, the data are plotted on a semi-log plot.

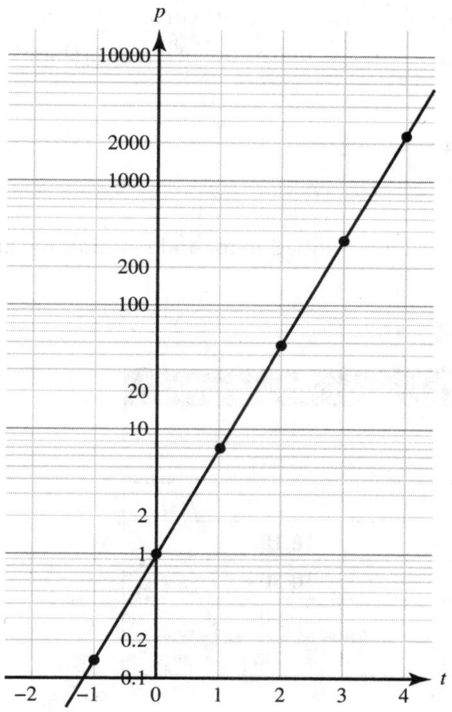

Figure 7.22

On this graph, a lot more information can be seen for smaller values of T and P.

Benefits of Semi-log Plots

Semi-log plots are helpful because exponential functions now have linear characteristics; more detail can be seen in graphs where there is a very wide range of values but some of the data are close together.

Linearizing Exponential Data

Now that it has been determined that exponential models appear linear on semi-log plots, this section discusses how to create a linear equation when given an exponential function.

Start with the general form of the exponential function:

$$y = a \cdot e^{kx}$$

Take the natural logarithm of both sides of the equation:

$$\ln y = \ln (a \cdot e^{kx})$$

Use log properties:

$$\ln y = \ln a + \ln e^{kx}$$

$$\ln y = \ln a + kx \cdot \ln e$$

$$\ln y = \ln a + kx$$

Rearrange the sum of the terms:

$$\ln y = kx + \ln a$$

Compare this equation to the general linear function, $y = mx + b$. Notice that the slope corresponds to k and the intercept corresponds to $\ln a$.

This is also true for the general form of the exponential function, $y = a \cdot b^x$. Its corresponding linear model for the semi-log plot is $y = (\log_n b)x + \log_n a$, where $n > 0$ and $n \neq 1$. The linear rate of change (slope) is $\log_n b$, and the initial linear value (intercept) is $\log_n a$.

Example

Given the data in Table 7.15 for the decay of a radioisotope, find an equation to model its half-life and then linearize the equation.

Table 7.15

t, Time in Seconds	*a*, Amount in Grams
0	100
100	56.657
200	32.100
300	18.187
400	10.304
500	5.838
600	3.308

Solution

The equation for decay based on half-life is $a = a_0\left(\frac{1}{2}\right)^{\frac{t}{n}}$, where:

$a =$ amount of the radioisotope
$a_0 =$ initial amount of the radioisotope
$t =$ time
$n =$ half-life

Linearize the equation using logarithms so that it can be plotted as a straight line.
Take log base 2 of both sides:

$$\log_2 a = \log_2\left(a_0\left(\frac{1}{2}\right)^{\frac{t}{n}}\right)$$

Use logarithmic properties to simplify:

$$\log_2 a = \log_2 a_0 + \log_2\left(\left(\frac{1}{2}\right)^{\frac{t}{n}}\right)$$

$$\log_2 a = \log_2 a_0 + \frac{t}{n}\log_2\left(\frac{1}{2}\right)$$

$$\log_2 a = \log_2 a_0 + \frac{t}{n}\log_2(2^{-1})$$

$$\log_2 a = \log_2 a_0 + \frac{t}{n}(-1)$$

$$\log_2 a = \log_2 a_0 - \frac{t}{n}$$

Therefore:

$$\log_2 a = -\frac{1}{n}t + \log_2 a_0$$

Compare this equation to $y = mx + b$:

$$y = \log_2 a$$

$$m = -\frac{1}{n}$$

$$x = t$$

$$b = \log_2 a_0$$

The slope of the line is $m = -\frac{1}{n}$, which can be used to find the half-life. Using values from Table 7.15, the line can either be graphed using a semi-log plot or the slope can be solved for algebraically.

The points used are (0, 100) since that is the initial amount. So $a_0 = 100$ and therefore the y-intercept is $b = \log_2 100$.

Substitute the point (100, 56.657) into the equation:

$$\log_2(56.657) = -\frac{1}{n}(100) + \log_2(100)$$

Solve for n:

$$\log_2(56.657) - \log_2(100) = -\frac{1}{n}(100)$$

$$\log_2\left(\frac{56.657}{100}\right) = -\frac{1}{n}(100)$$

$$\frac{\log_2(0.56657)}{100} = -\frac{1}{n}$$

$$n = -\frac{100}{\log_2(0.56657)} \approx 121.9997394$$

The half-life is approximately 122 days.

Multiple-Choice Questions

1. If $\log a = 2$ and $\log b = 4$, which of the following is the numerical value of $\log\left(\frac{a^3}{\sqrt{b}}\right)$?

   (A) 2
   (B) 4
   (C) 6
   (D) 8

2. The expression $3\log x - \frac{1}{3}\log y$ is equivalent to which of the following?

   (A) $\frac{\log x^3}{\log \sqrt[3]{x}}$
   (B) $\frac{3\log x}{\frac{1}{3}\log y}$
   (C) $\log \frac{x^3}{\sqrt[3]{y}}$
   (D) $\log \frac{x}{y}$

3. If $\log N = \frac{1}{2}(\log a - 2\log b) + \log c$, what does N equal?

   (A) $\frac{\sqrt{ac}}{b}$
   (B) $\frac{c\sqrt{a}}{b}$
   (C) $\sqrt{\frac{a+c}{b^2}}$
   (D) $\sqrt{\frac{a}{b^2}} + c$

4. If $f(x) = \log(x^2)$ and $g(x) = \frac{x}{10}$, then $f(g(x))$ equals which of the following?

   (A) $2(\log x - 1)$
   (B) $\frac{\log x}{50}$
   (C) $\frac{1}{2}(\log x + 1)$
   (D) $2\log x + 1$

5. What is the domain of the graph of $f(x) = c\log_b(x - a)$, where $a > 0$ and $c > 1$?

   (A) $(0, \infty)$
   (B) $(1, \infty)$
   (C) (a, ∞)
   (D) (a^c, ∞)

6. If $\log 30 = b$, then which of the following expressions is equivalent to $(b - 1)^2$?

   (A) $(\log 29)^2$
   (B) $2(\log(30) - 1)$
   (C) $(\log 3)^2$
   (D) $\log 899$

7. If $\log_3(x^2 + 5) = 2$, what is the value of x?

   (A) $x = -2, 2$
   (B) $x = -2$
   (C) $x = 2$
   (D) No solution

8. Which function is the inverse of $y = \log(x - 5)$?

   (A) $y = x^{10} - 5$
   (B) $y = 10^x - 5$
   (C) $y = 10^x + 5$
   (D) $y = \left(\frac{1}{10}\right)^x - 5$

9. A company tracked the sales of their app for the first 14 days after it was released. The data are shown in the table below. The company noticed that the sales slowed over time and wanted to write an equation to model this data to predict future sales. Which of the following regression equations best models the data in the given table?

x, day	1	2	3	4	5	6	7	8	9	10	11	12	13	14
y, total number of sales	217	552	807	897	1,001	1,080	1,126	1,162	1,218	1,275	1,290	1,335	1,367	1,393

(A) $y = 269.6 + 434.5 \ln(x)$
(B) $y = 297.3 + 331.7 \ln(x)$
(C) $y = 434.5 + 269.6 \ln(x)$
(D) $y = 331.7 + 297.3 \ln(x)$

10. Which scatterplot shows data that could be modeled by a logarithmic regression?

(A)
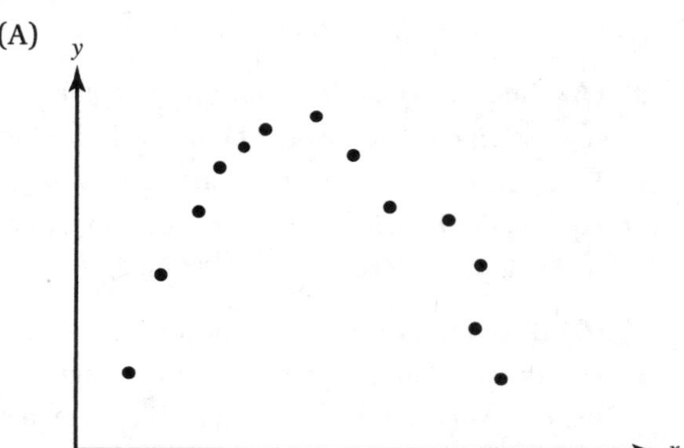

(B)
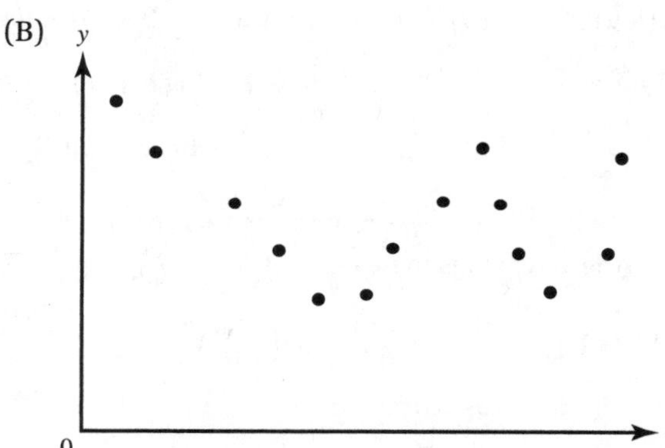

(C)
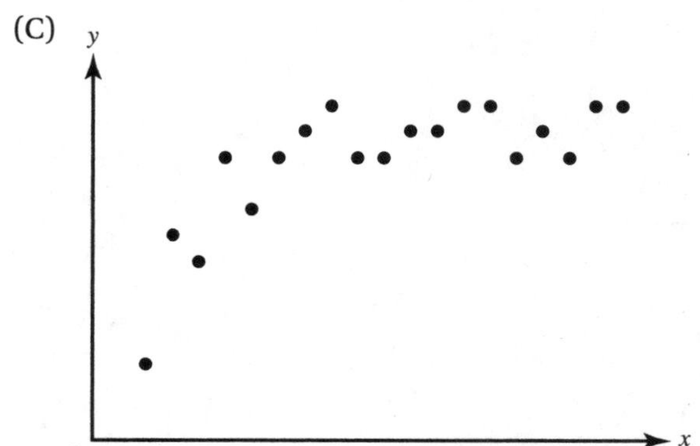

(D)
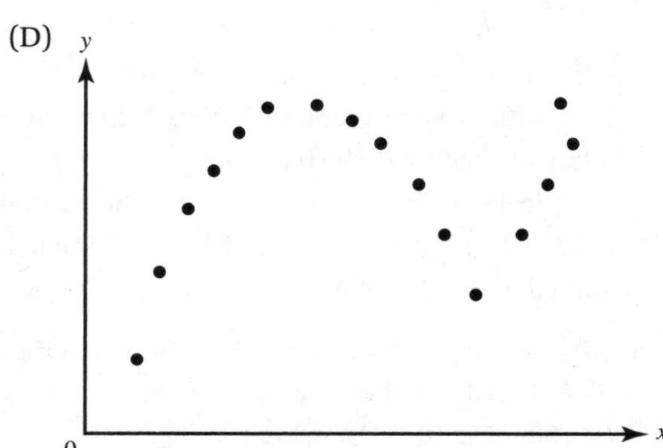

Answer Explanations

1. **(B)** Using logarithmic properties to expand the given expression,

$\log\left(\frac{a^3}{\sqrt{b}}\right) = \log a^3 - \log\sqrt{b} = 3\log a - \frac{1}{2}\log b.$

Substituting in the numerical values for each log expression gives:

$$3\log a - \frac{1}{2}\log b = 3(2) - \frac{1}{2}(4) = 6 - 2 = 4$$

2. **(C)** Use logarithmic properties to condense the given expression.

$$3\log x - \frac{1}{3}\log y = \log x^3 - \log\sqrt[3]{y} = \log\frac{x^3}{\sqrt[3]{y}}$$

3. **(B)** Use logarithmic properties to condense the right side of the given equation:

$$\frac{1}{2}(\log a - 2\log b) + \log c = \frac{1}{2}\log a - \frac{1}{2}\cdot 2\log b + \log c$$

$$= \log\sqrt{a} - \log b + \log c$$

$$= \log\frac{\sqrt{a}}{b} + \log c = \log\frac{c\sqrt{a}}{b}$$

So $\log N = \log\frac{c\sqrt{a}}{b}$, and therefore the only way for these two logs to be equal is if $N = \frac{c\sqrt{a}}{b}$.

4. **(A)** $f(g(x)) = f\left(\frac{x}{10}\right) = \log\left(\left(\frac{x}{10}\right)^2\right)$

Use log properties:

$$\log\left(\left(\frac{x}{10}\right)^2\right) = 2\log\left(\frac{x}{10}\right) = 2(\log x - \log 10)$$

$$= 2(\log x - 1)$$

5. **(C)** The domain of the basic parent function $\log x$ is $(0, \infty)$. The graph of $f(x) = c\log_b(x - a)$ is a transformation of the basic parent function with a vertical stretch by a factor of c and a horizontal shift to the right a units. Therefore, the domain is shifted to the right a units.

6. **(C)** If $\log 30 = b$ then $(b - 1)^2 = (\log 30 - 1)^2$. Since $\log 10 = 1$, then $(\log 30 - 1)^2 = (\log 30 - \log 10)^2$. Using log laws, $(\log 30 - \log 10)^2 = \left(\log\frac{30}{10}\right)^2 = (\log 3)^2$.

7. **(A)** Rewriting the logarithmic equation as an exponential equation shows that $\log_3(x^2 + 5) = 2$ is equivalent to $3^2 = x^2 + 5$. Solve the equation $x^2 + 5 = 9$. Isolating the x-term gives $x^2 = 4$, and so $x = \pm 2$. Check each solution in the original equation:

$$\log_3((-2)^2 + 5) = \log_3(4 + 5) = \log_3(9) = 2$$

$$\log_3((2)^2 + 5) = \log_3(4 + 5) = \log_3(9) = 2$$

Both solutions check in the original equation.

8. **(C)** To find the equation of an inverse, switch the x- and y-variables and solve for y. The inverse equation is $x = \log(y - 5)$. To solve for y, rewrite this equation in exponential form where the base is 10, $y - 5 = 10^x$. Isolating y gives $y = 10^x + 5$.

9. **(A)** Since the data slow over time, a logarithmic regression equation would be the best-fit model. Enter the data from the table into a graphing calculator. Find the logarithmic regression equation, which is $y = 269.639 + 434.4788\ln x$.

10. **(C)** The graphs are scatterplots on a Cartesian coordinate plane. To fit a logarithmic regression, the data should be increasing with the greatest increase in the beginning and then slowing down.

UNIT 3
Trigonometric and Polar Functions

8

Trigonometric Functions

Learning Objectives

In this chapter, you will learn:

- ➔ Periodic behavior through trigonometric functions
- ➔ Properties of trigonometric functions
- ➔ How to graph trigonometric functions
- ➔ How to model periodic behavior using trigonometric functions
- ➔ How to solve trigonometric equations and inequalities
- ➔ How to prove trigonometric identities

Before starting this chapter, it is important to have a good understanding of the necessary background knowledge in trigonometry. It is reviewed in the Prior Knowledge section of this book.

8.1 Periodic Phenomena

Periodic phenomena occur in the physical world. Examples include the seasonal variations in the climate, the number of daylight hours, the phases of the moon, and so on. These phenomena illustrate variable behavior that is repeated over time. This repetition may be called periodic, oscillatory, or cyclic in different situations.

The values of a periodic relationship repeat over regular intervals. Each complete pattern of values is known as a cycle.

Table 8.1 shows input and output values for a periodic relationship.

Table 8.1

x	0	1	2	3	4	5	6	7	8
y	0	1	0	-1	0	1	0	-1	0

Table 8.1 shows that with every increase of the x-value, the output values cycle from 0 to 1 to 0 to -1 and back to 0 and then repeat that same pattern. Without knowing more about the equation, a possible graph of the table of values is shown in Figure 8.1.

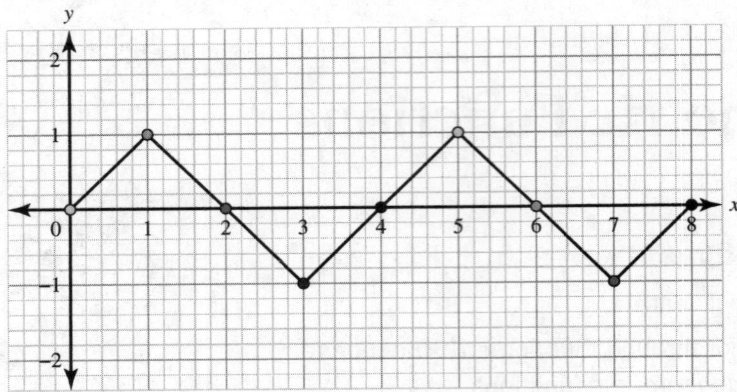

Figure 8.1

The graph indicates that one complete cycle of the relationship is in the interval [0, 4]. Then it repeats and another complete cycle of the relationship is in the interval [4, 8]. Recognizing the pattern allows the graph to be continued in either direction. Figure 8.2 shows the graph drawn over the interval from [−4, 12].

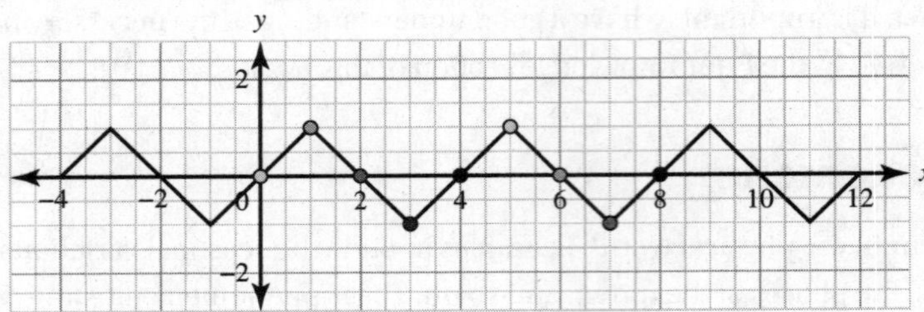

Figure 8.2

Figure 8.2 now has 4 cycles of the pattern shown in Figure 8.1.

The period of the function is defined as the interval in which one full cycle of a periodic relationship can be identified. The period of the function can also be defined as the smallest positive value k such that $f(x + k) = f(x)$ for all x in the domain.

In our previous example, the period is $k = 4$ because that is the length of the interval for one full cycle of the graph. It can also be seen that if the graph is shifted 4 units to the left, the transformed graph would look exactly like the original graph; therefore, $f(x + 4) = f(x)$.

> The graph of a periodic relationship can be constructed from the graph of a single cycle of the relationship.

> If k is the period of a function, the behavior of a periodic function is completely determined by any interval of width k.

Example

Given the graph of the function in Figure 8.3 over the domain [−4, 4], do the following.

(a) State whether the function has periodic behavior by identifying its period.
(b) State where the function is increasing.
(c) State where the function is decreasing.
(d) State where the function is concave up.
(e) State where the function is concave down.
(f) State the points of inflection.

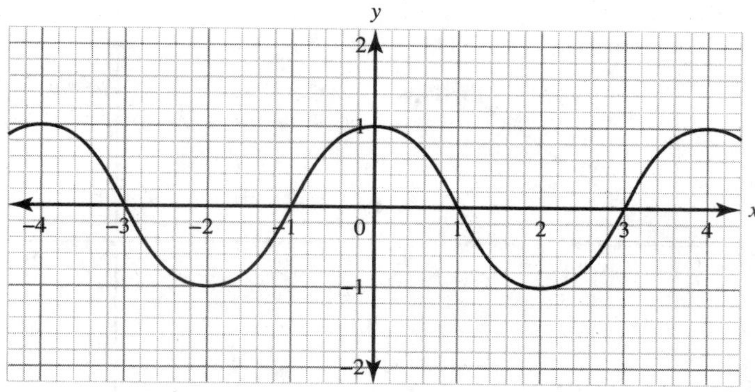

Figure 8.3

Solution

(a) The function has periodic behavior because it repeats its cycle every 4 units along the x-axis. Therefore, the period is 4.

(b) The function is increasing in the intervals $(-2, 0) \cup (2, 4)$.

(c) The function is decreasing in the intervals $(-4, -2) \cup (0, 2)$.

(d) The function is concave up in the intervals $(-3, -1) \cup (1, 3)$.

(e) The function is concave down in the intervals $[-4, -3) \cup (-1, 1) \cup (3, 4]$.

(f) The points of inflection are at $x = -3, -1, 1, 3$.

Behaviors of Periodic Functions

A few behaviors of periodic functions can be noted from the previous example:

- The period can be estimated by investigating successive equal-length output values and finding where the pattern begins to repeat.
- Intervals found in one period of a periodic function where the function increases, decreases, is concave up, or is concave down will be in every period of the function.

8.2 Sine, Cosine, and Tangent

Trigonometric functions are functions that have a periodic relationship. This family of functions is dependent on angle measures.

In the coordinate plane, an angle is in standard position when its vertex is at the origin and one of its sides, called the initial side, remains fixed on the positive x-axis. The side of the angle that rotates is called the terminal side.

- If the terminal side rotates in a counterclockwise direction, as shown in Figure 8.4 on the left, the angle of rotation is positive.
- If the terminal side rotates in a clockwise direction, as shown in Figure 8.4 on the right, the angle of rotation is negative.

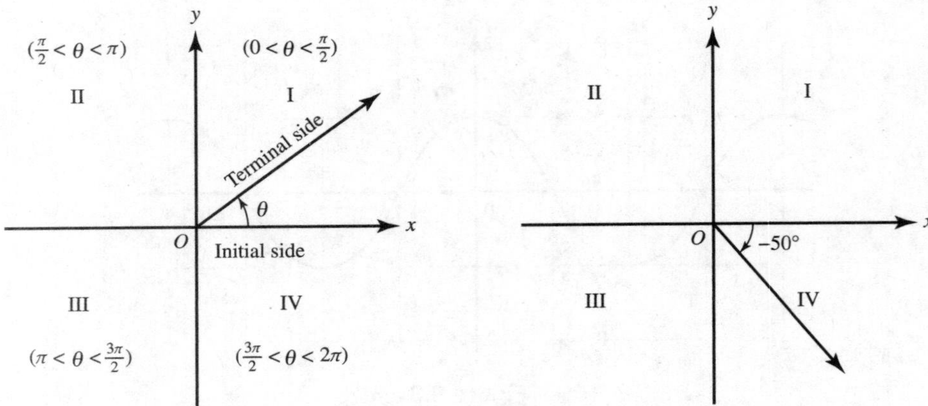

Figure 8.4

Angle measures can be expressed in units of degrees or in real-number units called radians. Degrees are based on fractional parts of a circular revolution. Radian measure compares the length of an arc that a central angle of a circle subtends (cuts off) with the radius of the circle. The Greek letter θ, read as theta, is commonly used to represent an angle of unknown measure.

In Figure 8.5, angle θ cuts off an arc of circle O that has the same length as the radius of the circle. Therefore, angle θ measures 1 radian.

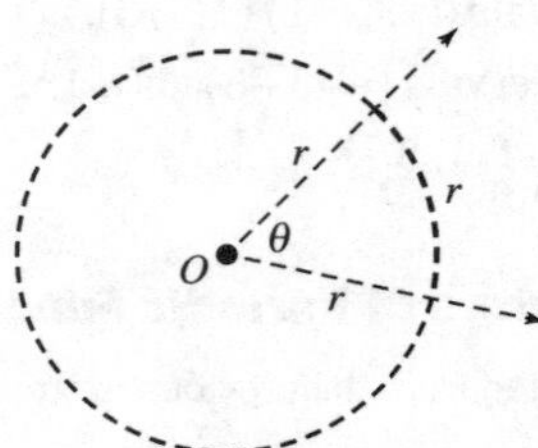

Figure 8.5 Defining a radian in a circle with radius r

Since the total number of radii that can be marked off along the circumference of the circle is 2π, the radian measure of a circle is 2π. The degree measure of a circle is 360°. Therefore, 2π radians $= 360°$, so π radians $= 180°$. This relationship helps to convert from one unit of angle measure to the other.

Radian and Degree Conversions

- To convert from degrees to radians, multiply the number of degrees by $\frac{\pi}{180°}$.
- To convert from radians to degrees, multiply the number of radians by $\frac{180°}{\pi}$.

Example

Convert 60° to radian measure.

Solution

To convert 60° to radian measure, multiply by $\frac{\pi}{180°}$:

$$60° \cdot \frac{\pi}{180°} = \frac{60\pi}{180} = \frac{\pi}{3} \text{ radians}$$

Example

Convert $\frac{7\pi}{12}$ radians to degrees.

Solution

To convert $\frac{7\pi}{12}$ radians to degrees, multiply by $\frac{180°}{\pi}$:

$$\frac{7\pi}{12} \cdot \frac{180}{\pi} = \frac{1{,}260\pi}{12\pi} = 105°$$

Important Angle Measures

Quadrant I:

$30° = \frac{\pi}{6}$ radians

$45° = \frac{\pi}{4}$ radians

$60° = \frac{\pi}{3}$ radians

Multiples:

$120° = 2(60°) = \frac{2\pi}{3}$ radians

$150° = 5(30°) = \frac{5\pi}{6}$ radians

$225° = 5(45°) = \frac{5\pi}{4}$ radians

Quadrantal angles: angles whose terminal ray is either the *x*- or *y*-axis.

$0° = 0$ radians

$90° = \frac{\pi}{2}$ radians

$180° = \pi$ radians

$360° = 2\pi$ radians

Coterminal Angles

Angles in standard position that share a terminal ray are known as coterminal angles. Angles of rotation of 50° and 410° are coterminal angles, as are angles 210° and −150°, as shown in Figures 8.6 and 8.7.

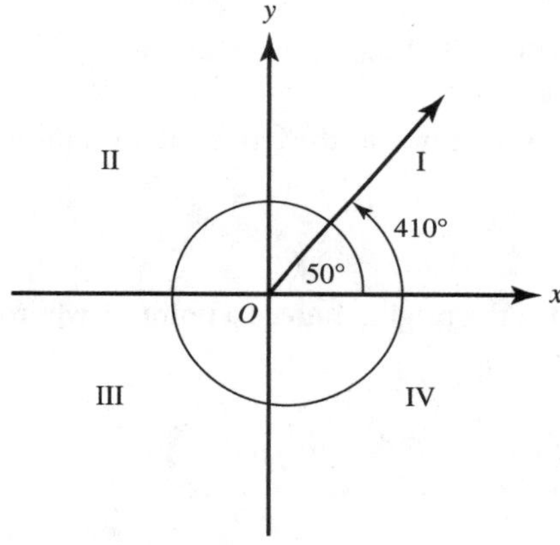

Figure 8.6 Coterminal angles

Converting the coterminal angles to radian measure gives $50° = \frac{5\pi}{18}$ and $410° = \frac{41\pi}{18}$. Taking the difference between the two angles in radian measure results in $\frac{41\pi}{18} - \frac{5\pi}{18} = 2\pi$, which equals 360°.

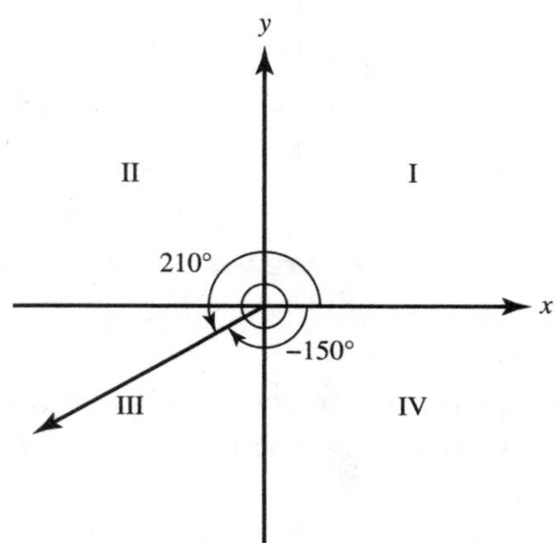

Figure 8.7 Coterminal angles

Converting the coterminal angles to radian measure gives $-150° = -\frac{5\pi}{6}$ and $210° = \frac{7\pi}{6}$. Taking the difference between the two angles in radian measure results in $\frac{7\pi}{6} - \frac{-5\pi}{6} = 2\pi$, which equals 360°.

The above shows that angles in standard position that share a terminal ray differ by an integer number of revolutions.

The unit circle is the circle whose center is at the origin, *O*, and whose radius is 1, as shown in Figure 8.8.

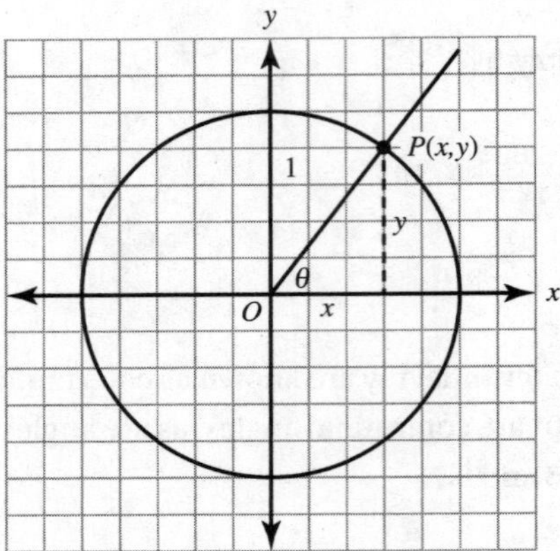

Figure 8.8 The unit circle

Since the unit circle has a radius of 1, the radian measure is the same as the length of the subtended arc.

Sine, Cosine, and Tangent of an Angle

Given an angle in standard position and a circle centered at the origin, there is a point, *P*, where the terminal ray intersects the circle, as shown in Figure 8.9.

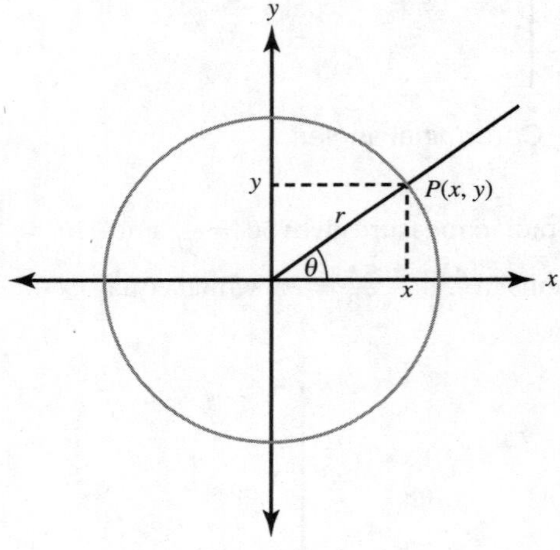

Figure 8.9

The number r denotes the distance from P to the origin.

The sine of the angle θ is the ratio of the vertical displacement of P from the x-axis (which is represented as the y-value in Figure 8.9) to the distance between the origin point and point P (which is represented as the r-value in Figure 8.9). In other words:

$$\sin\theta = \frac{y}{r}$$

The cosine of the angle θ is the ratio of the horizontal displacement of P from the y-axis (which is represented as the x-value in Figure 8.9) to the distance between the origin point and point P (which is represented as the r-value in Figure 8.9). In other words:

$$\cos\theta = \frac{x}{r}$$

The tangent of the angle θ is the slope, if it exists, of the terminal ray. Since the slope of the terminal ray is the ratio of the vertical displacement to the horizontal displacement over any interval, the tangent of the angle θ is the ratio of the y-coordinate to the x-coordinate of the point at which the terminal ray intersects the circle. In other words:

$$\tan\theta = \frac{y}{x}$$

In the case of the unit circle, the distance between the origin and the point on the circle is 1, as shown in Figure 8.10.

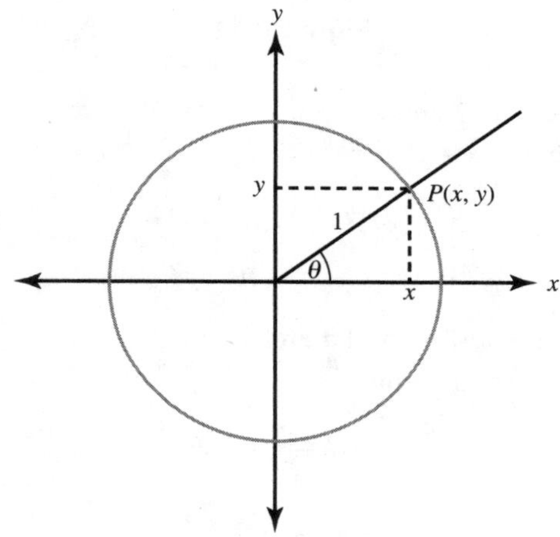

Figure 8.10

The unit circle changes the trigonometric relationships as follows:

$$\sin\theta = \frac{y}{r} = \frac{y}{1} = y$$

$$\cos\theta = \frac{x}{1} = x$$

$$\tan\theta = \frac{y}{x}$$

Therefore, for a unit circle, the sine of the angle is the y-coordinate of point P and the cosine of the angle is the x-coordinate of point P.

For a unit circle, the tangent of the angle is the ratio of the angle's sine to its cosine.

› Example

If $P(\sqrt{7}, -3)$ is a point on the terminal side of angle θ, find the values of $\sin\theta$ and $\cos\theta$.

✓ Solution

Since this was not given as a unit circle, create a model, as shown in Figure 8.11, to help determine the values of x, y, and r for the given point.

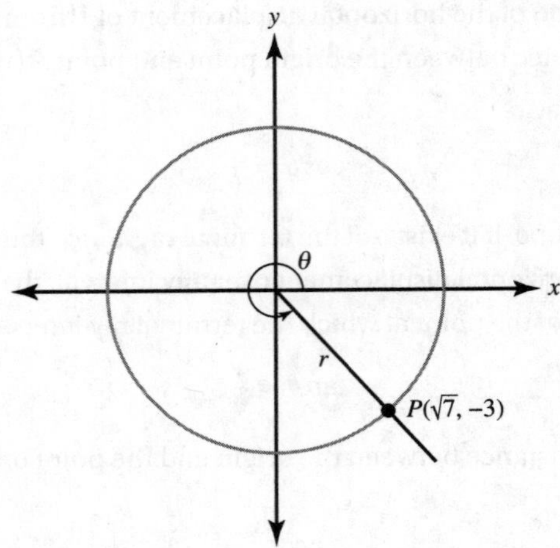

Figure 8.11

Based on the information provided, $x = +\sqrt{7}$ and $y = -3$.

Find r using the equation of the circle, $x^2 + y^2 = r^2$:

$$(\sqrt{7})^2 + (-3)^2 = r^2$$

$$16 = r^2$$

Since r represents a distance, it is always positive and therefore $r = 4$.

Use the coordinate definitions of sine and cosine:

$$\sin\theta = \frac{y}{r} = \frac{-3}{4}$$

$$\cos\theta = \frac{x}{r} = \frac{\sqrt{7}}{4}$$

Example

A point moves along the unit circle O in the counterclockwise direction from $A(1, 0)$ to $P(-0.6, -0.8)$. Find the values of $\sin\theta$, $\cos\theta$, and $\tan\theta$.

Solution

Since it was given that point P lies on the unit circle, $\cos\theta = x = -0.6$, $\sin\theta = y = -0.8$, and $\tan\theta = \frac{y}{x} = \frac{-0.8}{-0.6} = \frac{4}{3}$.

8.3 Sine and Cosine Function Values

The last section explained that given an angle of measure θ in standard position and a circle with radius r centered at the origin, there is a point, $P(x, y)$, where the terminal ray intersects the circle. As a result, $\sin\theta = \frac{y}{r}$ and $\cos\theta = \frac{x}{r}$. Solving these two equations for x and y, respectively, gives the following:

$$x = r\cos\theta$$

$$y = r\sin\theta$$

The coordinates of point $P(x, y)$ are $P(r\cos\theta, r\sin\theta)$.

In a unit circle since $r = 1$, the coordinates of point $P(x, y)$ are $P(\cos\theta, \sin\theta)$.

Example

Given the circles below, determine the coordinates of the indicated points.

1.

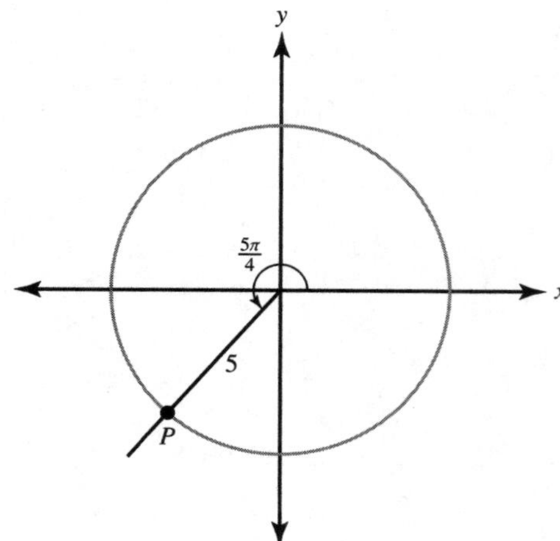

2.

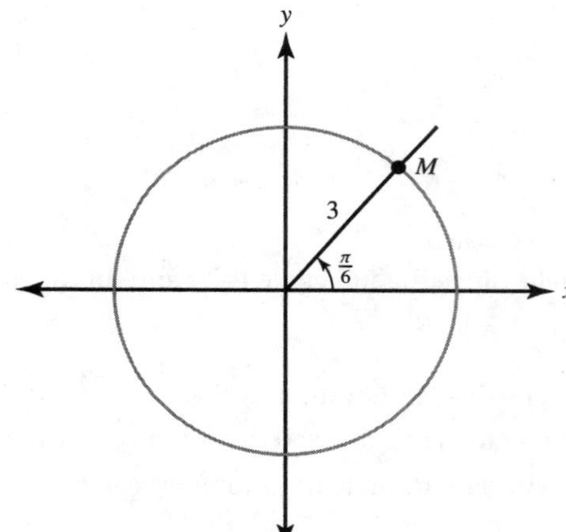

3.

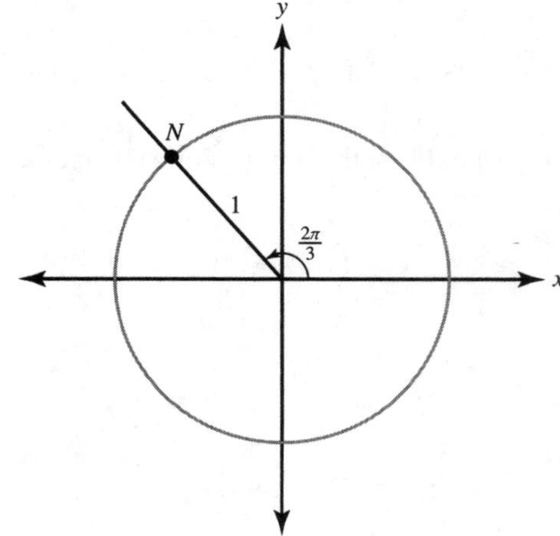

4.

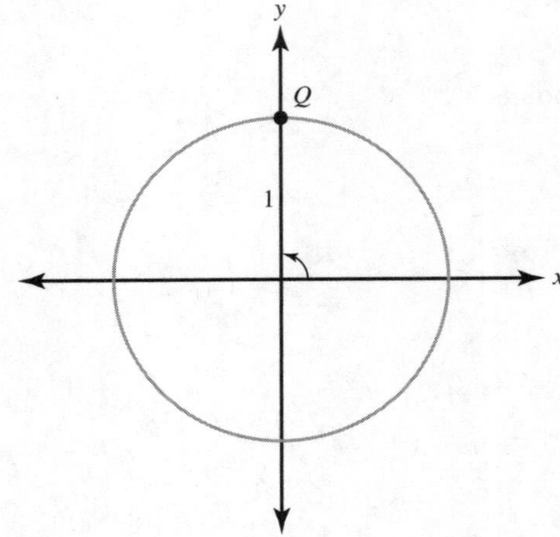

Solution

Since each of the angles is in standard position, the coordinates of the point where the terminal ray intersects the circle are in the form $(r\cos\theta, r\sin\theta)$.

1. $P(5\cos\frac{5\pi}{4}, 5\sin\frac{5\pi}{4})$
2. $M(3\cos\frac{\pi}{6}, 3\sin\frac{\pi}{6})$
3. $N(1\cos\frac{2\pi}{3}, 1\sin\frac{2\pi}{3}) = (\cos\frac{2\pi}{3}, \sin\frac{2\pi}{3})$
4. Point Q is located on the y-axis; this is known as a quadrantal angle. Since the full circle is 2π and this angle is a quarter of the way, $\theta = \frac{2\pi}{4} = \frac{\pi}{2}$. So $Q(1\cos\frac{\pi}{2}, 1\sin\frac{\pi}{2}) = (\cos\frac{\pi}{2}, \sin\frac{\pi}{2})$.

It is more common for the coordinates of points to be written as number values rather than as trigonometric expressions. The geometry and trigonometry of right triangles in the unit circle can be used to find the exact values for the cosine and sine of angles that are multiples of $\frac{\pi}{4}$ and $\frac{\pi}{6}$. A review of right triangle trigonometry can be found in the Prior Knowledge section of this book.

Using the Unit Circle to Evaluate Sine and Cosine

The unit circle is an important tool to find the exact values for the cosine and sine of angles since the coordinates of the points on the unit circle are always in the form $(\cos\theta, \sin\theta)$.

Since the unit circle is centered at the origin with a radius of 1, the coordinates of the points along the axes are known, as shown in Figure 8.12.

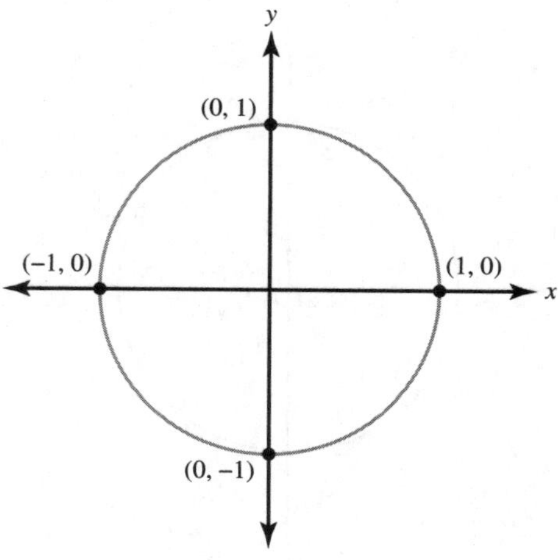

Figure 8.12

The angles associated with the points are also known as they are in standard position, as shown in Figure 8.13.

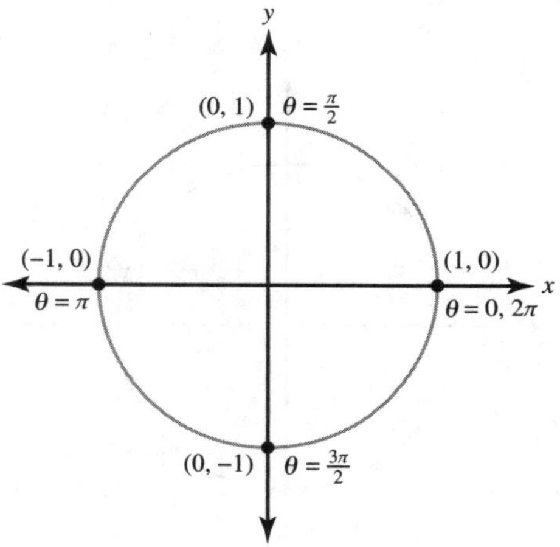

Figure 8.13

Using Figure 8.13 and knowing that the coordinates of the point on the unit circle are $(\cos\theta, \sin\theta)$ leads to the evaluations organized in Table 8.2.

Table 8.2

θ	0	$\frac{\pi}{2}$	π	$\frac{3\pi}{2}$	2π
cos θ	1	0	−1	0	1
sin θ	0	1	0	−1	0

To evaluate other angles, use the 30°-60°-90° and the 45°-45°-90° special right triangles. In radian measure, their angles are $\frac{\pi}{6}, \frac{\pi}{3}$, and $\frac{\pi}{2}$, respectively, and $\frac{\pi}{4}, \frac{\pi}{4}$, and $\frac{\pi}{2}$, respectively. Both triangles are reviewed in the Prior Knowledge section.

Since the side lengths of the special right triangles are known, graphing these triangles in the unit circle allows the coordinates of the point on the terminal side to be determined.

Starting with $\frac{\pi}{6}$, the triangle is constructed in Quadrant I as shown in Figure 8.14.

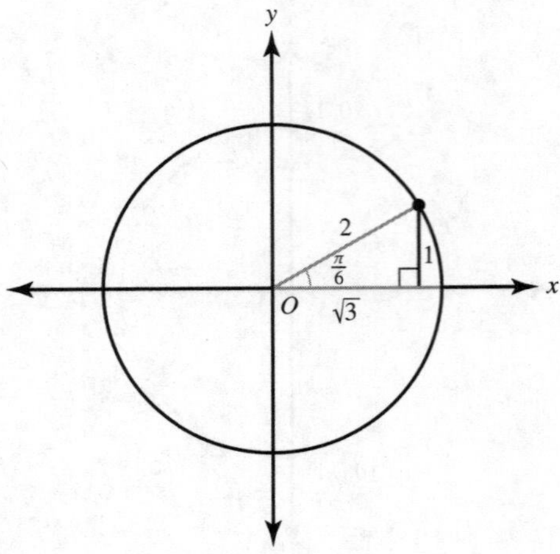

Figure 8.14

To have the circle become a unit circle, the radius needs to equal 1. Multiplying the sides of the triangle by $\frac{1}{2}$ will keep the triangle in proportion and the angle measures the same, as shown in Figure 8.15.

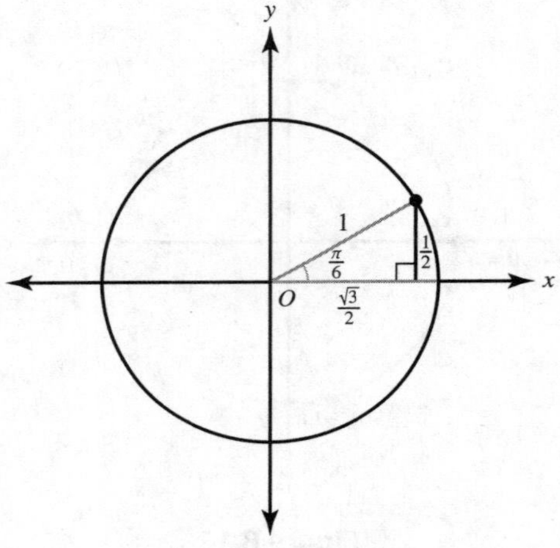

Figure 8.15

Using the sides of the triangle determines the coordinates of the point where the terminal side of the angle intersects the circle, as shown in Figure 8.16.

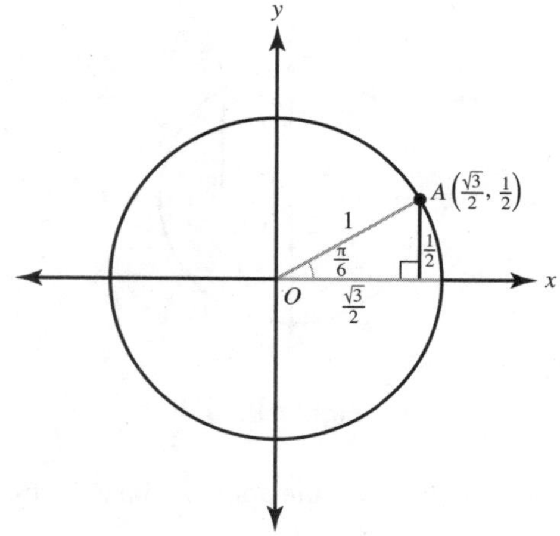

Figure 8.16

Using Figure 8.16 and knowing that the coordinates of the point on the unit circle are $(\cos\theta, \sin\theta)$, the following expressions can be evaluated:

$$\cos\frac{\pi}{6} = \frac{\sqrt{3}}{2}$$
$$\sin\frac{\pi}{6} = \frac{1}{2}$$

The same process will be followed for $\theta = \frac{\pi}{4}$ and $\theta = \frac{\pi}{3}$.

When $\theta = \frac{\pi}{4}$, the coordinates of the point where the terminal side intersects the unit circle are found in Figure 8.17.

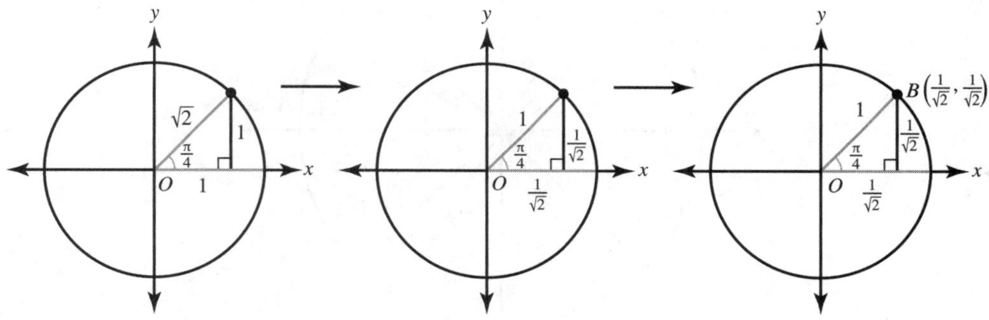

Figure 8.17

Using Figure 8.17 and knowing that the coordinates of the point on the unit circle are $(\cos\theta, \sin\theta)$, the following expressions can be evaluated:

$$\cos\frac{\pi}{4} = \frac{1}{\sqrt{2}} = \frac{\sqrt{2}}{2}$$
$$\sin\frac{\pi}{4} = \frac{1}{\sqrt{2}} = \frac{\sqrt{2}}{2}$$

When $\theta = \frac{\pi}{3}$, the coordinates of the point where the terminal side intersects the unit circle are found in Figure 8.18.

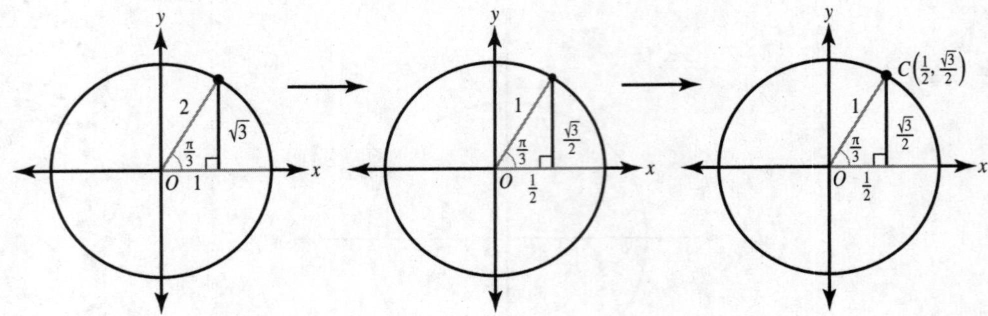

Figure 8.18

Using Figure 8.18 and knowing that the coordinates of the point on the unit circle are $(\cos\theta, \sin\theta)$, the following expressions can be evaluated:

$$\cos\frac{\pi}{3} = \frac{1}{2}$$

$$\sin\frac{\pi}{3} = \frac{\sqrt{3}}{2}$$

Multiples of $\frac{\pi}{4}$ and $\frac{\pi}{6}$ can be evaluated by reflecting the angles and their corresponding points from Quadrant I over the x- and y-axes. In Figure 8.19, the points and their corresponding angles are filled in for Quadrant I and the quadrantal angles, which are the angles that lie on the axes.

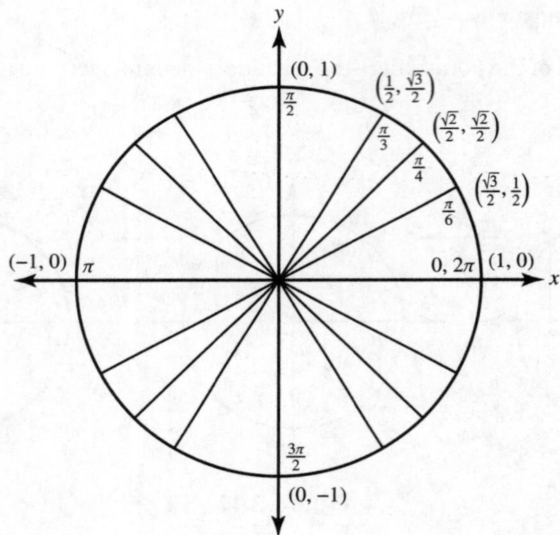

Figure 8.19

When the angles are reflected over the y-axis, the x-coordinates of the points are negated. Their corresponding angles can be found by subtracting their Quadrant I angle measure, which is referred to as the reference angle, from π radians as shown in Figure 8.20.

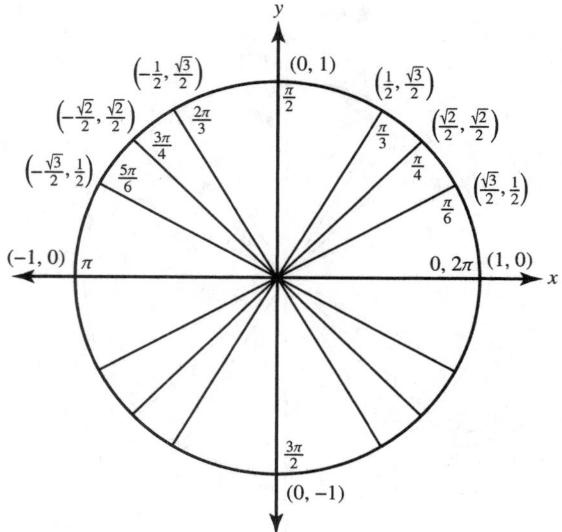

Figure 8.20

When the reference angles are reflected over the y-axis and then the x-axis, both the x- and y-coordinates of the points are negated. Their corresponding angles can be found by adding their reference angle to π radians as shown in Figure 8.21.

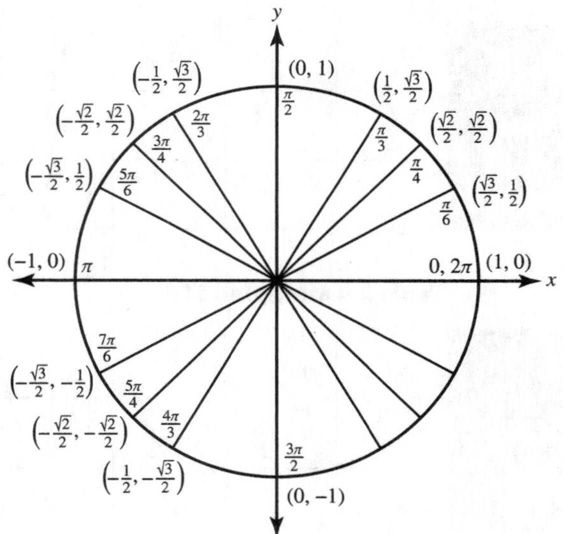

Figure 8.21

When the reference angles are reflected over the x-axis, the y-coordinates of the points are negated. Their corresponding angles can be found by subtracting their reference angle from 2π radians as shown in Figure 8.22.

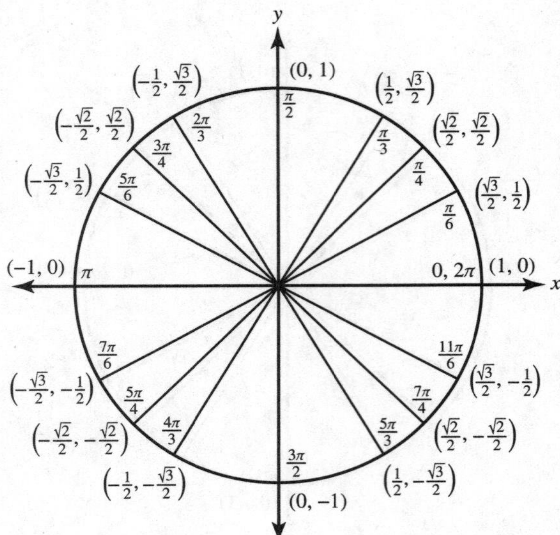

Figure 8.22

Using Figure 8.22 and knowing that the coordinates of the point on the unit circle are $(\cos\theta, \sin\theta)$ leads to the following evaluations organized in Tables 8.3, 8.4, and 8.5.

Table 8.3 Multiples of $\theta = \frac{\pi}{6}$

θ	$\frac{\pi}{6}$	$\frac{5\pi}{6}$	$\frac{7\pi}{6}$	$\frac{11\pi}{6}$
$\cos\theta$	$\frac{\sqrt{3}}{2}$	$-\frac{\sqrt{3}}{2}$	$-\frac{\sqrt{3}}{2}$	$\frac{\sqrt{3}}{2}$
$\sin\theta$	$\frac{1}{2}$	$\frac{1}{2}$	$-\frac{1}{2}$	$-\frac{1}{2}$

Table 8.4 Multiples of $\theta = \frac{\pi}{4}$

θ	$\frac{\pi}{4}$	$\frac{3\pi}{4}$	$\frac{5\pi}{4}$	$\frac{7\pi}{4}$
$\cos\theta$	$\frac{\sqrt{2}}{2}$	$-\frac{\sqrt{2}}{2}$	$-\frac{\sqrt{2}}{2}$	$\frac{\sqrt{2}}{2}$
$\sin\theta$	$\frac{\sqrt{2}}{2}$	$\frac{\sqrt{2}}{2}$	$-\frac{\sqrt{2}}{2}$	$-\frac{\sqrt{2}}{2}$

Table 8.5 Multiples of $\theta = \frac{\pi}{3}$

θ	$\frac{\pi}{3}$	$\frac{2\pi}{3}$	$\frac{4\pi}{3}$	$\frac{5\pi}{3}$
$\cos\theta$	$\frac{1}{2}$	$-\frac{1}{2}$	$-\frac{1}{2}$	$\frac{1}{2}$
$\sin\theta$	$\frac{\sqrt{3}}{2}$	$\frac{\sqrt{3}}{2}$	$-\frac{\sqrt{3}}{2}$	$-\frac{\sqrt{3}}{2}$

Example

Evaluate the following trigonometric expressions. Write the answers in simplest radical form.

1. $\cos\left(\frac{\pi}{3}\right)$
2. $\cos\left(\frac{7\pi}{3}\right)$
3. $\cos\left(\frac{13\pi}{3}\right)$
4. $\sin\left(\frac{7\pi}{4}\right)$
5. $\sin\left(-\frac{\pi}{6}\right)$
6. $\sin\left(\frac{13\pi}{3}\right)$

Solution

The trigonometric expressions can be evaluated using Figure 8.22 and knowing that the coordinates of the point on the unit circle are $(\cos\theta, \sin\theta)$.

1. $\cos\left(\frac{\pi}{3}\right)$ is the x-coordinate of the point on the unit circle whose $\theta = \frac{\pi}{3}$:
$$\cos\left(\frac{\pi}{3}\right) = \frac{1}{2}$$
2. $\cos\left(\frac{7\pi}{3}\right)$ is the x-coordinate of the point on the unit circle whose $\theta = \frac{7\pi}{3}$. Since $\frac{7\pi}{3} > 2\pi$, it involves more than one revolution around the unit circle. To evaluate this angle, find its coterminal angle by subtracting 2π from the given angle. The coterminal angle is $\frac{7\pi}{3} - 2\pi = \frac{\pi}{3}$:
$$\cos\left(\frac{7\pi}{3}\right) = \cos\left(\frac{\pi}{3}\right) = \frac{1}{2}$$
3. $\cos\left(\frac{13\pi}{3}\right)$ is the x-coordinate of the point on the unit circle whose $\theta = \frac{13\pi}{3}$. Since $\frac{13\pi}{3} > 2\pi$, it involves more than one revolution around the unit circle. To evaluate this angle, find its coterminal angle by subtracting 2π from the given angle. To find the coterminal angle, subtract $\frac{13\pi}{3} - 2\pi = \frac{7\pi}{3}$ and then subtract again to get $\frac{7\pi}{3} - 2\pi = \frac{\pi}{3}$:
$$\cos\left(\frac{13\pi}{3}\right) = \cos\left(\frac{7\pi}{3}\right) = \cos\left(\frac{\pi}{3}\right) = \frac{1}{2}$$
4. $\sin\left(\frac{7\pi}{4}\right)$ is the y-coordinate of the point on the unit circle whose $\theta = \frac{\pi}{4}$:
$$\sin\left(\frac{7\pi}{4}\right) = -\frac{\sqrt{2}}{2}$$
5. $\sin\left(-\frac{\pi}{6}\right)$ is the y-coordinate of the point on the unit circle whose $\theta = -\frac{\pi}{6}$. Since $-\frac{\pi}{6} < 0$, the angle is rotating in a clockwise direction. To evaluate this angle, find its coterminal angle by adding the angle to 2π. The coterminal angle is $2\pi + \left(-\frac{\pi}{6}\right) = \frac{11\pi}{6}$:
$$\sin\left(-\frac{\pi}{6}\right) = \sin\left(\frac{11\pi}{6}\right) = -\frac{1}{2}$$

6. $\sin\left(\frac{13\pi}{3}\right)$ is the y-coordinate of the point on the unit circle whose $\theta = \frac{13\pi}{3}$. Since $\frac{13\pi}{3} > 2\pi$, it involves more than one revolution around the unit circle. To evaluate this angle, find its coterminal angle by subtracting 2π from the angle. To find the coterminal angle, subtract $\frac{13\pi}{3} - 2\pi = \frac{7\pi}{3}$ and then subtract again to get $\frac{7\pi}{3} - 2\pi = \frac{\pi}{3}$:

$$\sin\left(\frac{13\pi}{3}\right) = \sin\left(\frac{7\pi}{3}\right) = \sin\left(\frac{\pi}{3}\right) = \frac{\sqrt{3}}{2}$$

It is not necessary to have all the angles and their corresponding coordinates in all four quadrants memorized. If the angles in Quadrant I are understood, the remaining angles can be figured out using reflections and the reference angles from Quadrant I.

8.4 Sine and Cosine Function Graphs

Trigonometric functions are periodic because the number of revolutions around the unit circle is infinite. The values of a periodic function repeat over regular intervals. Each complete pattern of values is a cycle. When θ varies from 0 to 2π, $\sin\theta$ and $\cos\theta$ take on values from -1 to 1. The previous section showed that an angle formed by adding or subtracting any multiple of 2π to θ is coterminal with θ and that the same pattern of values for $\sin\theta$ and $\cos\theta$ repeats every 2π. Therefore, sine and cosine are periodic functions with periods of 2π.

To discover the graphs of the sine and cosine curves requires "unwrapping" the unit circle to plot the values of sine and cosine on the coordinate plane.

As point $P(x, y)$ moves counterclockwise in a unit circle, its position is determined by angle θ as shown in Figure 8.22.

For any given θ, $\sin\theta$ specifies the height (y) of P while $\cos\theta$ gives its horizontal position (x). The process of creating the graphs of $f(\theta) = \sin\theta$ and $f(\theta) = \cos\theta$ is based on associating θ with its respective $\sin\theta$ or $\cos\theta$ evaluations and writing them as points in the coordinate plane, where the x-coordinate represents θ and the y-coordinate represents its trigonometric value.

Using the unit circle, Tables 8.6 and 8.7 have been created to graph the functions $f(\theta) = \cos\theta$ and $f(\theta) = \sin\theta$.

Table 8.6

$f(\theta) = \cos\theta$	
$x = \theta$	$y = \cos\theta$
0	1
$\frac{\pi}{4}$	$\frac{\sqrt{2}}{2}$
$\frac{\pi}{2}$	0
$\frac{3\pi}{4}$	$-\frac{\sqrt{2}}{2}$
π	-1
$\frac{5\pi}{4}$	$-\frac{\sqrt{2}}{2}$
$\frac{3\pi}{2}$	0
$\frac{7\pi}{4}$	$\frac{\sqrt{2}}{2}$
2π	1

Table 8.7

$f(\theta) = \sin\theta$	
$x = \theta$	$y = \sin\theta$
0	0
$\frac{\pi}{4}$	$\frac{\sqrt{2}}{2}$
$\frac{\pi}{2}$	1
$\frac{3\pi}{4}$	$\frac{\sqrt{2}}{2}$
π	0
$\frac{5\pi}{4}$	$-\frac{\sqrt{2}}{2}$
$\frac{3\pi}{2}$	-1
$\frac{7\pi}{4}$	$-\frac{\sqrt{2}}{2}$
2π	0

Plotting the points from each table onto the coordinate plane gives the graphs of $y = \cos x$ (or $f(\theta) = \cos\theta$) and $y = \sin x$ (or $f(\theta) = \sin\theta$), as shown in Figures 8.23 and 8.24.

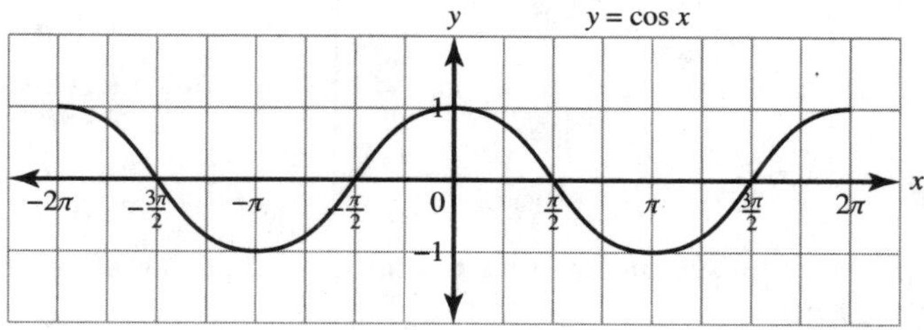

Figure 8.23

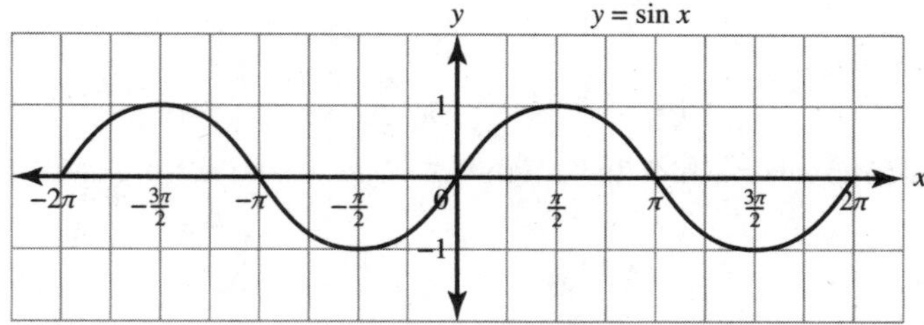

Figure 8.24

Notice the graphs have the same basic shape and that the cosine curve is a transformation of the sine curve to the left by $\frac{\pi}{2}$ units.

The domain of both the cosine and sine function graphs is all real numbers. The range of both the cosine and sine function graphs is $[-1, 1]$. The period of both the cosine and sine function graphs is 2π.

8.5 Sinusoidal Functions

A sinusoidal function is any function that involves additive and multiplicative transformations of $f(\theta) = \sin\theta$. The sine and cosine functions are examples of a sinusoidal function since $\cos\theta = \sin\left(\theta + \frac{\pi}{2}\right)$, which notes the horizontal shift left by $\frac{\pi}{2}$.

There are different key characteristics of the graphs of sinusoidal functions.

Frequency

The frequency of a sinusoidal function is the number of cycles that its graph completes in an interval of 2π radians. Since the sine curve and the cosine curve each complete one cycle every 2π radians, the frequency of each curve is 1.

> **Period and Frequency**
>
> In general, the period and frequency of a sinusoidal function are reciprocals. The period of $f(\theta) = \sin\theta$ and of $g(\theta) = \cos\theta$ is 2π, and the frequency of each is $\frac{1}{2\pi}$.

Amplitude

The amplitude of a sinusoidal function is half the difference between its maximum and minimum values. Since the maximum and minimum values of both the sine curve and cosine curve are 1 and -1, respectively, the amplitude of both $f(\theta) = \sin\theta$ and $g(\theta) = \cos\theta$ is $\frac{1-(-1)}{2} = 1$.

Midline

The midline of the graph of a sinusoidal function is determined by the average of the maximum and minimum values of the function. Since the maximum and minimum values of both the sine curve and cosine curve are 1 and -1, respectively, the midline of $f(\theta) = \sin\theta$ and of $g(\theta) = \cos\theta$ is $\frac{1+(-1)}{2} = 0$. The midline is the horizontal line $y = 0$.

Concavity

As input values increase, the graphs of sinusoidal functions oscillate between concave down and concave up.

Even/Odd Functions

Since the graph of $f(\theta) = \sin\theta$ has point symmetry with respect to the origin, it is an odd function. Therefore, $f(-\theta) = \sin(-\theta) = -\sin\theta = -f(\theta)$.

Since the graph of $f(\theta) = \cos\theta$ has line symmetry with respect to the y-axis, it is an even function. Therefore, $f(-\theta) = \cos(-\theta) = \cos\theta = f(\theta)$.

Example

If $0 \le x \le 2\pi$, determine the interval on which the graph of $y = \sin x$ is decreasing and, at the same time, the graph of $y = \cos x$ is increasing.

Solution

To get a visual, sketch the graphs of $y = \sin x$ and $y = \cos x$ on the same set of axes as shown in Figure 8.25.

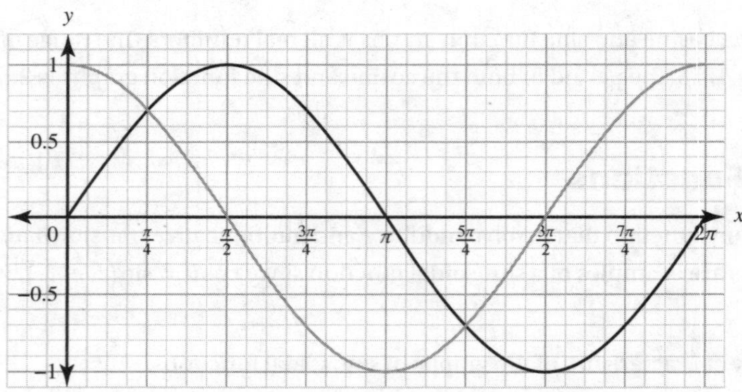

Figure 8.25

The sine curve is decreasing and the cosine curve is increasing on the interval $\left(\pi, \frac{3\pi}{2}\right)$.

Example

Use the graphs of $y = \sin x$ and $y = \cos x$ in the interval $[0, 2\pi]$ to graph $y = \sin x$ and $y = \cos x$ in the interval $[-2\pi, 2\pi]$.

Solution

The graph of $y = \sin x$ in the interval $[0, 2\pi]$ is shown in Figure 8.26.

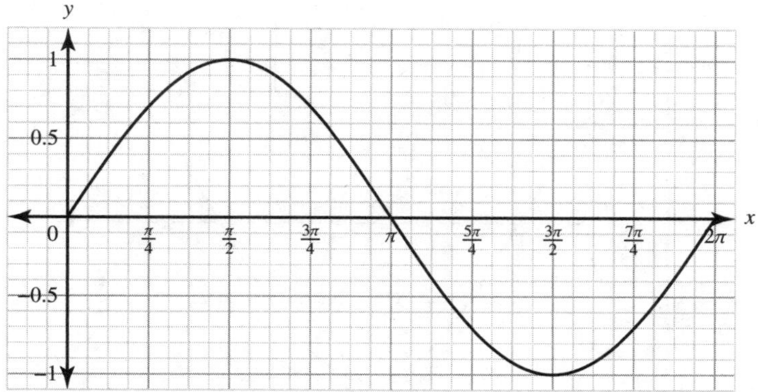

Figure 8.26

Since sine is an odd function, the points are reflected in the origin and transformed such that $(x, y) \rightarrow (-x, -y)$, as shown in Tables 8.8 and 8.9, which gives the results of following the transformation rule.

Table 8.8

x	0	$\frac{\pi}{2}$	π	$\frac{3\pi}{2}$	2π
y	0	1	0	-1	0

Table 8.9

x	0	$-\frac{\pi}{2}$	$-\pi$	$-\frac{3\pi}{2}$	-2π
y	0	-1	0	1	0

Plotting the points leads to the graph in Figure 8.27.

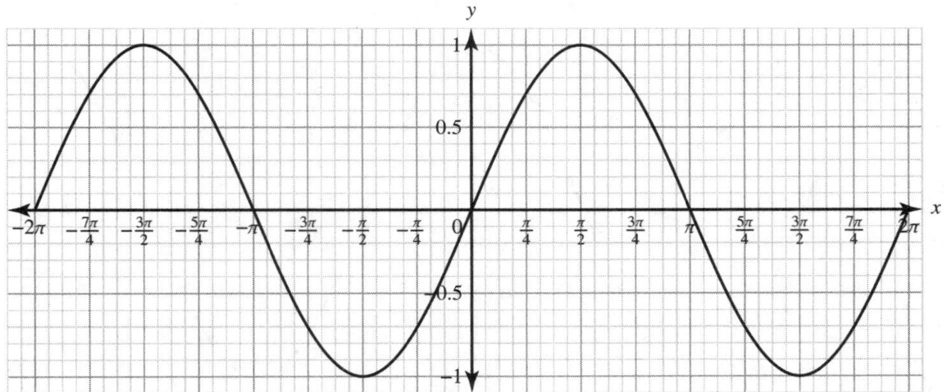

Figure 8.27

The graph of $y = \cos x$ in the interval $[0, 2\pi]$ is shown in Figure 8.28.

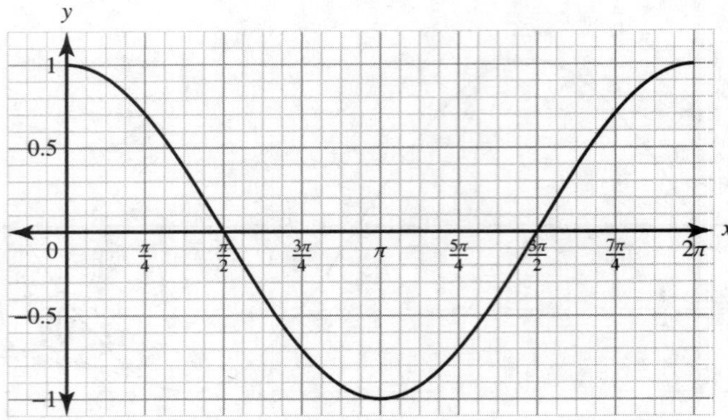

Figure 8.28

Since cosine is an even function, the points are reflected in the y-axis and transformed such that $(x, y) \to (-x, y)$, as shown in Tables 8.10 and 8.11, which gives the results of following the transformation rule.

Table 8.10

x	0	$\frac{\pi}{2}$	π	$\frac{3\pi}{2}$	2π
y	1	0	-1	0	1

Table 8.11

x	0	$-\frac{\pi}{2}$	$-\pi$	$-\frac{3\pi}{2}$	-2π
y	1	0	-1	0	1

Plotting the points leads to the graph in Figure 8.29.

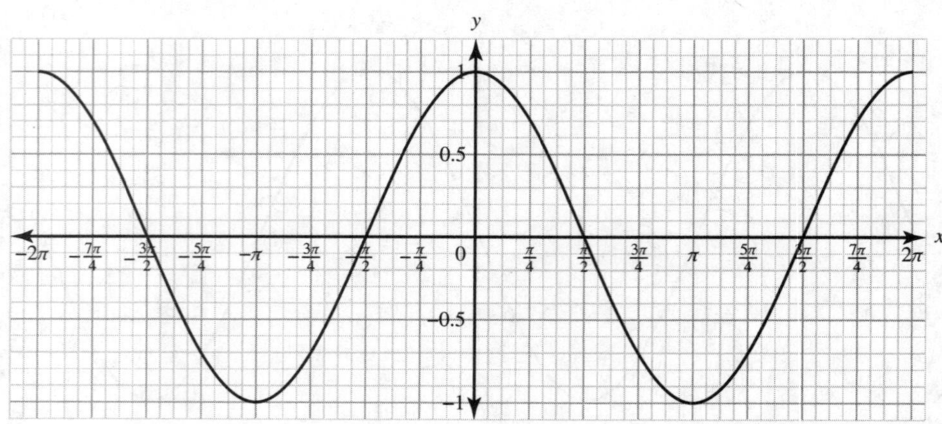

Figure 8.29

8.6 Sinusoidal Function Transformations

Functions that can be written in the form $f(\theta) = a\sin(b(\theta + c)) + d$ or in the form $g(\theta) = a\cos(b(\theta + c)) + d$, where a, b, c, and d are real numbers and $a \neq 0$, are sinusoidal functions. They are transformations of the sine and cosine functions. All the transformation properties discussed in Section 4.2 can be applied to the graphs of the sine and cosine functions.

Amplitude of $y = a\sin(x)$ and $y = a\cos(x)$

The value of a affects the amplitude of the sine and cosine graphs.

The amplitude is $|a|$ and is a vertical dilation of the graph of $f(\theta) = \sin\theta$ or $g(\theta) = \cos\theta$.

Example

Compare each of the following functions to their respective parent functions and list their transformations. Graph each function on the same set of axes as their parent function.

1. $f(x) = 2\sin x$
2. $g(x) = \frac{1}{2}\cos x$

Solution

1. The parent function of $f(x) = 2\sin x$ is $y = \sin x$. The graph of $f(x)$ is a vertical dilation by a factor of 2. Since the maximum and minimum values of its parent function are 1 and -1, the maximum and minimum values of $f(x) = 2\sin x$ are 2 and -2. Therefore, the amplitude is $\frac{2-(-2)}{2} = 2$. The graphs of the two functions over the domain $[0, 2\pi]$ are shown in Figure 8.30.

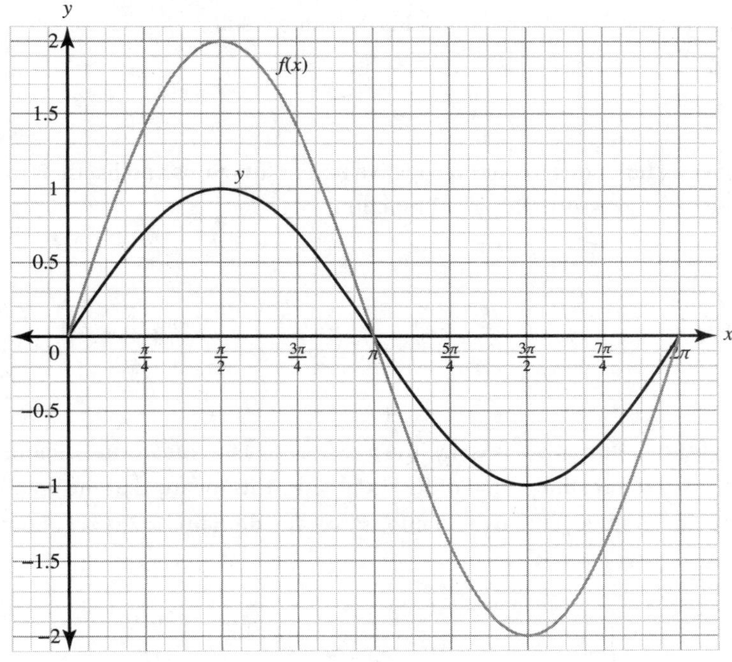

Figure 8.30

2. The parent function of $g(x) = \frac{1}{2}\cos x$ is $y = \cos x$. The graph of $g(x)$ is a vertical dilation by a factor of $\frac{1}{2}$. Since the maximum and minimum values of its parent function are 1 and -1, the maximum and minimum values of $g(x) = \frac{1}{2}\cos x$ are $\frac{1}{2}$ and $-\frac{1}{2}$. Therefore, the amplitude is $\frac{\frac{1}{2}-\left(-\frac{1}{2}\right)}{2} = \frac{1}{2}$. The graphs of the two functions over the domain $[0, 2\pi]$ are shown in Figure 8.31.

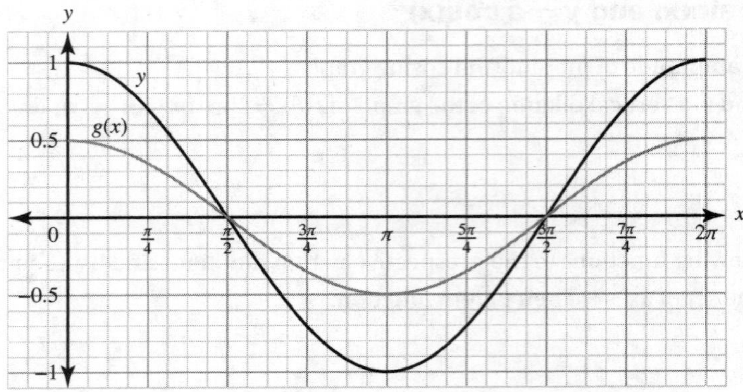

Figure 8.31

Frequency of $y = \sin(bx)$ and $y = \cos(bx)$

The value of b affects the frequency for the sine and cosine graphs. The b-value states the number of full cycles of the sine and cosine curve in the interval $[0, 2\pi]$.

The graph of the multiplicative transformation $y = \sin(bx)$ of the sine function $f(\theta) = \sin\theta$ is a horizontal dilation of the graph of f and differs in period by a factor of $\left|\frac{1}{b}\right|$. In general, the period of any sine or cosine curve is $\frac{2\pi}{|b|}$.

The same transformation of the cosine function yields the same result.

Example

Compare each of the following functions to their respective parent functions and list their transformations. Graph each function on the same set of axes as their parent function.

1. $f(x) = \cos 2x$
2. $g(x) = \cos\frac{1}{2}x$

Solution

1. The parent function of $f(x) = \cos 2x$ is $y = \cos x$. The graph of $f(x)$ is a horizontal dilation of y and differs in period by a factor of $\frac{1}{2}$. Therefore, the period of $f(x)$ is $\frac{2\pi}{2} = \pi$. This means that the graph of $f(x)$ completes one full cycle in π radians. The graphs of the two functions over the domain $[0, 2\pi]$ are shown in Figure 8.32.

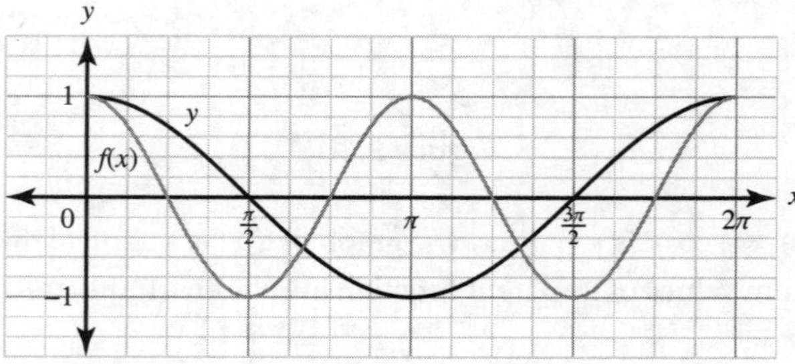

Figure 8.32

2. The parent function of $g(x) = \cos\frac{1}{2}x$ is $y = \cos x$. The graph of $g(x)$ is a horizontal dilation of y and differs in period by a factor of $\frac{1}{\frac{1}{2}} = 2$. Therefore, the period of $g(x)$ is $\frac{2\pi}{\frac{1}{2}} = 4\pi$. This means that the graph of $g(x)$ completes one-half of a full cycle in 2π radians. The graphs of the two functions over the domain $[0, 2\pi]$ are shown in Figure 8.33.

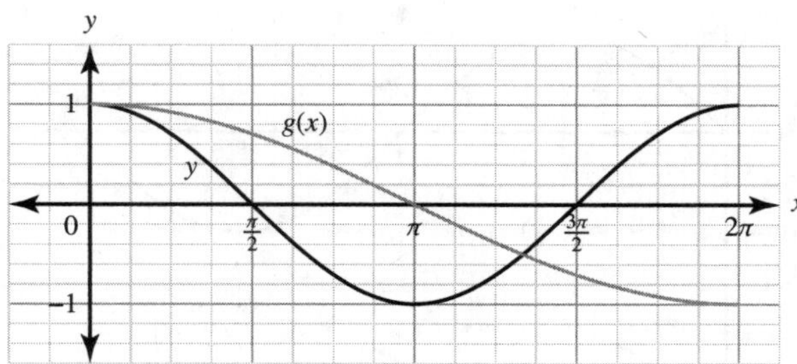

Figure 8.33

Vertical Translation of $y = \sin(x) + d$ and $y = \cos(x) + d$

Adding a constant, d, to the right side of the trigonometric parent functions $f(\theta) = \sin\theta$ and $g(\theta) = \cos\theta$ shifts the graphs of these functions up when $d > 0$ and down when $d < 0$. The midline of the equations also shifts, and the equation of the midline is the horizontal line $y = d$.

Example

Compare each of the following functions to their respective parent functions and list their transformations. Graph each function on the same set of axes as their parent function.

1. $f(x) = \cos x + 2$
2. $g(x) = \sin x - 3$

Solution

1. The parent function of $f(x) = \cos x + 2$ is $y = \cos x$. The graph of $f(x)$ is a vertical translation of y up 2 units. The equation of the midline is $y = 2$. This means that the maximum value of $f(x)$ is 3 and the minimum value is 1. The graphs of the two functions over the domain $[0, 2\pi]$ are shown in Figure 8.34.

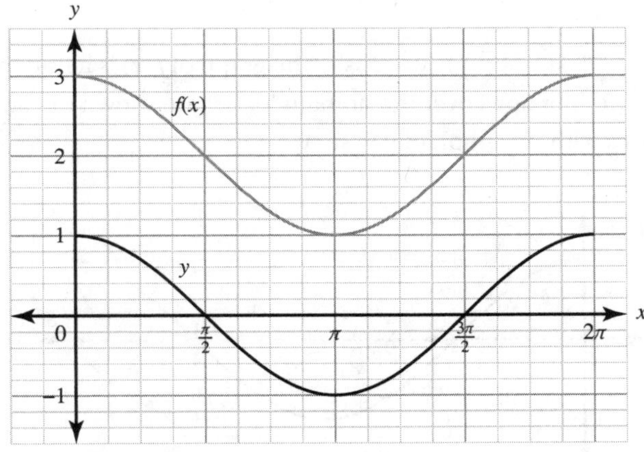

Figure 8.34

2. The parent function of $g(x) = \sin x - 3$ is $y = \sin x$. The graph of $g(x)$ is a vertical translation of y down 3 units. The equation of the midline is $y = -3$. This means that the maximum value of $g(x)$ is -2 and the minimum value is -4. The graphs of the two functions over the domain $[0, 2\pi]$ are shown in Figure 8.35.

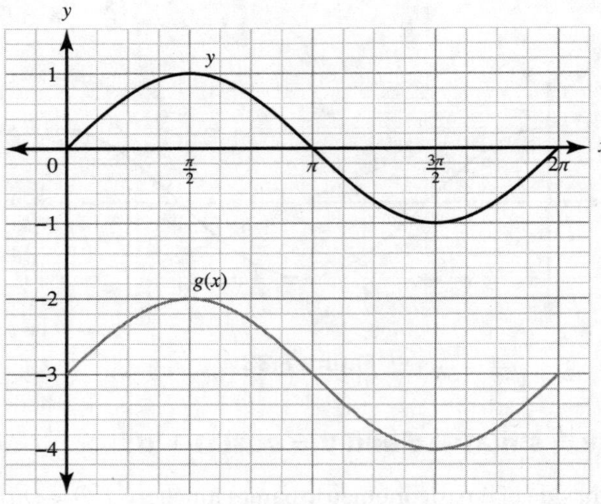

Figure 8.35

Horizontal Translation of $y = \sin(x + c)$ and $y = \cos(x + c)$

Adding a constant, c, to the input of the trigonometric parent functions $f(\theta) = \sin\theta$ and $g(\theta) = \cos\theta$ shifts the graphs of these functions left when $c > 0$ and right when $c < 0$. A horizontal translation of a periodic function is called a phase shift.

Example

Compare each of the following functions to their respective parent functions and list their transformations. Graph each function on the same set of axes as their parent function.

1. $f(x) = \cos\left(x - \frac{\pi}{2}\right)$
2. $g(x) = \sin(x + \pi)$

Solution

1. The parent function of $f(x) = \cos\left(x - \frac{\pi}{2}\right)$ is $y = \cos x$. The graph of $f(x)$ is a horizontal translation of y right $\frac{\pi}{2}$ units. The graphs of the two functions over the domain $[0, 2\pi]$ are shown in Figure 8.36. The graph of $f(x) = \cos\left(x - \frac{\pi}{2}\right)$ coincides with the graph of $y = \sin x$.

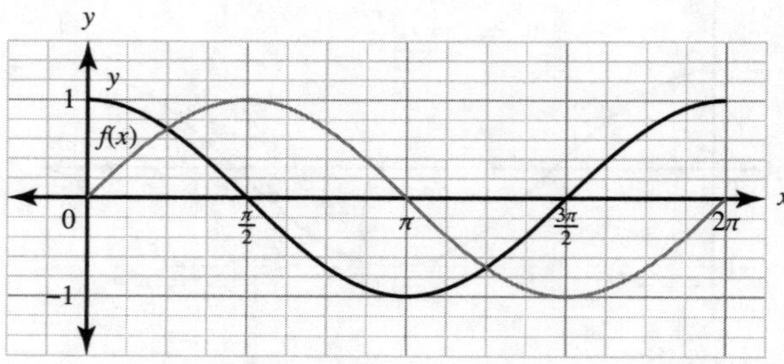

Figure 8.36

2. The parent function of $g(x) = \sin(x + \pi)$ is $y = \sin x$. The graph of $g(x)$ is a horizontal translation of y left π units. The graphs of the two functions over the domain $[-\pi, 2\pi]$ are shown in Figure 8.37.

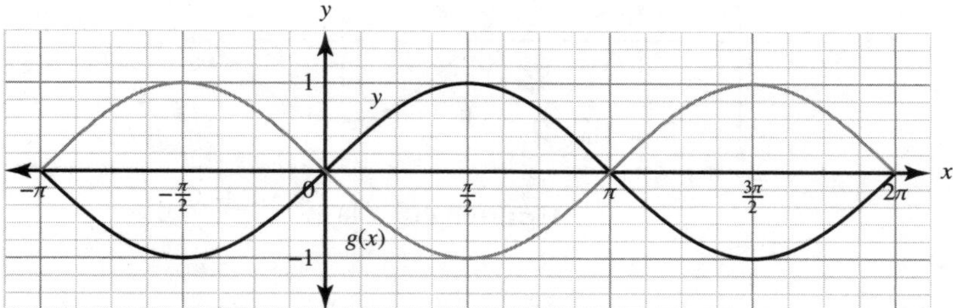

Figure 8.37

Reflecting Trigonometric Functions

The graphs of sine and cosine can be reflected over the x- and y-axes the same way that all parent functions can be reflected.

The graphs of $f(x) = -\sin x$ and $g(x) = -\cos x$ are the reflections of the graphs of the basic sine and cosine curves in the x-axis, respectively, as shown in Figures 8.38 and 8.39.

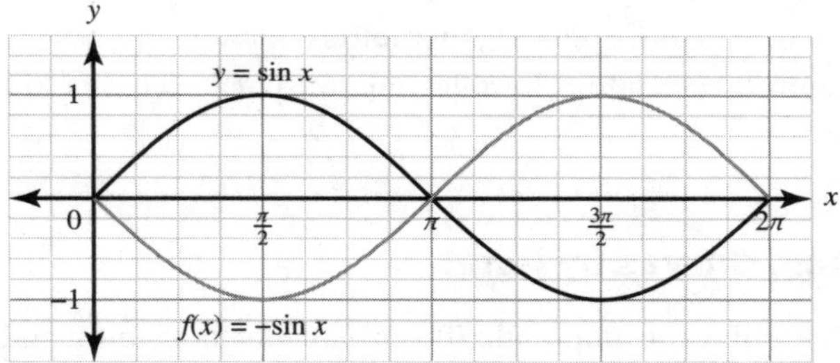

Figure 8.38

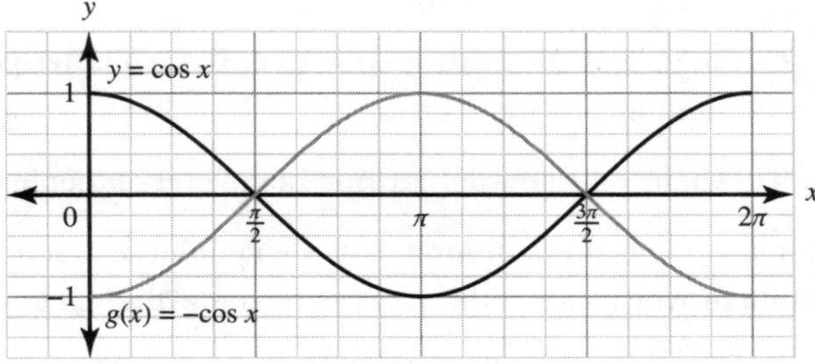

Figure 8.39

The graphs of $h(x) = \sin(-x)$ and $k(x) = \cos(-x)$ are the reflections of the graphs of the basic sine and cosine curves in the y-axis, as shown in Figures 8.40 and 8.41. Note that the cosine curve is symmetric with respect to the y-axis, so the graphs of $y = \cos x$ and $k(x) = \cos(-x)$ coincide.

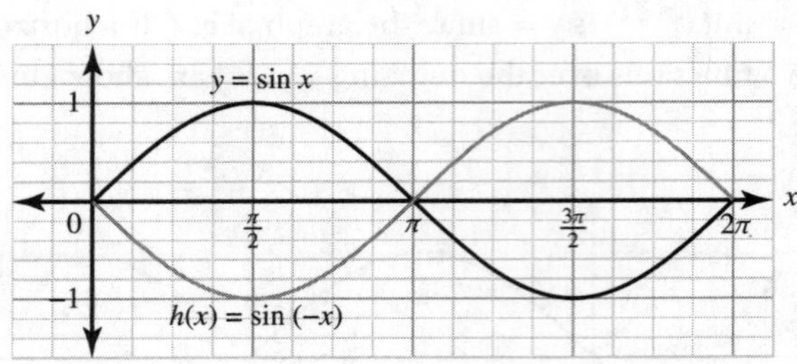

Figure 8.40

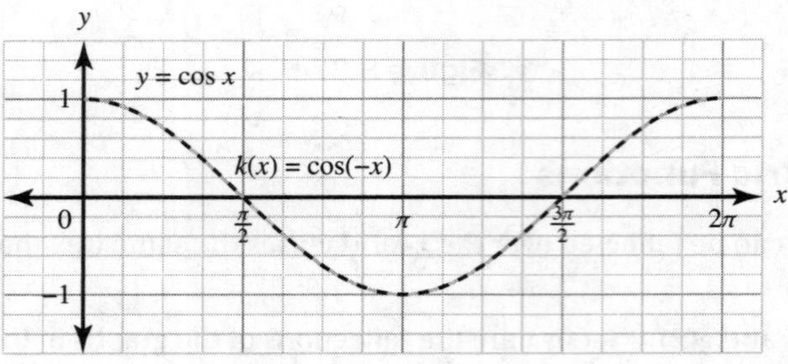

Figure 8.41

All the transformations can be found in one single trigonometric equation and can be graphed by hand or graphed using technology, such as a graphing calculator. The following steps make it easier to graph a sine or cosine curve by hand.

Graphing Sine and Cosine Curves by Hand

STEP 1 Using the a-value, determine the amplitude. This will tell the maximum and minimum points of the graph.

STEP 2 Using the b-value, find the period $= \frac{2\pi}{b}$. This will give the interval for one full cycle of the sine or cosine curve.

STEP 3 To figure out how to scale the x-axis, divide the period by 4. These quarter-period marks plot key points such as the maximum, minimum, and intercepts.

STEP 4 Using the c-value, shift the points the appropriate number of units either left or right.

STEP 5 Using the d-value, shift the points the appropriate number of units either down or up. The d-value gives the equation of the midline, which can be drawn as a horizontal dashed line.

STEP 6 Look for negations of the equation and reflect over the x- or y-axes where appropriate.

Example

Graph one full cycle for each of the following functions.

1. $f(x) = 3\sin 2x$
2. $g(x) = -2\cos\frac{1}{2}x$
3. $h(x) = 3\cos(x + \pi) + 1$
4. $m(x) = \sin\left(2\pi x - \frac{\pi}{2}\right) - 1$

✓ Solution

For each of the graphs, follow the 6 steps listed previously to graph sine and cosine curves by hand.

1. $f(x) = 3\sin 2x$

   STEPS:

   1. $a = 3$
   2. $b = 2$, so period $= \frac{2\pi}{2} = \pi$. The b-value means there are two full cycles from $[0, 2\pi]$. So to ensure one full cycle, the domain of the graph will be $[0, \pi]$.
   3. $x\text{-scale} = \frac{\text{period}}{4} = \frac{\pi}{4}$
   4. $c = 0$, no phase shift
   5. $d = 0$, no vertical shift
   6. No negations, no reflections

   First scale the x-axis by $\frac{\pi}{4}$, and then ensure the y-axis has a maximum value of 3 and a minimum value of -3. The sine curve starts at the origin and then increases, decreases, and increases again to complete 1 cycle. Follow the overall parent curve, ensuring that the key points are plotted on the quarter-period marks. The graph of $f(x)$ is shown in Figure 8.42.

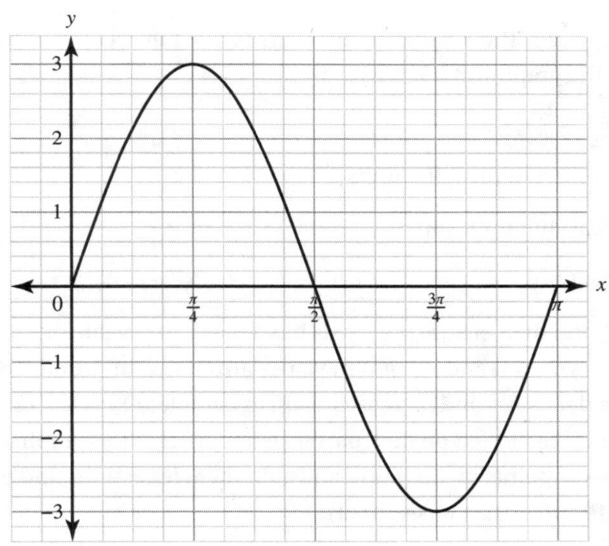

Figure 8.42

2. $g(x) = -2\cos\frac{1}{2}x$

   1. $a = -2$, amplitude $= |a| = 2$
   2. $b = \frac{1}{2}$, so period $= \frac{2\pi}{\frac{1}{2}} = 4\pi$. The b-value means there is half of a full cycle from $[0, 2\pi]$. So to ensure one full cycle, the domain of the graph will be $[0, 4\pi]$.
   3. $x\text{-scale} = \frac{\text{period}}{4} = \frac{4\pi}{4} = \pi$
   4. $c = 0$, no phase shift

5. $d = 0$, no vertical shift

6. The output is negated, reflection over the x-axis

First scale the x-axis by π, and then ensure the y-axis has a maximum value of 2 and a minimum value of -2. The cosine curve starts at its maximum value and then decreases and increases again to complete 1 cycle. Follow the overall parent curve, ensuring that the key points are plotted on the quarter-period marks. This graph is shown as a dashed line on Figure 8.43. Then take the key points of this graph and reflect over the x-axis, which is the solid graph shown in Figure 8.43 and is the final solution.

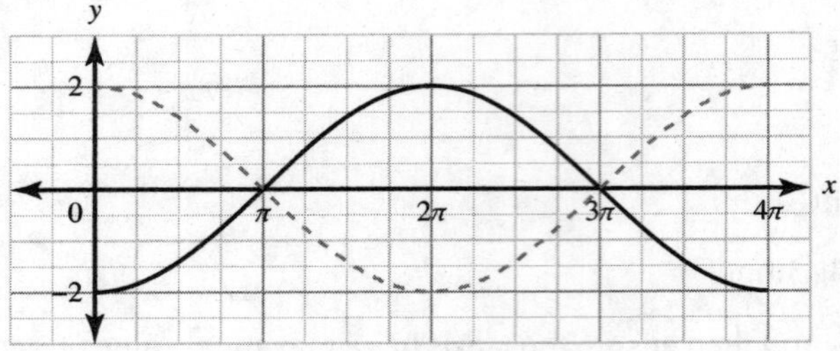

Figure 8.43

3. $h(x) = 3\cos(x + \pi) + 1$

1. $a = 3$

2. $b = 1$, so period $= \frac{2\pi}{1} = 2\pi$. The b-value means there is one full cycle from $[0, 2\pi]$. So to ensure one full cycle, the domain of the graph will be $[0, 2\pi]$.

3. x-scale $= \frac{\text{period}}{4} = \frac{2\pi}{4} = \frac{\pi}{2}$

4. $c = \pi$, there is a phase shift to the left π units

5. $d = 1$, there is a vertical shift up 1 unit

6. No negations, no reflections

First scale the x-axis by $\frac{\pi}{2}$, and then ensure the y-axis has a maximum value of $3 + 1 = 4$ and a minimum value of $-3 + 1 = -2$ since the maximum and minimum values are shifted up 1 unit. The cosine curve starts at the maximum value and then decreases and increases again to complete 1 cycle. Follow the overall parent curve, ensuring that the key points are plotted on the quarter-period marks. This graph is shown as a dashed line on Figure 8.44. Then take the key points of this graph and shift them left π units and up 1 unit, which is the solid graph shown in Figure 8.44 and is the final solution.

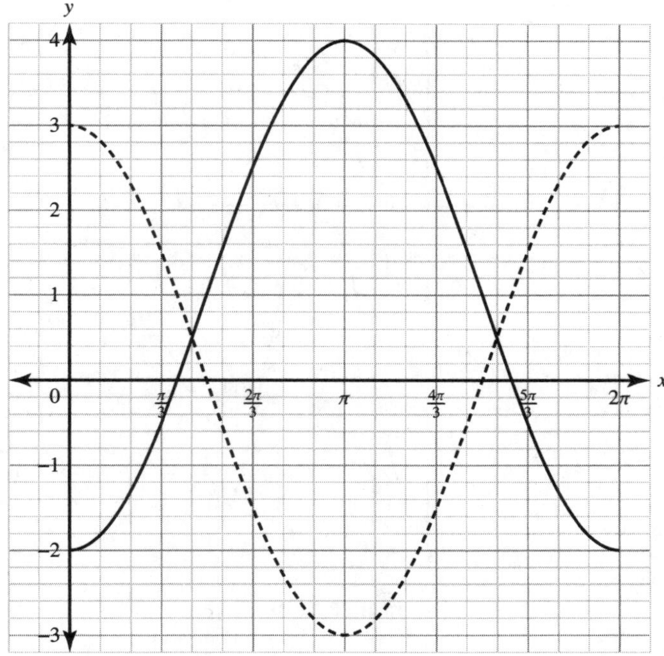

Figure 8.44

4. $m(x) = \sin\left(2\pi x - \frac{\pi}{2}\right) - 1$
   1. $a = 1$
   2. To find the b-value, the coefficient in front of the x in the parentheses must be 1. Factor out 2π from each term in the parentheses to rewrite the equation as:

      $$m(x) = \sin 2\pi\left(x - \frac{1}{4}\right) - 1$$

      This equation will be used for the remaining steps of the problem.

      $b = 2\pi$, so period $= \frac{2\pi}{2\pi} = 1$. The b-value means there will be 2π full cycles from $[0, 2\pi]$. So to ensure one full cycle, the domain of the graph will be $[0, 1]$.
   3. x-scale $= \frac{\text{period}}{4} = \frac{1}{4}$
   4. $c = -\frac{1}{4}$, there is a phase shift to the right $\frac{1}{4}$ unit
   5. $d = -1$, there is a vertical shift down 1 unit
   6. No negations, no reflections

   First scale the x-axis by $\frac{1}{4}$, and ensure the y-axis has a maximum value of $1 - 1 = 0$ and a minimum value of $-1 - 1 = -2$ since the maximum and minimum values are shifted down 1. The sine curve starts at the origin and then increases, decreases, and increases again to complete 1 cycle. Follow the overall parent curve, ensuring that the key points are plotted on the quarter-period marks. This graph is shown as a dashed line on Figure 8.45. Then take the key points of this graph and shift them right $\frac{1}{4}$ unit and down 1 unit, which is the solid graph shown in Figure 8.45 and is the final solution.

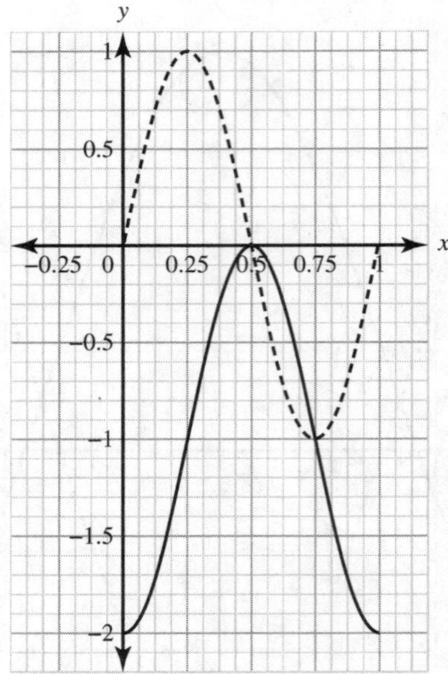

Figure 8.45

It is important to note that in the last example, the x-scale was by rational numbers that were not multiples of π. This is not uncommon, especially when trigonometric equations are used to model phenomena.

Example

In each of the following functions, identify its parent function and describe the transformations that take place.

1. $f(x) = -3\sin(x - \pi) + 5$
2. $g(x) = \frac{1}{2}\cos\left(2x - \frac{\pi}{2}\right) - 7$

Solution

1. The parent function for $f(x) = -3\sin(x - \pi) + 5$ is $y = \sin x$. The graph of $f(x)$ has a vertical dilation by a factor of 3, is horizontally translated to the right π units, is vertically translated up 5 units, and is reflected over the x-axis.
2. The parent function for $g(x) = \frac{1}{2}\cos\left(2x - \frac{\pi}{2}\right) - 7$ is $y = \cos x$. To determine the appropriate transformations, the given equation must be written in the form $y = a\cos b(x - c) + d$. Factoring out 2 from the input, the equation can be rewritten in the proper form as:
$$g(x) = \frac{1}{2}\cos 2\left(x - \frac{\pi}{4}\right) - 7$$
The graph of $g(x)$ has a vertical dilation by a factor of $\frac{1}{2}$, a horizontal dilation by a factor of $\frac{1}{2}$, is horizontally translated to the right $\frac{\pi}{4}$ unit, and is vertically translated down 7 units.

Since the graphs of the sine and cosine curves are equivalent because of a phase shift, when given a periodic function and its data, there are infinitely many equations that can act as a model for the data. Each solution presented in this section is just an example of one possible equation, but that equation has the smallest phase shift possible.

Example

1. What is an equation of the trigonometric function shown in Figure 8.46?

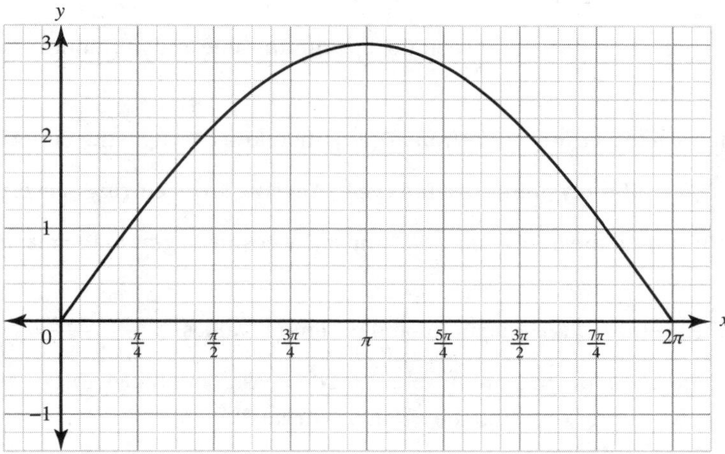

Figure 8.46

2. What is an equation of the trigonometric function found in Figure 8.47?

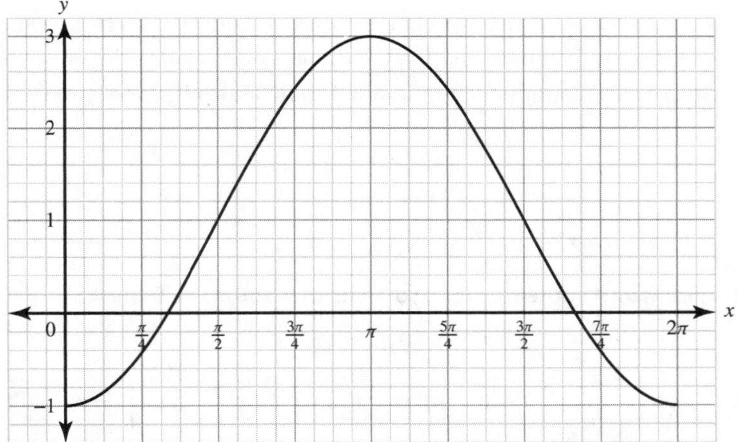

Figure 8.47

Solution

For each of the equations, follow the 6 steps to graph sine and cosine curves by hand.

1. The curve has the basic shape of a sine curve.
    1. The amplitude is the distance from the maximum value to the midline $= 3 - 0 = 3 = a$. There is half of a full cycle of a sine curve from $[0, 2\pi]$; therefore, $b = \frac{1}{2}$. So the period $= \frac{2\pi}{\frac{1}{2}} = 4\pi$.
    2. The x-scale $= \frac{4\pi}{4} = \pi$. The key points of half of a full cycle of sine are occurring at every multiple of π.
    3. The sine curve begins at the origin; there is no phase shift. So $c = 0$.
    4. The sine curve begins at the origin; there is no vertical shift. So $d = 0$.

5. The sine curve begins at the origin and increases. Since there are no reflections, there are no negations.

6. An equation of the curve is $y = 3\sin\frac{1}{2}x$.

2. The curve has the basic shape of a cosine curve.

1. The amplitude is half the distance between the maximum value and minimum value $= \frac{3-(-1)}{2} = \frac{4}{2} = 2 = a$.

2. There is one full cycle of a cosine curve from $[0, 2\pi]$, therefore $b = 1$. So the period $= \frac{2\pi}{1} = 2\pi$.

3. The x-scale $= \frac{2\pi}{4} = \frac{\pi}{2}$. The key points of a full cycle of cosine are occurring at every multiple of $\frac{\pi}{2}$.

4. The given curve begins at the minimum value and since cosine begins at the maximum value, there is no phase shift. So $c = 0$.

5. The midline is the average of the maximum and minimum values $= \frac{3+(-1)}{2} = \frac{2}{2} = 1 = d$.

6. The cosine curve begins at the maximum value and decreases, but the given curve does the opposite. There is a reflection over the x-axis, so the a-value is negated.

An equation of the curve is $y = -2\cos x + 1$.

8.7 Sinusoidal Function Context and Data Modeling

As in the previous chapters, trigonometric functions can be used to model data. Phenomena that are cyclic in nature can be represented by sinusoidal functions. The techniques discussed in the previous section can be applied to determine or estimate the amplitude, frequency, period, and shifts needed to write the equation of the sinusoidal function.

Example

The mean monthly maximum temperature (°C) for a small town is shown in Table 8.12.

Table 8.12

Month	Jan	Feb	Mar	April	May	June	July	Aug	Sept	Oct	Nov	Dec
Temp (°C)	28	27	25.5	22	18.5	16	15	16	18	21.5	24	26

Write an equation to model this data.

Solution

Since temperature is a periodic phenomenon, a sine curve could model the given data. Using the equation $y = a\sin b(x - c) + d$, the 6 steps from the previous section will be used to determine a, b, c, and d.

1. The amplitude $= \frac{\text{maximum} - \text{minimum}}{2} = \frac{28 - 15}{2} = 6.5 = a$.

2. The period is 12 months since the cycle of temperature repeats itself after each year. Since period $= \frac{2\pi}{b}$, then $12 = \frac{2\pi}{b}$ and so $b = \frac{\pi}{6}$.

3. The x-scale is not necessary here to find the equation. If the data was graphed, the x-axis would use each month to plot the given data point.

4. The vertical shift should be found before looking for a possible phase shift. The midline $= \frac{\text{maximum} + \text{minimum}}{2} = \frac{28 + 15}{2} = 21.5 = d$.

5. The model so far is $y = 6.5 \sin \frac{\pi}{6}(x - c) + 21.5$. The sine curve begins at the midline, which in this case is 21.5. Table 8.12 shows that in the month of October, the maximum temperature was 21.5°C. This can be written as the point (10, 21.5). This point can be used as a starting point for the sine curve, meaning that the curve is shifted to the right 10 units. Therefore, $c = 10$.

6. After the phase shift, the curve begins at the midline and increases, following the behavior of a general sine curve. There are no reflections.

The equation $y = 6.5 \sin \frac{\pi}{6}(x - 10) + 21.5$ is the model.

Example

The number of visitors at a resort rises and falls during the year according to the graph in Figure 8.48. Determine an equation of this graph in terms of the month number, t.

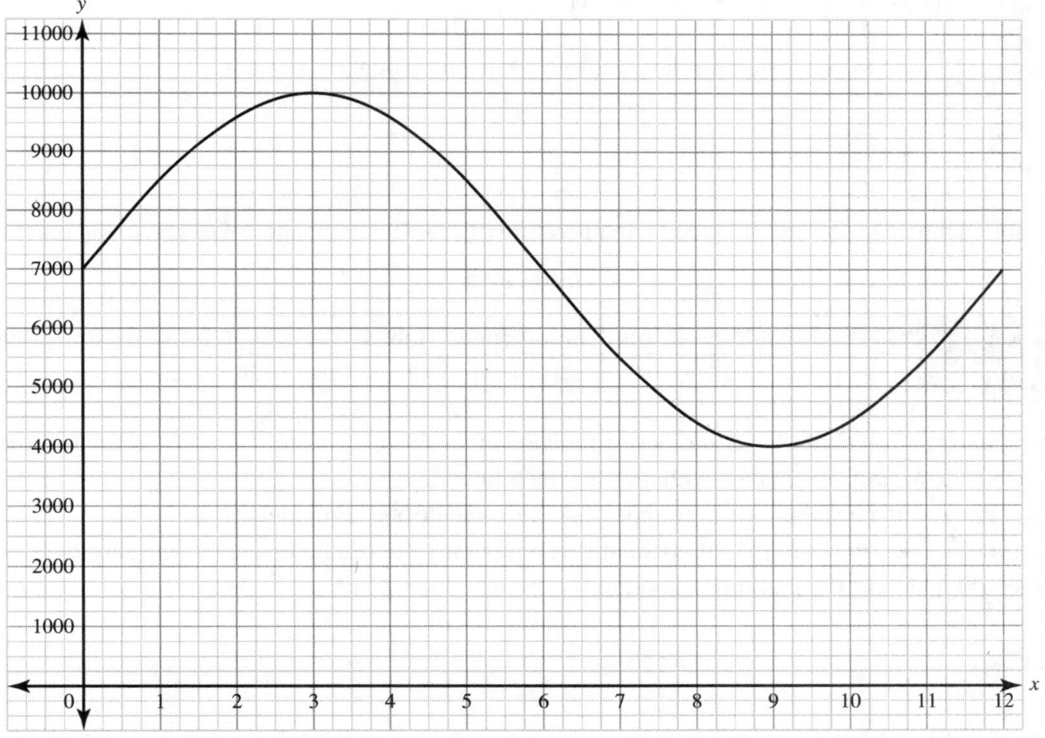

Figure 8.48

✓ Solution

Since the curve has the basic shape of the sine curve, its equation has the form $y = a \sin b(t + c) + d$. The 6 steps from the previous section will be used to determine a, b, c, and d.

1. The amplitude $= \dfrac{\text{maximum} - \text{minimum}}{2} = \dfrac{10{,}000 - 4{,}000}{2} = 3{,}000 = a$.
2. The period is 12 months since that is the interval of time shown in the graph to get one full cycle of sine. Since period $= \frac{2\pi}{b}$, then $12 = \frac{2\pi}{b}$ and so $b = \frac{\pi}{6}$.
3. The x-scale is not necessary here to find the equation.
4. The vertical shift should be found before a possible phase shift. The midline $= \dfrac{\text{maximum} + \text{minimum}}{2}$ $= \dfrac{10{,}000 + 4{,}000}{2} = 7{,}000 = d$.
5. The model has no phase shift.
6. The curve begins at the midline and increases, following the behavior of a general sine curve. There are no reflections.

The equation $y = 3{,}000 \sin\left(\frac{\pi}{6}t\right) + 7{,}000$ is the model.

Technology can be used to find an appropriate sinusoidal function model for a data set.

› Example

Table 8.13 shows the mean monthly maximum temperature (°C) for Old Bethpage, New York, in the year 2022.

Table 8.13

Month	Jan	Feb	Mar	April	May	June	July	Aug	Sept	Oct	Nov	Dec
Temp (°C)	15	14	15	18	21	25	27	26	24	20	18	16

(a) Using January = 1, February = 2, etc., a sine function of the form $T = a \sin b(m - c) + d$ can be used to model the data. Find an equation without using technology.

(b) Use technology to check your answer to part (a). How well does your model fit?

(c) Using your model from part (b), approximate the temperature in Old Bethpage for February 2023. Comment on the reasonableness of this approximation.

✓ Solution

(a) Since the equation has the form $T = a \sin b(m - c) + d$, the 6 steps from the previous section along with a scatterplot of the data can be used to determine a, b, c, and d. Figure 8.49 shows the scatterplot.

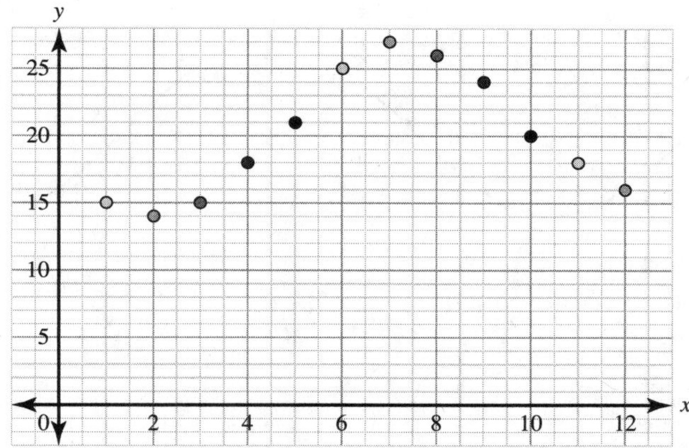

Figure 8.49

1. The amplitude $= \dfrac{\text{maximum} - \text{minimum}}{2} = \dfrac{27 - 14}{2} = 6.5 = a$.
2. The period is 12 months since the cycle of temperature repeats itself after each year. Since period $= \dfrac{2\pi}{b}$, then $12 = \dfrac{2\pi}{b}$ and so $b = \dfrac{\pi}{6}$.
3. The x-scale is not necessary here to find the equation.
4. The vertical shift should be found before a possible phase shift. The midline $= \dfrac{\text{maximum} + \text{minimum}}{2} = \dfrac{27 + 14}{2} = 20.5 = d$.
5. The model so far is $T = 6.5 \sin\frac{\pi}{6}(m - c) + 20.5$. The sine curve begins at the midline, which in this case is 20.5. Table 8.13 shows that in the month of May, the maximum temperature was 21°C, which is very close to 20.5°C. This can be written as the point (5, 21). This point can be used as a starting point for the sine curve, meaning that the curve is shifted to the right approximately 5 units. Therefore, $c = 5$.
6. After the phase shift, the curve begins at the midline and increases, following the behavior of a general sine curve. There are no reflections.

   The equation $T = 6.5 \sin\frac{\pi}{6}(m - 5) + 20.5$ is the model.

(b) To find the equation of a sinusoidal curve using technology, enter the data into your calculator. If using a TI-84, press STAT, EDIT, and list the months numerically into L1 and their respective temperatures into L2. After the data have been entered, press STAT and then CALC. Select C:SinReg. The number of iterations is the amount of full cycles wanted. In this case, enter 1. Enter in the period of 12, and select Calculate. Approximating the coefficients to 3 decimal places gives the equation $T = 6.129 \sin(0.597m - 2.840) + 20.557$.

To compare this equation to the one found in (a), factor out the b-value from the input:

$$T = 6.129 \sin 0.597(m - 4.757) + 20.557$$

The a- and d-values are close to the values found in part (a). The b-value found in part (a) was $\frac{\pi}{6} \approx 0.5235$, which is close to the calculated value of 0.597. The phase shift values differ by 0.243. The calculated equation opted to shift the function right 4.757 instead of by 5. Using the data in the table, the temperatures between months 4 and 5 are between 18°C and 21°C, which fall around the midline value of 20.5. So this is a reasonable phase shift. The equation calculated by hand is graphed along with the scatterplot in Figure 8.50, and the equation calculated by the calculator is graphed along with the scatterplot in Figure 8.51. Both seem to approximate the data reasonably. However, the equation found using technology has a closer fit to more of the data points.

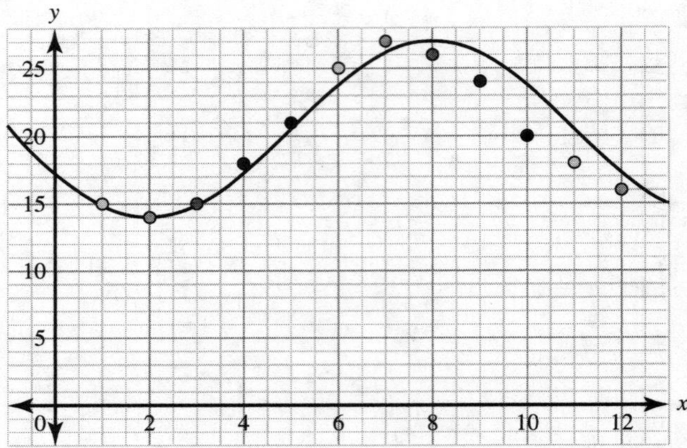

Figure 8.50

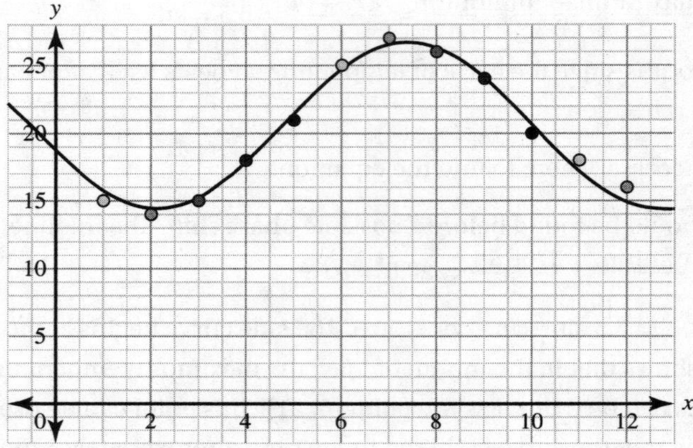

Figure 8.51

(c) To approximate the temperature in Old Bethpage for February 2023, substitute 14 for m in the equation $T = 6.129 \sin 0.597(m - 4.757) + 20.557$ and solve for T. Be sure that your calculator is set to radian mode:

$$T = 6.129 \sin 0.597((14) - 4.757) + 20.557 \approx 16.3119°\text{C}$$

When compared with the February temperature in Table 8.13, this temperature is slightly higher. However, looking at the graph in Figure 8.51 shows that the equation was graphed above the data point, explaining the possible overestimate.

Example

The height $h(t)$ meters of the tide above mean sea level on May 26 at Tobay Beach is modeled by the equation $h(t) = 4 \sin\left(\frac{\pi t}{6}\right)$, where t is the number of hours after midnight.

(a) Graph $y = h(t)$ for $0 \leq t \leq 24$.

(b) When was high tide, and what was the maximum height?

(c) What was the height at 2 P.M.?

(d) If a ship can cross the harbor provided the tide is at least 2 meters above mean sea level, when is crossing possible on May 26?

Solution

(a) To graph, follow the 6 steps and check with a graphing calculator set to radian mode.

1. Since $a = 4$, the amplitude $= 4$. The maximum value is 4, and the minimum value is -4.
2. Since $b = \frac{\pi}{6}$, the period $= \frac{2\pi}{b} = \frac{2\pi}{\frac{\pi}{6}} = 12$.
3. x-scale $= \frac{\text{period}}{4} = \frac{12}{4} = 3$.
4. Since $c = 0$, there is no phase shift.
5. Since $d = 0$, there is no vertical shift, so there is no midline.
6. Since there are no negations, there are no reflections.

The graph is shown in Figure 8.52.

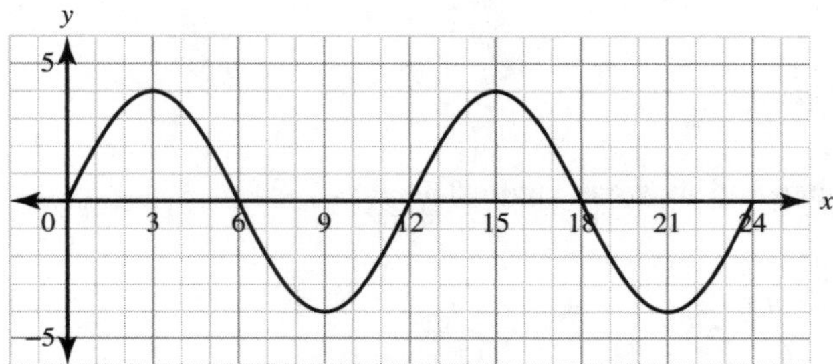

Figure 8.52

(b) High tide is at $t = 3$ and $t = 15$, which are 3 A.M. and 3 P.M. The maximum height is 4 meters above mean sea level.

(c) At 2 P.M., $t = 14$. To find the height, evaluate $h(14) = 4\sin\left(\frac{\pi(14)}{6}\right) \approx 3.464$ meters. The tide is 3.464 meters above mean sea level at 2 P.M.

(d) To find the time when crossing is possible, set $h(t) = 2$ and solve for t. To find the solution, use the graphing calculator. Let $Y1 = 4\sin\left(\frac{\pi t}{6}\right)$ and $Y2 = 2$. Then find their points of intersection as shown in Figure 8.53.

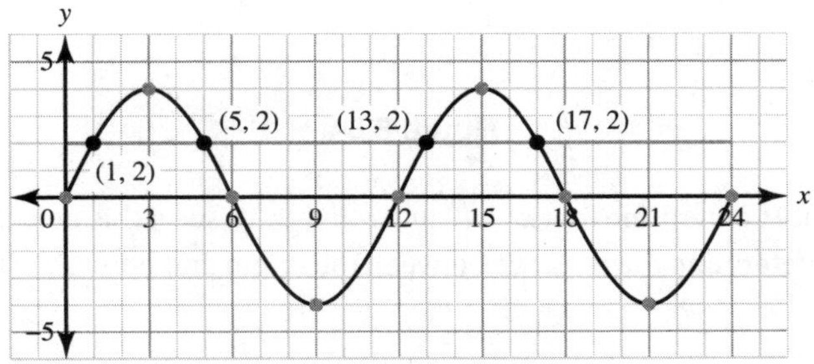

Figure 8.53

From the graph, there are four points of intersection: $t_1 = 1$, $t_2 = 5$, $t_3 = 13$, and $t_4 = 17$. Since it was given that a ship can cross the harbor when the tide is at least 2 meters above mean sea level, the ship can cross between 1 A.M. and 5 A.M. or between 1 P.M. and 5 P.M.

8.8 The Tangent Function

There are other trigonometric functions aside from the sine and cosine function, including the tangent function. In Section 8.2, the following relationship was determined between the tangent of an angle and the sine and cosine of that angle:

$$\sin\theta = \frac{y}{r} = \frac{y}{1} = y$$

$$\cos\theta = \frac{x}{r} = \frac{x}{1} = x$$

$$\tan\theta = \frac{y}{x}$$

$$\tan\theta = \frac{\sin\theta}{\cos\theta},\ \cos\theta \neq 0$$

Example

If $P(\sqrt{7}, -3)$ is a point on the terminal side of $\angle\theta$, find the value of $\tan\theta$.

Solution

The given information is graphed and shown in Figure 8.54.

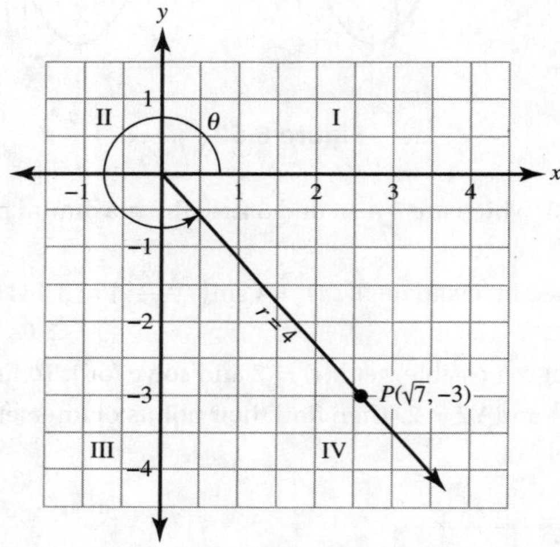

Figure 8.54

Using the coordinates of Point P, determine that $x = \sqrt{7}$ and $y = -3$. To see if this point lies on the unit circle, use the Pythagorean theorem to determine the distance between the origin and point P, notated as r:

$$x^2 + y^2 = r^2$$

$$(\sqrt{7})^2 + (-3)^2 = r^2$$

$$16 = r^2$$

Since r represents a distance, it is always positive:

$$r = \sqrt{16} = 4$$

Point *P* does not lie on the unit circle. However, the trigonometric values can still be determined:

$$\sin\theta = \frac{y}{r} = \frac{-3}{4}$$
$$\cos\theta = \frac{x}{r} = \frac{\sqrt{7}}{4}$$
$$\tan\theta = \frac{y}{x} = \frac{-3}{\sqrt{7}} = \frac{-3\sqrt{7}}{7}$$

Graph of $f(\theta) = \tan\theta$

The graph of the tangent function looks very different from the graphs of the sine and cosine functions. To determine points on the tangent graph, take the ratio of the sine and cosine values of selected angles, as shown in Table 8.14.

Table 8.14

θ	0	$\frac{\pi}{6}$	$\frac{\pi}{4}$	$\frac{\pi}{3}$	$\frac{\pi}{2}$	$\frac{3\pi}{4}$	π	$\frac{5\pi}{4}$	$\frac{3\pi}{2}$	$\frac{7\pi}{4}$	2π
sin θ	0	$\frac{1}{2}$	$\frac{1}{\sqrt{2}} = \frac{\sqrt{2}}{2}$	$\frac{\sqrt{3}}{2}$	1	$\frac{1}{\sqrt{2}} = \frac{\sqrt{2}}{2}$	0	$-\frac{1}{\sqrt{2}} = -\frac{\sqrt{2}}{2}$	-1	$-\frac{1}{\sqrt{2}} = -\frac{\sqrt{2}}{2}$	0
cos θ	1	$\frac{\sqrt{3}}{2}$	$\frac{1}{\sqrt{2}} = \frac{\sqrt{2}}{2}$	$\frac{1}{2}$	0	$-\frac{1}{\sqrt{2}} = -\frac{\sqrt{2}}{2}$	-1	$-\frac{1}{\sqrt{2}} = -\frac{\sqrt{2}}{2}$	0	$\frac{1}{\sqrt{2}} = \frac{\sqrt{2}}{2}$	1
tan θ	$\frac{0}{1} = 0$	$\frac{\frac{1}{2}}{\frac{\sqrt{3}}{2}} = \frac{1}{\sqrt{3}} = \frac{\sqrt{3}}{3}$	$\frac{\frac{1}{\sqrt{2}}}{\frac{1}{\sqrt{2}}} = 1$	$\frac{\frac{\sqrt{3}}{2}}{\frac{1}{2}} = \sqrt{3}$	$\frac{1}{0}$ undefined	$\frac{\frac{1}{\sqrt{2}}}{-\frac{1}{\sqrt{2}}} = -1$	$\frac{0}{-1} = 0$	$\frac{-\frac{1}{\sqrt{2}}}{-\frac{1}{\sqrt{2}}} = 1$	$\frac{-1}{0}$ undefined	$\frac{-\frac{1}{\sqrt{2}}}{\frac{1}{\sqrt{2}}} = -1$	$\frac{0}{1} = 0$

From Table 8.14, the function $f(\theta) = \tan\theta$ is not defined at $\theta = \frac{\pi}{2}$ or at any odd-integer multiple of $\frac{\pi}{2}$. So the graph has vertical asymptotes through these *x*-values as shown in Figure 8.55.

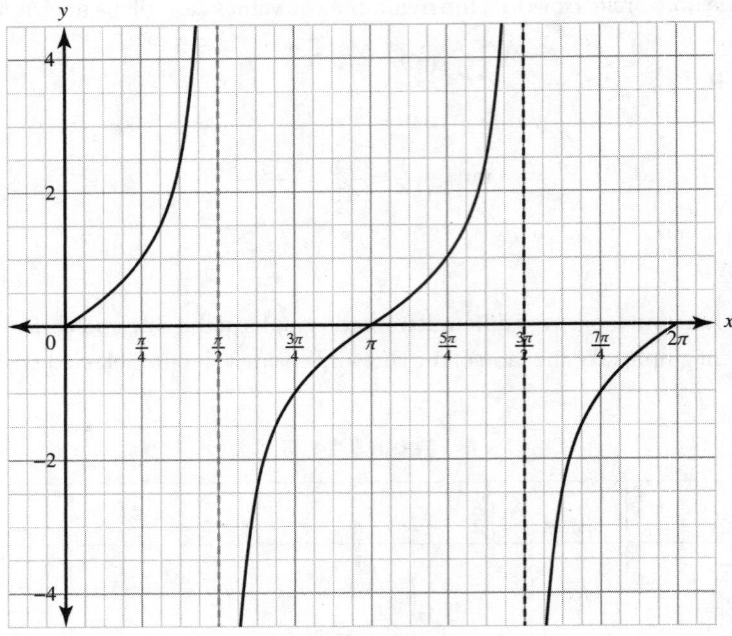

Figure 8.55

The Graph of the Tangent Function

Listed below are key facts about the graph of the tangent function:

- The period $= \pi$. One full cycle of tangent can be observed from $\left(\frac{\pi}{2}, \frac{3\pi}{2}\right)$. The length of this interval is π.
- There is no amplitude since there is no absolute maximum or absolute minimum values.
- The domain is all real numbers except for odd multiples of $\pm\frac{\pi}{2}$.
- The range is all real numbers.
- The x-intercepts occur at integer multiples of π.
- The vertical asymptotes occur at odd-integer multiples of $\pm\frac{\pi}{2}$. This means that the tangent function demonstrates periodic asymptotic behavior at input values $\theta = \frac{\pi}{2} + k\pi$ for integer values of k. The graph completes one full cycle between consecutive vertical asymptotes.
- The tangent function increases.
- The graph changes from concave down to concave up between consecutive asymptotes. The point of inflection exists at x-intercepts.

Transformations of $f(\theta) = \tan\theta$

The transformations that were applied to functions in the previous sections can also be applied to $f(\theta) = \tan\theta$ with the same results.

Additive Transformations

The graph of the additive transformation $g(\theta) = \tan\theta + d$ of the tangent function $f(\theta) = \tan\theta$ is a vertical translation of the graph of f by d units.

The graph of the additive transformation $h(\theta) = \tan(\theta + c)$ of the tangent function $f(\theta) = \tan\theta$ is a horizontal translation, or phase shift, of the graph of f by c units.

Example

Describe the transformation of $g(\theta) = \tan\theta + 4$ and $h(\theta) = \tan\theta - 2$ when compared with the parent function $f(\theta) = \tan\theta$. Graph each function along with its parent function on the same set of axes.

✓ Solution

The graph of $g(\theta) = \tan\theta + 4$ is a vertical shift 4 units up of the graph of $f(\theta) = \tan\theta$. The graph of $g(\theta)$ is solid and the graph of $f(\theta)$ is dashed as shown in Figure 8.56.

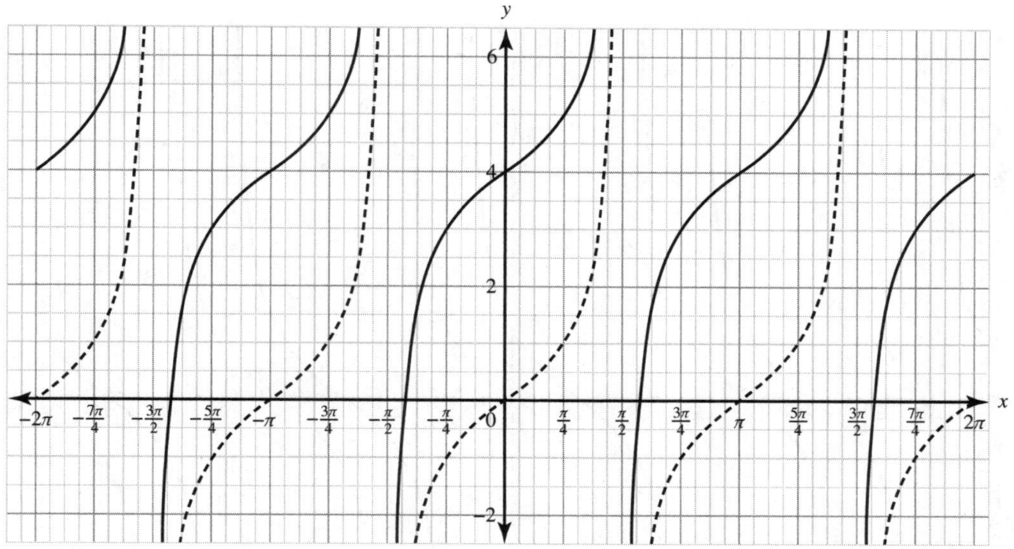

Figure 8.56

The graph of $h(\theta) = \tan\theta - 2$ is a vertical shift 2 units down of the graph of $f(\theta) = \tan\theta$. The graph of $h(\theta)$ is solid and the graph of $f(\theta)$ is dashed as shown in Figure 8.57.

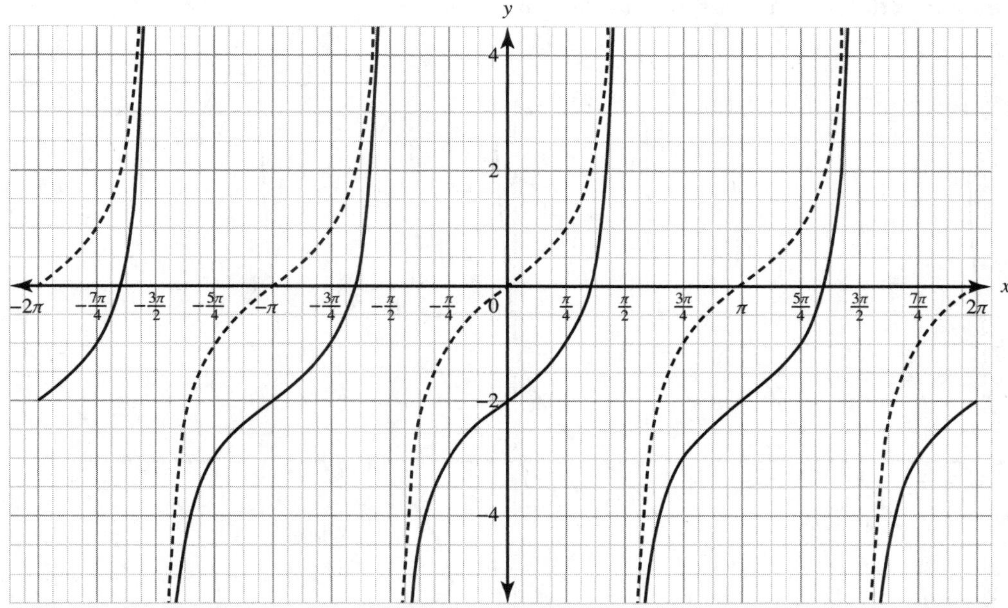

Figure 8.57

› Example

Describe the transformation of $g(\theta) = \tan(\theta + 4)$ and $h(\theta) = \tan(\theta - 2)$ when compared with the parent function $f(\theta) = \tan\theta$. Graph each function along with its parent function on the same set of axes.

Solution

The graph of $g(\theta) = \tan(\theta + 4)$ is a horizontal shift 4 units left of the graph of $f(\theta) = \tan\theta$. The graph of $g(\theta)$ is solid and the graph of $f(\theta)$ is dashed as shown in Figure 8.58.

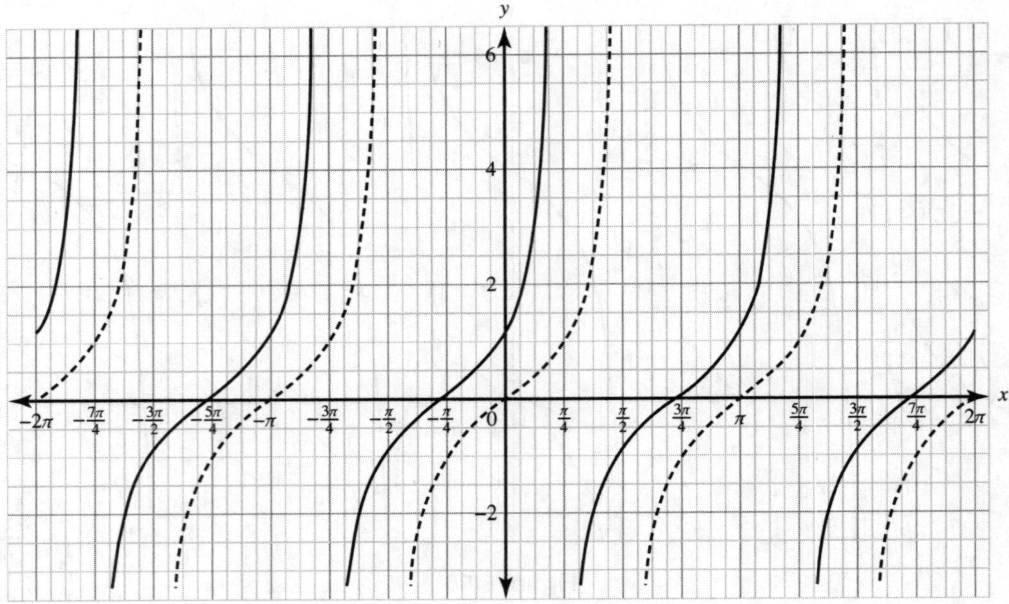

Figure 8.58

The graph of $h(\theta) = \tan(\theta - 2)$ is a horizontal shift 2 units right of the graph of $f(\theta) = \tan\theta$. The graph of $h(\theta)$ is solid and the graph of $f(\theta)$ is dashed as shown in Figure 8.59.

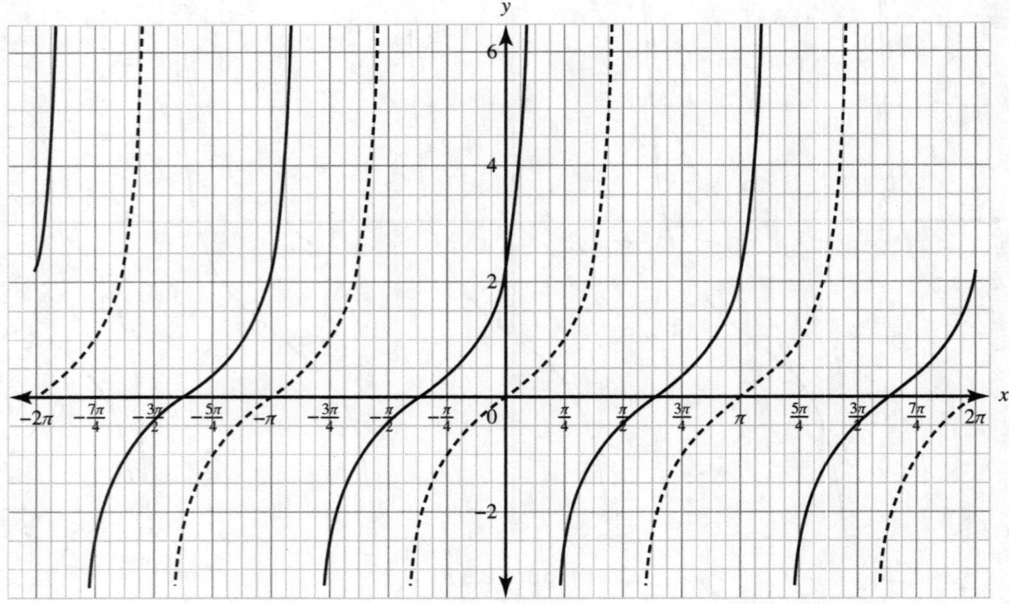

Figure 8.59

Multiplicative Transformations

The graph of the multiplicative transformation $g(\theta) = a \tan\theta$ of the tangent function $f(\theta) = \tan\theta$ is a vertical dilation of the graph of f by a factor of $|a|$.

If $a < 0$, the transformation involves a reflection over the x-axis.

The graph of the multiplicative transformation $h(\theta) = \tan(b\theta)$ of the tangent function $f(\theta) = \tan\theta$ is a horizontal dilation of the graph of f and differs in period by a factor of $\left|\frac{1}{b}\right|$.

If $b < 0$, the transformation involves a reflection over the y-axis.

Example

Describe the transformation of $g(\theta) = 4\tan\theta$ and $h(\theta) = \tan(2\theta)$ when compared to the parent function $f(\theta) = \tan\theta$. Graph each function along with its parent function on the same set of axes.

Solution

The graph of $g(\theta) = 4\tan\theta$ is a vertical dilation by a factor of 4. The graph of $g(\theta)$ is solid and the graph of $f(\theta)$ is dashed as shown in Figure 8.60.

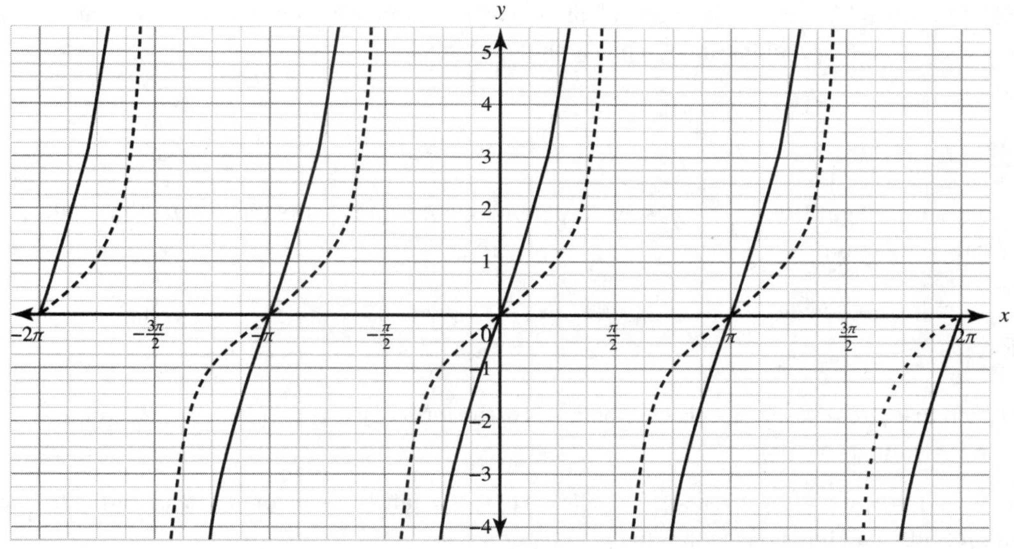

Figure 8.60

The graph of $h(\theta) = \tan(2\theta)$ is a horizontal dilation and differs in period by a factor of $\frac{1}{2}$. The graph of $h(\theta)$ is solid and the graph of $f(\theta)$ is dashed as shown in Figure 8.61.

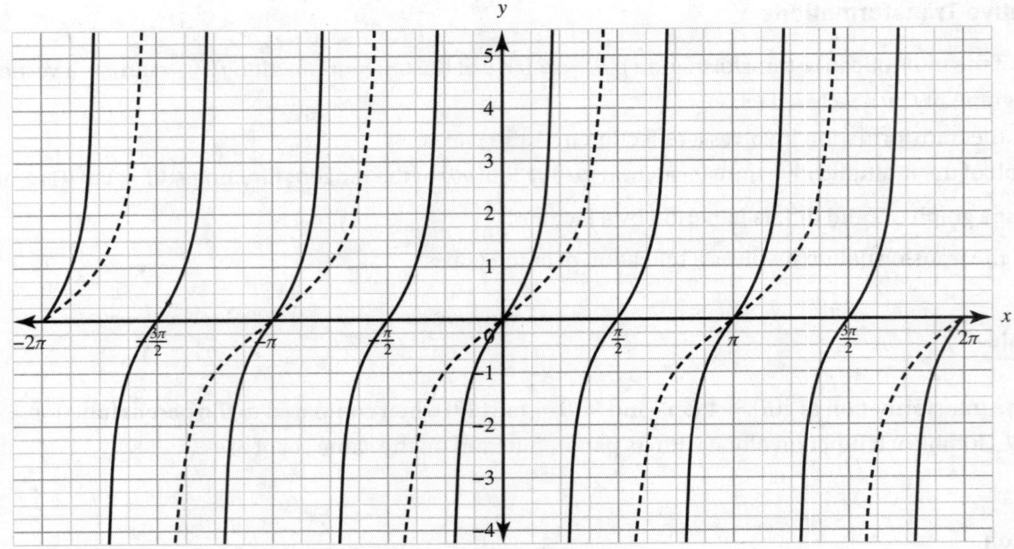

Figure 8.61

Graph of $f(\theta) = a \tan b(x + c) + d$

The graph of $f(\theta) = a \tan b(x + c) + d$ is a vertical dilation of the graph of $y = \tan\theta$ by a factor of $|a|$, has a period of $\left|\frac{1}{b}\right|\pi$ units, is a vertical shift by d units, and is a phase shift of c units.

Example

Given the equation $y = 2\tan 3\left(x + \frac{\pi}{4}\right) + 1$, state its parent function and list the transformations of the given equation from its parent function.

Solution

The parent function is $f(\theta) = \tan\theta$. The given equation is of the form of $y = a \tan b(x + c) + d$, where $a = 2$, $b = 3$, $c = \frac{\pi}{4}$, and $d = 1$.

The graph of $y = 2\tan 3\left(x + \frac{\pi}{4}\right) + 1$ is a vertical dilation of $f(\theta) = \tan\theta$ by a factor of 2 has a period of $\frac{\pi}{3}$ units, has a vertical shift up 1 unit, and has a phase shift to the left $\frac{\pi}{4}$ unit.

8.9 Inverse Trigonometric Functions

Since the graphs of the sine, cosine, and tangent functions do not pass the horizontal line test, none of these functions has an inverse function. Inverse trigonometric functions can be formed by restricting the domains of the original trigonometric functions so that the graphs of the restricted functions pass the horizontal line test.

For inverse trigonometric functions, the input and output values are switched from their corresponding trigonometric functions. The output value of an inverse trigonometric function is often interpreted as an angle measure, and the input is a value in the range of the corresponding function.

Inverse Sine Function

Since sine is a periodic function, it has an inverse only if its domain is restricted. In the interval $\left[-\frac{\pi}{2}, \frac{\pi}{2}\right]$, the graph of $y = \sin x$ passes the horizontal line test while $y = \sin x$ takes on its full range of values from $y = -1$ to $y = 1$, as shown in Figure 8.62.

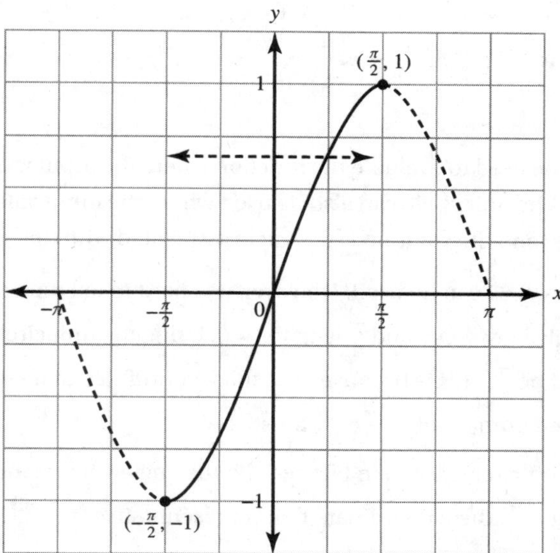

Figure 8.62 Graph of $y = \sin x$ on its restricted domain $\left[-\frac{\pi}{2}, \frac{\pi}{2}\right]$

Over the restricted domain $\left[-\frac{\pi}{2}, \frac{\pi}{2}\right]$, the sine function has an inverse function represented as $y = \sin^{-1} x$ or called $y = \arcsin x$, which can also be read as "the angle whose sine is x."

To graph $y = \arcsin x$, switch the input and output values for the graph of $y = \sin x$ and plot as shown in Figure 8.63.

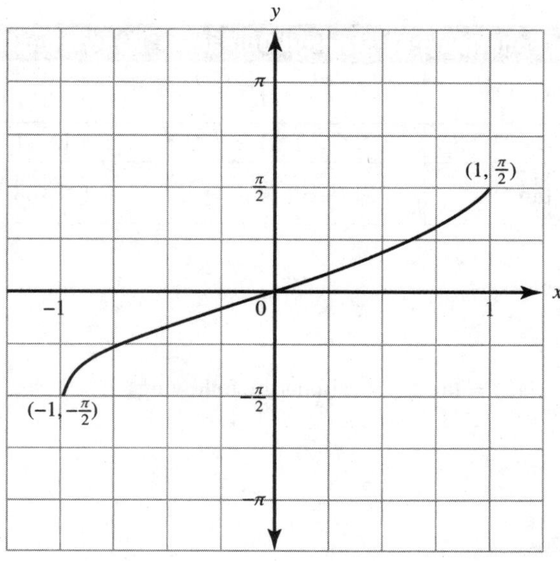

Figure 8.63 Graph of $y = \arcsin x$

The graph in Figure 8.63 can be used to evaluate inverse sine functions at specific values, and so can the unit circle.

Example

Given $f(\theta) = \arcsin\theta$, evaluate the following.

1. $f\left(\frac{1}{2}\right)$
2. $f\left(-\frac{\sqrt{3}}{2}\right)$

Solution

The graph of $f(\theta) = \arcsin\theta$ can be used to evaluate the function where the input value is found on the x-axis and the solution is the output value. The unit circle can also be used where the input value is located as the y-coordinate of the point of the unit circle and the solution is the angle associated with that given point.

1. For $f\left(\frac{1}{2}\right)$, the graph in Figure 8.63 can be used. However, it is hard to tell the output value specifically other than estimating it as an angle between 0 and $\frac{\pi}{4}$ when $x = \frac{1}{2}$. Using the unit circle shown in Figure 8.22, the angle whose sine is $\frac{1}{2}$ would be $\frac{\pi}{6}$. This is because at $\frac{\pi}{6}$, the y-coordinate that represents $\sin\theta$ is $\frac{1}{2}$. This angle also falls within the restricted domain of $\left[-\frac{\pi}{2}, \frac{\pi}{2}\right]$ for sine.
2. For $f\left(-\frac{\sqrt{3}}{2}\right)$, utilizing the unit circle shown in Figure 8.22, the angle whose sine is $-\frac{\sqrt{3}}{2}$ is $\frac{5\pi}{3}$ or its coterminal angle $-\frac{\pi}{3}$. This is because at $-\frac{\pi}{3}$, the y-coordinate that represents $\sin\theta$ is $-\frac{\sqrt{3}}{2}$. The angle $-\frac{\pi}{3}$ also falls within the restricted domain of $\left[-\frac{\pi}{2}, \frac{\pi}{2}\right]$ for sine.

Inverse Cosine and Inverse Tangent Functions

Since cosine and tangent are also periodic functions, they have an inverse only if their domains are restricted. Table 8.15 lists the key properties of the three inverse functions.

Table 8.15 Inverse Trigonometric Functions

Inverse Function	Restricted Domain	Restricted Range
$y = \sin^{-1}x$	$[-1, 1]$	$\left[-\frac{\pi}{2}, \frac{\pi}{2}\right]$
$y = \cos^{-1}x$	$[-1, 1]$	$[0, \pi]$
$y = \tan^{-1}x$	$(-\infty, \infty)$	$\left(-\frac{\pi}{2}, \frac{\pi}{2}\right)$

Example

If $f(x) = \sin x$, $g(x) = \cos^{-1}x$, and $h(x) = \tan^{-1}x$, evaluate the following.

1. $g\left(\frac{\sqrt{3}}{2}\right)$
2. $g(-1)$
3. $h(-1)$
4. $f\left(g\left(\frac{\sqrt{3}}{2}\right)\right)$

Solution

1. $g\left(\frac{\sqrt{3}}{2}\right) = \cos^{-1}\left(\frac{\sqrt{3}}{2}\right) = \frac{\pi}{6}$ since $\cos\left(\frac{\pi}{6}\right) = \frac{\sqrt{3}}{2}$.
2. $g(-1) = \cos^{-1}(-1) = \pi$ since $\cos(\pi) = -1$.
3. $h(-1) = \tan^{-1}(-1) = -\frac{\pi}{4}$ since $\tan\left(-\frac{\pi}{4}\right) = \tan\left(\frac{7\pi}{4}\right) = \frac{-\frac{\sqrt{2}}{2}}{\frac{\sqrt{2}}{2}} = -1$.
4. $f\left(g\left(\frac{\sqrt{3}}{2}\right)\right) = \sin\left(\cos^{-1}\left(\frac{\sqrt{3}}{2}\right)\right) = \sin\left(\frac{\pi}{6}\right) = \frac{1}{2}$.

Example

If $f(x) = \sin(\arctan x)$, what is the value of $f(1)$?

Solution

$$f(1) = \sin(\arctan(1)) = \sin\left(\frac{\pi}{4}\right) = \frac{\sqrt{2}}{2}$$

Example

What is the value of $\tan\left(\arccos\left(-\frac{\sqrt{3}}{2}\right)\right)$?

Solution

First evaluate $\arccos\left(-\frac{\sqrt{3}}{2}\right)$. The restricted range of arccosine is $[0, \pi]$. To find the angle whose $\cos\theta = -\frac{\sqrt{3}}{2}$, use the unit circle to find the reference angle whose x-coordinate is $\frac{\sqrt{3}}{2}$. This reference angle is $\frac{\pi}{6}$. Since the cosine value is negative, the angle must lie in Quadrant II. Therefore, the angle is $\theta = \pi - \frac{\pi}{6} = \frac{5\pi}{6}$. Now evaluate $\tan\left(\frac{5\pi}{6}\right)$. Using the unit circle, $\tan\left(\frac{5\pi}{6}\right) = \frac{\frac{1}{2}}{-\frac{\sqrt{3}}{2}} = -\frac{1}{\sqrt{3}} = -\frac{\sqrt{3}}{3}$.

Technology

A graphing calculator can also be used to evaluate the inverse trigonometric functions, although the answers will be decimal approximations and not exact values. Be sure the calculator is set to radian mode.

Example

Approximate $\sin^{-1}\left(\frac{3}{7}\right)$ to the nearest thousandth.

Solution

Since there is no point on the unit circle with a *y*-coordinate of $\frac{3}{7}$, the only way to find the angle whose sine is $\frac{3}{7}$ is to use a calculator. After checking to see that the calculator is set to radian mode, type 2ND SIN 3/7 and press ENTER. The calculator should display .4429110441. This approximation can be rounded to 0.443 radians.

8.10 Trigonometric Equations and Inequalities

Inverse trigonometric functions are useful in solving equations and inequalities involving trigonometric functions. It is important to keep in mind that since the inverse trigonometric functions have restricted domains, solutions to equations and inequalities may need to be modified.

To solve a trigonometric equation, first solve for the trigonometric function. Then find the angle, keeping in mind that the trigonometric function is positive or negative in more than one quadrant. Since trigonometric functions are periodic, trigonometric equations often have infinite solutions.

Example

Solve $2\sin(x) + 1 = 0$ for *x*, where $0 \leq x \leq 2\pi$.

Solution

First solve for the trigonometric function, $\sin x$:

$$2\sin(x) + 1 = 0$$

$$\sin x = -\frac{1}{2}$$

Next, solve for the angle *x* using inverse trigonometric functions and its reference angle:

$$\sin x = -\frac{1}{2}$$

$$x = \sin^{-1}\left(-\frac{1}{2}\right)$$

The reference angle is $\frac{\pi}{6}$ since that is the angle whose *y*-coordinate is $\frac{1}{2}$.

Since sine is negative in Quadrants III and IV, there are two possible solutions, as shown in Table 8.16.

Table 8.16

Quadrant III:	Quadrant IV:
$x = \pi + \frac{\pi}{6} = \frac{7\pi}{6}$	$x = 2\pi - \frac{\pi}{6} = \frac{11\pi}{6}$

Both solutions check out when looking at the complete unit circle where $0 \leq x \leq 2\pi$. However, using the reference angle method allows one to memorize only the angles in Quadrant I to figure out the angles in the other quadrants.

A Graphing Calculator and Inverse Trig Functions

A graphing calculator can be used to find the angle measure of inverse trigonometric functions. However, the calculator does not take into account all of the different angles that would fit the trigonometric equation. It is better to find the reference angle, which is the angle in Quadrant I, and place it in the appropriate quadrant to find the angle of the trigonometric equation that fits the value and the sign. To find the reference angle using the calculator, always evaluate the inverse trigonometric function using a positive input value, even if the input value is negative.

Example

Solve $\tan^2 x = \tan x + 2$, where $0 \le x \le 2\pi$, for x to the nearest tenth of a degree.

Solution

First solve for the trigonometric function, $\tan x$, by rewriting the given quadratic equation in standard form and then factoring:

$$\tan^2 x - \tan x - 2 = 0$$
$$(\tan x + 1)(\tan x - 2) = 0$$
$$\tan x + 1 = 0 \text{ or } \tan x - 2 = 0$$
$$\tan x = -1 \text{ or } \tan x = 2$$

Next, solve for angle x using inverse trigonometric functions and reference angles, as shown in Table 8.17.

Table 8.17

$x = \tan^{-1}(-1)$	$x = \tan^{-1}(2)$
Using the unit circle, the reference angle is $\frac{\pi}{4}$ since that is the angle whose ratio of $\frac{y}{x} = 1$. Since tangent is negative in Quadrants II and IV, there are two possible solutions. Quadrant II: $x = \pi - \frac{\pi}{4} = \frac{3\pi}{4}$ Quadrant IV: $x = 2\pi - \frac{\pi}{4} = \frac{7\pi}{4}$	Using the graphing calculator, the reference angle is approximately 1.107148718. Since tangent is positive in Quadrants I and III, there are two possible solutions. Quadrant I: $x = \text{reference angle} = 1.107148718$ Quadrant III: $x = \pi + 1.107148718 \approx 4.248741371$

The four possible solutions within the given domain are $x = \frac{3\pi}{4}, \frac{7\pi}{4}$, 1.1, 4.2 radians.

Example

Find the exact solutions of $\sqrt{2}\cos\left(x - \frac{3\pi}{4}\right) + 1 = 0$.

Solution

First solve for the trigonometric function $\cos\left(x - \frac{3\pi}{4}\right)$:

$$\cos\left(x - \frac{3\pi}{4}\right) = -\frac{1}{\sqrt{2}}$$

Next, solve for the angle $\left(x - \frac{3\pi}{4}\right)$ using inverse trigonometric functions and its reference angle:

$$\left(x - \frac{3\pi}{4}\right) = \cos^{-1}\left(-\frac{1}{\sqrt{2}}\right)$$

Using the unit circle, the reference angle is $\frac{\pi}{4}$ since that is the angle whose x-coordinate is $\frac{1}{\sqrt{2}}$.

Since cosine is negative in Quadrants II and III, there are two possible solutions in the interval $[0, 2\pi]$, as shown in Table 8.18.

Table 8.18

Quadrant II:	Quadrant III:
$x - \frac{3\pi}{4} = \pi - \frac{\pi}{4} = \frac{3\pi}{4}$	$x - \frac{3\pi}{4} = \pi + \frac{\pi}{4} = \frac{5\pi}{4}$
Solving for x:	Solving for x:
$x - \frac{3\pi}{4} = \frac{3\pi}{4}$	$x - \frac{3\pi}{4} = \frac{5\pi}{4}$
$x = \frac{3\pi}{4} + \frac{3\pi}{4} = \frac{6\pi}{4} = \frac{3\pi}{2}$	$x = \frac{5\pi}{4} + \frac{3\pi}{4} = \frac{8\pi}{4} = 2\pi$

Since trigonometric functions are periodic, there are infinitely many solutions to trigonometric equations since they repeat every period. This example did not include a specific domain for solutions. In that case, all solutions can be included for every 2π period. This can be represented by $k2\pi$.

The solutions are $x = \frac{3\pi}{2} + k2\pi$ and $x = 2\pi + k2\pi$ where k is an integer.

Example

Find the exact solutions of $\sqrt{3}\sin x = \cos x$.

Solution

Isolate the trigonometric functions on one side of the equation:

$$\frac{\sin x}{\cos x} = \frac{1}{\sqrt{3}}$$

Rewrite the trigonometric expression:

$$\tan x = \frac{1}{\sqrt{3}}$$

Solve for the angle x using inverse trigonometric functions and its reference angle:

$$x = \tan^{-1}\left(\frac{1}{\sqrt{3}}\right)$$

Using the unit circle, the reference angle is $\frac{\pi}{6}$ since that is the angle whose ratio $\frac{y}{x} = \frac{1}{\sqrt{3}}$.

Since tangent is positive in Quadrants I and III, there are two possible solutions in the interval $[0, 2\pi]$, as shown in Table 8.19.

Table 8.19

Quadrant I:	Quadrant III:
$x = \text{reference angle} = \frac{\pi}{6}$	$x = \pi + \frac{\pi}{6} = \frac{7\pi}{6}$

Since trigonometric functions are periodic, there are infinitely many solutions to trigonometric equations since they repeat every period. This example did not include a specific domain for solutions. In that case, all solutions can be included for every 2π period. This can be represented by $k2\pi$.

The solutions are $x = \frac{\pi}{6} + k2\pi$ and $x = \frac{7\pi}{6} + k2\pi$ where k is an integer.

Example

Solve for x given the inequality $\cos\left(2x + \frac{\pi}{4}\right) < \frac{1}{2}$ where $0 \leq x \leq 2\pi$.

Solution

Inequalities are different from equations in that there are multiple solutions in a solution set. First find where the cosine of an angle is less than $\frac{1}{2}$.

Consider the graphs of $y = \cos x$ and $y = \frac{1}{2}$ as shown in Figure 8.64.

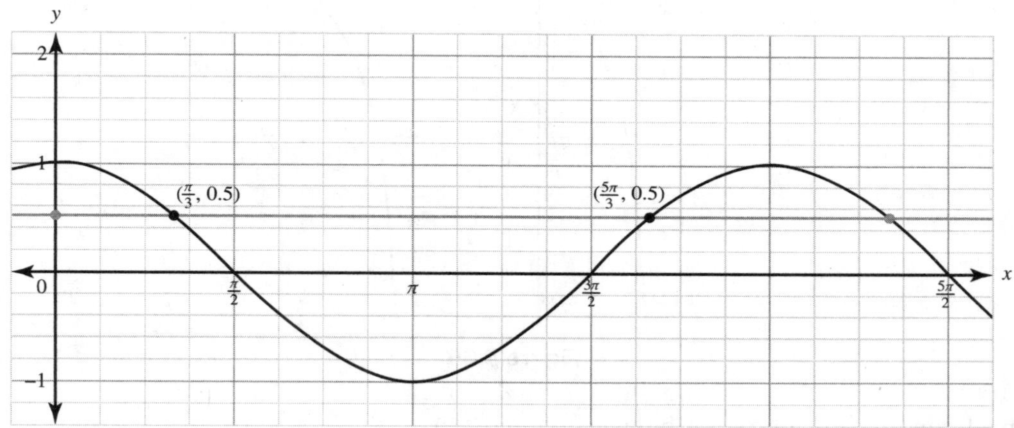

Figure 8.64

The graphs intersect when $x = \frac{\pi}{3}$ and $\frac{5\pi}{3}$. The graph of cosine is under the graph of $y = \frac{1}{2}$ between those values.

Therefore, the following inequality can be stated:

$$\frac{\pi}{3} < 2x + \frac{\pi}{4} < \frac{5\pi}{3}$$

Solve for x:

$$\frac{\pi}{3} - \frac{\pi}{4} < 2x < \frac{5\pi}{3} - \frac{\pi}{4}$$

$$\frac{\pi}{12} < 2x < \frac{17\pi}{12}$$

$$\frac{\pi}{12} \cdot \frac{1}{2} < x < \frac{17\pi}{12} \cdot \frac{1}{2}$$

$$\frac{\pi}{24} < x < \frac{17\pi}{24}$$

Example

Solve for x given the inequality $\tan(x) < 0.414$.

Solution

First, find where $\tan(x) < 0.414$ by seeing where the graph of the tangent curve is below the horizontal line $y = 0.414$, as shown in Figure 8.65.

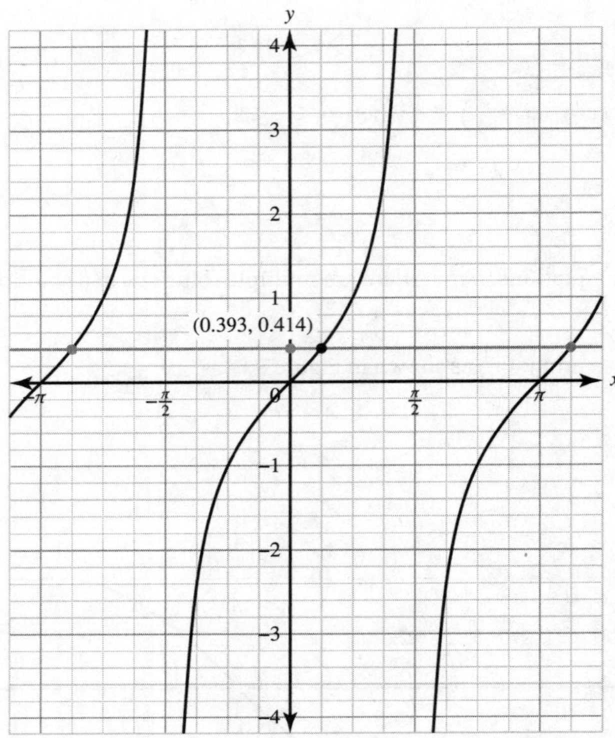

Figure 8.65

Using the graph, $\tan x < 0.414$ is between the angles $-\frac{\pi}{2}$ and 0.393.

Therefore, the following inequality can be stated:

$$-\frac{\pi}{2} < x < 0.393$$

Since no domain was specified, there are infinitely many answers that are repeated for every cycle of π.

Therefore, the angles that satisfy the given inequality are $-\frac{\pi}{2} + k\pi < x < 0.393 + k\pi$, where k is an integer.

8.11 The Secant, Cosecant, and Cotangent Functions

There are a total of six trigonometric functions. The remaining three are reciprocals of the basic trigonometric functions sine, cosine, and tangent.

The Secant Function

The secant function, $f(\theta) = \sec\theta$, is the reciprocal of the cosine function, where $\cos\theta \neq 0$. To determine the graph of $f(\theta) = \sec\theta$, consider Table 8.20, which shows values of both the cosine and secant functions.

Table 8.20

θ	-2π	$-\frac{3\pi}{2}$	$-\pi$	$-\frac{\pi}{2}$	0	$\frac{\pi}{2}$	π	$\frac{3\pi}{2}$	2π
cos θ	1	0	-1	0	1	0	-1	0	1
sec θ	$\frac{1}{1}=1$	$\frac{1}{0}$ undefined	$\frac{1}{-1}=-1$	$\frac{1}{0}$ undefined	$\frac{1}{1}=1$	$\frac{1}{0}$ undefined	$\frac{1}{-1}=-1$	$\frac{1}{0}$ undefined	$\frac{1}{1}=1$

At the x-values where the secant function is undefined, there are vertical asymptotes. The graphs of $f(\theta) = \sec\theta$ and its reciprocal $f(\theta) = \cos\theta$ are displayed on the same set of axes in Figure 8.66.

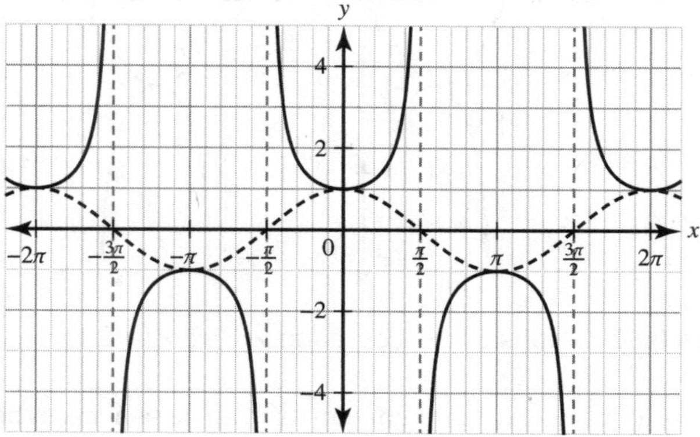

Figure 8.66 Graphs of $y = \sec x$ and $y = \cos x$ $(-2\pi \le x \le 2\pi)$

The following are key characteristics of the secant function that can be seen from the graph:

- The period of the secant function, like that of the cosine function, is 2π.
- Whenever $\cos x = \pm 1$, $\sec x = \pm 1$.
- The graph of $y = \sec x$ has asymptotes at the values of x for which $\cos x = 0$.
- A local maximum of $y = \cos x$ coincides with a local minimum of $y = \sec x$, and a local minimum of $y = \cos x$ coincides with a local maximum of $y = \sec x$.
- The range of the secant function is $(-\infty, -1] \cup [1, \infty)$.

The Cosecant Function

The cosecant function, $f(\theta) = \csc\theta$, is the reciprocal of the sine function, where $\sin\theta \neq 0$. To determine the graph of $f(\theta) = \csc\theta$, consider Table 8.21, which shows values of both the sine and cosecant functions.

Table 8.21

θ	-2π	$-\frac{3\pi}{2}$	$-\pi$	$-\frac{\pi}{2}$	0	$\frac{\pi}{2}$	π	$\frac{3\pi}{2}$	2π
sin θ	0	1	0	-1	0	1	0	-1	0
csc θ	$\frac{1}{0}$ undefined	$\frac{1}{1}=1$	$\frac{1}{0}$ undefined	$\frac{1}{-1}=-1$	$\frac{1}{0}$ undefined	$\frac{1}{1}=1$	$\frac{1}{0}$ undefined	$\frac{1}{-1}=-1$	$\frac{1}{0}$ undefined

At the x-values where the cosecant function is undefined, there are vertical asymptotes. The graphs of $f(\theta) = \csc\theta$ and its reciprocal $f(\theta) = \sin\theta$ are displayed on the same set of axes in Figure 8.67.

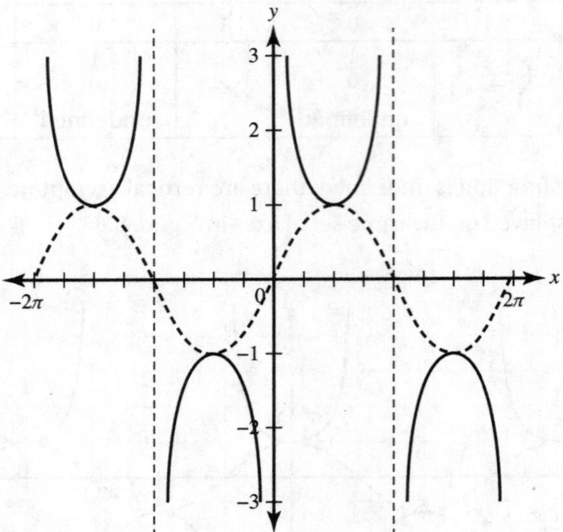

Figure 8.67 Graphs of $y = \csc x$ and $y = \sin x$ $(-2\pi \leq x \leq 2\pi)$

The following are key characteristics of the cosecant function that can be seen from the graph:

- The period of the cosecant function, like that of the sine function, is 2π.
- Whenever $\sin x = \pm 1$, $\csc x = \pm 1$.
- The graph of $y = \csc x$ has asymptotes at the values of x for which $\sin x = 0$.
- A local maximum of $y = \sin x$ coincides with a local minimum of $y = \csc x$, and a local minimum of $y = \sin x$ coincides with a local maximum of $y = \csc x$.
- The range of the cosecant function is $(-\infty, -1] \cup [1, \infty)$.

The Cotangent Function

The cotangent function, $f(\theta) = \cot\theta$, is the reciprocal of the tangent function, where $\tan\theta \neq 0$. Equivalently, $\cot\theta = \dfrac{\cos\theta}{\sin\theta}$, $\sin\theta \neq 0$. To determine the graph of $f(\theta) = \cot\theta$, consider Table 8.22, which shows values of the sine, cosine, and cotangent functions.

Table 8.22

θ	-2π	$-\frac{3\pi}{2}$	$-\pi$	$-\frac{\pi}{2}$	0	$\frac{\pi}{2}$	π	$\frac{3\pi}{2}$	2π
cos θ	1	0	−1	0	1	0	−1	0	1
sin θ	0	1	0	−1	0	1	0	−1	0
cot θ	$\frac{1}{0}$ undefined	$\frac{0}{1} = 0$	$\frac{-1}{0}$ undefined	$\frac{0}{-1} = 0$	$\frac{1}{0}$ undefined	$\frac{0}{1} = 0$	$\frac{-1}{0}$ undefined	$\frac{0}{-1} = 0$	$\frac{1}{0}$ undefined

At the x-values where the cotangent function is undefined, there are vertical asymptotes. The graphs of $f(\theta) = \cot\theta$ and $f(\theta) = \sin\theta$ are displayed on the same set of axes in Figure 8.68.

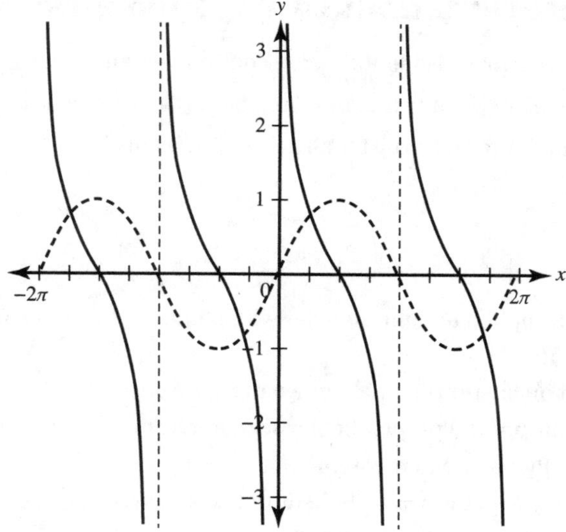

Figure 8.68 Graphs of $y = \cot x$ and $y = \sin x\ (-2\pi \leq x \leq 2\pi)$

The following are key characteristics of the cotangent function that can be seen from the graph:

- The period of the cotangent function, like that of the tangent function, is π.
- The graph of $y = \cot x$ has vertical asymptotes at the values of x for which $\sin x = 0$ since $\cot x = \frac{\cos x}{\sin x}$.
- The graph of the cotangent function is decreasing between consecutive vertical asymptotes.
- The cotangent function, like the tangent function, has no amplitude.

To evaluate the reciprocal trigonometric functions, first evaluate their reciprocal functions: sine, cosine, and tangent.

Example

Find the exact value of each of the following:

1. $\sec\left(\frac{\pi}{6}\right)$
2. $\csc\left(\frac{\pi}{2}\right)$
3. $\cot\left(\frac{7\pi}{4}\right)$
4. $\cos^{-1}(\sec\pi)$

Solution

1. $\sec\left(\frac{\pi}{6}\right) = \frac{1}{\cos\left(\frac{\pi}{6}\right)} = \frac{1}{\frac{\sqrt{3}}{2}} = \frac{2}{\sqrt{3}} = \frac{2\sqrt{3}}{3}$
2. $\csc\left(\frac{\pi}{2}\right) = \frac{1}{\sin\left(\frac{\pi}{2}\right)} = \frac{1}{1} = 1$
3. $\cot\left(\frac{7\pi}{4}\right) = \frac{1}{\tan\left(\frac{7\pi}{4}\right)} = \frac{1}{-1} = -1$
4. $\cos^{-1}(\sec\pi) = \cos^{-1}\left(\frac{1}{\cos(\pi)}\right) = \cos^{-1}\left(\frac{1}{-1}\right) = \cos^{-1}(-1) = \pi + k2\pi$

8.12 Equivalent Representations of Trigonometric Functions

Important trigonometric formulas, known as identities, are common when working with trigonometric expressions and equations. An example of an identity that has already been discussed is $\tan\theta = \frac{\sin\theta}{\cos\theta}$. This equation is an example of a trigonometric identity since it is true for all possible substitutions of θ for which the expressions are defined.

Pythagorean Identities

The Pythagorean theorem can be applied to right triangles with points on the unit circle at coordinates $(\cos\theta, \sin\theta)$, resulting in the Pythagorean identity.

Let $P(x, y)$ represent any point on the unit circle as shown in Figure 8.10. Constructing a right triangle by dropping a perpendicular segment from point P to the x-axis gives, by the Pythagorean theorem, $x^2 + y^2 = 1$. Since P is on the unit circle, $x = \cos\theta$ and $y = \sin\theta$. Substituting $\cos\theta$ for x and $\sin\theta$ for y produces the Pythagorean identity: $\sin^2\theta + \cos^2\theta = 1$.

Pythagorean Trigonometric Identity

$\sin^2\theta + \cos^2\theta = 1$

Dividing each term of the Pythagorean identity by either $\sin^2\theta$ or $\cos^2\theta$ leads to two additional Pythagorean identities that are listed in Table 8.23. Each of the three Pythagorean trigonometric identities may be written in more than one way.

Table 8.23 Pythagorean Trigonometric Identities

Three Pythagorean Trigonometric Identities	Equivalent Forms
▪ $\sin^2\theta + \cos^2\theta = 1$	▪ $\sin^2\theta = 1 - \cos^2\theta$ or $\cos^2\theta = 1 - \sin^2\theta$
▪ $\tan^2\theta + 1 = \sec^2\theta$	▪ $\tan^2\theta = \sec^2\theta - 1$ or $\sec^2\theta - \tan^2\theta = 1$
▪ $1 + \cot^2\theta = \csc^2\theta$	▪ $\cot^2\theta = \csc^2\theta - 1$ or $\csc^2\theta - \cot^2\theta = 1$

Example

Rewrite the expression in terms of $\sin\theta$: $\cos\theta(\sec\theta - \cos\theta)$.

Solution

A common strategy to rewrite trigonometric expressions is to break them into the basic trigonometric functions of $\sin\theta$ and $\cos\theta$:

$$\cos\theta(\sec\theta - \cos\theta) = \cos\theta\left(\frac{1}{\cos\theta} - \cos\theta\right)$$

Distribute:

$$= 1 - \cos^2\theta$$

Use the Pythagorean identity:

$$= \sin^2\theta$$

The Pythagorean identities can also be algebraically manipulated to establish relationships among trigonometric functions and the inverse trigonometric functions.

Consider the right triangle in Figure 8.69, whose hypotenuse is 1 and legs are x and y.

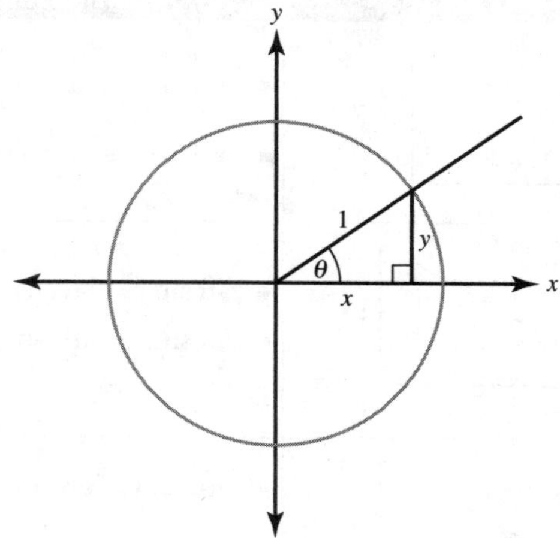

Figure 8.69

Instead of referring to the other leg as y, use the Pythagorean theorem to solve for the leg in terms of x:

$$x^2 + y^2 = 1^2$$
$$y^2 = 1 - x^2$$
$$y = \pm\sqrt{1 - x^2}$$

Since this triangle is drawn in Quadrant I, $y = +\sqrt{1 - x^2}$.

The following two trigonometric equations can be written:

$$\sin\theta = \frac{\sqrt{1 - x^2}}{1}$$

and

$$\cos\theta = \frac{x}{1}$$

Using inverse trigonometric functions, each can be written for θ:

$$\theta = \arcsin\sqrt{1 - x^2}$$

and

$$\theta = \arccos x$$

Since both are referring to the same angle θ, the two equations can be set equal:

$$\arccos x = \arcsin\sqrt{1 - x^2}$$

In the right triangle, the sides designated with lengths 1 and x can be rearranged, allowing for a different third side to be found in terms of x. This shows different inverse trigonometric relationships that are outlined in Table 8.24. The table assumes that x is positive and, depending on which quadrant the triangle is drawn in, will restrict the domain of certain functions.

Table 8.24

Diagram	Relationships
Right triangle with angle θ, hypotenuse 1, adjacent side x, opposite side $\sqrt{1-x^2}$	▪ $\arccos x = \arcsin\sqrt{1-x^2}$ ▪ $\arccos x = \arctan\dfrac{\sqrt{1-x^2}}{x}$
Right triangle with angle θ, hypotenuse 1, opposite side x, adjacent side $\sqrt{1-x^2}$	▪ $\arcsin x = \arccos\sqrt{1-x^2}$ ▪ $\arcsin x = \arctan\dfrac{x}{\sqrt{1-x^2}}$
Right triangle with angle θ, hypotenuse $\sqrt{1+x^2}$, opposite side x, adjacent side 1	▪ $\arctan x = \arcsin\dfrac{x}{\sqrt{1+x^2}}$ ▪ $\arctan x = \arccos\dfrac{x}{\sqrt{1+x^2}}$

Sine and Cosine Sum Identities

There are identities that allow trigonometric functions of the sum or difference of two angles to be expressed in terms of combinations of trigonometric functions of the individual angles.

> **Sum Identities**
>
> $\sin(\alpha+\beta) = \sin\alpha\cos\beta + \cos\alpha\sin\beta$
>
> $\cos(\alpha+\beta) = \cos\alpha\cos\beta - \sin\alpha\sin\beta$

Example

Find the exact value of $\sin\left(\frac{\pi}{9}\right)\cos\left(\frac{2\pi}{9}\right) + \cos\left(\frac{\pi}{9}\right)\sin\left(\frac{2\pi}{9}\right)$.

Solution

Since neither angle, $\frac{\pi}{9}$ nor $\frac{2\pi}{9}$, has known coordinates on the unit circle, use a trigonometric identity to write an equivalent expression that will result in an angle with known coordinates on the unit circle.

Use the sum identity for sine:

$$\sin\left(\frac{\pi}{9}\right)\cos\left(\frac{2\pi}{9}\right) + \cos\left(\frac{\pi}{9}\right)\sin\left(\frac{2\pi}{9}\right) = \sin\left(\frac{\pi}{9}+\frac{2\pi}{9}\right) = \sin\left(\frac{3\pi}{9}\right) = \sin\left(\frac{\pi}{3}\right) = \frac{\sqrt{3}}{2}$$

Example

Find the exact value of $\cos\left(\frac{7\pi}{12}\right)$.

Solution

Since the angle $\frac{7\pi}{12}$ is not an angle with known coordinates on the unit circle, use a trigonometric identity to write an equivalent expression that will result in an angle with known coordinates on the unit circle.

Since $\frac{\pi}{3} + \frac{\pi}{4} = \frac{4\pi}{12} + \frac{3\pi}{12} = \frac{7\pi}{12}$, use the formula for $\cos(\alpha + \beta)$, where $\alpha = \frac{\pi}{3}$ and $\beta = \frac{\pi}{4}$:

$$\begin{aligned}\cos\left(\tfrac{\pi}{3} + \tfrac{\pi}{4}\right) &= \cos\left(\tfrac{\pi}{3}\right)\cos\left(\tfrac{\pi}{4}\right) - \sin\left(\tfrac{\pi}{3}\right)\sin\left(\tfrac{\pi}{4}\right)\\ &= \left(\tfrac{1}{2}\right)\left(\tfrac{\sqrt{2}}{2}\right) - \left(\tfrac{\sqrt{3}}{2}\right)\left(\tfrac{\sqrt{2}}{2}\right)\\ &= \frac{\sqrt{2}}{4} - \frac{\sqrt{6}}{4}\\ &= \frac{\sqrt{2} - \sqrt{6}}{4}\end{aligned}$$

Sine and Cosine Difference Identities

The sum identities for sine and cosine can also be used as difference identities by adding a negative angle.

The derivation of the difference identity for sine is as follows:

$$\sin(\alpha - \beta) = \sin(\alpha + -\beta) = \sin\alpha\cos(-\beta) + \cos\alpha\sin(-\beta)$$

Since the graph of sine is symmetric with respect to the origin, sine is an odd function, meaning that $\sin(-\alpha) = -\sin\alpha$. Since the graph of cosine is symmetric with respect to the y-axis, cosine is an even function, meaning that $\cos(-\alpha) = \cos\alpha$.

Substitute this into the previous equation:

$$\begin{aligned}\sin(\alpha - \beta) &= \sin\alpha\cos(-\beta) + \cos\alpha\sin(-\beta)\\ \sin(\alpha - \beta) &= \sin\alpha\cos(\beta) + \cos\alpha\cdot - \sin(\beta)\\ \sin(\alpha - \beta) &= \sin\alpha\cos\beta - \cos\alpha\sin\beta\end{aligned}$$

The derivation of the difference identity for cosine is as follows:

$$\cos(\alpha - \beta) = \cos(\alpha + -\beta) = \cos\alpha\cos(-\beta) - \sin\alpha\sin(-\beta)$$

Substitute the even and odd function identities:

$$\begin{aligned}\cos(\alpha - \beta) &= \cos\alpha\cos(\beta) - \sin\alpha\cdot - \sin(\beta)\\ \cos(\alpha - \beta) &= \cos\alpha\cos\beta + \sin\alpha\sin\beta\end{aligned}$$

Difference Identities

$\sin(\alpha - \beta) = \sin\alpha\cos\beta - \cos\alpha\sin\beta$

$\cos(\alpha - \beta) = \cos\alpha\cos\beta + \sin\alpha\sin\beta$

Example

If $\alpha = \frac{\pi}{6}$ and $\beta = \arccos\frac{3}{5}$, what is the exact value of $\sec(\alpha - \beta)$?

Solution

Since $\sec(\alpha - \beta) = \dfrac{1}{\cos(\alpha - \beta)}$, first find the exact value of $\cos(\alpha - \beta)$. Since $\alpha = \frac{\pi}{6}$, that means $\sin\alpha = \frac{1}{2}$ and $\cos\alpha = \frac{\sqrt{3}}{2}$.

If $\beta = \arccos\frac{3}{5}$, then $\angle\beta$ is in Quadrant I where $\cos\beta = \frac{3}{5} = \frac{x}{r}$. Therefore, $x = 3$, $y = 4$, and $r = 5$. So $\sin\beta = \frac{y}{r} = \frac{4}{5}$.

Use the difference identity for cosine:

$$\cos(\alpha - \beta) = \cos\alpha\cos\beta + \sin\alpha\sin\beta$$
$$= \left(\frac{\sqrt{3}}{2}\right)\left(\frac{3}{5}\right) + \left(\frac{1}{2}\right)\left(\frac{4}{5}\right)$$
$$= \frac{3\sqrt{3}+4}{10}$$

Since $\cos(\alpha - \beta) = \frac{3\sqrt{3}+4}{10}$, $\sec(\alpha - \beta) = \frac{10}{3\sqrt{3}+4}$.

Double-Angle Identities

The sum identities for sine and cosine can also be used to derive double-angle identities for sine and cosine. The derivation of the double-angle identity for sine is as follows:

$$\sin(2\alpha) = \sin(\alpha + \alpha) = \sin\alpha\cos\alpha + \cos\alpha\sin\alpha$$

Combine like terms:

$$\sin(2\alpha) = 2\sin\alpha\cos\alpha$$

The derivation of the double-angle identity for cosine is as follows:

$$\cos(2\alpha) = \cos(\alpha + \alpha) = \cos\alpha\cos\alpha - \sin\alpha\sin\alpha$$
$$\cos(2\alpha) = \cos^2\alpha - \sin^2\alpha$$

Use the Pythagorean identities to substitute for $\cos^2\alpha$:

$$\cos(2\alpha) = 1 - \sin^2\alpha - \sin^2\alpha = 1 - 2\sin^2\alpha$$

Use the Pythagorean identities to substitute for $\sin^2\alpha$:

$$\cos(2\alpha) = \cos^2\alpha - (1 - \cos^2\alpha) = \cos^2\alpha - 1 + \cos^2\alpha = 2\cos^2\alpha - 1$$

Double-Angle Identities

$\sin(2\alpha) = 2\sin\alpha\cos\alpha$

$\cos(2\alpha) = \cos^2\alpha - \sin^2\alpha$

$\cos(2\alpha) = 1 - 2\sin^2\alpha$

$\cos(2\alpha) = 2\cos^2\alpha - 1$

Example

Given $\sin x = \frac{3}{4}$, find the exact value of $\cos(2x)$.

Solution

Since the value of $\sin x$ is given, choose the identity for $\cos(2x)$ that involves only sine, $\cos(2x) = 1 - 2\sin^2 x$:

$$\cos(2x) = 1 - 2\left(\frac{3}{4}\right)^2 = -\frac{1}{8}$$

Proving Trigonometric Identities

Properties of trigonometric functions, known trigonometric identities, and other algebraic properties can be used to verify additional trigonometric identities. Proving that a trigonometric equation is an identity involves showing that the two sides of the equation can be made to look exactly alike.

Strategies to Prove Trigonometric Identities

A few strategies can be applied when proving trigonometric identities:

- Start with the more complicated side of the equation, and express it using only sines and cosines.
- Factor.
- Combine and simplify fractional terms.
- Make a substitution using a known trigonometric identity such as a quotient, reciprocal, or Pythagorean identity.

If none of the above strategies work, it may help to change the other side of the equation to sines and cosines as well.

Example

Prove that the following equation is an identity for all values of θ for which the expressions are defined:

$$(\cot\theta + \csc\theta)(1 - \cos\theta) = \sin\theta$$

Solution

Starting with the left side of the equation, which is more complicated, rewrite the terms using only sine and cosine:

$$(\cot\theta + \csc\theta)(1 - \cos\theta) = \left(\frac{\cos\theta}{\sin\theta} + \frac{1}{\sin\theta}\right)(1 - \cos\theta)$$

Combine and simplify the fractional terms:

$$= \left(\frac{\cos\theta + 1}{\sin\theta}\right)(1 - \cos\theta)$$

Multiply the fractions:

$$= \frac{1 - \cos^2\theta}{\sin\theta}$$

Substitute using a Pythagorean identity:

$$= \frac{\sin^2\theta}{\sin\theta}$$

Simplify:

$$= \sin\theta$$

Example

Prove that the following equation is an identity for all values of θ for which the expressions are defined:

$$\sec\theta - \cos\theta = \sin\theta\tan\theta$$

Solution

Rewrite the terms on both sides of the equation using sine and cosine:

$$\frac{1}{\cos\theta} - \cos\theta = \sin\theta \cdot \frac{\sin\theta}{\cos\theta}$$

Simplify the right side of the equation using multiplication:

$$\frac{1}{\cos\theta} - \cos\theta = \frac{\sin^2\theta}{\cos\theta}$$

The new goal is to show the left side of the equation can equal the right side, $\frac{\sin^2\theta}{\cos\theta}$.

Combine and simplify the fractional terms on the left side of the equation:

$$\frac{1}{\cos\theta} - \cos\theta = \frac{1}{\cos\theta} - \frac{\cos^2\theta}{\cos\theta}$$

$$= \frac{1 - \cos^2\theta}{\cos\theta}$$

Substitute using a Pythagorean identity:

$$= \frac{\sin^2\theta}{\cos\theta}$$

Both sides of the equation are equal to $\frac{\sin^2\theta}{\cos\theta}$; the proof is complete.

Solving Trigonometric Equations

Previously, it was shown that solving trigonometric equations is done in two stages. First the trigonometric function is solved for, and then the angle is solved for. There may be more than one solution since a trigonometric function is periodic.

When presented with a trigonometric equation with mixed trigonometric functions or different angles, a substitution using a Pythagorean identity or reciprocal identity may be needed.

Example

Solve for x, to the nearest tenth of a radian, given that $3\cos^2 x + 5\sin x = 4$, where $0 \le x \le 2\pi$.

✓ Solution

Since the equation contains two different trigonometric functions, a trigonometric identity is needed to rewrite the equation in terms of one trigonometric function.

Using a Pythagorean identity, the equation can be rewritten into the following equivalent form:

$$3\cos^2 x + 5\sin x = 4$$

$$3(1 - \sin^2 x) + 5\sin x = 4$$

$$3 - 3\sin^2 x + 5\sin x = 4$$

This is now a quadratic equation that can be rewritten in standard form:

$$3\sin^2 x - 5\sin x + 1 = 0$$

First solve for the trigonometric function $\sin x$ using the quadratic formula:

$$\sin x = \frac{-(-5) \pm \sqrt{(-5)^2 - 4(3)(1)}}{2(3)}$$

$$\sin x = \frac{5 \pm \sqrt{13}}{6}$$

Table 8.25 shows how to solve for angle x.

Table 8.25

$x = \sin^{-1}\left(\frac{5+\sqrt{13}}{6}\right)$ $x \approx \sin^{-1}(1.4343)$ There is no possible angle solution here since the maximum value of $\sin x$ is 1.	$x = \sin^{-1}\left(\frac{5-\sqrt{13}}{6}\right)$ $x \approx \sin^{-1}(0.2324)$ Use the calculator to find the reference angle: reference angle $= 0.2346$ radians. Sine is positive in Quadrants I and II. Quadrant I: $x =$ reference angle ≈ 0.2346 radians Quadrant II: $x \approx \pi - 0.2346 = 2.9070$ radians

The equation has two solutions: 0.2 radians and 2.9 radians.

Example

Find all values of x in the interval $0 \leq x \leq 2\pi$ that satisfy the equation $\cos 2x - \cos x = 0$.

Solution

Since the equation contains two different types of angles, a trigonometric identity is needed to rewrite the equation in terms of angle x.

Use a double-angle identity to rewrite the equation into an equivalent form:

$$\cos 2x - \cos x = 0$$
$$(2\cos^2 x - 1) - \cos x = 0$$
$$2\cos^2 x - \cos x - 1 = 0$$

First solve for the trigonometric function $\cos x$ by factoring:

$$(2\cos x + 1)(\cos x - 1) = 0$$
$$\cos x = -\frac{1}{2},\ \cos x = 1$$

Table 8.26 shows how to solve for angle x.

Table 8.26

$x = \cos^{-1}\left(-\frac{1}{2}\right)$	$x = \cos^{-1}(1)$
The reference angle $= \frac{\pi}{3}$. Cosine is negative in Quadrants II and III. Quadrant II: $x = \pi - \frac{\pi}{3} = \frac{2\pi}{3}$ Quadrant III: $x = \pi + \frac{\pi}{3} = \frac{4\pi}{3}$	This happens at the quadrantal angles $x = 0$ and $x = 2\pi$.

The solutions are $x = 0, \frac{2\pi}{3}, \frac{4\pi}{3}, 2\pi$.

Multiple-Choice Questions

1. Which of the following choices is equivalent to the expression $\cos\alpha(\sec\alpha - \cos\alpha)$?

(A) $\cos^2\alpha$
(B) $\cos\alpha - \sin\alpha$
(C) $\sin^2\alpha$
(D) $\cot\alpha - 1$

2. If the terminal side of angle A in standard position is in Quadrant I and $\cos A = \frac{5}{13}$, what is the value of $\sin A \cdot \tan A$?

(A) $\frac{144}{65}$
(B) $\frac{60}{65}$
(C) $\frac{144}{169}$
(D) $\frac{60}{156}$

3. If x is a positive acute angle and $\cos x = a$, which of the following is an expression for $\tan x$ in terms of a?

(A) $\frac{1-a}{a}$
(B) $\sqrt{1-a^2}$
(C) $\frac{\sqrt{1-a^2}}{a}$
(D) $\frac{1}{1-a}$

4. If $\tan x = \frac{\sqrt{3}}{3}$ and $\cos x = -\frac{\sqrt{3}}{2}$, find the measure of $\angle x$.

(A) $\frac{2\pi}{3}$
(B) $\frac{5\pi}{6}$
(C) $\frac{7\pi}{6}$
(D) $\frac{5\pi}{3}$

5. The graph of which equation has an amplitude of $\frac{1}{2}$ and a period of π?

(A) $y = 2\sin 2x$
(B) $y = 2\sin\frac{1}{2}x$
(C) $y = \frac{1}{2}\sin 2x$
(D) $y = \frac{1}{2}\sin\frac{1}{2}x$

6. Which of the following expressions is equivalent to $\sin\theta$?

(A) $\cos(\theta + 1)$
(B) $\cos\left(\theta + \frac{\pi}{2}\right)$
(C) $\cos\left(\theta - \frac{\pi}{2}\right)$
(D) $\cos(\theta - 1)$

7. Which of the following equations could represent the graph shown below?

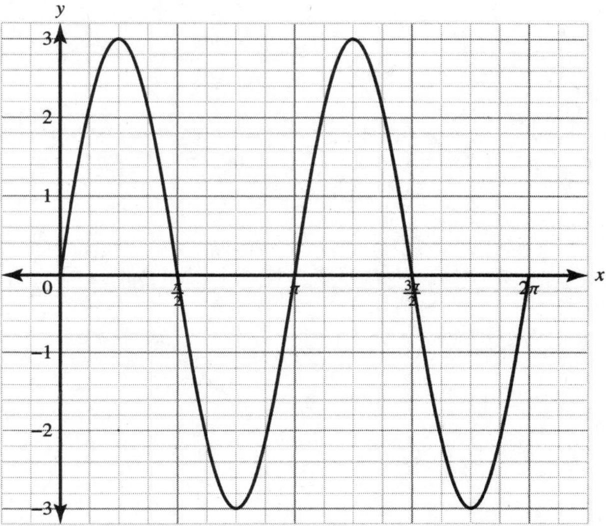

(A) $y = 3\sin 2x$
(B) $y = 2\sin 3x$
(C) $y = 3\sin x$
(D) $y = 2\sin 4x$

8. At which value of x does the graph of $f(\theta) = \tan\left(\theta - \frac{\pi}{4}\right)$ have a vertical asymptote?

(A) $-\frac{\pi}{4}$
(B) 0
(C) $\frac{\pi}{4}$
(D) $\frac{\pi}{2}$

9. If $f(x) = \sin(\cos^{-1} x)$, what is the value of $f\left(\frac{1}{2}\right)$?

(A) 1
(B) $\frac{1}{2}$
(C) $\frac{\sqrt{2}}{2}$
(D) $\frac{\sqrt{3}}{2}$

10. If $\cos x = \frac{12}{13}$, $\sin y = \frac{4}{5}$, and x and y are acute angles, what is the value of $\sin(x - y)$?

(A) $\frac{72}{65}$
(B) $\frac{56}{65}$
(C) $-\frac{16}{65}$
(D) $-\frac{33}{65}$

Answer Explanations

1. **(C)** Rewrite the expression in terms of sine and cosine:

$$\begin{aligned}\cos\alpha(\sec\alpha - \cos\alpha) &= \cos\alpha\left(\frac{1}{\cos\alpha} - \cos\alpha\right)\\ &= 1 - \cos^2\alpha = \sin^2\alpha\end{aligned}$$

2. **(A)** By the Pythagorean identity, $\sin^2 A + \cos^2 A = 1$, which gives $\sin^2 A + \left(\frac{5}{13}\right)^2 = 1$. So $\sin^2 A = 1 - \frac{25}{169} = \frac{144}{169}$. Since angle A terminates in Quadrant I, $\sin A = +\frac{12}{13}$. Since $\tan A = \frac{\sin A}{\cos A} = \frac{\frac{12}{13}}{\frac{5}{13}} = \frac{12}{5}$, that means $\sin A \cdot \tan A = \left(\frac{12}{13}\right)\left(\frac{12}{5}\right) = \frac{144}{65}$.

3. **(C)** Angle x is in Quadrant I. Since $\cos x = a$ and the Pythagorean identity, $\sin^2 x + \cos^2 x = 1$, then $\sin^2 x + (a)^2 = 1$. Solving for $\sin x$, $\sin x = +\sqrt{1 - a^2}$. Since $\tan x = \frac{\sin x}{\cos x} = \frac{\sqrt{1 - a^2}}{a}$.

4. **(C)** Since $\tan x = \frac{\sqrt{3}}{3} = \frac{\sin x}{\cos x}$ and $\cos x = -\frac{\sqrt{3}}{2}$, then $\sin x$ must be negative as well. So angle x must terminate in Quadrant III where both sine and cosine are negative. Using the unit circle, the angle in Quadrant III whose cosine value is $-\frac{\sqrt{3}}{2}$ is $\frac{7\pi}{6}$.

5. **(C)** If the amplitude of a sine curve is $\frac{1}{2}$, that means $a = \frac{1}{2}$. If the period is π, then $\pi = \frac{2\pi}{b}$ and so $b = 2$. A possible equation is $y = \frac{1}{2}\sin 2x$.

6. **(C)** The graph of a cosine curve can translate to a sine curve if shifted to the right $\frac{\pi}{2}$ unit. That means that $\sin\theta = \cos\left(\theta - \frac{\pi}{2}\right)$.

7. **(A)** The amplitude of the graph is 3, so $a = 3$. Since there are 2 full sine curves in the interval $[0, 2\pi]$, $b = 2$. A possible equation is $y = 3\sin 2x$.

8. **(A)** The graph of $f(\theta) = \tan\left(\theta - \frac{\pi}{4}\right)$ is the graph of tangent shifted to the right $\frac{\pi}{4}$ unit. The tangent graph has a vertical asymptote at $x = \frac{\pi}{2}$. So the graph of $f(\theta)$ will have a vertical asymptote at $x = -\frac{\pi}{2} + \frac{\pi}{4} = -\frac{\pi}{4}$.

9. **(D)** To evaluate $f\left(\frac{1}{2}\right)$, use the unit circle:

$$f\left(\frac{1}{2}\right) = \sin\left(\cos^{-1}\left(\frac{1}{2}\right)\right) = \sin\left(\frac{\pi}{3}\right) = \frac{\sqrt{3}}{2}$$

10. **(D)** Since x and y are given as acute angles, both terminate in Quadrant I. To determine $\sin x$ and $\cos y$, use the Pythagorean identity $\sin^2 x + \cos^2 x = 1$. When $\cos x = \frac{12}{13}$, then $\sin x = \frac{5}{13}$. When $\sin y = \frac{4}{5}$, then $\cos y = \frac{3}{5}$. To evaluate $\sin(x - y)$, use the identity $\sin(x - y) = \sin x\cos y - \cos x\sin y$. Substitute the values for the trigonometric expressions:

$$\sin(x - y) = \left(\frac{5}{13}\right)\left(\frac{3}{5}\right) - \left(\frac{12}{13}\right)\left(\frac{4}{5}\right) = -\frac{33}{65}$$

9

Polar Functions

Learning Objectives

In this chapter, you will learn:

- ➔ The different properties of the polar plane
- ➔ How to graph polar functions
- ➔ The different properties of polar functions

9.1 Trigonometry and Polar Coordinates

The polar coordinate system is graphed on a polar grid, which is a series of concentric circles radiating out from the origin, like a bull's-eye. This is different from the previously used rectangular coordinate system that is based on a grid of rectangles with ordered pairs (x, y). In the polar coordinate system, a point P is located relative to a fixed point, O, called the pole; a horizontal ray whose endpoint is the pole, called the polar axis; and the directed angle, θ, whose initial side is on the polar axis and whose vertex is on the pole. Polar coordinates are defined as an ordered pair, (r, θ), such that $|r|$ represents the radius of the circle on which the point lies and θ represents the measure of an angle in standard position whose terminal ray includes the point $P(r, \theta)$. This is shown in Figure 9.1.

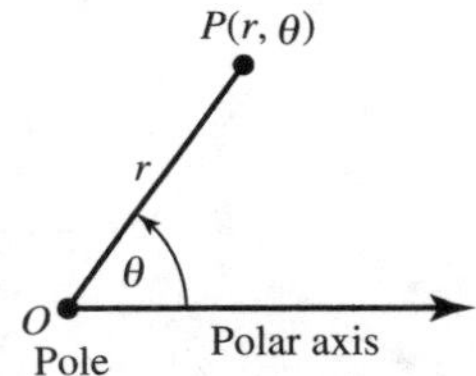

Figure 9.1

When locating a point, $P(r, \theta)$, in the polar system:

- If $\angle\theta$ is positive, the terminal side of the angle is rotated counterclockwise. If $\angle\theta$ is negative, the terminal side of the angle is rotated clockwise. These are shown in Figures 9.2 and 9.3.

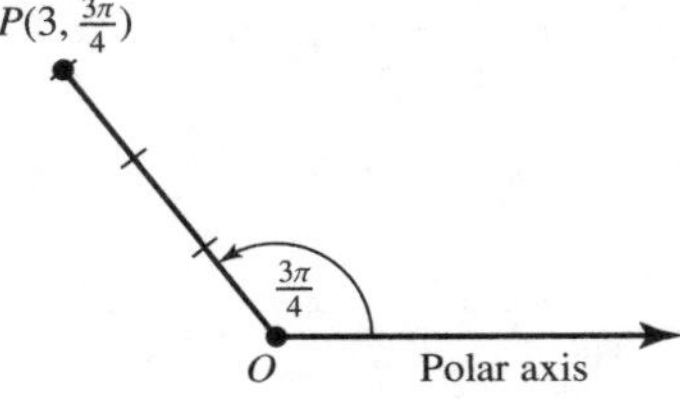

Figure 9.2 Graph of $P\left(3, \frac{3\pi}{4}\right)$

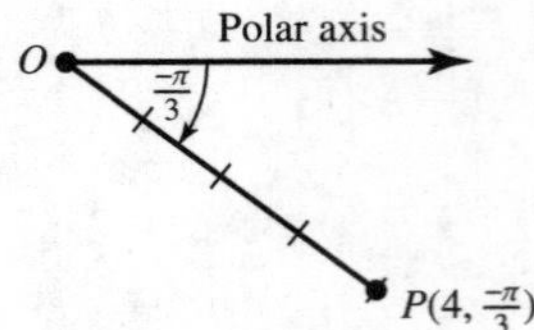

Figure 9.3 Graph of $P\left(4, \frac{-\pi}{3}\right)$

- If r is positive, P is r units along the terminal side of $\angle\theta$, as shown in Figure 9.4. If r is negative, first locate $(|r|, \theta)$. Then extend the terminal side of $\angle\theta$ its own length but in the opposite direction, as shown in Figure 9.5.

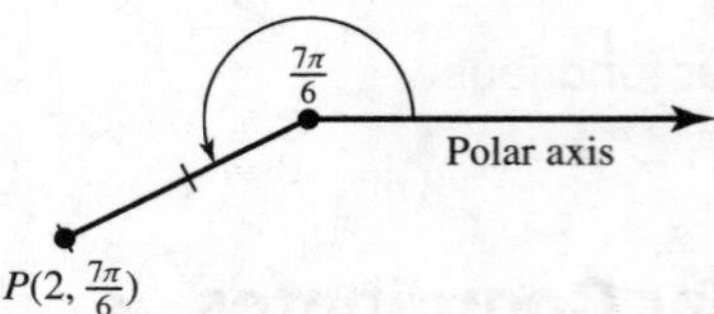

Figure 9.4 Graph of $P\left(2, \frac{7\pi}{6}\right)$

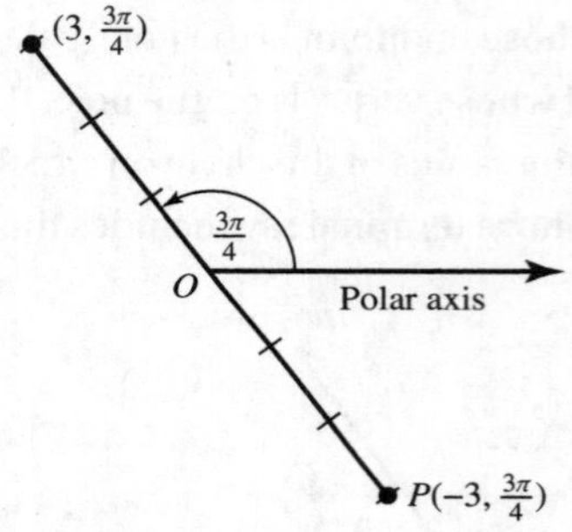

Figure 9.5 Graph of $P\left(-3, \frac{3\pi}{4}\right)$

In the polar coordinate system, the same point can be represented in many ways.

Example

Plot each of the following points in the polar coordinate system.

1. $C\left(5, \frac{7\pi}{6}\right)$
2. $M\left(5, \frac{19\pi}{6}\right)$
3. $P\left(-5, \frac{\pi}{6}\right)$

✓ Solution

1. Since $\theta = \dfrac{7\pi}{6}$ is positive, the angle is rotated counterclockwise and is 5 units along the terminal side, as shown in Figure 9.6.

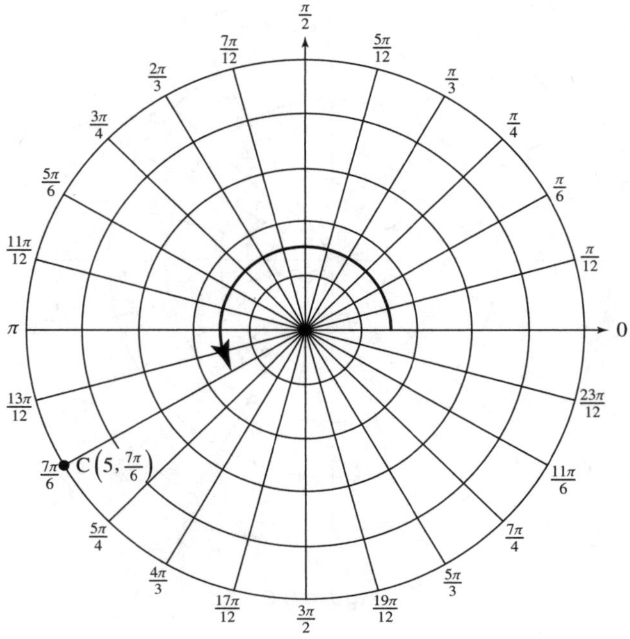

Figure 9.6

2. Since $\theta = \dfrac{19\pi}{6}$ is positive, the angle is rotated counterclockwise twice around the unit circle to land on its coterminal angle, $\dfrac{7\pi}{6}$, and is 5 units along the terminal side. This is seen in Figure 9.7.

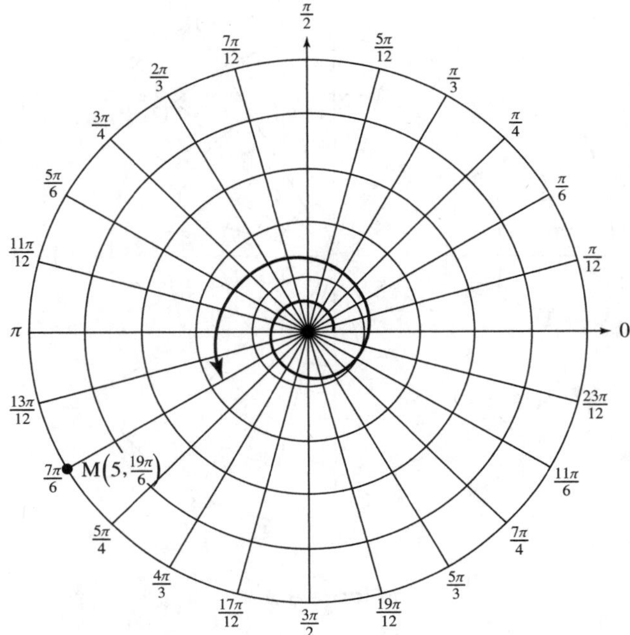

Figure 9.7

3. Since r is negative, first locate the point $\left(|-5|, \frac{\pi}{6}\right) = \left(5, \frac{\pi}{6}\right)$. Since $\theta = \frac{\pi}{6}$ is positive, the angle is rotated counterclockwise around the unit circle and is 5 units along the terminal side. This is shown in Figure 9.8.

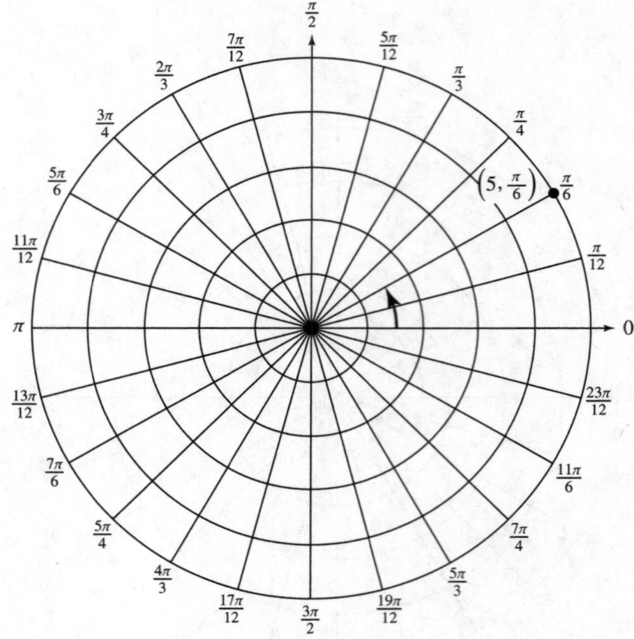

Figure 9.8

Then extend the terminal side of θ 5 units in the opposite direction, which will be along angle $\frac{7\pi}{6}$, and plot point P. This is seen in Figure 9.9.

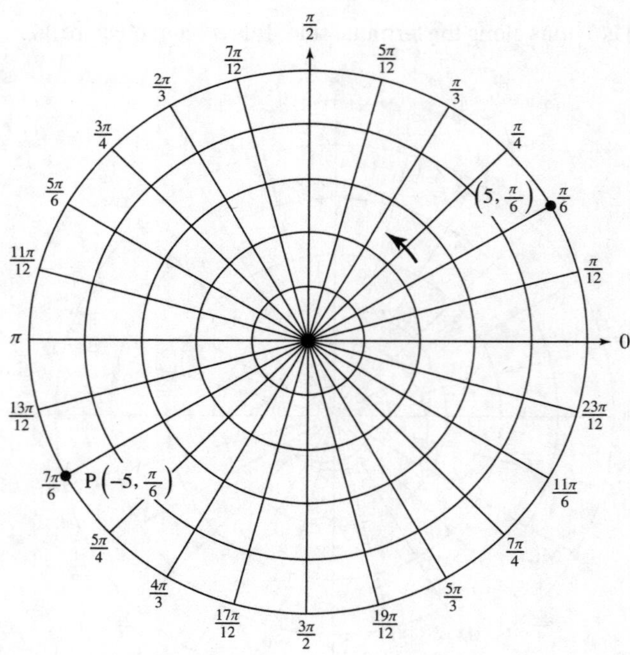

Figure 9.9

All three points in the previous example refer to the same point.

Coordinate System Conversions

The polar coordinate system can be converted to the rectangular coordinate system by aligning the pole with the origin and the polar axis with the positive x-axis, as shown in Figure 9.10. When the same point is represented in rectangular and polar coordinates, formulas that relate the two coordinate systems can be derived. Using the sine and cosine ratios in right triangle OAP gives:

$$\cos\theta = \frac{x}{r}, \text{ so } x = r\cos\theta$$

$$\sin\theta = \frac{y}{r}, \text{ so } y = r\sin\theta$$

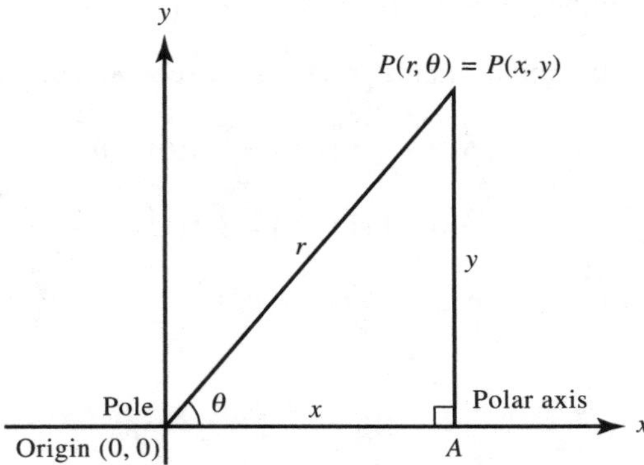

Figure 9.10 Rectangular and polar coordinate systems

Polar-Rectangular Conversion Formulas

Let point P have rectangular coordinates (x, y) and polar coordinates (r, θ).

- To go from polar (r, θ) to rectangular (x, y), use these relationships:

$$x = r\cos\theta \text{ and } y = r\sin\theta$$

- To go from rectangular (x, y) to polar (r, θ), use these relationships:

$$x^2 + y^2 = r^2 \rightarrow r = \sqrt{x^2 + y^2}$$

$$\tan\theta = \frac{y}{x} \rightarrow \theta = \arctan\left(\frac{y}{x}\right) \text{ for } x > 0, y > 0$$

$$\theta = \pi - \arctan\left(\left|\frac{y}{x}\right|\right) \text{ for } x < 0, y > 0$$

$$\theta = \arctan\left(\left|\frac{y}{x}\right|\right) + \pi \text{ for } x < 0, y < 0$$

$$\theta = 2\pi - \arctan\left(\left|\frac{y}{x}\right|\right) \text{ for } x > 0, y < 0$$

Example

Convert each point from polar to rectangular coordinates.

1. $\left(5, \frac{4\pi}{3}\right)$
2. $\left(-1, \frac{\pi}{2}\right)$

Solution

1. To convert $\left(5, \frac{4\pi}{3}\right)$ into rectangular coordinates (x, y), use the conversion formulas with $r = 5$ and $\theta = \frac{4\pi}{3}$:

$$x = r\cos\theta = 5\cos\frac{4\pi}{3} = 5\left(-\cos\left(\frac{\pi}{3}\right)\right) = 5\left(-\frac{\sqrt{3}}{2}\right) = -\frac{5\sqrt{3}}{2}$$

$$y = r\sin\theta = 5\sin\left(\frac{4\pi}{3}\right) = 5\left(-\sin\left(\frac{\pi}{3}\right)\right) = 5\left(-\frac{1}{2}\right) = -\frac{5}{2}$$

The equivalent rectangular coordinates are $(x, y) = \left(-\frac{5\sqrt{3}}{2}, -\frac{5}{2}\right)$.

2. To convert $\left(-1, \frac{\pi}{2}\right)$ into rectangular coordinates (x, y), use the conversion formulas with $r = -1$ and $\theta = \frac{\pi}{2}$:

$$x = r\cos\theta = -1\cos\frac{\pi}{2} = -1(0) = 0$$

$$y = r\sin\theta = -1\sin\left(\frac{\pi}{2}\right) = -1(1) = -1$$

The equivalent rectangular coordinates are $(x, y) = (0, -1)$.

Example

Convert each point from rectangular to polar form (r, θ), where θ is the smallest possible positive angle when $r > 0$.

1. $(3, \sqrt{3})$
2. $(-5, -12)$
3. $(2, -2)$
4. $(-1, 0)$

Solution

1. To convert $(3, \sqrt{3})$ into polar coordinates (r, θ), find θ and r when $x = 3$ and $y = \sqrt{3}$:

$$r = \sqrt{x^2 + y^2} = \sqrt{(3)^2 + (\sqrt{3})^2} = \sqrt{12} = 2\sqrt{3}$$

$$\theta = \arctan\left(\frac{y}{x}\right) \text{ for } x > 0, y > 0 \rightarrow \theta = \arctan\left(\frac{\sqrt{3}}{3}\right) = \frac{\pi}{6}$$

The equivalent polar coordinates are $(r, \theta) = \left(2\sqrt{3}, \frac{\pi}{6}\right)$.

2. To convert $(-5, -12)$ into polar coordinates (r, θ), find θ and r when $x = -5$ and $y = -12$:

$$r = \sqrt{x^2 + y^2} = \sqrt{(-5)^2 + (-12)^2} = \sqrt{169} = 13$$

$$\theta = \arctan\left(\frac{y}{x}\right) + \pi \text{ for } x < 0, y < 0 \rightarrow \theta = \arctan\left(\left|\frac{-12}{-5}\right|\right) + \pi \approx 1.176 + \pi \approx 4.318$$

The equivalent polar coordinates are $(r, \theta) = (13, 4.318)$.

3. To convert $(2, -2)$ into polar coordinates (r, θ), find θ and r when $x = 2$ and $y = -2$:

$$r = \sqrt{x^2 + y^2} = \sqrt{(2)^2 + (-2)^2} = \sqrt{8} = 2\sqrt{2}$$

$$\theta = 2\pi - \arctan\left(\left|\frac{y}{x}\right|\right) \text{ for } x > 0, y < 0 \rightarrow \theta = 2\pi - \arctan\left(\left|\frac{-2}{2}\right|\right) = 2\pi - \frac{\pi}{4} = \frac{7\pi}{4}$$

The equivalent polar coordinates are $(r, \theta) = \left(2\sqrt{2}, \frac{7\pi}{4}\right)$.

4. To convert $(-1, 0)$ into polar coordinates (r, θ), find θ and r when $x = -1$ and $y = 0$:

$$r = \sqrt{x^2 + y^2} = \sqrt{(-1)^2 + (0)^2} = \sqrt{1} = 1$$

Since $(-1, 0)$ lies on an axis, θ is a quadrantal angle. So $\theta = \pi$.

The equivalent polar coordinates are $(r, \theta) = (1, \pi)$.

Polar Representation of a Complex Number

When complex numbers of the form $a + bi$ are graphed, the rectangular coordinates (a, b) are plotted where a is measured along the horizontal real axis and b is measured along the vertical imaginary axis. The set of all such points forms the complex plane.

In the previous section, it was discovered that it is possible to represent the same point in both polar and rectangular coordinate systems. In a similar way, the complex plane can be aligned with the polar coordinate system and produce formulas that allow the conversion of a complex number into polar form.

Let point (a, b) in the complex plane correspond to the complex number $a + bi$, and let (r, θ) represent the polar coordinates of the same point. This is shown in Figure 9.11. Using the right triangle and the trigonometric ratios:

$$\cos\theta = \frac{a}{r} \rightarrow a = r\cos\theta$$

$$\sin\theta = \frac{b}{r} \rightarrow b = r\sin\theta$$

Therefore, the complex number, $a + bi$, can be written in polar form:

$$a + bi = (r\cos\theta) + (r\sin\theta)i = r(\cos\theta + i\sin\theta)$$

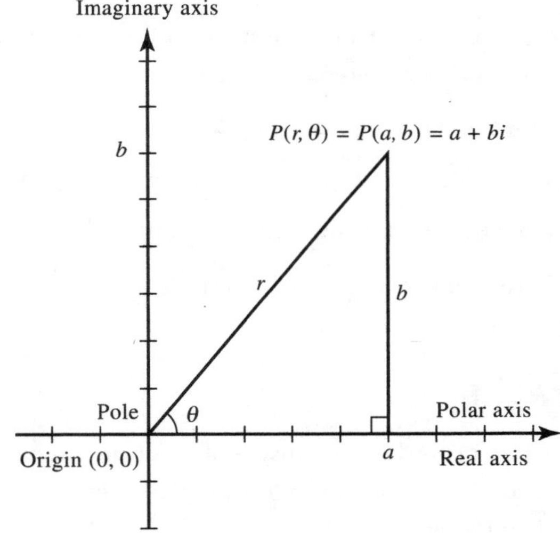

Figure 9.11

Polar Form of a Complex Number

The polar form of the complex number $z = a + bi$ is written as:

$$z = r(\cos\theta + i\sin\theta)$$

where $r^2 = a^2 + b^2 \rightarrow r = \sqrt{a^2 + b^2}$ and $\tan\theta = \frac{b}{a} \rightarrow \theta = \tan^{-1}\left(\frac{b}{a}\right)$.

The r-value is called the modulus of z, and θ is referred to as the argument of z.

The polar form of the complex number, $r(\cos\theta + i\sin\theta)$, is sometimes referred to as the trigonometric form or the "rcisθ" form of a complex number. For example, the shorthand notation $2\text{ cis }\frac{\pi}{3}$ means $2\left(\cos\frac{\pi}{3} + i\sin\frac{\pi}{3}\right)$.

Example

Convert each complex number into polar form.

1. $1 - \sqrt{3}\,i$
2. $5 + 0i$
3. $0 - 4i$

Solution

1. To write $1 - \sqrt{3}\,i$ in polar form, find θ and r when $a = 1$ and $b = -\sqrt{3}$:

$$r = \sqrt{a^2 + b^2} = \sqrt{(1)^2 + (-\sqrt{3})^2} = \sqrt{1+3} = \sqrt{4} = 2$$

Since a is positive and b is negative, θ is a Quadrant IV angle. To find θ, first find its reference angle and then subtract it from 2π:

$$\tan^{-1}\left(\left|\frac{-\sqrt{3}}{1}\right|\right) = \tan^{-1}\sqrt{3} = \frac{\pi}{3}$$

$$\theta = 2\pi - \frac{\pi}{3} = \frac{5\pi}{3}$$

The polar form of the complex number is $z = 2\left(\cos\frac{5\pi}{3} + i\sin\frac{5\pi}{3}\right)$.

2. To write $5 + 0i$ in polar form, find θ and r when $a = 5$ and $b = 0$:

$$r = \sqrt{a^2 + b^2} = \sqrt{(5)^2 + (0)^2} = \sqrt{25+0} = \sqrt{25} = 5$$

Since a is positive and b is 0, θ is a quadrantal angle on the positive x-axis, so $\theta = 0$. The polar form of the complex number is $z = 5(\cos 0 + i\sin 0)$.

3. To write $0 - 4i$ in polar form, find θ and r when $a = 0$ and $b = -4$:

$$r = \sqrt{a^2 + b^2} = \sqrt{(0)^2 + (-4)^2} = \sqrt{0+16} = \sqrt{16} = 4$$

Since a is 0 and b is negative, θ is a quadrantal angle on the negative y-axis. So $\theta = \frac{3\pi}{2}$.

The polar form of the complex number is $z = 4\left(\cos\frac{3\pi}{2} + i\sin\frac{3\pi}{2}\right)$.

9.2 Polar Function Graphs

The graph of the function $r = f(\theta)$ in polar coordinates consists of input-output pairs of values where the input values are angle measures and the output values are radii. The graphs of some polar coordinate equations can be easily seen from their appropriate definitions.

Example

Graph the following equations.

1. $r = 3$
2. $\theta = \frac{\pi}{6}$

Solution

1. For $r = 3$, the graph consists of all points (r, θ), where $r = 3$. This is all points whose distance from the origin is 3. Therefore, the graph is a circle with center O and radius 3, as shown in Figure 9.12.

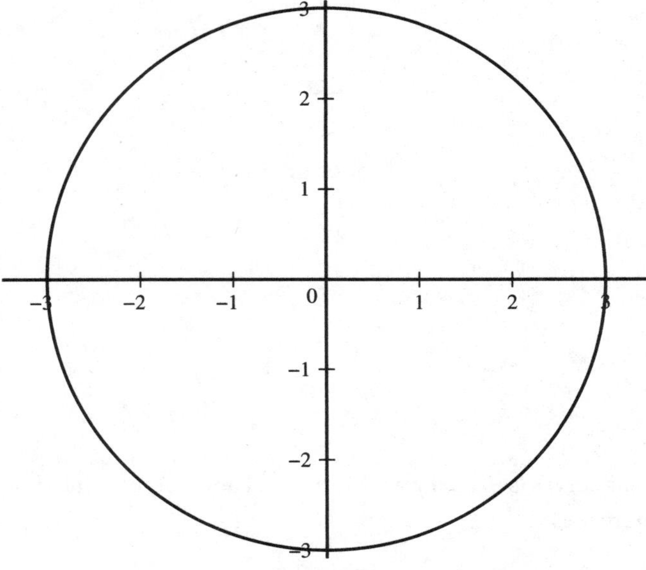

Figure 9.12

2. For $\theta = \frac{\pi}{6}$, the graph consists of all points (r, θ), where $\theta = \frac{\pi}{6}$. If $r \geq 0$, the points $\left(r, \frac{\pi}{6}\right)$ lie on the terminal side of an angle of $\frac{\pi}{6}$ radians whose initial side is the polar axis. If $r < 0$, the points $\left(r, \frac{\pi}{6}\right)$ lie on the extension of the terminal side of an angle of $\frac{\pi}{6}$ radians across the origin. The graph is a straight line as shown in Figure 9.13.

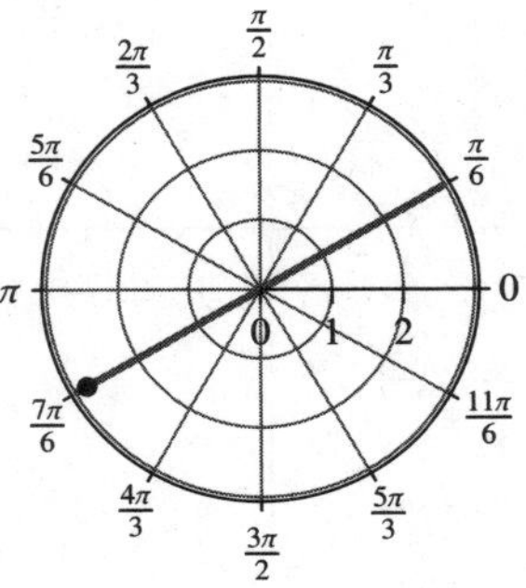

Figure 9.13

The domain of the polar function $r = f(\theta)$, given graphically, can be restricted to a desired portion of the function by selecting endpoints corresponding to the desired angle and radius. In the second part of the previous example if the full line was not the intended graph, the domain could have been restricted to having $r \geq 0$. Then the graph of $\theta = \frac{\pi}{6}$ would be the set of points $(r, \theta) = \left(r, \frac{\pi}{6}\right)$ with $r \geq 0$. This is the ray starting at the origin and going in the direction $\theta = \frac{\pi}{6}$ as shown in Figure 9.14.

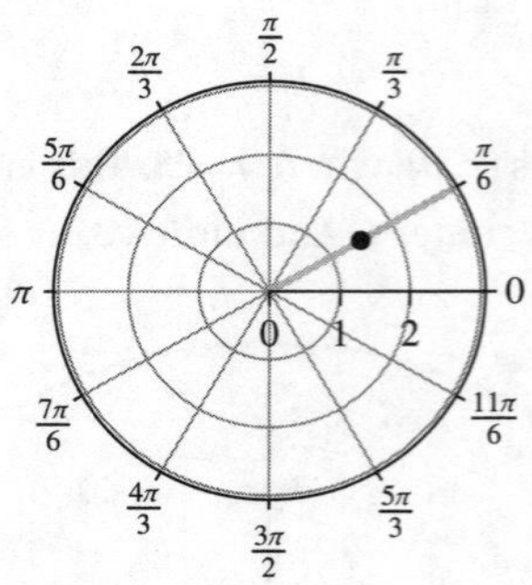

Figure 9.14

Example

Graph: $r = 6\cos\theta$

Solution

One strategy for graphing polar curves is to create a table of input and output values and then plot the points. This is shown in Table 9.1 and Figure 9.15.

Table 9.1

θ	0	$\frac{\pi}{4}$	$\frac{\pi}{2}$	$\frac{3\pi}{4}$	π	$\frac{5\pi}{4}$	$\frac{3\pi}{2}$	$\frac{7\pi}{4}$	2π
r	6	$3\sqrt{2}$	0	$-3\sqrt{2}$	-6	$-3\sqrt{2}$	0	$3\sqrt{2}$	6

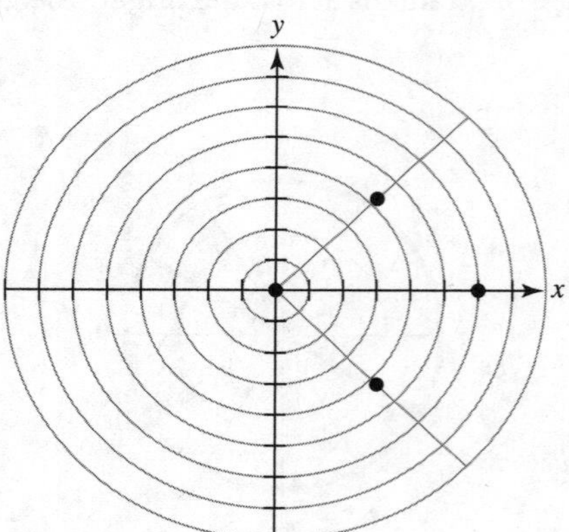

Figure 9.15

Although Table 9.1 has 9 ordered pairs, there are only 4 distinct points on the graph in Figure 9.15. To have a better understanding of the graph, an additional strategy must be used.

Graph $r = 6\cos\theta$ on the θr-plane and use it as a guide for graphing the equation on the xy-plane as shown in Figures 9.16 and 9.17. As θ ranges from 0 to $\frac{\pi}{2}$, r ranges from 6 to 0. In the xy-plane, this means that the curve starts 6 units from the origin on the positive x-axis, when $\theta = 0$, and gradually returns to the origin, at $\theta = \frac{\pi}{2}$.

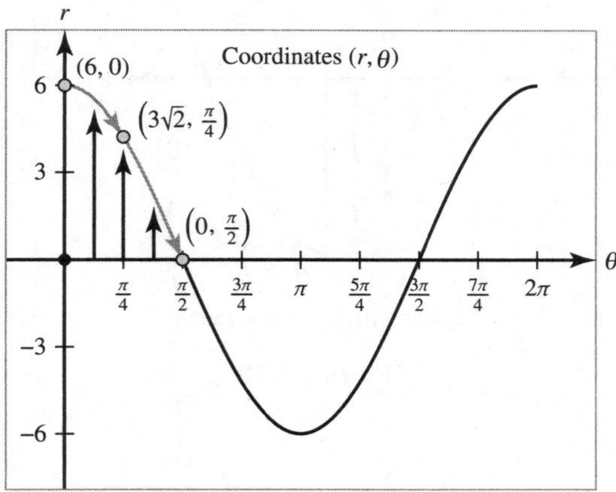

Figure 9.16

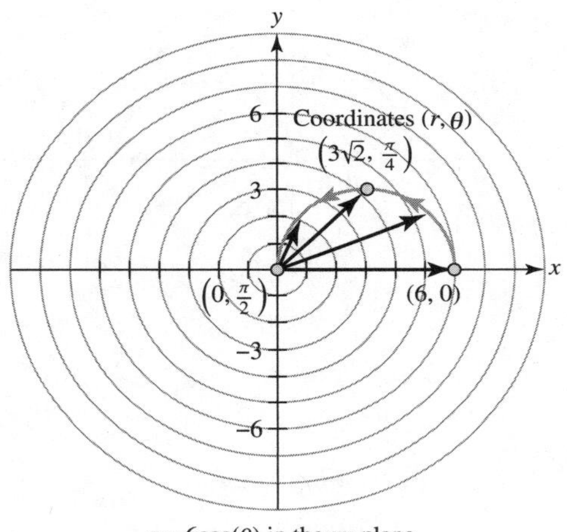

$r = 6\cos(\theta)$ in the xy-plane

Figure 9.17

In Figure 9.18 in the θr-plane, the arrows are drawn from the θ-axis to the curve $r = 6\cos\theta$. In Figure 9.19 in the xy-plane, each of these arrows starts at the origin and is rotated through the corresponding angle θ, in accordance with how polar coordinates are plotted.

Repeat the process as θ ranges from $\frac{\pi}{2}$ to π. As shown in Figure 9.18, the r-values are all negative. This means that in the xy-plane, instead of graphing in Quadrant II, graph in Quadrant IV, with all the angle rotations starting from the negative x-axis, as shown in Figure 9.19.

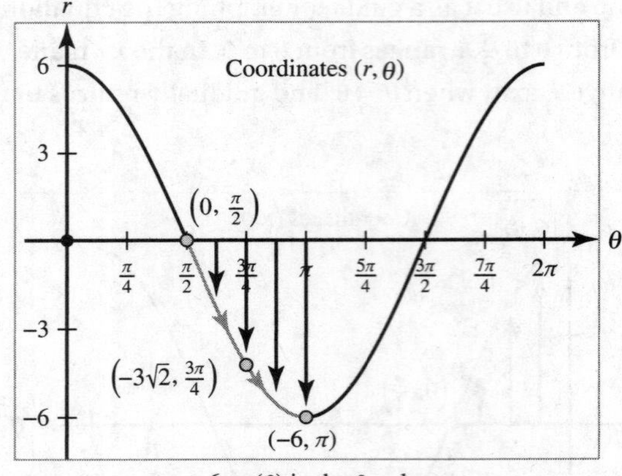

$r = 6\cos(\theta)$ in the θr-plane

Figure 9.18

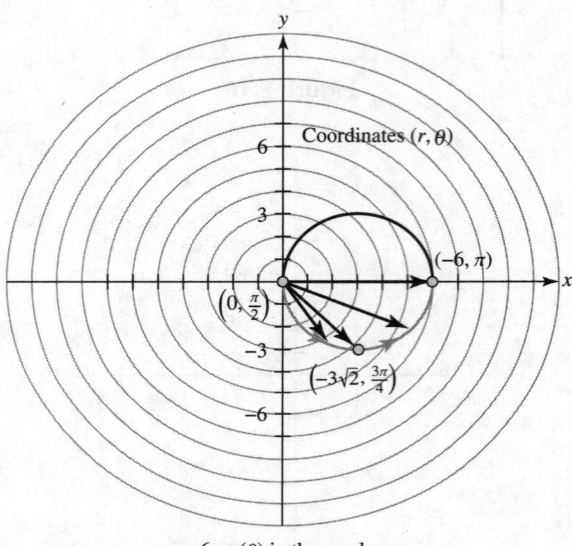

$r = 6\cos(\theta)$ in the xy-plane

Figure 9.19

As θ ranges from π to $\frac{3\pi}{2}$, the r-values are still negative, as shown in Figure 9.20. This means the graph is traced out in Quadrant I instead of Quadrant III. Since the $|r|$ for these values of θ match the r-values for θ in $\left[0, \frac{\pi}{2}\right]$, the curve begins to retrace itself at this point. The final graph is shown in Figure 9.21.

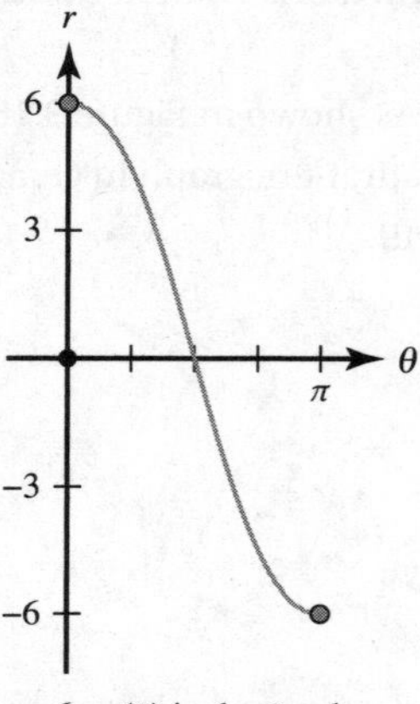

$r = 6\cos(\theta)$ in the θr-plane

Figure 9.20

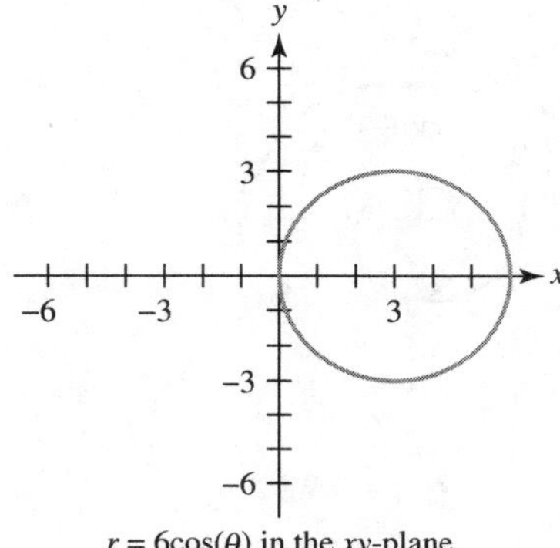

$r = 6\cos(\theta)$ in the xy-plane

Figure 9.21

Example

Graph: $r = \theta$

Solution

First graph the part of the curve with $\theta \geq 0$. Start at $r = \theta = 0$, which is the origin. As $r = \theta$ gets larger, the graph travels around counterclockwise and the radius also gets larger. The graph traces out a spiral as shown in Figure 9.22.

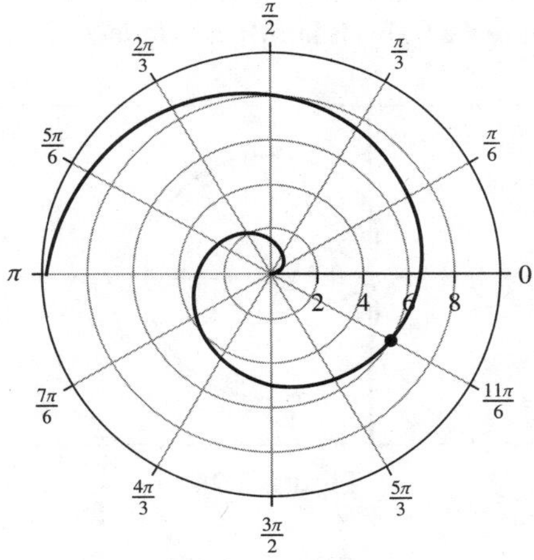

Figure 9.22

As $r = \theta$ gets negative, the graph travels around clockwise. Since r is also negative, the graph plots points with length $|r|$ in the reverse of the terminal side of the angle. This traces out a spiral that is the mirror image through the y-axis of the first half of the graph, as shown in Figure 9.23.

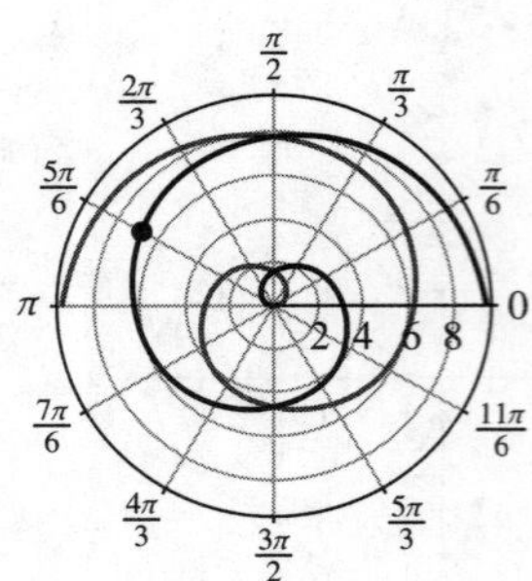

Figure 9.23

Polar graphs can also be graphed using technology such as a graphing calculator.

> When graphing polar functions in the form $r = f(\theta)$, changes in input values correspond to changes in angle measure from the positive x-axis and changes in output values correspond to changes in distance from the origin.

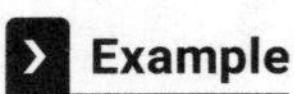

Example

Graph: $r = 2 + 4\cos\theta$

✓ Solution

Using a graphing calculator, press MODE and select POLAR. Go to Y= and notice that it no longer shows y but instead shows r. Enter $r = 2 + 4\cos\theta$. Go to WINDOW and enter the appropriate minimum and maximum values for x, y, and θ. Since cosine has a period of 2π, a complete graph can be obtained by taking $0 \le \theta \le 2\pi$. Press GRAPH, which is what is shown in Figure 9.24. This is known as a limaçon.

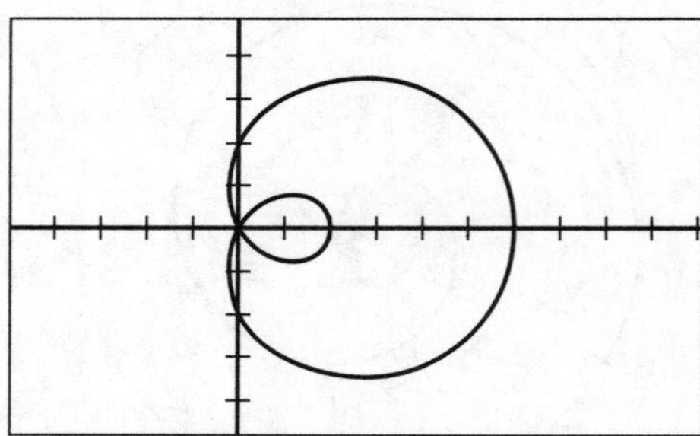

Figure 9.24

Table 9.2 shows common polar graphs, where a and b are constants.

Table 9.2

Name	Equation	Example	
Circle	$r = a\sin\theta$ $r = a\cos\theta$	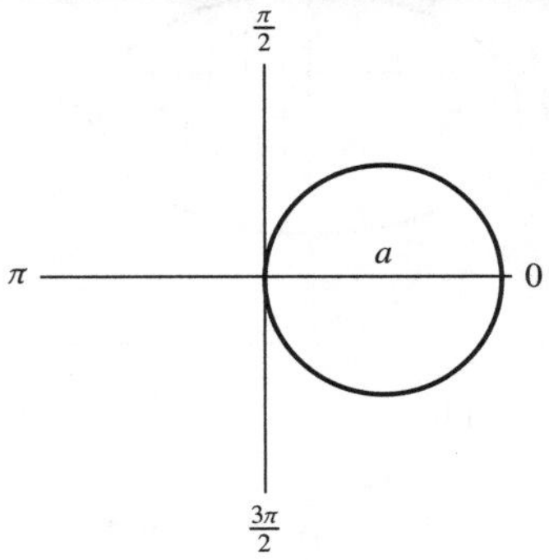 $r = a\sin\theta$	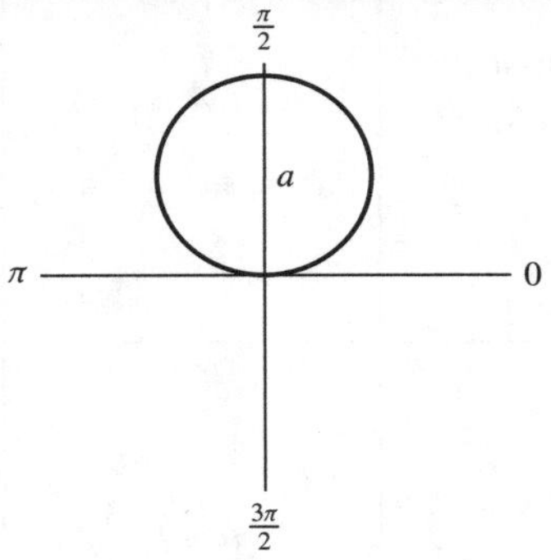 $r = a\cos\theta$
Cardioid	$r = a(1 \pm \sin\theta)$ $r = a(1 \pm \cos\theta)$	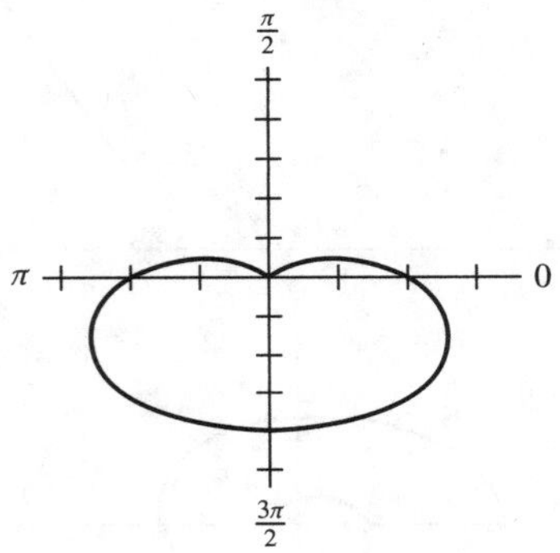 $r = a(1 - \sin\theta)$	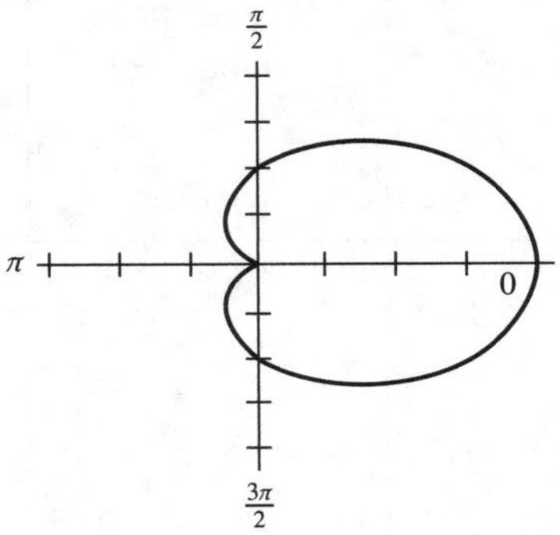 $r = a(1 + \cos\theta)$
Limaçon	$r = a \pm b\sin\theta$ $r = a \pm b\cos\theta$ $(a, b > 0; a \neq b)$	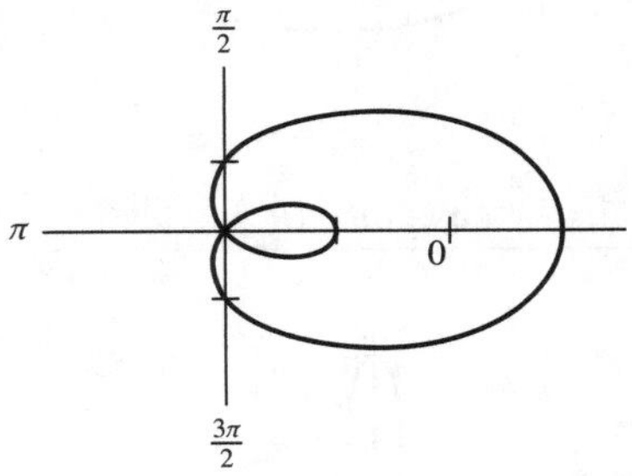 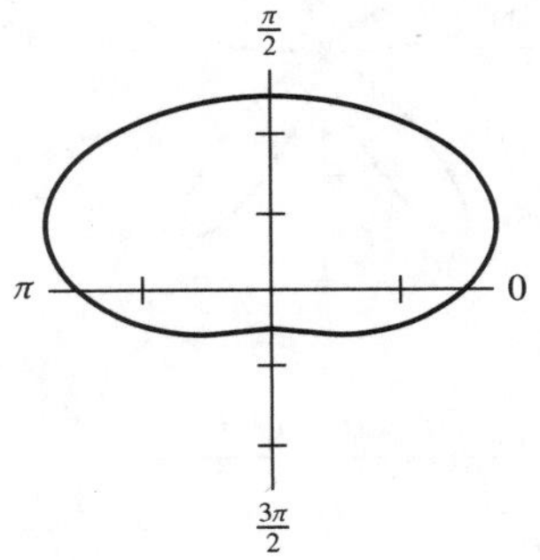	$a < b$ $r = a + b\cos\theta$
			$b < a < 2b$ $r = a + b\sin\theta$

(*Continued*)

Name	Equation	Example
Limaçon (continued)		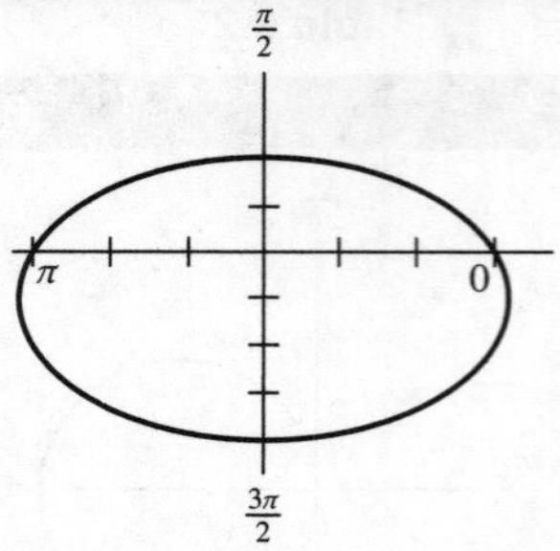 $a \geq 2b$ $r = a - b \sin \theta$
Line	$\theta = K$ passes through pole slope $= K$	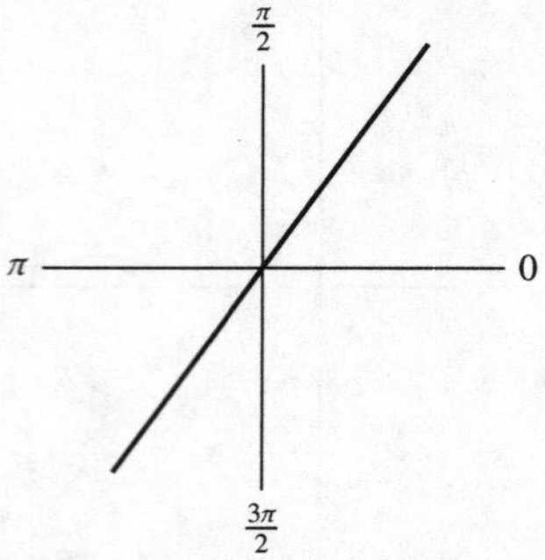
Spiral	$r = a\theta \quad (\theta \geq 0)$ $r = a\theta \quad (\theta \leq 0)$	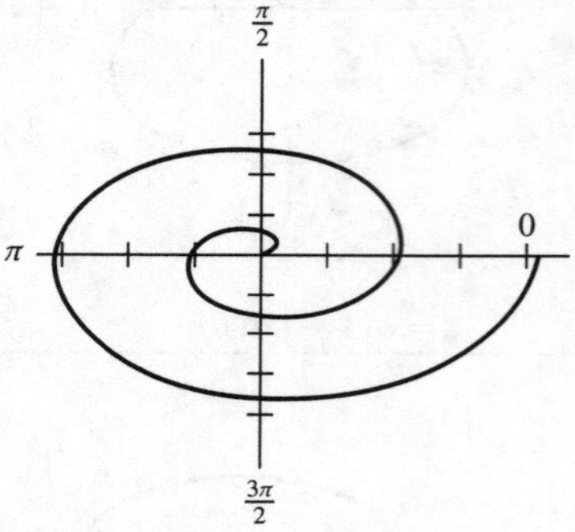 $r = a\theta \quad (\theta \geq 0)$ 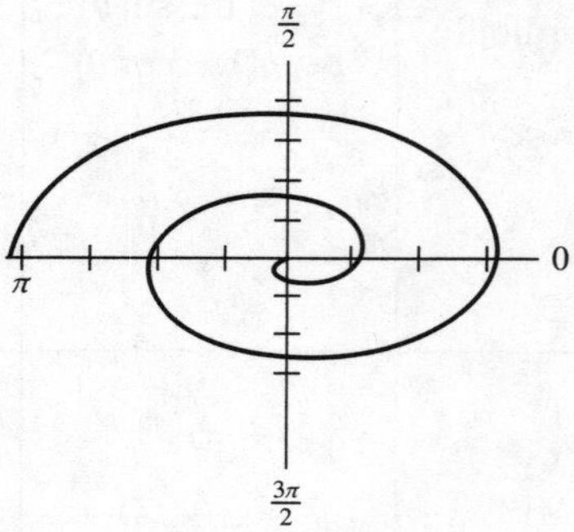 $r = a\theta \quad (\theta \leq 0)$
Rose	$r = a \sin n\theta$ $r = a \cos n\theta$ $(n \geq 2)$ There are n petals when n is odd. There are $2n$ petals when n is even.	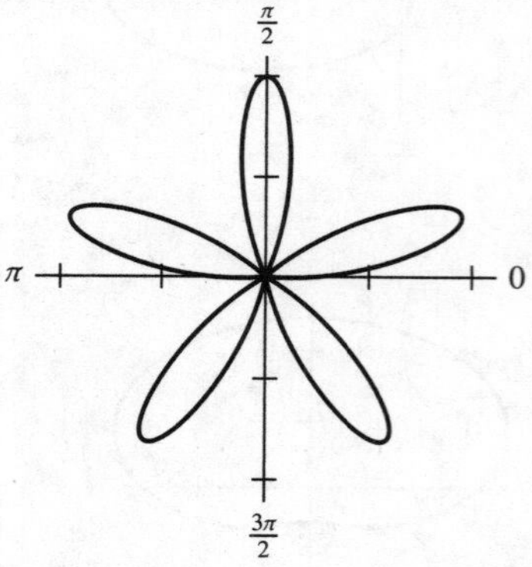 $r = a \sin n\theta$ 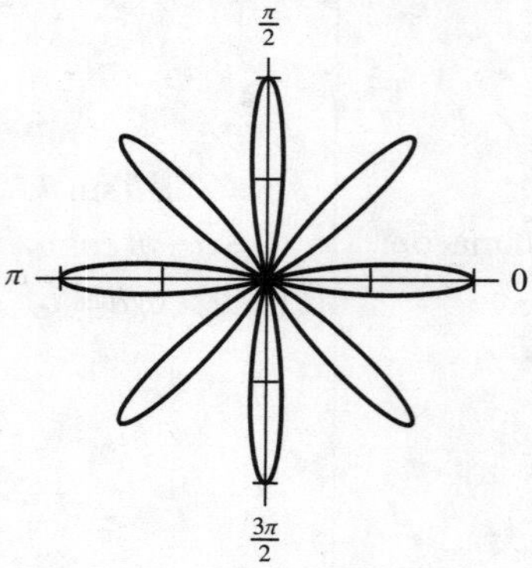 $r = a \cos n\theta$

9.3 Rates of Change in Polar Functions

In previous units, functions in the rectangular coordinate plane had interesting behaviors that were unique to their different graphs. The same is true for graphs of polar functions.

Example

Graph the polar equation $r = 1 + \cos\theta$ and describe the behavior of the graph over the domain $0 \le \theta \le 2\pi$.

Solution

To examine the different characteristics of this graph, first graph $r = 1 + \cos\theta$ in the θr-plane by shifting the graph of the cosine curve up 1 unit as shown in Figure 9.25.

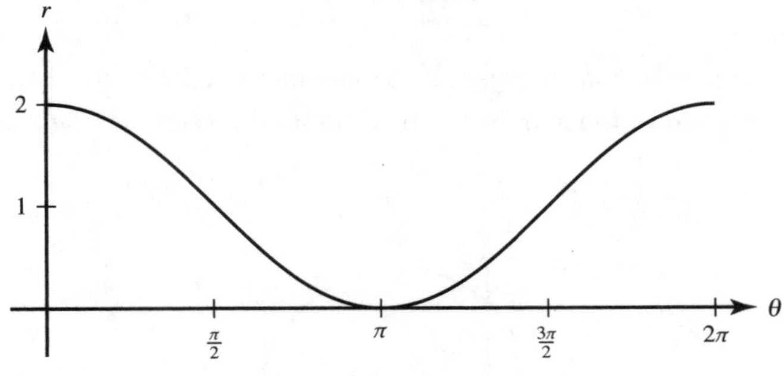

Figure 9.25

As θ increases from 0 to $\frac{\pi}{2}$, r decreases from 2 to 1. This part of the polar curve in the polar coordinate system is shown in Figure 9.26. The distance between the origin and the polar function is decreasing.

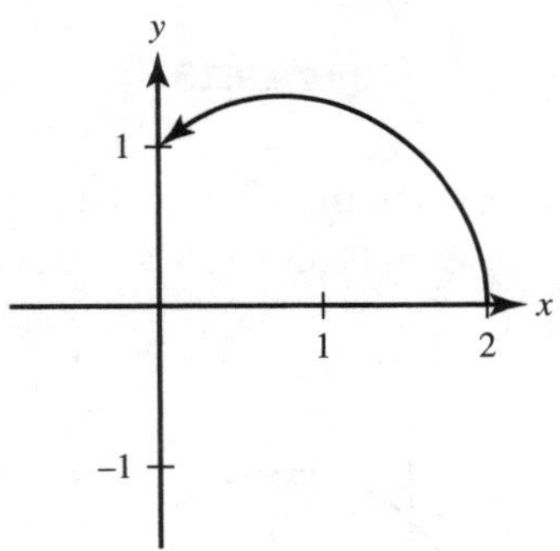

Figure 9.26

As θ increases from $\frac{\pi}{2}$ to π, it can be seen from Figure 9.25 that r decreases from 1 to 0. This next part of the curve in the polar coordinate system is shown in Figure 9.27. The distance between the origin and the polar function is decreasing.

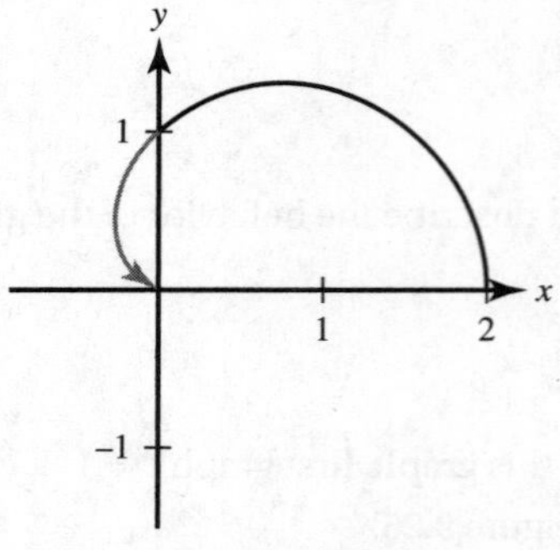

Figure 9.27

As θ increases from π to $\frac{3\pi}{2}$, it can be seen from Figure 9.25 that r increases from 0 to 1. This next part of the curve in the polar coordinate system is shown in Figure 9.28. The distance between the origin and the polar function is increasing.

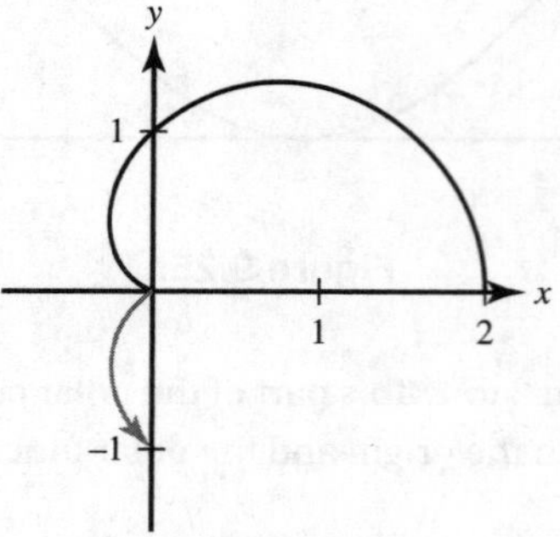

Figure 9.28

Finally, as θ increases from $\frac{3\pi}{2}$ to 2π, it can be seen from Figure 9.25 that r increases from 1 to 2. This final part of the curve in the polar coordinate system is shown in Figure 9.29. The distance between the origin and the polar function is increasing.

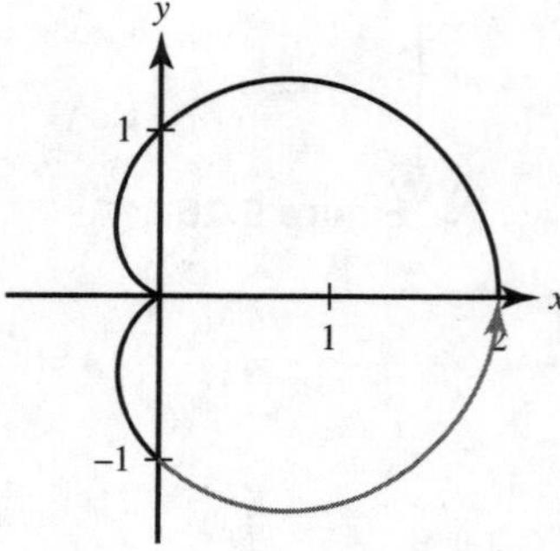

Figure 9.29

Put together the parts of the curve from Figures 9.26–9.29; the complete curve is graphed and shown in Figure 9.30. It is called a cardioid because it is shaped like a heart.

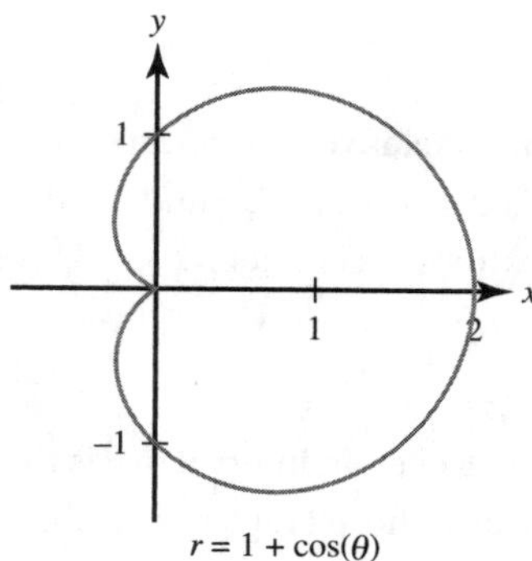

Figure 9.30

It is also important to note that in Figure 9.25 at $\theta = \pi$, the function changes from decreasing to increasing, resulting in a relative minimum. In the graph shown in Figure 9.30, at $\theta = \pi$ the function is at the origin, which is the time it is the closest to the origin. In Figure 9.25, at $\theta = 0$ and $\theta = 2\pi$, the graph is at a relative maximum. In the graph shown in Figure 9.30, at $\theta = 0$ and $\theta = 2\pi$ the function is at a point that is farthest from the origin.

This example leads to the following generalizations of the graph of a polar function:

- If a polar function, $r = f(\theta)$, is positive and increasing or is negative and decreasing, the distance between $f(\theta)$ and the origin is increasing.
- If a polar function, $r = f(\theta)$, is positive and decreasing or is negative and increasing, the distance between $f(\theta)$ and the origin is decreasing.
- For a polar function, $r = f(\theta)$, if the function changes from increasing to decreasing or from decreasing to increasing on an interval, the function has a relative extremum on the interval that corresponds to a point relatively closest to or farthest from the origin.

Example

Using the graph of $r = \cos(2\theta)$ shown in Figure 9.31, state the relative extrema of the curve.

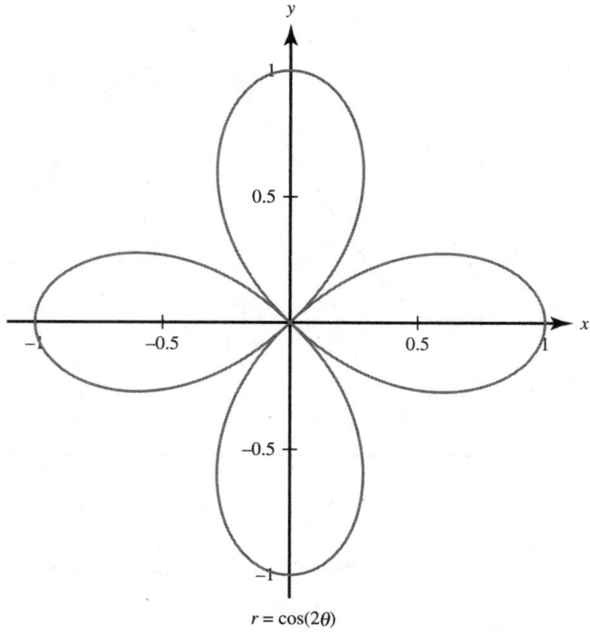

Figure 9.31

Solution

For a polar function, $r = f(\theta)$, the function has a relative extremum on the interval that corresponds to a point relatively closest to or farthest from the origin. At $\theta = 0, \frac{\pi}{2}, \pi, \frac{3\pi}{2}$, and 2π, the curve is farthest from the origin, having the greatest distance of 1 unit. The relative extrema occur at $0, \frac{\pi}{2}, \pi, \frac{3\pi}{2}$, and 2π.

Average Rate of Change

The average rate of change of r with respect to θ over an interval of θ is the ratio of the change in the radius values to the change in θ over an interval of θ. Graphically, the average rate of change indicates the rate at which the radius is changing per radian.

Example

For the polar equation $r = 4 - 2\sin\theta$, find the average rate of change over the following intervals.

1. $\left[0, \frac{\pi}{2}\right]$
2. $\left[\frac{\pi}{2}, \pi\right]$
3. $\left[\pi, \frac{3\pi}{2}\right]$
4. $\left[\frac{3\pi}{2}, 2\pi\right]$

Solution

1. As θ ranges from 0 to $\frac{\pi}{2}$, r decreases from 4 to 2 as shown in Figure 9.32. This means that the curve in the xy-plane starts 4 units from the origin on the positive x-axis and gradually pulls in toward a point 2 units from the origin on the positive y-axis as shown in Figure 9.33.

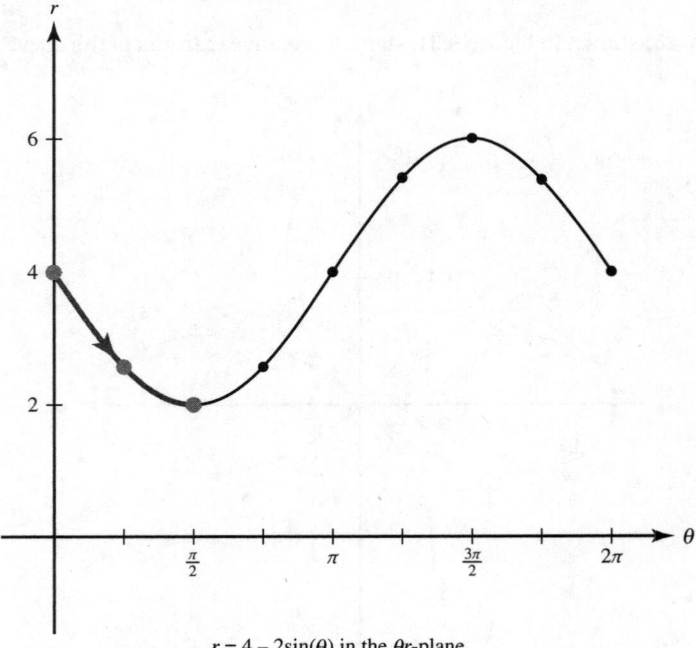

$r = 4 - 2\sin(\theta)$ in the θr-plane

Figure 9.32

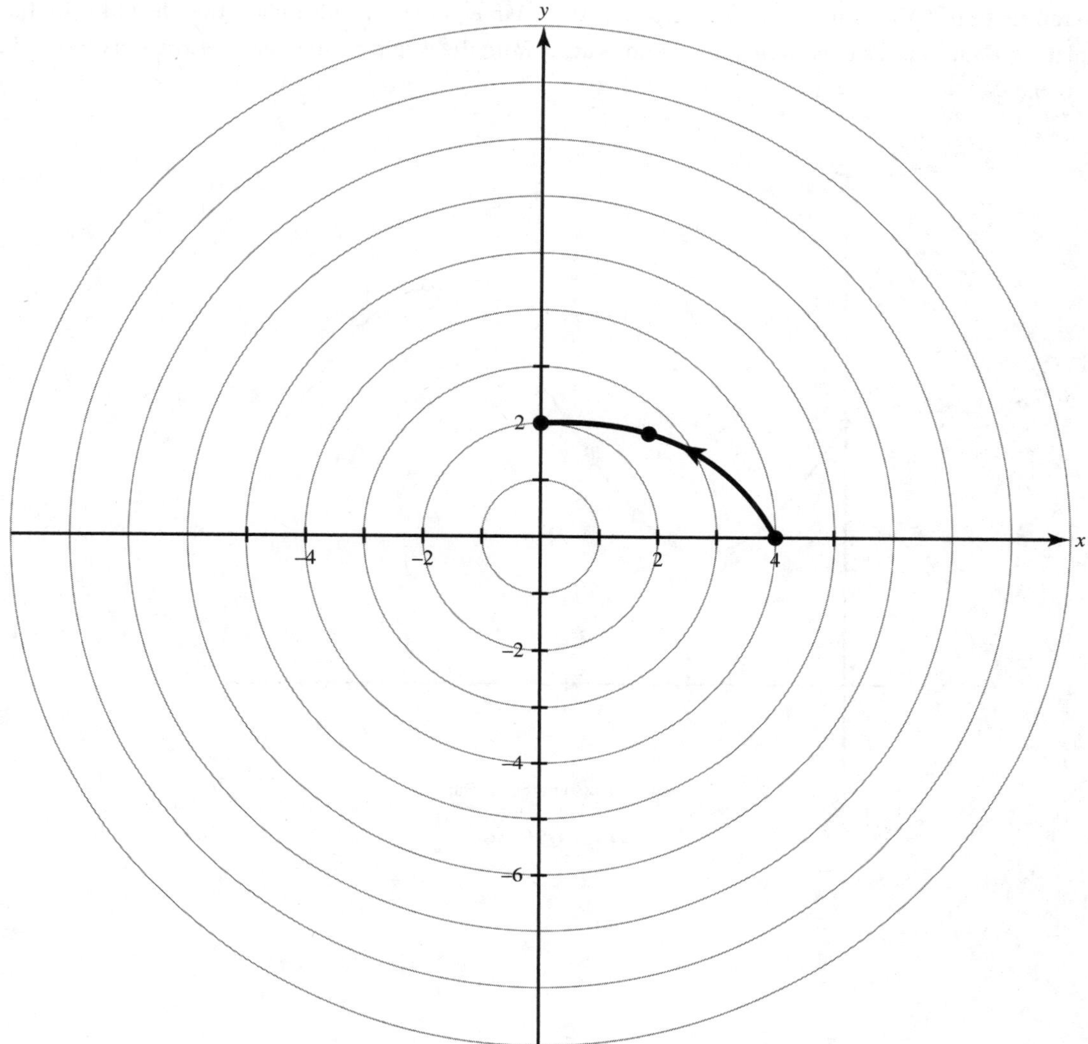

$r = 4 - 2\sin(\theta)$ in the xy-plane

Figure 9.33

The average rate of change $= \dfrac{4-2}{0-\frac{\pi}{2}} = \dfrac{2}{-\frac{\pi}{2}} = -\frac{4}{\pi}$ radius/radian.

2. As θ ranges from $\frac{\pi}{2}$ to π, r increases from 2 to 4 as shown in Figure 9.34. This means that the curve in the xy-plane gradually pulls toward the point 4 units away from the origin on the negative x-axis as shown in Figure 9.35.

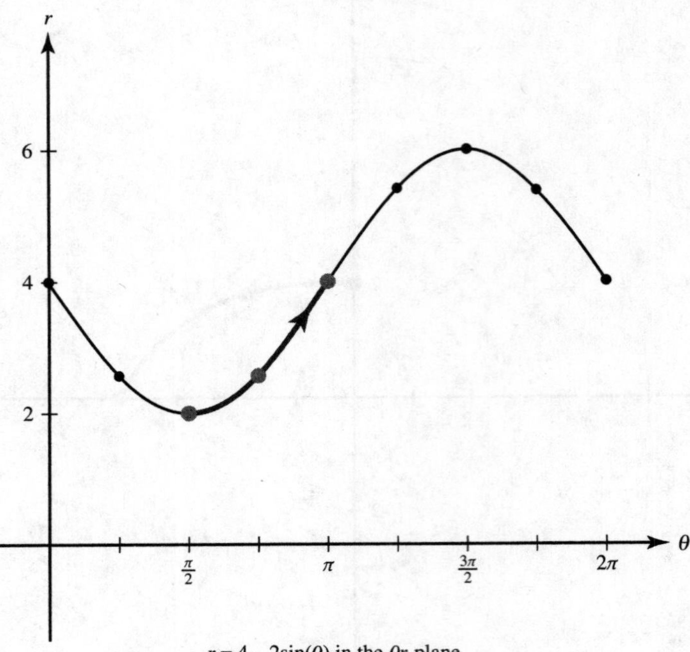

$r = 4 - 2\sin(\theta)$ in the θr-plane

Figure 9.34

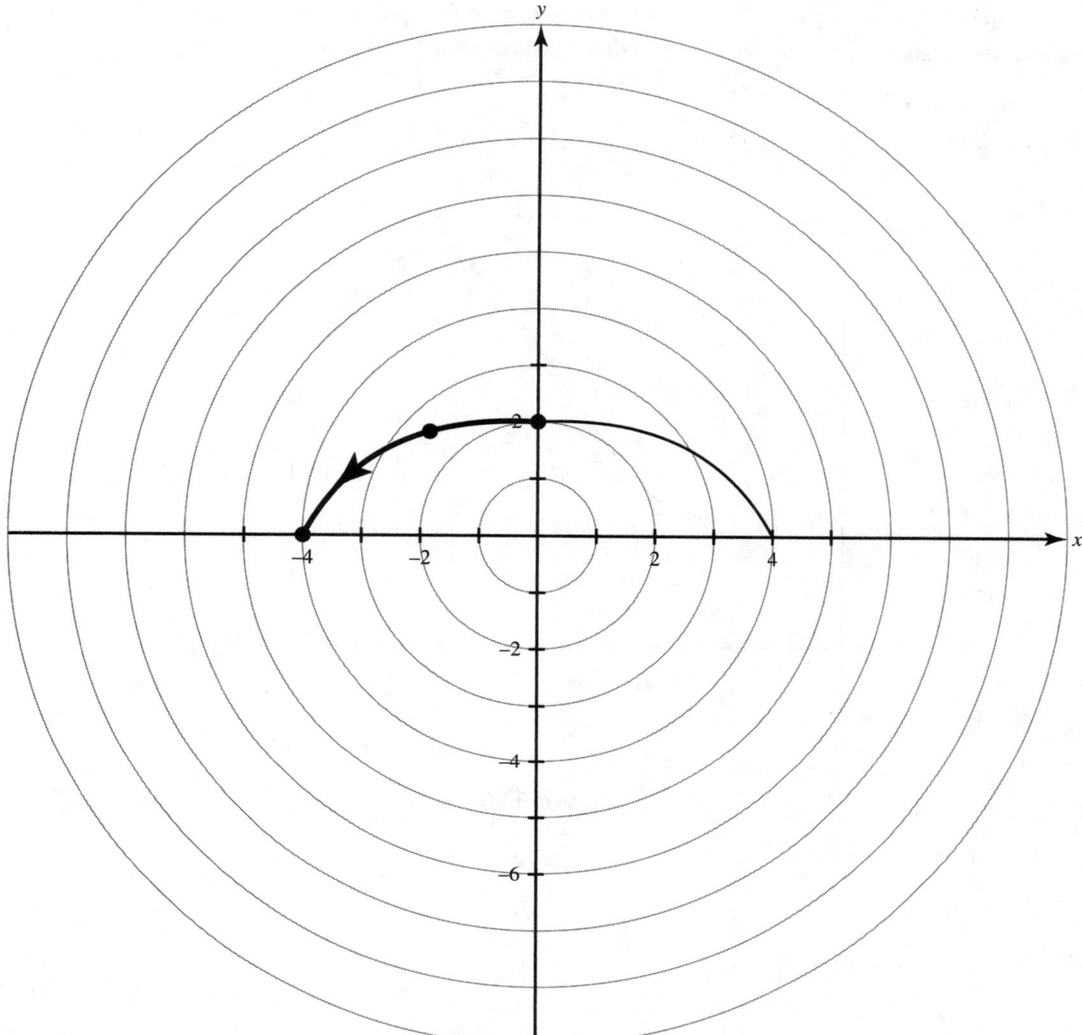

$r = 4 - 2\sin(\theta)$ in the xy-plane

Figure 9.35

The average rate of change $= \dfrac{2-4}{\frac{\pi}{2}-\pi} = \dfrac{-2}{-\frac{\pi}{2}} = \frac{4}{\pi}$ radius/radian.

3. Over the interval $\left[\pi, \frac{3\pi}{2}\right]$, r increases from 4 to 6 as shown in Figure 9.36. On the xy-plane, the curve moves away from the negative x-axis toward the negative y-axis as shown in Figure 9.37.

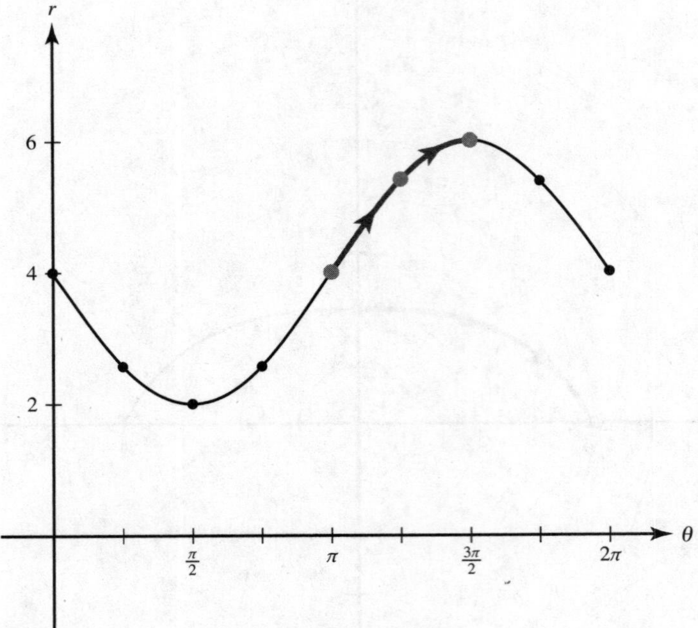

$r = 4 - 2\sin(\theta)$ in the θr-plane

Figure 9.36

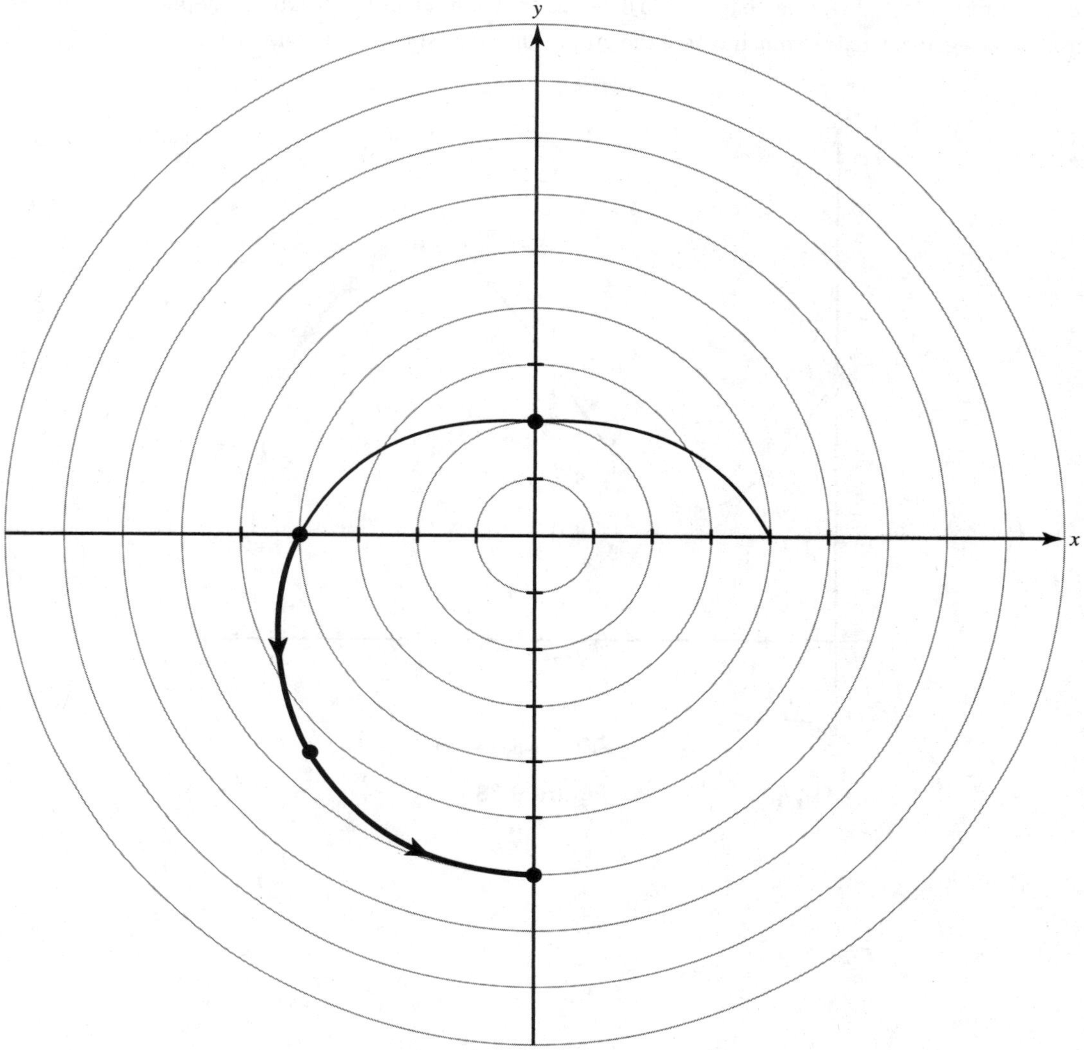

$r = 4 - 2\sin(\theta)$ in the xy-plane

Figure 9.37

The average rate of change $= \dfrac{4-6}{\pi - \dfrac{3\pi}{2}} = \dfrac{-2}{-\dfrac{\pi}{2}} = \frac{4}{\pi}$ radius/radian.

4. Over the interval $\left[\frac{3\pi}{2}, 2\pi\right]$, r decreases from 6 to 4 as shown in Figure 9.38. On the xy-plane, the curve moves in from the negative y-axis to finish where the graph started as shown in Figure 9.39.

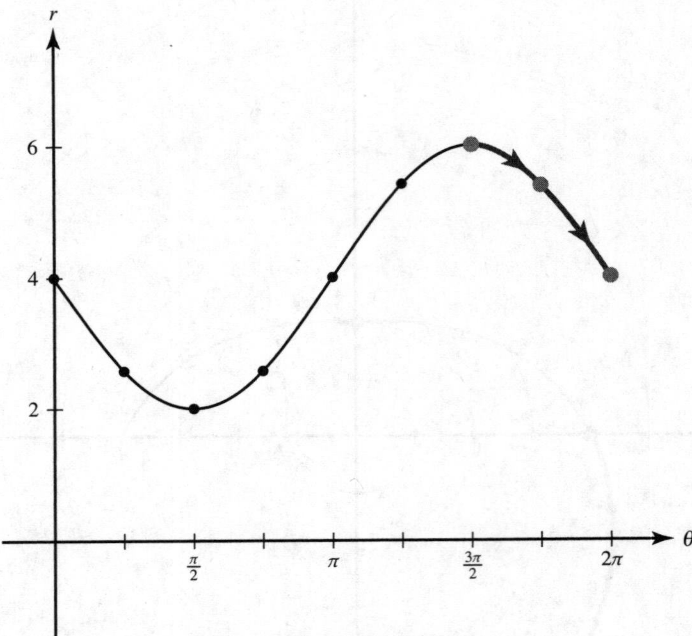

$r = 4 - 2\sin(\theta)$ in the θr-plane

Figure 9.38

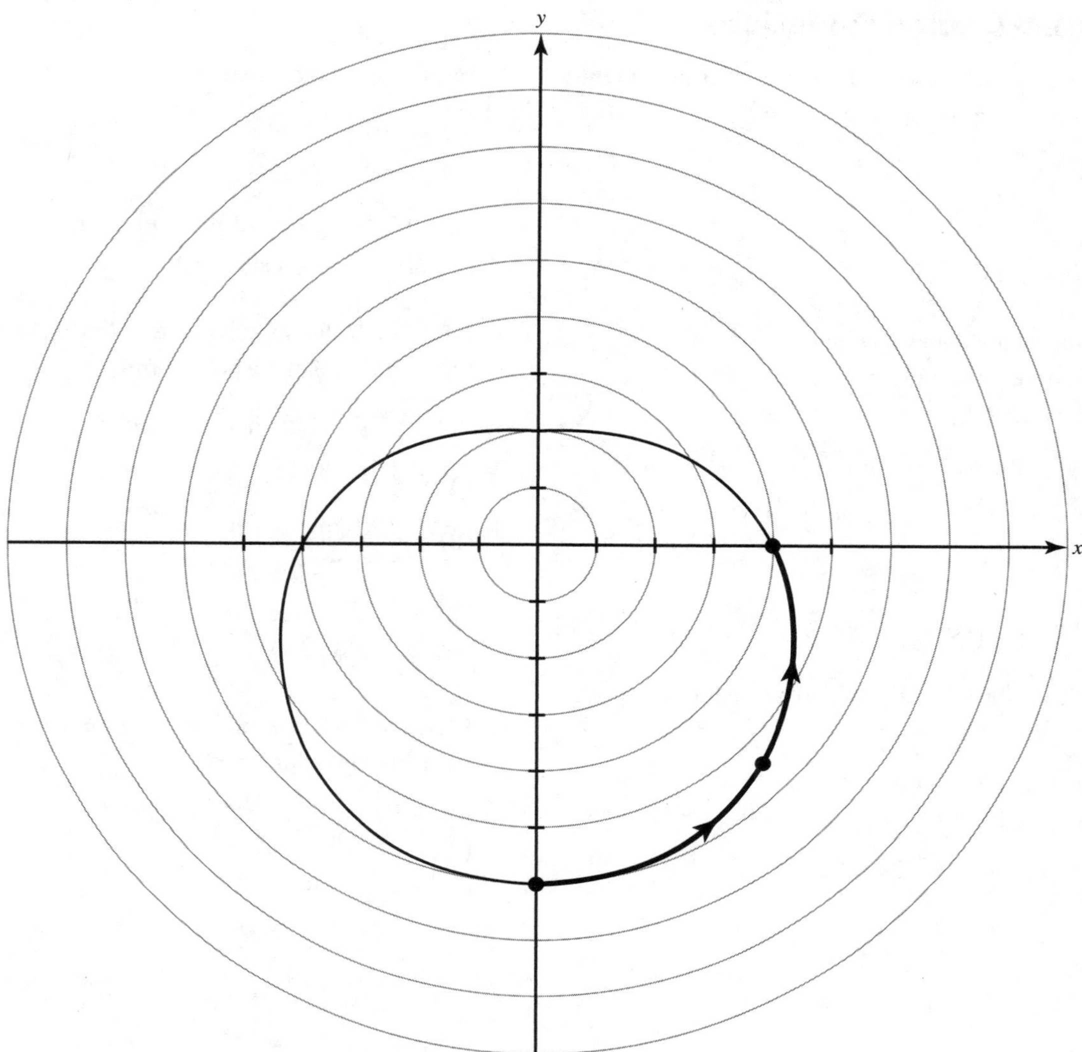

$r = 4 - 2\sin(\theta)$ in the xy-plane

Figure 9.39

The average rate of change $= \dfrac{6-4}{\frac{3\pi}{2} - 2\pi} = \dfrac{2}{-\frac{\pi}{2}} = -\frac{4}{\pi}$ radius/radian.

Multiple-Choice Questions

1. Which of the following choices is equivalent to the complex number $2\left(\cos\frac{\pi}{3} + i\sin\frac{\pi}{3}\right)$?

   (A) $1 + i\sqrt{3}$
   (B) $(1, \sqrt{3})$
   (C) $\sqrt{3} + i$
   (D) $(\sqrt{3}, 1)$

2. Which of the following choices represents the complex number $k - ki$, where $k < 0$, in polar form?

   (A) $k\sqrt{2}\left(\cos\frac{3\pi}{4} + i\sin\frac{3\pi}{4}\right)$
   (B) $-k\sqrt{2}\left(\cos\frac{3\pi}{4} + i\sin\frac{3\pi}{4}\right)$
   (C) $k\sqrt{2}\left(\cos\frac{7\pi}{4} + i\sin\frac{7\pi}{4}\right)$
   (D) $-k\sqrt{2}\left(\cos\frac{7\pi}{4} + i\sin\frac{7\pi}{4}\right)$

3. Which of the following choices is an equivalent polar form of the complex number $2 - 2i\sqrt{3}$?

   (A) $4\left(\cos\frac{\pi}{6} + i\sin\frac{\pi}{6}\right)$
   (B) $4\left(\cos\frac{\pi}{3} + i\sin\frac{\pi}{3}\right)$
   (C) $4\left(\cos\left(-\frac{\pi}{6}\right) + i\sin\left(-\frac{\pi}{6}\right)\right)$
   (D) $4\left(\cos\left(-\frac{\pi}{3}\right) + i\sin\left(-\frac{\pi}{3}\right)\right)$

4. Write $-5i$ in polar form.

   (A) $5\left(\cos\frac{\pi}{2} + i\sin\frac{\pi}{2}\right)$
   (B) $5(\cos\pi + i\sin\pi)$
   (C) $5\left(\cos\left(-\frac{\pi}{2}\right) + i\sin\left(-\frac{\pi}{2}\right)\right)$
   (D) $5(\cos(-\pi) + i\sin(-\pi))$

5. Which of the following choices is the equivalent Cartesian form of the polar form $\sqrt{3}\left(\cos\frac{5\pi}{6} + i\sin\frac{5\pi}{6}\right)$?

   (A) $-\frac{3}{2} + \frac{\sqrt{3}}{2}i$
   (B) $-\frac{3}{2} - \frac{\sqrt{3}}{2}i$
   (C) $\frac{3}{2} + \frac{\sqrt{3}}{2}i$
   (D) $\frac{3}{2} - \frac{\sqrt{3}}{2}i$

6. Which of the following complex numbers cannot be written in polar form?

   (A) $-i$
   (B) 0
   (C) $-1 - i$
   (D) -1

7. Which of the following choices shows the point $\left(-2, -\frac{5\pi}{4}\right)$ in polar coordinates plotted on the polar coordinate system?

(A)

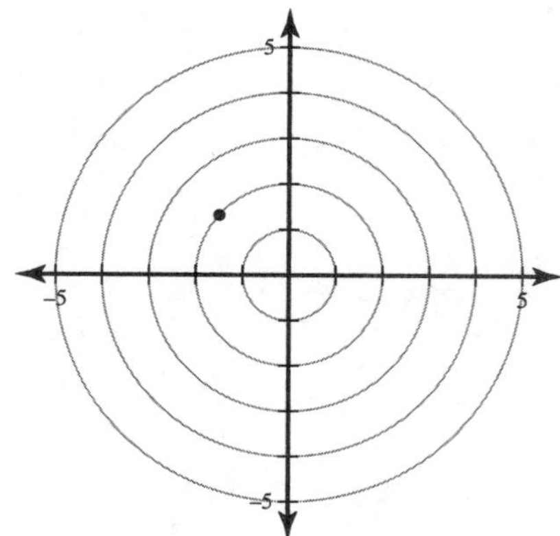

(B)

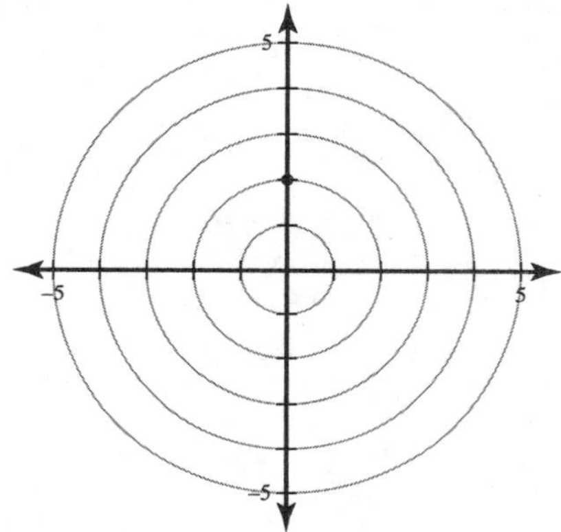

(C)

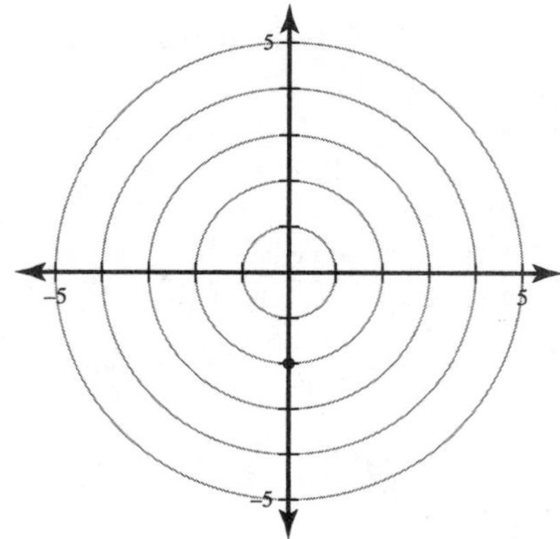

(D)

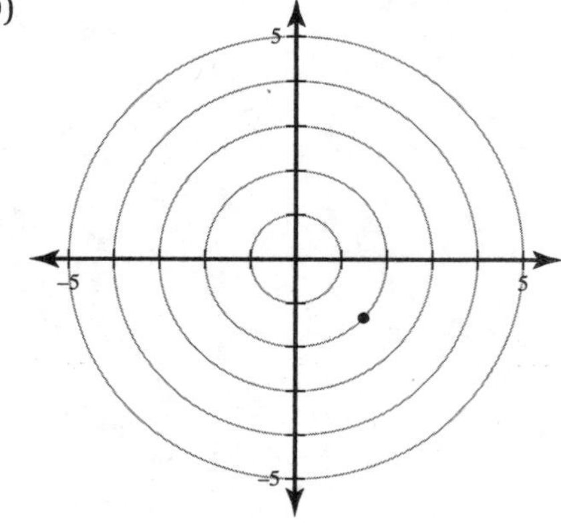

8. Which of the following depicts the graph of $r = 2 + 2\sin\theta$?

(A)

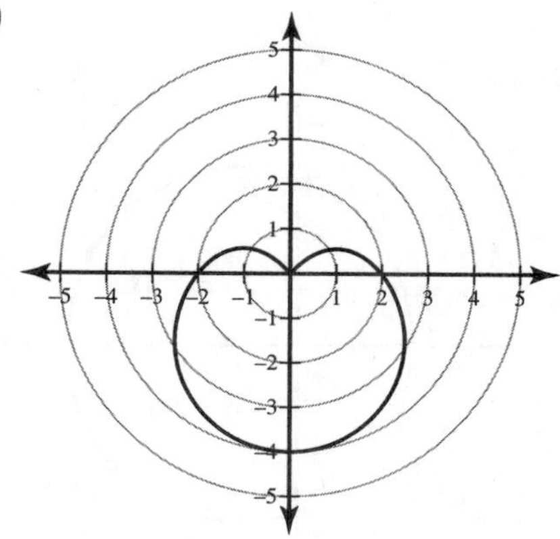

(B)

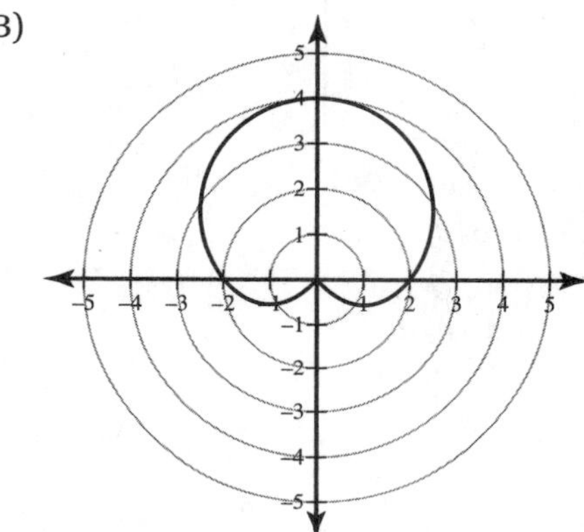

(C)

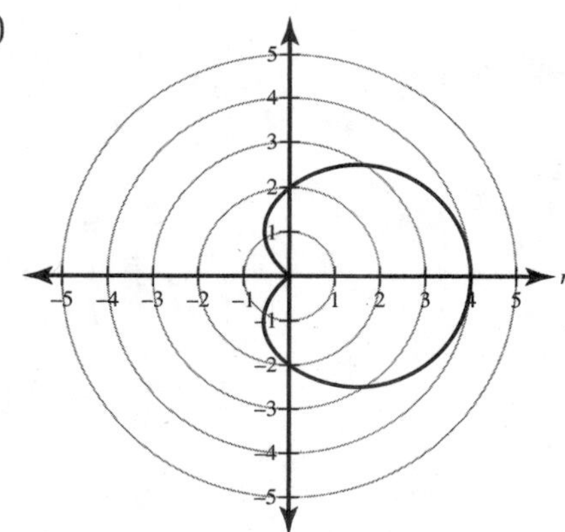

(D)

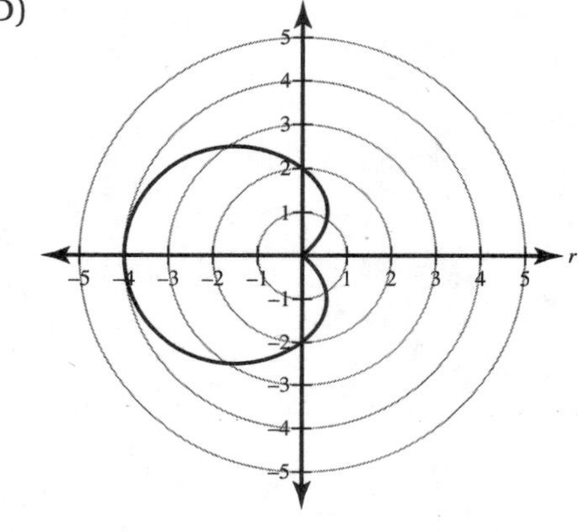

9. Which of the following polar equations represents the graph given below?

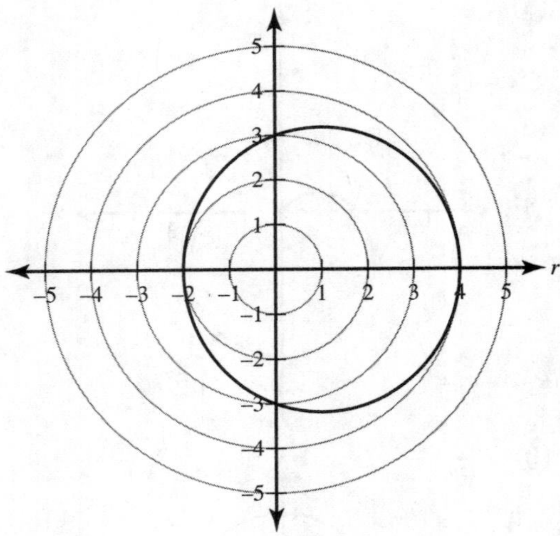

(A) $r = 3 + \sin\theta$
(B) $r = 6\cos\theta$
(C) $r = 6\sin\theta$
(D) $r = 3 + \cos\theta$

10. Find the average rate of change for the polar curve $r = 2 + 4\cos\theta$ in the interval $\left[0, \frac{\pi}{2}\right]$.

(A) $-\frac{8}{\pi}$
(B) $\frac{4}{\pi}$
(C) $\frac{8}{\pi}$
(D) 6

Answer Explanations

1. **(A)** The given number is a polar representation of a complex number, where $r = 2$, $\theta = \frac{\pi}{3}$. An equivalent representation would be to write it as a complex number in the form $a + bi$:

$$a = r\cos\theta = 2\cos\frac{\pi}{3} = 2\left(\frac{1}{2}\right) = 1$$

$$b = r\sin\theta = 2\sin\frac{\pi}{3} = 2\left(\frac{\sqrt{3}}{2}\right) = \sqrt{3}$$

An equivalent representation is $1 + i\sqrt{3}$.

2. **(B)** To convert the complex number $k - ki$, where $k < 0$, into polar form, find r and θ. First calculate r:

$$r = \sqrt{(k)^2 + (-k)^2} = \sqrt{2k^2} = k\sqrt{2}$$

However, since $k < 0$ and $r > 0$, then $r = -k\sqrt{2}$. To find θ, first find the reference angle:

$$\tan^{-1}\left(\left|\frac{-k}{k}\right|\right) = \tan^{-1}(1) = \frac{\pi}{4}$$

Since $k < 0$, the complex number $k - ki$ would be graphed in Quadrant II. Therefore:

$$\theta = \pi - \frac{\pi}{4} = \frac{3\pi}{4}$$

The polar form of the complex number $k - ki$ is $-k\sqrt{2}\left(\cos\frac{3\pi}{4} + i\sin\frac{3\pi}{4}\right)$.

3. **(D)** To convert the complex number $2 - 2i\sqrt{3}$ into polar form, find r and θ. First calculate r:

$$r = \sqrt{(2)^2 + (-2\sqrt{3})^2} = \sqrt{4 + 12} = \sqrt{16} = 4$$

To find θ, first find the reference angle:

$$\tan^{-1}\left(\left|\frac{-2\sqrt{3}}{2}\right|\right) = \tan^{-1}(\sqrt{3}) = \frac{\pi}{3}$$

Since the complex number $2 - 2i\sqrt{3}$ would be graphed in Quadrant IV:

$$\theta = 2\pi - \frac{\pi}{3} = \frac{5\pi}{3}$$

or

$$\theta = -\frac{\pi}{3}$$

The polar form of the complex number $2 - 2i\sqrt{3}$ is $4\left(\cos\left(-\frac{\pi}{3}\right) + i\sin\left(-\frac{\pi}{3}\right)\right)$.

4. **(C)** To convert the complex number $-5i$ into polar form, find r and θ. First calculate r:

$$r = \sqrt{(0)^2 + (-5)^2} = \sqrt{0 + 25} = \sqrt{25} = 5$$

Since the complex number $-5i$ would be graphed on the $-y$-axis, θ is a quadrantal angle. So $\theta = \frac{3\pi}{2}$ or $\theta = -\frac{\pi}{2}$. The polar form of the complex number $-5i$ is $5\left(\cos\left(-\frac{\pi}{2}\right) + i\sin\left(-\frac{\pi}{2}\right)\right)$.

5. **(A)** Evaluate each of the trigonometric expressions using the unit circle:

$$\sqrt{3}\left(\cos\frac{5\pi}{6} + i\sin\frac{5\pi}{6}\right) = \sqrt{3}\left(-\frac{\sqrt{3}}{2} + i\frac{1}{2}\right)$$

$$= \sqrt{3}\cdot -\frac{\sqrt{3}}{2} + \sqrt{3}\cdot\frac{1}{2}i = -\frac{3}{2} + \frac{\sqrt{3}}{2}i$$

6. **(B)** For a complex number to be graphed and written in polar form, it must lie on a circle with $r > 0$. The complex number 0 has an r-value $= 0$ and therefore cannot be written in polar form.

7. **(D)** To plot the point $\left(-2, -\frac{5\pi}{4}\right)$, first locate the angle $-\frac{5\pi}{4}$, which is in Quadrant II. Since $|r| = 2$, move 2 units along the terminal side of the angle. Since $r = -2$, extend the terminal side of the angle in the opposite direction and move 2 units, which places the point in Quadrant IV.

8. **(B)** Use technology, create a table of values, or refer to the list of polar graphs in Table 9.2. The graph of $r = 2 + 2\sin\theta$ is a cardioid.

9. **(D)** The graph depicted is a limaçon, which narrows the options to Choices (A) or (D) since Choices (B) and (C) are the equations of a circle. The graph can be found using technology, creating a table of values, or referring to the list of polar graphs in Table 9.2. It can be seen on the graph that at $\theta = 0$, $r = 4$. Substituting $\theta = 0$ into the two remaining choices can determine the correct equation. For Choice (D), when $\theta = 0$, $r = 3 + \cos(0) = 3 + 1 = 4$. This input-output pair from Choice (D) is found on the graph.

10. **(A)** To find the average rate of change over the interval, evaluate $r(0) = 2 + 4\cos(0) = 6$ and $r\left(\frac{\pi}{2}\right) = 2 + 4\cos\left(\frac{\pi}{2}\right) = 2$. The average rate of change $= \frac{\Delta r}{\Delta\theta} = \frac{2 - 6}{\frac{\pi}{2} - 0} = \frac{-4}{\frac{\pi}{2}} = -\frac{8}{\pi}$.

UNIT 4

Functions Involving Parameters, Vectors, and Matrices

The Unit 4 topics covered in the following three chapters will not be tested on the AP Precalculus exam; however, they may be part of your school's course curriculum. Reviewing these chapters may help build your understanding of topics that you will encounter on the exam.

10

Parametric Functions

Learning Objectives

In this chapter, you will learn:

- ➔ The different properties of parametric equations
- ➔ How to graph parametric equations
- ➔ Properties of implicit equations
- ➔ How to graph implicit equations
- ➔ Properties and graphs of conic sections

NOTE

These Unit 4 topics are **not** assessed on the AP exam but may be part of your school's course curriculum.

All the previous chapters dealt with functions and graphs as defined by a single equation in two variables. In certain situations, however, these types of functions and graphs may not provide a complete picture of the real-world situation it is modeling. A plane curve can be described by a function that uses three variables instead of two. The extra variable, typically represented by t or θ, is called the parameter and provides additional information about the process or function represented by the curve.

10.1 Parametric Functions

Suppose that a particle is moving within the coordinate plane in such a way as to trace out the graph of $y = x^2 - 3x$. From this function, when $x = 5$, $y = (5)^2 - 3(5) = 10$. However, this function does not say when the particle was at (5, 10). Although the function determines where the particle has been, it does not tell when the particle was at those locations.

It is helpful in this situation to introduce a third variable that represents time, t. By expressing both x and y as a function of time, t, such as $x(t) = 3t - 4$ and $y(t) = t^2 + 1$, the position of the particle can be obtained when a specific time is given. Calculate where the particle is located at time $t = 3$:

$$x(3) = 3(3) - 4 = 5,\ y(3) = (3)^2 + 1 = 10 \rightarrow (5, 10)$$

Over any specified interval of t-values, the pair of functions $x(t) = 3t - 4$ and $y(t) = t^2 + 1$ defines a plane curve consisting of the set of ordered pairs $(3t - 4,\ t^2 + 1)$.

Parametric Equations

Let x and y represent functions of the same variable, t, called the parameter, defined on some specified interval of t-values. The functions $x(t)$ and $y(t)$ are the parametric equations for the plane curve that consists of the set of ordered pairs $(x(t), y(t))$. This can also be expressed as a single parametric function $f(t) = (x(t), y(t))$.

Graphing Parametric Equations

A numerical table of values can be generated for the parametric function $f(t) = (x(t), y(t))$ by evaluating x_i and y_i at several values of t_i within the domain.

A graph of a parametric function can be sketched by connecting several points from the numerical table of values in order of increasing value of t.

Example

Create a table of values and sketch the parametric curve for the parametric function $f(t) = (t^2 + t, 2t - 1)$.

Solution

To create a table of values, pick values of t and substitute them into the parametric equations, as shown in Table 10.1. Since no specific domain was given, the values of t are arbitrarily chosen.

Table 10.1

t	-2	-1	$-\frac{1}{2}$	0	1
x	2	0	$-\frac{1}{4}$	0	2
y	-5	-3	-2	-1	1

Parametric curves have direction of motion. The direction of motion is given by increasing t. So when plotting parametric curves, include arrows that show the direction of motion.

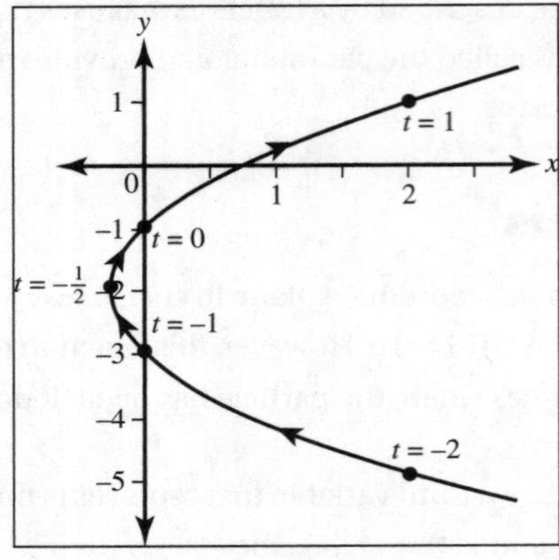

Figure 10.1

In this example, since there was no domain provided or limits on time, the graph will continue in both directions as shown in Figure 10.1.

> The domain of the parametric function $f(t)$ is often restricted, which results in start and end points on the graph of $f(t)$.

Example

Create a table of values and sketch the parametric curve for the parametric function $f(t) = \left(t^2 - 4, \frac{t}{2}\right)$ over the time interval $-2 \le t \le 3$.

Solution

The parametric function is given by $f(t) = \left(t^2 - 4, \frac{t}{2}\right) = (x(t), y(t))$. To create the table of values, as shown in Table 10.2, find the x- and y-coordinates using the equations $x(t) = t^2 - 4$ and $y(t) = \frac{t}{2}$.

Table 10.2

t	-2	-1	0	1	2	3
x	0	-3	-4	-3	0	5
y	-1	$-\frac{1}{2}$	0	$\frac{1}{2}$	1	$\frac{3}{2}$

To graph the parametric function, plot the set of ordered pairs as points and trace out the path as t is increasing. The graph is shown in Figure 10.2.

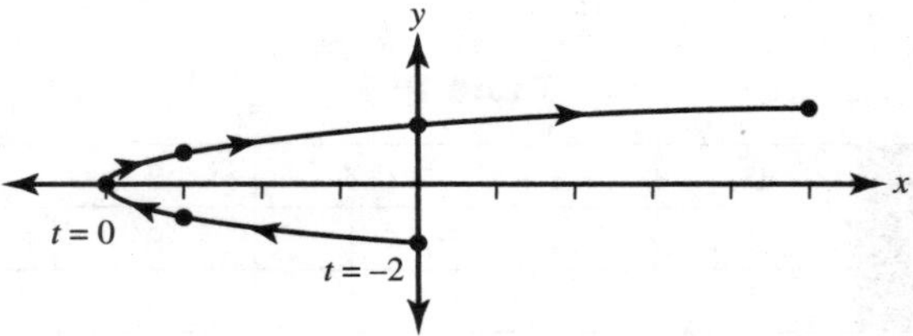

Figure 10.2

Most graphing calculators have a parametric model that can graph a pair of parametric equations.

Example

Using technology, graph $x = 2t + 1$ and $y = 4t^2 - 1$ for $-3 \leq t \leq 4$.

Solution

Change the mode of the calculator to parametric by pressing MODE and selecting PARAMETRIC. Then select Y=. In X_{1T} type $2T + 1$ and in Y_{1T} type $4T^2 - 1$. Select WINDOW and let Tmin = -3 and Tmax = 4. Possible dimensions for x and y can be Xmin = -10, Xmax = 10, Ymin = -15, and Ymax = 70. To see the parametric function being traced out, for the TI-84 Plus calculator, select Y= and move the cursor to the left of X_{1T}. Then press ENTER twice to have a key symbol appear. Select GRAPH, and the graph of the parametric function will trace out as t increases. The finished graph is shown in Figure 10.3.

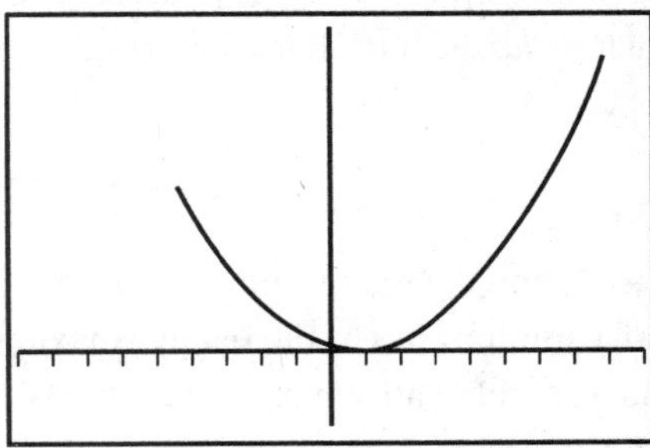

Figure 10.3

10.2 Parametric Functions Modeling Planar Motion

A parametric function given by $f(t) = (x(t), y(t))$ can be used to model particle motion in the plane. The graph of this function indicates the position of a particle at time t.

Example

A particle moves in the xy-plane so that at any time t, its position is given by $x = t^2$ and $y = t$.

(a) Graph the path of the particle over the time interval $0 \leq t \leq 4$.

(b) Find the position of the particle when $t = 10$.

Solution

(a) To help graph, first create a table of values, as shown in Table 10.3. Figure 10.4 shows the graph.

Table 10.3

t	0	1	2	3	4
x	0	1	4	9	16
y	0	1	2	3	4

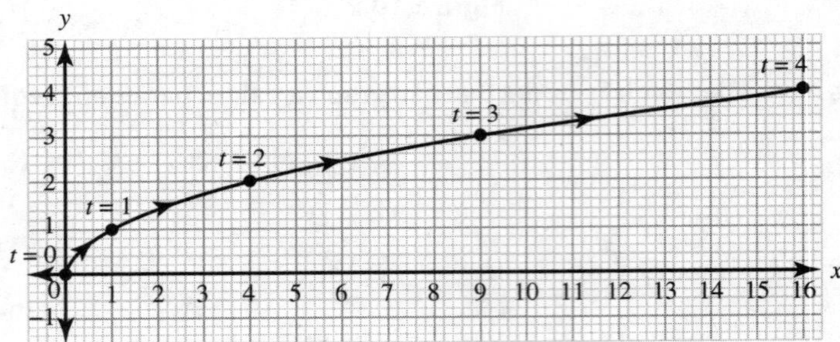

Figure 10.4

(b) The position of the particle when t is $10 = (x(10), y(10)) = (10^2, 10) = (100, 10)$.

Extrema of Particle Motion

The horizontal and vertical extrema of a particle's motion can be determined by identifying the maximum and minimum values of the functions $x(t)$ and $y(t)$, respectively.

Example

Find the relative maximum and relative minimum values of the parametric curve defined by the parametric equations $x = 3t^2$, $y = 2t$, $-2 \leq t \leq 2$. Include a sketch to check your results.

Solution

The relative maximum value of $x = 3t^2$ is 12, and the relative minimum value is 0. Therefore, the horizontal relative maximum is 12 and the horizontal relative minimum is 0. The relative maximum value of $y = 2t$ is 4, and the relative minimum value is -4. Therefore, the vertical relative maximum value is 4 and the vertical relative minimum value is -4. This can be verified using the graph of the parametric curve shown in Figure 10.5.

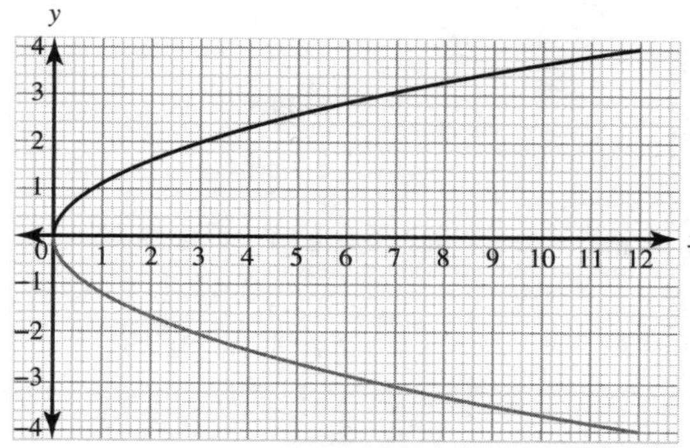

Figure 10.5

Zeros of Parametric Curves

The real zeros of the function $x(t)$ correspond to y-intercepts, and the real zeros of the function $y(t)$ correspond to x-intercepts.

Example

Algebraically find the x- and y-intercepts of the parametric curve defined by the equations $x = t^2 + t$, $y = 2t - 1$. Include a sketch to check your results.

Solution

First find the zeros of $x = t^2 + t$:

$$0 = t^2 + t$$
$$0 = t(t + 1)$$
$$t = 0,\ t = -1$$

At these t-values, calculate y:

$$y = 2(0) - 1 = -1$$
$$y = 2(-1) - 1 = -3$$

Then find the zeros of $y = 2t - 1$:

$$0 = 2t - 1$$
$$t = \frac{1}{2}$$

At this t-value, calculate x:

$$x = \left(\frac{1}{2}\right)^2 + \left(\frac{1}{2}\right) = \frac{3}{4}$$

The y-intercepts are $(0, -1)$ and $(0, -3)$. The x-intercept is $\left(\frac{3}{4}, 0\right)$. This can be verified on the sketch of the graph in Figure 10.6.

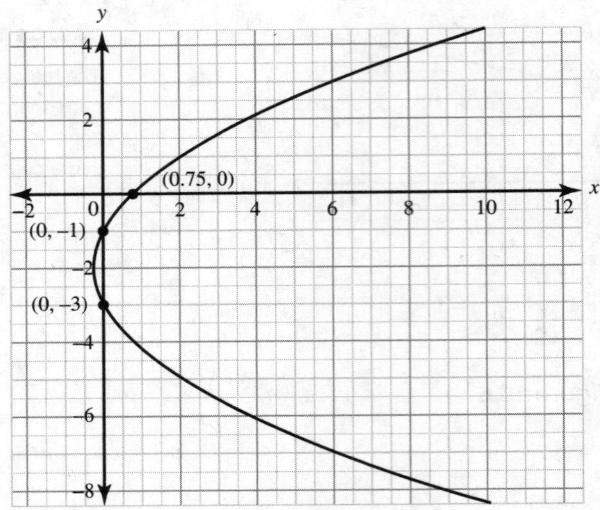

Figure 10.6

10.3 Parametric Functions and Rates of Change

The most common application of parametric equations is to describe particle motion in a plane. This is known as parametric planar motion. This section will identify key characteristics of parametric planar motion functions that are related to direction and rate of change.

As we have seen in the previous sections, as the parameter, usually t, increases, the direction of planar motion of a particle can be analyzed in terms of x and y independently.

> **Direction of Planar Motion Functions**
>
> If $x(t)$ is increasing or decreasing, the direction of motion is to the right or left, respectively. If $y(t)$ is increasing or decreasing, the direction of motion is up or down, respectively.

Example

Describe the path of the particle whose position is given by the parametric equations $x = 3\cos t$, $y = 2\sin t$, $0 \le t \le 2\pi$. Include a sketch to check your results.

Solution

To describe the path of the particle and include a sketch, create a table of values over the given domain. See Table 10.4 and Figure 10.7.

Table 10.4

t	0	$\frac{\pi}{4}$	$\frac{\pi}{2}$	$\frac{3\pi}{4}$	π	$\frac{5\pi}{4}$	$\frac{3\pi}{2}$	$\frac{7\pi}{4}$	2π
x	3	$\frac{3\sqrt{2}}{2}$	0	$-\frac{3\sqrt{2}}{2}$	-3	$-\frac{3\sqrt{2}}{2}$	0	$\frac{3\sqrt{2}}{2}$	3
y	0	$\sqrt{2}$	2	$\sqrt{2}$	0	$-\sqrt{2}$	-2	$-\sqrt{2}$	0

At $t = 0$, the direction of the particle is right 3 units.

At $t = \frac{\pi}{4}$, the direction of the particle is right $\frac{3\sqrt{2}}{2}$ units and up $\sqrt{2}$ units.

At $t = \frac{\pi}{2}$, the direction of the particle is up 2 units.

At $t = \frac{3\pi}{4}$, the direction of the particle is left $\frac{3\sqrt{2}}{2}$ units and up $\sqrt{2}$ units.

At $t = \pi$, the direction of the particle is left 3 units.

At $t = \frac{5\pi}{4}$, the direction of the particle is left $\frac{3\sqrt{2}}{2}$ units and down $\sqrt{2}$ units.

At $t = \frac{3\pi}{2}$, the direction of the particle is down 2 units.

At $t = \frac{7\pi}{4}$, the direction of the particle is right $\frac{3\sqrt{2}}{2}$ units and down $\sqrt{2}$ units.

At $t = 2\pi$, the direction of the particle is right 3 units.

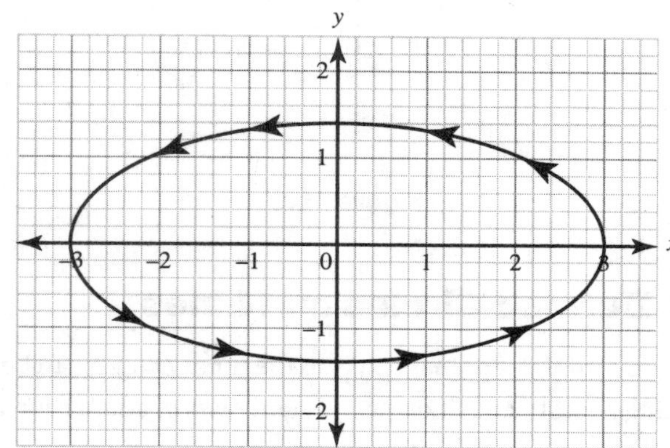

Figure 10.7

The previous example shows that at any given point in the plane, the direction of planar motion may be different for different values of t.

> **Average Rate of Change**
>
> Over a given interval $[t_1, t_2]$ within the domain, the average rate of change can be computed for $x(t)$ and $y(t)$ independently.
>
> The ratio of the average rate of change of y to the average rate of change of x gives the slope of the graph between the points on the curve corresponding to t_1 and t_2 if the average rate of change of $x(t) \neq 0$.

Example

For the particle whose position is determined by the equations $x = t$, $y = \sqrt{1 - t^2}$, over the given interval $-1 \leq t \leq 1$, find the following.

(a) The average rate of change for $x(t)$

(b) The average rate of change for $y(t)$

(c) The slope of the graph between the points on the curve corresponding to $t = -1$ and $t = 1$

Solution

(a) Calculate the average rate of change for $x(t)$:

$$\frac{\Delta x}{\Delta t} = \frac{x(1) - x(-1)}{1 - (-1)} = \frac{1 - (-1)}{1 + 1} = \frac{2}{2} = 1$$

(b) Calculate the average rate of change for $y(t)$:

$$\frac{\Delta y}{\Delta t} = \frac{y(1) - y(-1)}{1 - (-1)} = \frac{0 - 0}{1 + 1} = \frac{0}{2} = 0$$

(c) Calculate the slope of the graph:

$$\frac{\frac{\Delta y}{\Delta t}}{\frac{\Delta x}{\Delta t}} = \frac{\Delta y}{\Delta x} = \frac{0}{1} = 0$$

The points that correspond to the curve at $t = -1$ and $t = 1$ are $(-1, 0)$ and $(1, 0)$, respectively. Connecting those points results in a horizontal secant segment whose slope is 0, as shown in Figure 10.8. This figure confirms the calculated slope.

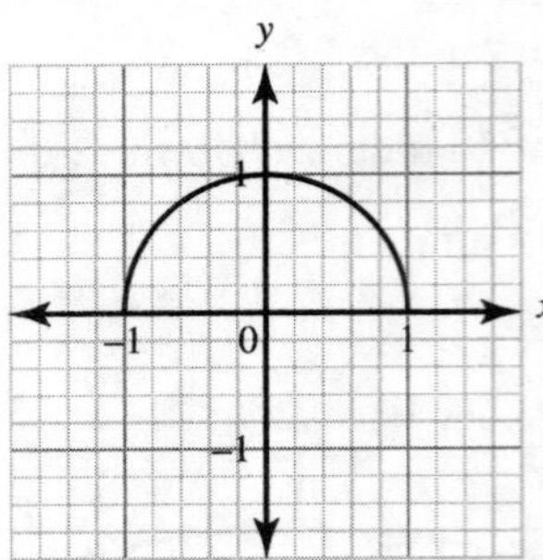

Figure 10.8

10.4 Parametrically Defined Circles and Lines

A linear path along the line segment from the point (x_1, y_1) to the point (x_2, y_2) can be parameterized in many ways. One such way includes using an initial position (x_1, y_1) and rates of change for x with respect to t and for y with respect to t.

Example

Find a parameterization of the line through the points $A = (-2, 3)$ and $B = (3, 6)$.

Solution

Let $A = (-2, 3)$ represent the initial position, meaning the position at $t = 0$. Let $B = (3, 6)$ represent the position at $t = 1$.

Calculate the rate of change for x with respect to t:

$$\frac{\Delta x}{\Delta t} = \frac{3 - (-2)}{1 - 0} = 5$$

Calculate the rate of change for y with respect to t:

$$\frac{\Delta y}{\Delta t} = \frac{6 - 3}{1 - 0} = 3$$

The parametric equations are $x = -2 + 5t$ and $y = 3 + 3t$.

Parameterization of a Line

1. Let one point represent the initial position when $t = 0$, (x_0, y_0).
2. Let the other point represent the position when $t = 1$, (x_1, y_1).
3. Calculate $\frac{\Delta x}{\Delta t} = \frac{x_1 - x_0}{1 - 0}$ and $\frac{\Delta y}{\Delta t} = \frac{y_1 - y_0}{1 - 0}$.
4. Write the parametric equations where $x = x_0 + \frac{\Delta x}{\Delta t} \cdot t$ and $y = y_0 + \frac{\Delta y}{\Delta t} \cdot t$.

It is also possible to parameterize a circle. The simplest case would be to parameterize the unit circle. The unit circle is defined by the equation $x^2 + y^2 = 1$. The Pythagorean trigonometric identity from Section 8.12 is $\cos^2 t + \sin^2 t = 1$ with the domain $(0, 2\pi)$. This directly leads to the parameterization of the unit circle with the parametric equations being $x(t) = \cos t$ and $y(t) = \sin t$ where $0 \leq t \leq 2\pi$.

This can be verified by creating a table of values and a sketch of the parametric equations. These are shown in Table 10.5 and Figure 10.9.

Table 10.5

t	0	$\frac{\pi}{2}$	π	$\frac{3\pi}{2}$	2π
x	1	0	-1	0	1
y	0	1	0	-1	0

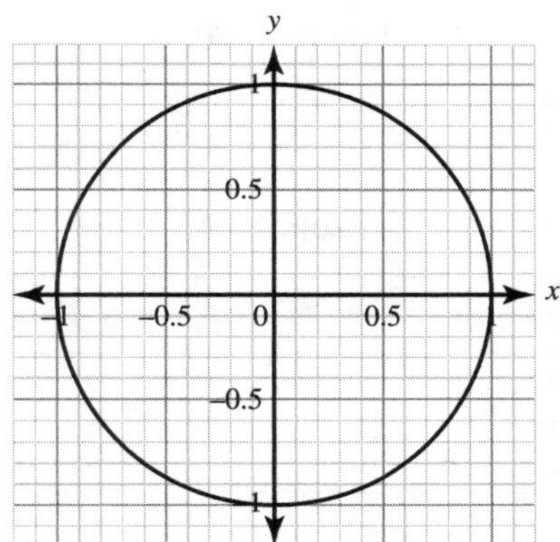

Figure 10.9

It is also possible to parameterize a circle of any given radius, a, where $a \geq 0$, centered at the origin with the equation $x^2 + y^2 = a^2$. The parametric equations are $x(t) = a \cos t$ and $y(t) = a \sin t$.

A circle of radius a centered at (h, k) with the equation $(x - h)^2 + (y - k)^2 = a^2$ shifts the x-coordinates of the circle centered at the origin horizontally by h and shifts the y-coordinates vertically by k. Since $x(t)$ defines the x-coordinates and $y(t)$ defines the y-coordinates, the parametric equations of the circle centered at (h, k) with radius a are $x(t) = h + a \cos t$ and $y(t) = k + a \sin t$.

Example

Find the parametric equations of the circle with the Cartesian equation $(x - 1)^2 + (y + 2)^2 = 9$. Graph a sketch of the circle, and include a table of values of the parametric equations.

Solution

The center of the circle is (1, −2), and the radius is 3. A sketch of the circle is shown in Figure 10.10.

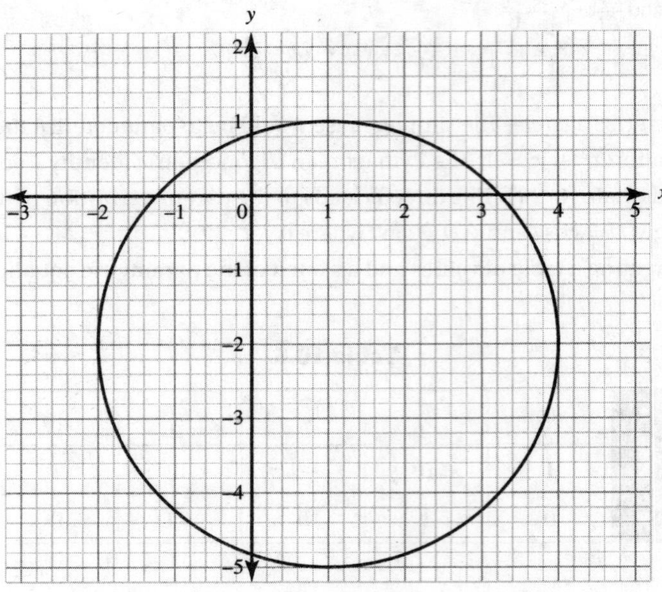

Figure 10.10

The parametric equations are $x(t) = 1 + 3\cos t$ and $y(t) = -2 + 3\sin t$. A table of values is given in Table 10.6.

Table 10.6

t	0	$\frac{\pi}{2}$	π	$\frac{3\pi}{2}$
x	4	1	−2	1
y	−2	1	−2	−5

All the ordered pairs found in the table are also ordered pairs of the graph.

10.5 Implicitly Defined Functions

An implicitly defined function is an equation in which the variable and output are not explicitly expressed in terms of one another. The equation of the unit circle, $x^2 + y^2 = 1$, is an example of an implicitly defined function.

In the example that follows, $x^3 + y^3 = 1$, y can be solved explicitly for x: $y = (1 - x^3)^{\frac{1}{3}}$. However, it is not always possible to write an implicit equation explicitly as seen in this example, $y + \tan^{-1} y = x$.

An implicit equation involving two variables can be graphed by finding solutions to the equation. This can be done by substituting different values for a variable and solving for the other variable. The graph is formed by plotting the different ordered pairs.

Example

Given the equation $x^3 + y^3 = 1$, do the following.

(a) Graph the equation as an implicit equation.
(b) Using technology, graph the explicit equation $y = (1 - x^3)^{\frac{1}{3}}$.

Solution

(a) To graph implicitly, substitute different values for x, and solve the equation for the other variable, as shown in Table 10.7. Figures 10.11 and 10.12 show how to graph the results.

Table 10.7

Substituting and Solving	x	y
If $x = 0$: $(0)^3 + y^3 = 1$ $y^3 = 1$ $y = 1$	0	1
If $x = 1$: $(1)^3 + y^3 = 1$ $y^3 = 0$ $y = 0$	1	0
If $x = 2$: $(2)^3 + y^3 = 1$ $y^3 = -7$ $y = \sqrt[3]{-7}$	2	$\sqrt[3]{-7} \approx -1.913$
If $x = 3$: $(3)^3 + y^3 = 1$ $y^3 = -26$ $y = \sqrt[3]{-26}$	3	$\sqrt[3]{-26} \approx -2.962$
If $x = -1$: $(-1)^3 + y^3 = 1$ $y^3 = 2$ $y = \sqrt[3]{2}$	-1	$\sqrt[3]{2} \approx 1.2599$
If $x = -2$: $(-2)^3 + y^3 = 1$ $y^3 = 9$ $y = \sqrt[3]{9}$	-2	$\sqrt[3]{9} \approx 2.0801$
If $x = -3$: $(-3)^3 + y^3 = 1$ $y^3 = 28$ $y = \sqrt[3]{28}$	-3	$\sqrt[3]{28} \approx 3.0366$

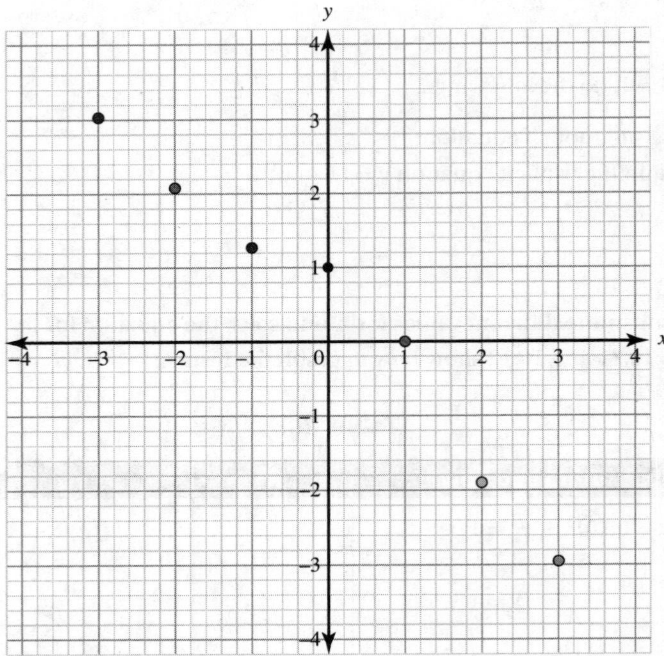

Figure 10.11

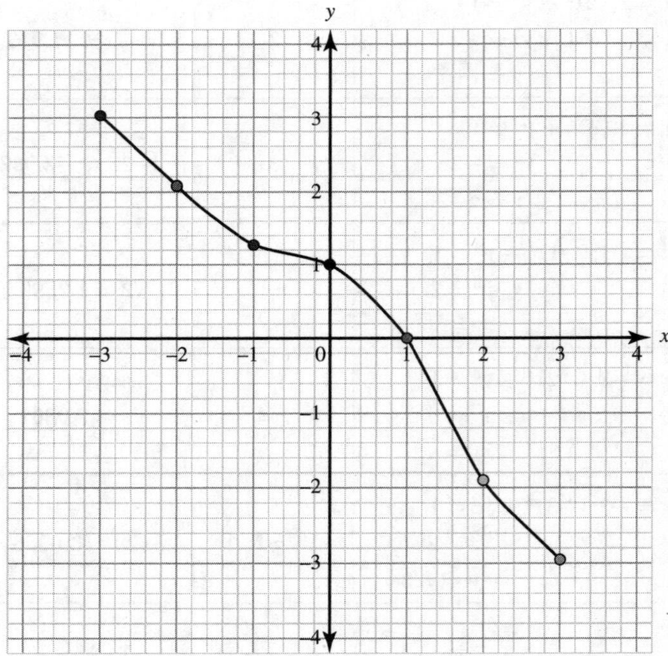

Figure 10.12

The drawback to this method is that many more solutions to points need to be found to have a better understanding of the graph.

(b) Using technology, the graph of the explicit equation $y = (1 - x^3)^{\frac{1}{3}}$ is shown in Figure 10.13.

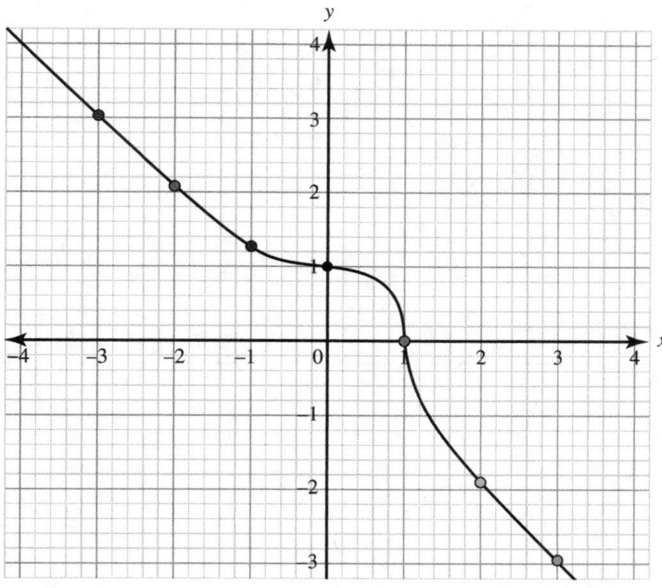

Figure 10.13

As a check, the graph in part (b) was plotted on the same set of axes as the points and did pass through all the points. What is important to note is the behavior of the graph when x is in the interval [0, 1]. More points would have been needed to be found in that interval to have seen that behavior in part (a).

Example

Given the equation of the circle $x^2 + y^2 = 16$, do the following.

(a) Graph the equation of the given circle.
(b) Write the given equation as a set of functions that defines y explicitly in terms of x.
(c) Graph the equations found in part (b).
(d) Write the given equation as a set of functions that defines x explicitly in terms of y.
(e) Graph the equations found in part (d).

Solution

(a) The given circle is centered at the origin and has a radius of 4. The graph is shown in Figure 10.14.

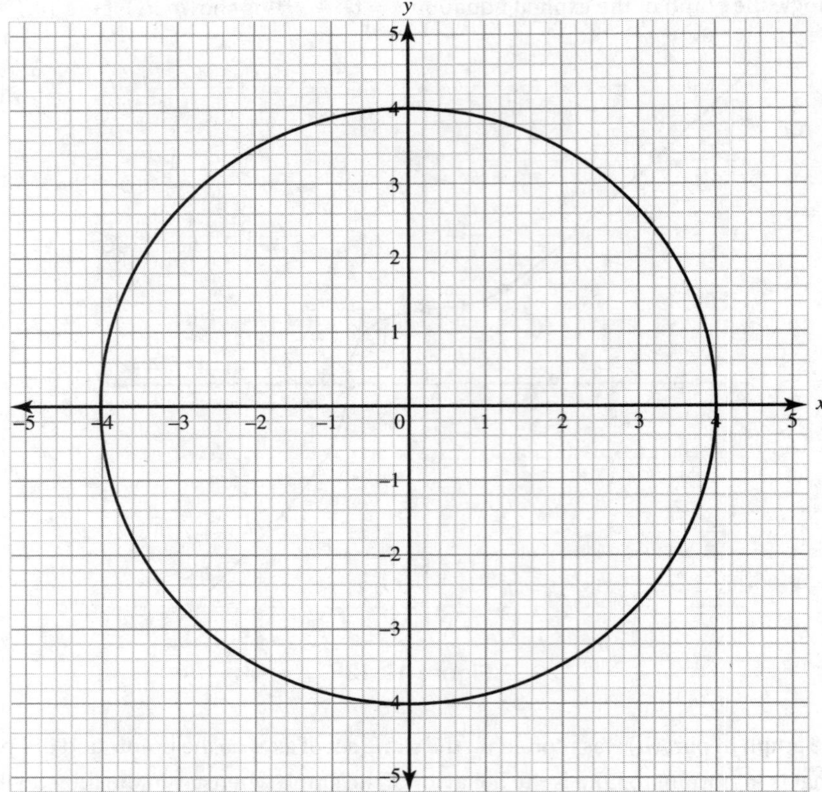

Figure 10.14

(b) Solve for y in terms of x:

$$x^2 + y^2 = 16$$
$$y^2 = 16 - x^2$$
$$y = \pm\sqrt{16 - x^2}$$

(c) When y is written explicitly in terms of x, two different equations are produced. Figure 10.15 shows the graph of $y = +\sqrt{16 - x^2}$. This semicircle is the top half of the circle graphed in part (a).

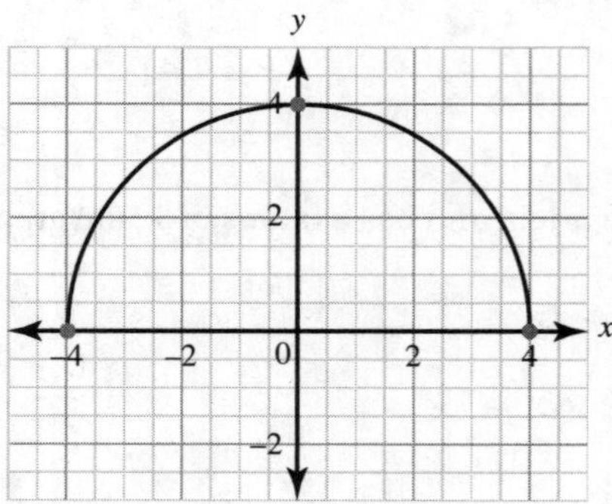

Figure 10.15

Figure 10.16 shows the graph of $y = -\sqrt{16 - x^2}$. This semicircle is the bottom half of the circle graphed in part (a).

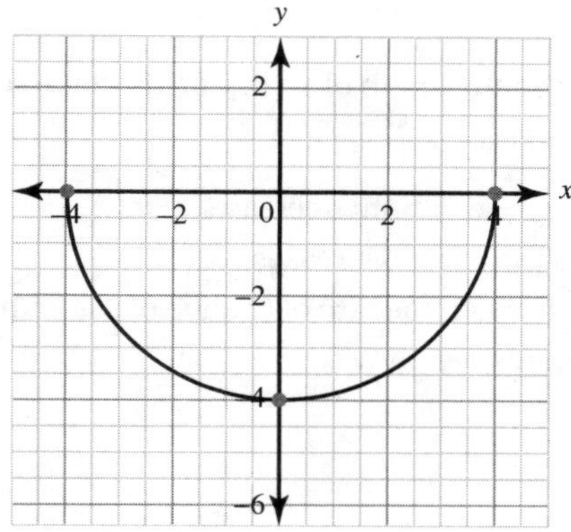

Figure 10.16

(d) Solve for x in terms of y:

$$x^2 + y^2 = 16$$
$$x^2 = 16 - y^2$$
$$x = \pm\sqrt{16 - y^2}$$

(e) When x is written explicitly in terms of y, two different equations are produced. Figure 10.17 is the graph of $x = +\sqrt{16 - y^2}$. This semicircle is the right half of the circle graphed in part (a).

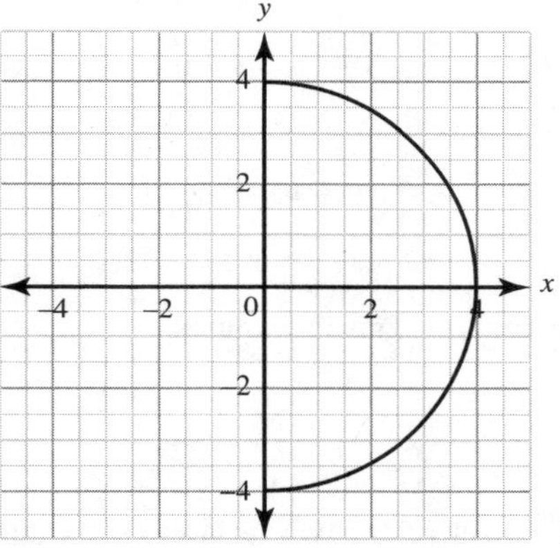

Figure 10.17

Figure 10.18 is the graph of $x = -\sqrt{16 - y^2}$. This semicircle is the left half of the circle graphed in part (a).

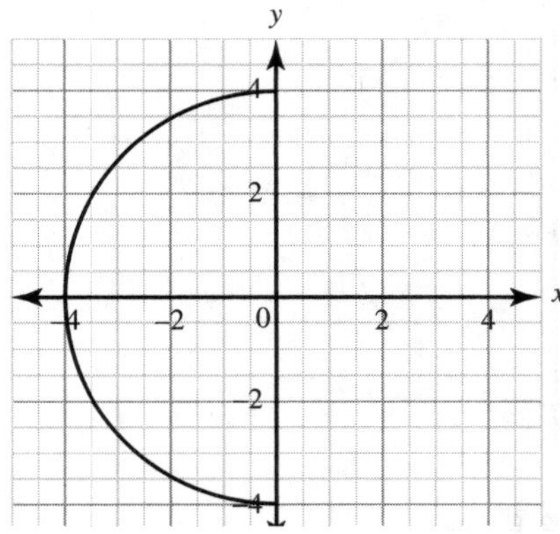

Figure 10.18

In this example, solving for y in terms of x graphed the top and bottom halves of the curve. However, solving for x in terms of y graphed the right and left halves of the curve.

Solving for one of the variables in an equation involving two variables can define a function whose graph is part or all of the graph of the implicit equation.

Rate of Change

For ordered pairs that are close together on the graph of an implicitly defined function, if the rate of change in the two variables is

- positive, the two variables both increase or both decrease.
- negative, as one variable increases the other decreases.
- zero, this indicates vertical intervals when the rate of change of x is with respect to y or horizontal intervals when the rate of change of y is with respect to x.

Example

Given the implicitly defined equation $x^2 + y^2 = 4$, find the following.

(a) The average rate of change when $-2 \leq x \leq 0$ and $0 \leq y \leq 2$

(b) The average rate of change when $0 \leq x \leq 2$ and $0 \leq y \leq 2$

Solution

(a) To calculate the average rate of change, first find the corresponding y-values when $x = -2$ and $x = 0$.

When $x = -2$:

$$(-2)^2 + y^2 = 4$$
$$y = 0$$

When $x = 0$:

$$(0)^2 + y^2 = 4$$
$$y = \pm 2$$

Reject the -2 since it is not in the interval of given y-values.

Calculate the average rate of change:

$$\frac{\Delta y}{\Delta x} = \frac{0 - 2}{-2 - 0} = \frac{-2}{-2} = 1$$

Since this average rate of change is positive, the two variables simultaneously increase or decrease. In Figure 10.19, the graph for $-2 \leq x \leq 0$ shows both variables increasing.

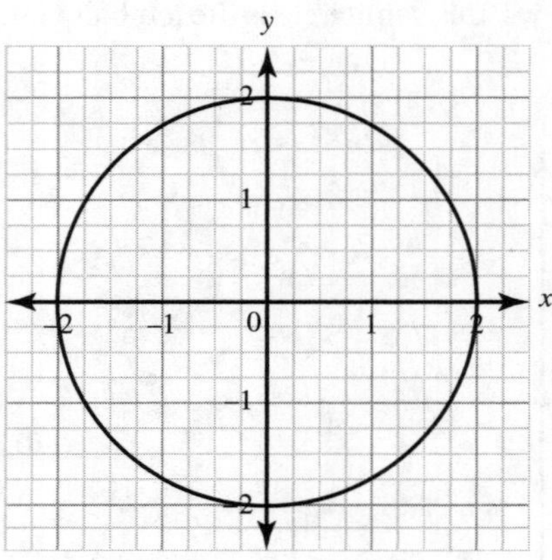

Figure 10.19

(b) To calculate the average rate of change, first find the corresponding y-values when $x = 0$ and $x = 2$.

When $x = 0$:

$$(0)^2 + y^2 = 4$$
$$y = \pm 2$$

Reject the -2 since it is not in the interval of given y-values.

When $x = 2$:

$$(2)^2 + y^2 = 4$$
$$y = 0$$

Calculate the average rate of change:

$$\frac{\Delta y}{\Delta x} = \frac{2 - 0}{0 - 2} = \frac{2}{-2} = -1$$

Since this average rate of change is negative, as one variable increases, the other decreases. In Figure 10.19, the graph for $0 \le x \le 2$ shows that as x is increasing, y is decreasing.

10.6 Conic Sections

Circles, ellipses, parabolas, and hyperbolas are all curves in the plane and are called conic sections. As shown in Figure 10.20, they can be formed by cutting a double cone with a plane at different angles:

- A circle is formed when a plane intersects one cone and is perpendicular to the axis.
- An ellipse is formed when a plane intersects one cone and is not perpendicular to the axis.
- A parabola is formed when a plane intersects one cone and is parallel to the edge of the cone.
- A hyperbola is formed when a plane intersects both cones.

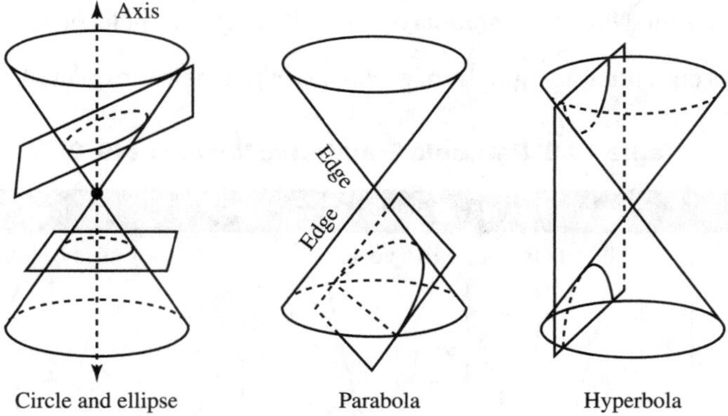

Figure 10.20

The general equations of these four conic sections are written implicitly in terms of x and y, as already seen with the circle equation in the previous section. Equations for the other conic sections can be developed from their geometric definitions.

The Parabola

As previously discussed, a parabola is the graph of a quadratic function. Another way to develop the equation of a parabola is from its geometric definition.

Geometric Definition of a Parabola

A parabola is the set of all points in a plane that are the same distance from both a fixed point and a given line. The fixed point is the focus, and the given line is the directrix. The line that contains the focus and is perpendicular to the directrix is the focal axis.

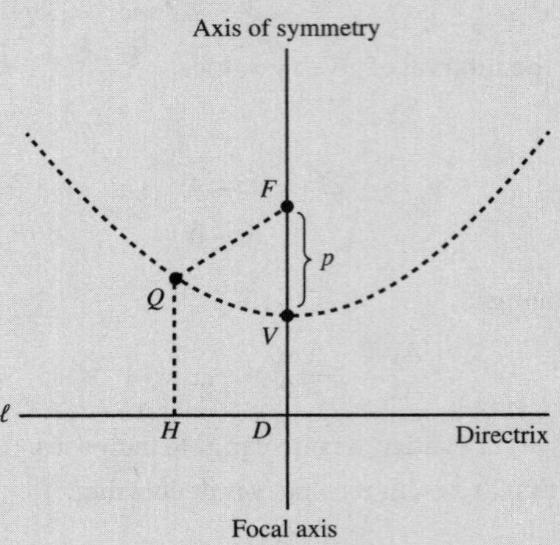

Figure 10.21 Finding all points Q such that $QF = QH$

Figure 10.21 shows a parabola formed by locating all points that are the same distance from point F (the focus) as they are from line ℓ (the directrix). Point Q represents all the points that are equidistant from F and H, meaning that $QF = QH$.

There are a few points to recognize about parabolas that are illustrated in Figure 10.21:

- The focal axis is a line of symmetry and is perpendicular to the directrix. The vertex is the point at which the parabola intersects the focal axis.
- The vertex is midway between the focus and the point at which the focal axis intersects the directrix.
- If the focal axis is a vertical line, the parabola opens up or opens down.
- If the focal axis is a horizontal line, the parabola opens to the right or opens to the left.

Table 10.8 shows the various equations and graphs of parabolas when the vertex is at (0, 0).

Table 10.8 Parabola Equations: Vertex at (0, 0)

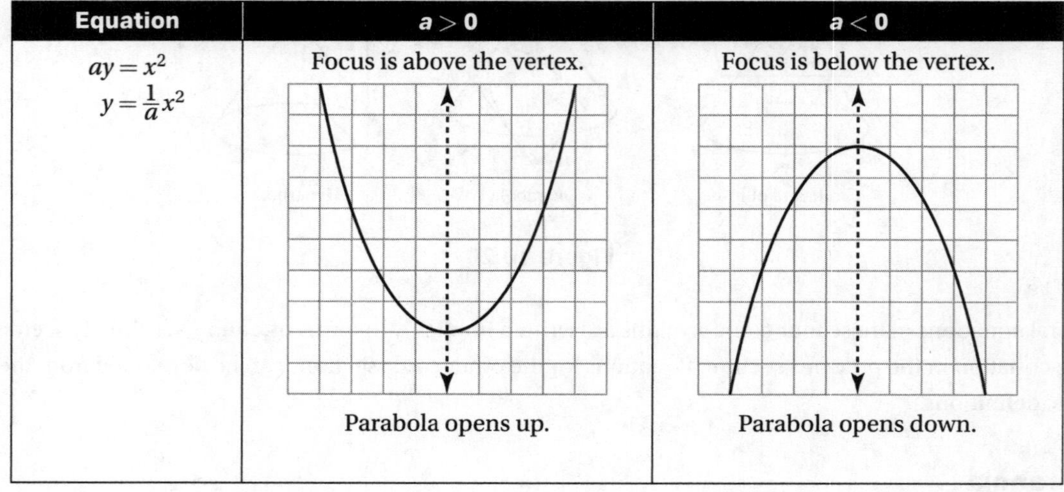

Equation	$a > 0$	$a < 0$
$ay = x^2$ $y = \frac{1}{a}x^2$	Focus is above the vertex. Parabola opens up.	Focus is below the vertex. Parabola opens down.

(Continued)

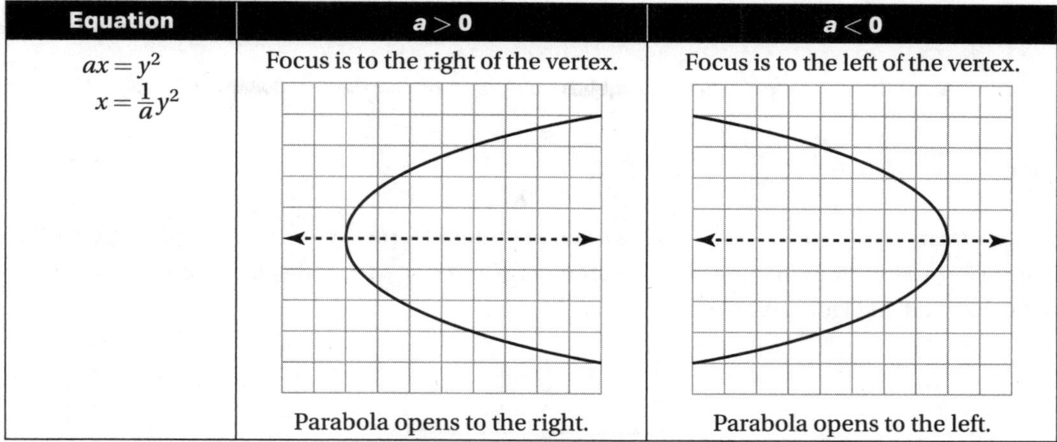

Equation	$a > 0$	$a < 0$
$ax = y^2$ $x = \frac{1}{a}y^2$	Focus is to the right of the vertex. Parabola opens to the right.	Focus is to the left of the vertex. Parabola opens to the left.

Parabola Equations: Vertex at (*h*, *k*)

To find the standard equation of a parabola whose vertex has the general coordinates (h, k), use the rules of horizontal and vertical translations. Shift $ay = x^2$ and $ax = y^2$ by $|h|$ units horizontally and $|k|$ units vertically as shown in Table 10.9.

Table 10.9

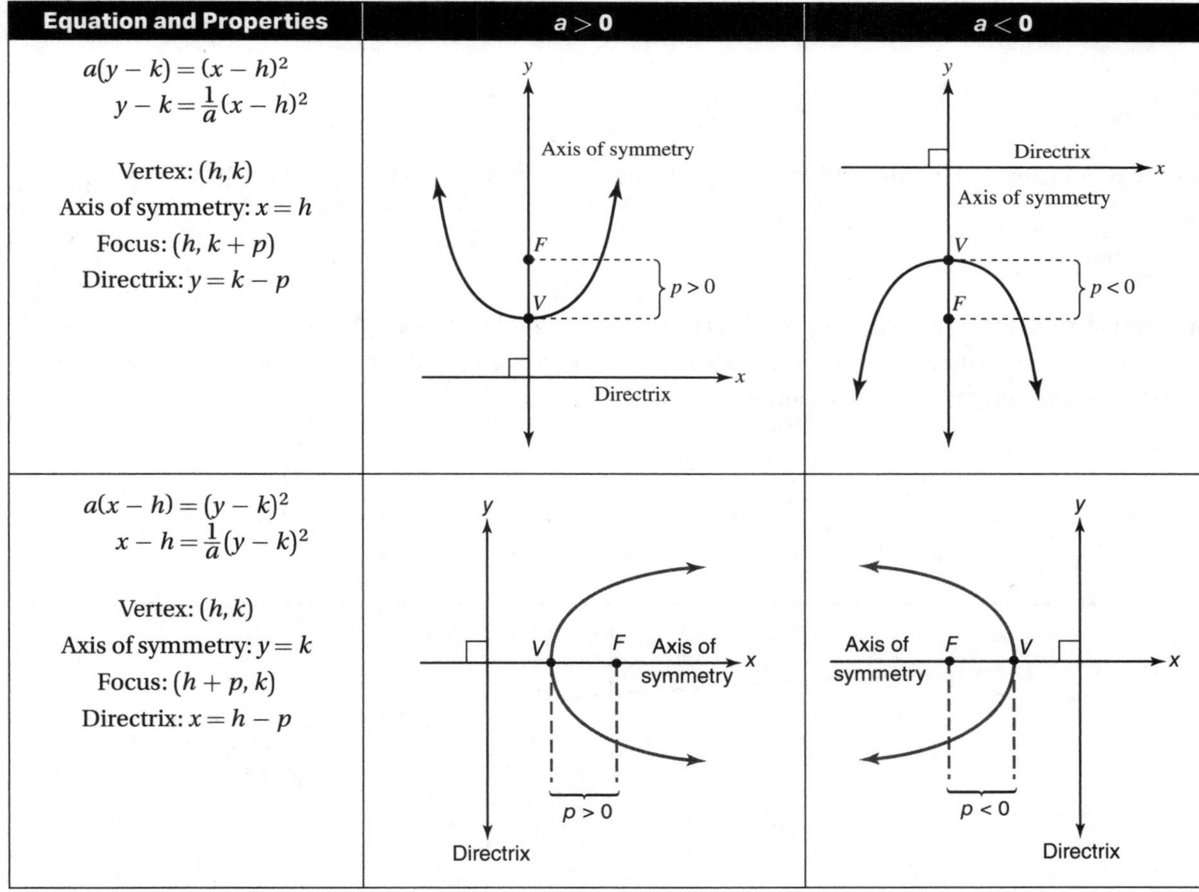

Equation and Properties	$a > 0$	$a < 0$
$a(y-k) = (x-h)^2$ $y - k = \frac{1}{a}(x-h)^2$ Vertex: (h, k) Axis of symmetry: $x = h$ Focus: $(h, k+p)$ Directrix: $y = k - p$	y Axis of symmetry F V $p > 0$ Directrix x	y Directrix x Axis of symmetry V F $p < 0$
$a(x-h) = (y-k)^2$ $x - h = \frac{1}{a}(y-k)^2$ Vertex: (h, k) Axis of symmetry: $y = k$ Focus: $(h+p, k)$ Directrix: $x = h - p$	y V F Axis of symmetry x $p > 0$ Directrix	y Axis of symmetry F V x $p < 0$ Directrix

By inspecting a parabola's equation when written in standard form, it is easy to determine the coordinates of the vertex and the orientation of the parabola.

Example

State the coordinates of the vertex and orientation of the parabola whose equation is given by $(x-1)^2 = -4(y+8)$.

Solution

Since the given equation $(x-1)^2 = -4(y+8)$ is written in standard form $a(y-k) = (x-h)^2$, the coordinates of the vertex are $(1, -8)$. Since the equation contains an x^2-term and $a = -4 < 0$, the parabola opens down. This can be verified by the graph in Figure 10.22.

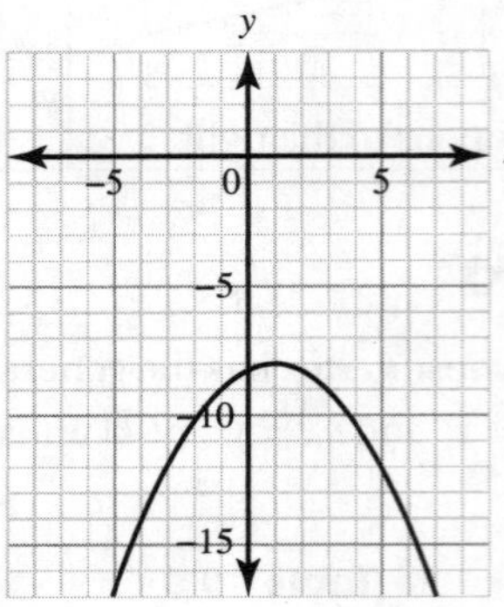

Figure 10.22

Example

State the coordinates of the vertex and orientation of the parabola whose equation is given by $(y+2)^2 = -\frac{1}{8}(x-5)$.

Solution

Since the given equation $(y+2)^2 = -\frac{1}{8}(x-5)$ is written in standard form $a(x-h) = (y-k)^2$, the coordinates of the vertex are $(5, -2)$. Since the equation contains a y^2-term and $a = -\frac{1}{8} < 0$, the parabola opens to the left. This can be verified by the graph shown in Figure 10.23.

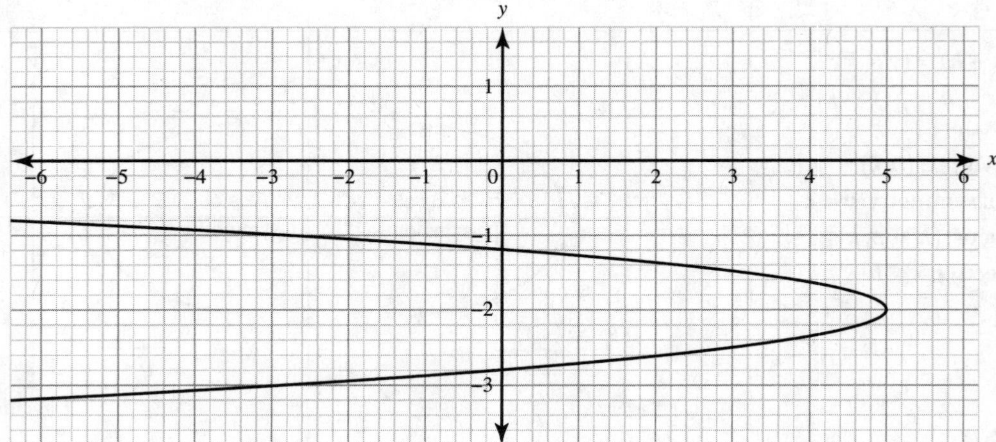

Figure 10.23

Example

Write the equation of a parabola whose vertex is at $(-2, 4)$ and passes through the origin as shown in the graph in Figure 10.24.

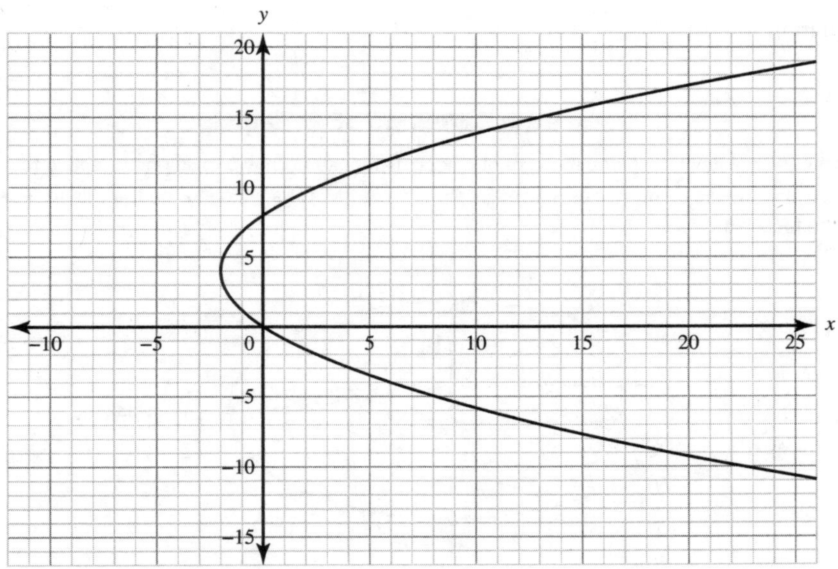

Figure 10.24

Solution

Since the graph is a horizontal parabola that opens to the right, its equation is written in the standard form $a(x - h) = (y - k)^2$. Since the vertex is $(-2, 4)$, substitute into the equation $h = -2$, $k = 4$. To find the a-value, substitute 0 for x and y:

$$a(0 + 2) = (0 - 4)^2$$

$$2a = 16$$

$$a = 8$$

The equation is $8(x + 2) = (y - 4)^2$.

The Ellipse

In the equation of the circle, $x^2 + y^2 = 16$, the coefficients of the x^2-term and y^2-term are equal. However, if one coefficient is made larger than the other, such as $x^2 + 8y^2 = 16$, the curve the equation describes becomes more oval and is called an ellipse. The equation of an ellipse is usually written so that the number 1 appears alone on the right side of the equation, such as $\frac{x^2}{16} + \frac{y^2}{2} = 1$.

Geometric Definition of an Ellipse

An ellipse is the set of all points in the plane the sum of whose distances from two fixed points, called foci, is constant. In Figure 10.25, the foci are labeled F and F'. The sum $PF + PF'$ is the same for any point P on the ellipse. The line that contains F and F' is called the focal axis.

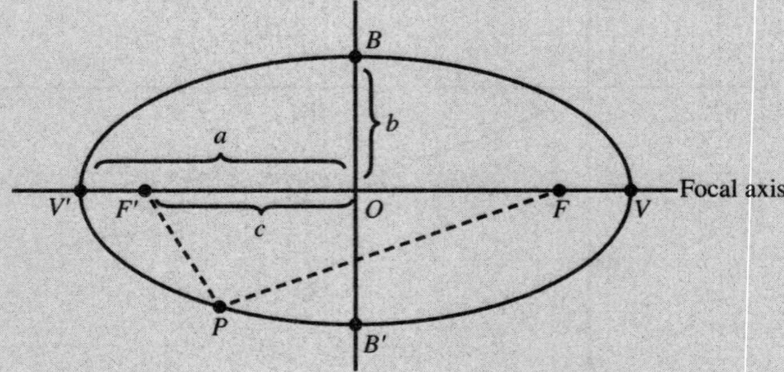

Figure 10.25 Terms related to an ellipse

There are a few points to recognize about ellipses that are illustrated in Figure 10.25:

- Point O, the midpoint of segment FF', is the center of the ellipse. The center of an ellipse is the midpoint of the segment whose endpoints are the foci. Each of the foci is c units from the center.
- Points V and V' are the vertices of the ellipse. The vertices are the two points where the focal axis intersects the ellipse.
- Line segment VV' is the major axis. The major axis coincides with the focal axis and is an axis of symmetry. Since each of the vertices is a units from the center, the length of the major axis is $2a$.
- Points B and B' are the endpoints of the minor axis. The minor axis is a line segment through the center of the ellipse that is perpendicular to the major axis and whose endpoints are points on the ellipse. The length of the minor axis is $2b$. The major axis of an ellipse is always longer than the minor axis.

Table 10.10 shows the various equations and graphs of ellipses when the center is at (0, 0).

Table 10.10 Ellipse Equations: Center at (0, 0) Where $a > b > 0$

Equation	Graph	Properties
$\frac{x^2}{a^2}+\frac{y^2}{b^2}=1$	Horizontal ellipse 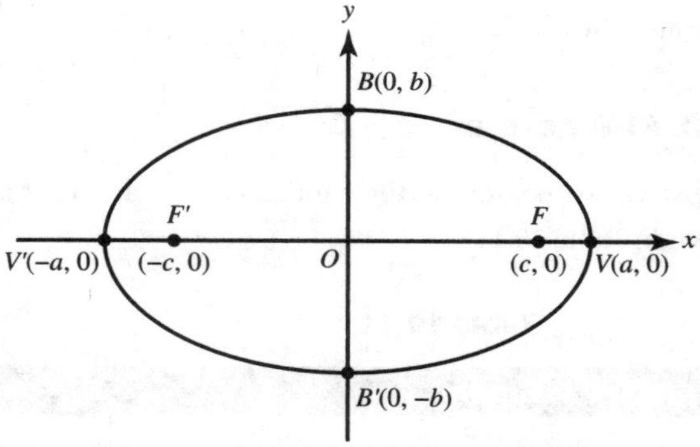	Foci: $(\pm c, 0)$ Vertices: $(\pm a, 0)$ Length of major axis: $2a$
$\frac{x^2}{b^2}+\frac{y^2}{a^2}=1$	Vertical ellipse 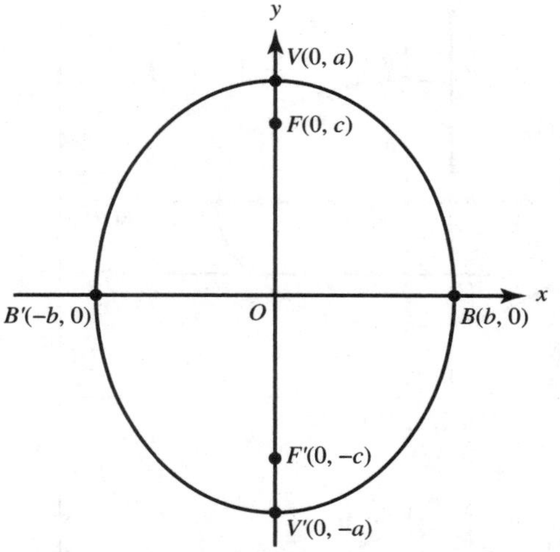	Foci: $(0, \pm c)$ Vertices: $(0, \pm a)$ Length of major axis: $2a$

When the equation of an ellipse is in standard form, it can be determined whether the ellipse is horizontal or vertical by comparing the denominators of the variable terms:

- If the denominator of the x^2-term is greater than the denominator of the y^2-term, the major axis is horizontal.
- If the denominator of the y^2-term is greater than the denominator of the x^2-term, the major axis is vertical.

In an ellipse, the numbers a, b, and c are related by the equation $a^2 = b^2 + c^2$. When finding the foci, the value of c can be found with $c = \pm\sqrt{a^2 - b^2}$.

› Example

Describe the key properties of the graph of $\frac{x^2}{9}+\frac{y^2}{4}=1$.

Solution

The ellipse has the form $\frac{x^2}{a^2} + \frac{y^2}{b^2} = 1$, where $a^2 = 9$ and $b^2 = 4$. Since the larger denominator is associated with the x^2-term, the ellipse equation describes a horizontal ellipse whose center is at (0, 0). The horizontal radius is 3, and the length of the major axis is 6. The vertical radius is 2, and the length of the minor axis is 4. To find the coordinates of the foci, solve for c, $c = \pm\sqrt{9-4} = \pm\sqrt{5}$. The coordinates of the foci are $(-\sqrt{5}, 0)$ and $(\sqrt{5}, 0)$. The coordinates of the vertices are $(-3, 0)$ and $(3, 0)$, which are also the x-intercepts.

Ellipse Equations: Center at (*h*, *k*) Where $a > b > 0$

To find the standard equation of an ellipse whose center has the general coordinates (h, k), use the rules of horizontal and vertical translations, as shown in Table 10.11.

Table 10.11

Equation	Graph	Properties
$\frac{(x-h)^2}{a^2} + \frac{(y-k)^2}{b^2} = 1$	Horizontal ellipse 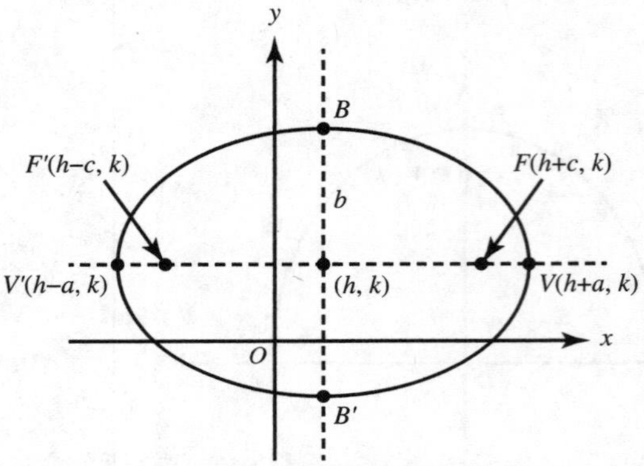 	Foci: $(h \pm c, k)$ Focal axis: $y = k$ Vertices: $(h \pm a, k)$ Length of major axis: $2a$
$\frac{(x-h)^2}{b^2} + \frac{(y-k)^2}{a^2} = 1$	Vertical ellipse 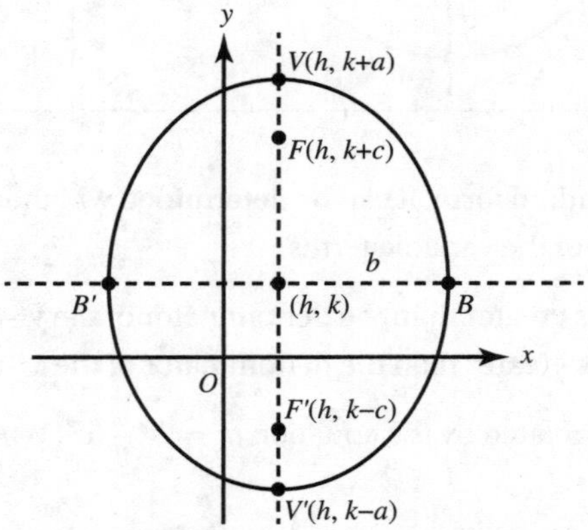 	Foci: $(h, k \pm c)$ Focal axis: $x = h$ Vertices: $(h, k \pm a)$ Length of major axis: $2a$

Example

Determine the coordinates of the center and the foci of the ellipse whose equation is $\frac{(x-1)^2}{4} + \frac{(y+7)^2}{17} = 1$.

Solution

Since the denominator of the y^2-term is greater than the denominator of the x^2-term, the ellipse is vertical with $a^2 = 17$ and $b^2 = 4$. The center is at $(1, -7)$. To write the coordinates of the foci, first calculate $c = \pm\sqrt{17 - 4} = \pm\sqrt{13}$. The coordinates of the foci are $(h, k \pm c) = (1, -7 \pm \sqrt{13}) = (1, -7 + \sqrt{13})$ and $(1, -7 - \sqrt{13})$.

Example

The foci of an ellipse are located at $(-1, 2)$ and $(7, 2)$. If the length of the minor axis is 6 units, find the following.

(a) The coordinates of the center and the vertices

(b) The standard form of the equation of the ellipse

Solution

(a) Since the foci have the same y-coordinate but different x-coordinates, the foci lie on a horizontal line. So this is a horizontal ellipse.
The center is on the same line as the foci and acts as the midpoint. Find the coordinates of the center:

$$(h, k) = \left(\frac{-1+7}{2}, \frac{2+2}{2}\right) = (3, 2)$$

To locate the vertices, find the value of a. Since the distance from the center $(3, 2)$ to the focus $(7, 2)$ is 4 units, $c = 4$. The length of the vertical minor axis is given as 6, so $2b = 6$ and $b = 3$. Therefore:

$$a = \pm\sqrt{b^2 + c^2} = \sqrt{3^2 + 4^2} = 5$$

Find the coordinates of the vertices:

$$(h \pm a, k) = (3 \pm 5, 2) = (-2, 2) \text{ and } (8, 2)$$

(b) The equation of this horizontal ellipse is $\frac{(x-h)^2}{a^2} + \frac{(y-k)^2}{b^2} = 1$, where $h = 3$, $k = 2$, $a^2 = 25$, and $b^2 = 9$:

$$\frac{(x-3)^2}{25} + \frac{(y-2)^2}{9} = 1$$

The Hyperbola

A hyperbola consists of two disconnected branches that are mirror images of each other as shown in Figure 10.26. A hyperbola, like an ellipse, can be defined in terms of two fixed points called foci.

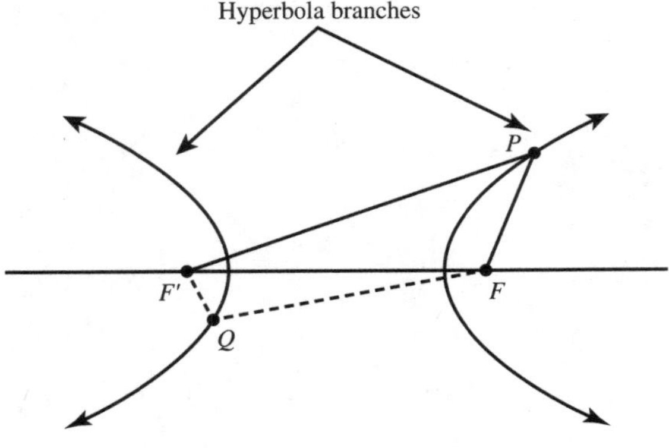

Figure 10.26

For any point on a hyperbola, the difference of the distances from that point to the foci is the same.

Geometric Definition of a Hyperbola

A hyperbola, as shown in Figure 10.27, is the set of all points in the plane the difference of whose distances from two fixed points, the foci, is the same. The focal axis is the line that contains the foci. The foci are always located in the interiors of the branches of the hyperbola.

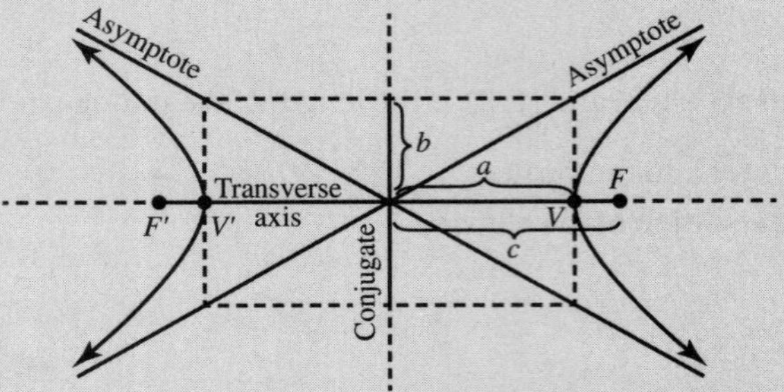

Figure 10.27 Axes and asymptotes of a hyperbola

There are a few points to recognize about hyperbolas that are illustrated in Figure 10.27:

- The terms foci, focal axis, center, and vertices used for ellipses have the same meanings for hyperbolas. The same is true for the distances a and c.
- The center of the hyperbola is the midpoint of segment FF'. The letter c represents the distance from the center to either of the two foci.
- The vertices are labeled as V and V'. The vertices are closer to the center than are the foci. The letter a represents the distance from the center to either of the two vertices. For a hyperbola, $c > a$.
- The transverse axis is the line segment whose endpoints are V and V'. The length of the transverse axis is $2a$.
- The conjugate axis is the line segment through the center that is perpendicular to the transverse axis and whose endpoints are b units from the center, where $b^2 = c^2 - a^2$. The length of the conjugate axis is $2b$. The conjugate axis, like the transverse axis, is an axis of symmetry.
- The Pythagorean relation for a hyperbola is $c^2 = a^2 + b^2$.
- The asymptotes are the two lines through the center of the hyperbola that the branches of the hyperbola approach as the curve extends away from the center. The asymptotes contain the diagonals of the rectangle that has the same center as the hyperbola and whose adjacent sides measure $2a$ and $2b$.

Table 10.12 shows the various equations and graphs of hyperbolas when the center is at (0, 0).

Table 10.12 Hyperbola Equations: Center at (0, 0)

Equation	Graph	Properties
$\frac{x^2}{a^2}-\frac{y^2}{b^2}=1$	Horizontal hyperbola 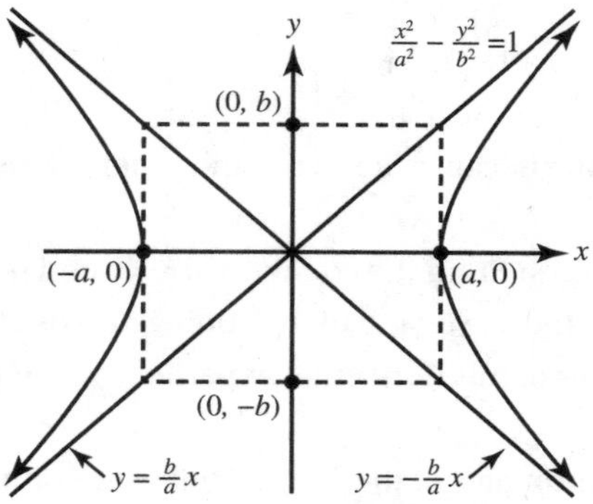 Equation of a horizontal hyperbola with center at (0,0)	Foci: $(\pm c, 0)$ Vertices: $(\pm a, 0)$ Asymptote: $y=\pm\frac{b}{a}x$
$\frac{y^2}{a^2}-\frac{x^2}{b^2}=1$	Vertical hyperbola 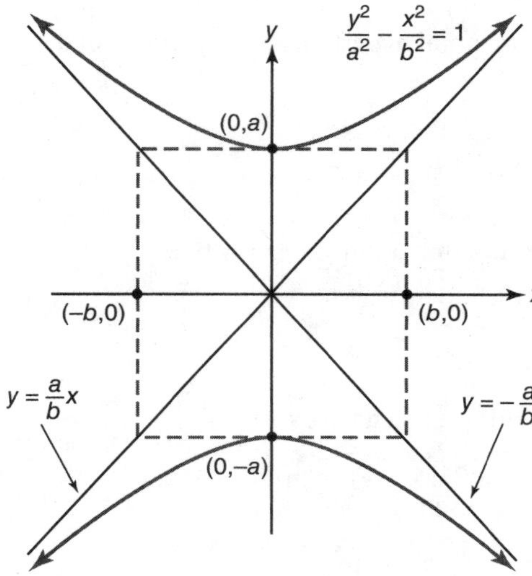 Equation of a vertical hyperbola with center at (0,0)	Foci: $(0, \pm c)$ Vertices: $(0, \pm a)$ Asymptote: $y=\pm\frac{a}{b}x$

When the equation of a hyperbola is in standard form, there are a few important properties to note:

- The Pythagorean relation is $c^2 = a^2 + b^2$, so c is always the greatest of the three distances.
- The order in which the x^2-term and y^2-term are written matters. In the equation for a horizontal hyperbola, the x^2-term comes before the y^2-term. In the equation for a vertical hyperbola, the y^2-term comes first.
- There is no requirement that a be greater than b or that a cannot equal b. Either of the two axes of a hyperbola may be longer than the other axis, or both axes may have the same length.

Example

List the key features of the hyperbola whose equation is $9x^2 - y^2 = 81$.

✓ Solution

Rewrite the equation in standard form by dividing each term by 81:

$$\frac{9x^2}{81} - \frac{y^2}{81} = \frac{81}{81}$$

$$\frac{x^2}{9} - \frac{y^2}{81} = 1$$

1. **Center and orientation of the hyperbola:** Because the sign of the x^2-term is positive, the hyperbola has a horizontal focal axis with its center at (0, 0).
2. **Lengths of transverse and conjugate axes:** From the equation, $a^2 = 9$ and $b^2 = 81$. So $a = 3$ and $b = 9$. The length of the transverse axis is $2 \cdot 3 = 6$, and the length of the conjugate axis is $2 \cdot 9 = 18$.
3. **Coordinates of the vertices:** The vertices of this horizontal hyperbola are on the x-axis with coordinates $(\pm a, 0) = (\pm 3, 0)$.
4. **Coordinates of the foci:** To find the coordinates of the foci, calculate the value of c:

$$c = \pm\sqrt{a^2 + b^2} = \pm\sqrt{9 + 81} = \pm\sqrt{90} = \pm 3\sqrt{10}$$

   The foci of this horizontal hyperbola are on the x-axis with coordinates $(\pm c, 0) = (\pm 3\sqrt{10}, 0)$.

5. **Equations of the asymptotes:** For a horizontal hyperbola, the equations of the asymptotes are $y = \pm\frac{b}{a}x = \pm\frac{9}{3}x = 3x$.

› Example

List the key features of the hyperbola whose equation is $25y^2 - 4x^2 = 100$.

✓ Solution

Rewrite the equation in standard form by dividing each term by 100:

$$\frac{25y^2}{100} - \frac{4x^2}{100} = \frac{100}{100}$$

$$\frac{y^2}{4} - \frac{x^2}{25} = 1$$

1. **Center and orientation of the hyperbola:** Because the sign of the y^2-term is positive, the hyperbola has a vertical focal axis with its center at (0, 0).
2. **Lengths of transverse and conjugate axes:** From the equation, $a^2 = 4$ and $b^2 = 25$. So $a = 2$ and $b = 5$. The length of the transverse axis is $2 \cdot 2 = 4$, and the length of the conjugate axis is $2 \cdot 5 = 10$.
3. **Coordinates of the vertices:** The vertices of this vertical hyperbola are on the y-axis with coordinates $(0, \pm a) = (0, \pm 2)$.
4. **Coordinates of the foci:** To find the coordinates of the foci, calculate the value of c:

$$c = \pm\sqrt{a^2 + b^2} = \pm\sqrt{4 + 25} = \pm\sqrt{29}$$

   The foci of this vertical hyperbola are on the y-axis with coordinates $(0, \pm c) = (0, \pm 3\sqrt{29})$.

5. **Equations of the asymptotes:** For a vertical hyperbola, the equations of the asymptotes are $y = \pm\frac{a}{b}x = \pm\frac{2}{5}x$.

Hyperbola Equations: Center at (*h*, *k*)

To find the standard equation of a hyperbola whose center has the general coordinates (h, k) use the rules of horizontal and vertical translations, as shown in Table 10.13.

Table 10.13

Equation	Graph	Properties
$\frac{(x-h)^2}{a^2} - \frac{(y-k)^2}{b^2} = 1$	Horizontal hyperbola Translating a horizontal hyperbola	Foci: $(h \pm c, k)$ Focal axis: $y = k$ Vertices: $(h \pm a, k)$ Asymptotes: $y = \pm\frac{b}{a}(x-h) + k$
$\frac{(y-k)^2}{a^2} - \frac{(x-h)^2}{b^2} = 1$	Vertical hyperbola Translating a vertical hyperbola	Foci: $(h, k \pm c)$ Focal axis: $x = h$ Vertices: $(h, k \pm a)$ Asymptotes: $y = \pm\frac{a}{b}(x-h) + k$

Example

Discuss the key features of the hyperbola whose equation is $\frac{(x-1)^2}{36} - \frac{(y+4)^2}{13} = 1$.

✓ Solution

1. **Center and orientation of the hyperbola:** Because the sign of the x^2-term is positive, the hyperbola has a horizontal focal axis with its center at $(1, -4)$.
2. **Lengths of transverse and conjugate axes:** From the equation, $a^2 = 36$ and $b^2 = 13$. So $a = 6$ and $b = \sqrt{13}$. The length of the transverse axis is $2 \cdot 6 = 12$, and the length of the conjugate axis is $2 \cdot \sqrt{13} = 2\sqrt{13}$.
3. **Coordinates of the vertices:** The vertices of this horizontal hyperbola are on the horizontal focal axis with coordinates $(h \pm a, k) = (1 \pm 6, -4) = (7, -4)$ and $(-5, -4)$.
4. **Coordinates of the foci:** To find the coordinates of the foci, calculate the value of c:

$$c = \pm\sqrt{a^2 + b^2} = \pm\sqrt{36 + 13} = \pm\sqrt{49} = \pm 7$$

   The foci of this horizontal hyperbola are on the horizontal focal axis with coordinates $(h \pm c, k) = (1 \pm 7, -4) = (8, -4)$ and $(-6, -4)$.

5. **Equations of the asymptotes:** For a horizontal hyperbola, the equations of the asymptotes are $y = \pm\frac{b}{a}(x - h) + k = \pm\frac{\sqrt{13}}{6}(x - 1) - 4.$

10.7 Parameterization of Functions

A parameterization $(x(t), y(t))$ for an implicitly defined function will, when $x(t)$ and $y(t)$ are substituted for x and y, respectively, satisfy the corresponding equation for every value of t in the domain.

If a curve is defined by the equation $y = f(x)$, a parametric equation is one where the x- and y-coordinates of the curve are both written as functions of another variable called a parameter; this is usually given by the letter t or θ. This can be done by letting $x = t$ and then $y = f(t)$.

Example

Find parametric equations for the equation $y = x^2 - 4$.

Solution

Let $x = t$. Then the parametric equations are $x = t$, $y = t^2 - 4$, where the domain of t is $(-\infty, \infty)$.

Example

Find two different pairs of parametric equations to represent the graph of $y = 2x^2 - 3$.

Solution

One possible pair of equations can be found by letting $x(t) = t$ and then replacing x with t in the equation for $y(t)$. This gives the parameterization $x(t) = t$, $y(t) = 2t^2 - 3$. Since there is no restriction on the domain in the original equation, there is no restriction on the values of t.

For the second parameterization, let $x(t) = 3t - 2$. The only thing that is necessary to do is to check that there are no restrictions imposed on x, meaning that the range of $x(t)$ is all real numbers. This is true for $x(t) = 3t - 2$. Substituting $3t - 2$ for x gives the following parameterization:

$$x(t) = 3t - 2$$

$$y(t) = 2(3t - 2)^2 - 3 = 2(9t^2 - 12t + 4) - 3 = 18t^2 - 24t + 5$$

The same curve in the plane can be parameterized in different ways and can be traversed in different directions with different parametric functions. In other words, a parametric representation need not be unique.

Inverses

If f is a function of x, then $y = f(x)$ can be parameterized as $(x(t), y(t)) = (t, f(t))$. If f is invertible, its inverse can be parameterized as $(x(t), y(t)) = (f(t), t)$ for an appropriate interval of t.

Example

For $f(x) = x^3 - 7$, do the following.

(a) Find the equation for the inverse of $f(x)$, $f^{-1}(x)$.
(b) Find parametric equations for $f(x)$, where $x = t$.
(c) Find parametric equations for $f^{-1}(x)$.

Solution

(a) To find the equation of the inverse, switch the x and y variables and solve for y:

$$x = y^3 - 7$$
$$x + 7 = y^3$$
$$y = f^{-1}(x) = \sqrt[3]{x+7}$$

(b) To find the parametric equations for $f(x)$, let $x = t$. Then $y = f(t) = t^3 - 7$. The parametric equations are $(x(t), y(t)) = (t, t^3 - 7)$.

(c) There are two ways to parameterize the inverse. One method is to have the parametric equations $f^{-1}(t) = (x(t), y(t)) = (t^3 - 7, t)$. A second method is to parameterize the equation of the inverse. To do that, let $x = t$. Then $f^{-1}(t) = \sqrt[3]{t+7}$. The parametric equations are $(x(t), y(t)) = \left(t, \sqrt[3]{t+7}\right)$.

To verify that both methods would produce equivalent equations, create tables of values, such as Tables 10.14 and 10.15.

Table 10.14

$x(t) = t$	$y(t) = \sqrt[3]{t+7}$
-2	$\sqrt[3]{-2+7} = \sqrt[3]{5}$
-1	$\sqrt[3]{-1+7} = \sqrt[3]{6}$
0	$\sqrt[3]{0+7} = \sqrt[3]{7}$
1	$\sqrt[3]{1+7} = \sqrt[3]{8}$
2	$\sqrt[3]{2+7} = \sqrt[3]{9}$

Table 10.15

$x(t) = t^3 - 7$	$y(t) = t$
$\left(\sqrt[3]{5}\right)^3 - 7 = -2$	$\sqrt[3]{5}$
$\left(\sqrt[3]{6}\right)^3 - 7 = -1$	$\sqrt[3]{6}$
$\left(\sqrt[3]{7}\right)^3 - 7 = 0$	$\sqrt[3]{7}$
$\left(\sqrt[3]{8}\right)^3 - 7 = 1$	$\sqrt[3]{8}$
$\left(\sqrt[3]{9}\right)^3 - 7 = 2$	$\sqrt[3]{9}$

Parametric to Rectangular

To change from parametric to rectangular form, eliminate the parameter by solving for t in one of the parametric equations. Then use that equation to replace t in the other equation.

Example

Write the parametric equations $x(t) = 2t + 1$, $y(t) = 4t^2 - 1$, $-3 \leq t \leq 4$, in rectangular form.

Solution

The pair of parametric equations can be changed to rectangular form using the following steps.

1. Solve for t in the first equation: $t = \frac{x-1}{2}$.
2. Substitute $\frac{x-1}{2}$ for t in the second equation:

$$y = 4\left(\frac{x-1}{2}\right)^2 - 1$$
$$y = 4\left(\frac{x^2 - 2x + 1}{4}\right) - 1$$
$$y = x^2 - 2x$$

3. Determine the domain and range of $y = x^2 - 2x$.

   Graphing the parametric equations provides a great visual to help determine the domain and range. This can be done using technology or by creating a table of values. A graph of the parametric equations is shown in Figure 10.28.

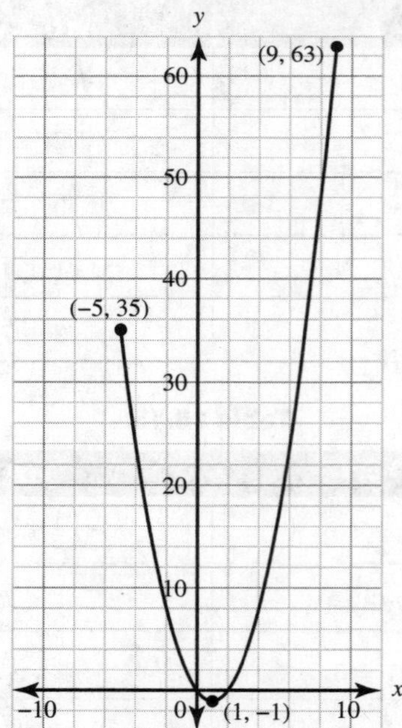

Figure 10.28

When tracing along the parametric curve from $t = -3$ to $t = 4$, x increases from -5 to 9, while y varies from a low of -1 to a high of 63. Hence, the domain of $y = x^2 - 2x$ is restricted to $-5 \leq x \leq 9$, while the range is limited to $-1 \leq y \leq 63$.

Changing from parametric to rectangular form can help identify the type of curve that a set of parametric equations represents.

Example

Find the rectangular equation of the plane curve defined parametrically by the equations $x = 4\cos t$, $y = 3\sin t$, $0 \leq t \leq 2\pi$.

Solution

Here a different technique will be used to isolate the parameter t. Solve for $\cos t$ and $\sin t$:

$$\cos t = \frac{x}{4}, \ \sin t = \frac{y}{3}$$

Using the Pythagorean trigonometric identity $\cos^2 t + \sin^2 t = 1$:

$$\left(\frac{x}{4}\right)^2 + \left(\frac{y}{3}\right)^2 = 1$$

$$\frac{x^2}{16} + \frac{y^2}{9} = 1$$

This is an equation of an ellipse centered at the origin.

Representing Conic Sections Parametrically

In the previous example, parametric equations that were transformed into a rectangular equation graphed an ellipse. This leads to the following parameterizations of conic sections.

Parametric Form of the Parabola

A parabola can be parameterized in the same way that any equation that can be solved for x or y can be parameterized.

Equations that can be solved for x can be parameterized as $(x(t), y(t)) = (f(t), t)$ by solving for x and replacing y with t.

Equations that can be solved for y can be parameterized as $(x(t), y(t)) = (t, f(t))$ by solving for y and replacing x with t.

Example

Parameterize the rectangular equation of the parabola $y^2 = ax$.

Solution

It is easier to solve this equation for x:

$$x = \frac{y^2}{a}$$

Let $y = t$ and replace t for y in the previous equation:

$$x = \frac{t^2}{a}$$

The parametric equations are $(x(t), y(t)) = \left(\frac{t^2}{a}, t\right)$.

Example

Parameterize the rectangular equation of the parabola $-8y = x^2$.

Solution

It is easier to solve this equation for y:

$$y = -\frac{1}{8}x^2$$

Let $x = t$ and replace t for x in the previous equation:

$$y = -\frac{1}{8}t^2$$

The parametric equations are $(x(t), y(t)) = \left(t, -\frac{t^2}{8}\right)$.

Parametric Form of the Ellipse

An ellipse can be parameterized using trigonometric functions.

The equation of an ellipse centered at (h, k) in standard form is $\frac{(x-h)^2}{a^2} + \frac{(y-k)^2}{b^2} = 1$. To express in parametric form, first multiply the equation by a^2b^2:

$$b^2(x-h)^2 + a^2(y-k)^2 = a^2b^2$$

Solve for $y - k$:

$$a^2(y-k)^2 = a^2b^2 - b^2(x-h)^2$$
$$a^2(y-k)^2 = b^2(a^2 - (x-h)^2)$$
$$a(y-k) = \pm\sqrt{b^2(a^2 - (x-h)^2)}$$
$$a(y-k) = \pm\sqrt{b^2}\sqrt{(a^2 - (x-h)^2)}$$

The constant b absorbs the appropriate sign for the equation depending on the points it passes through:

$$a(y-k) = b\sqrt{(a^2 - (x-h)^2)}$$
$$(y-k) = \frac{b}{a}\sqrt{(a^2 - (x-h)^2)}$$

Let $x - h = a\cos t$:

$$(y-k) = \frac{b}{a}\sqrt{(a^2 - (a\cos t)^2)}$$
$$(y-k) = \frac{b}{a}\sqrt{(a^2 - a^2\cos^2 t)}$$
$$(y-k) = \frac{b}{a}\sqrt{a^2(1 - \cos^2 t)}$$
$$(y-k) = \frac{b}{a} \cdot a\sqrt{\sin^2 t}$$
$$y - k = b\sin t$$

Solve for y:

$$y = k + b\sin t$$

Therefore, the equation of an ellipse $\frac{(x-h)^2}{a^2} + \frac{(y-k)^2}{b^2} = 1$ centered at (h, k) in parametric form is

$$x(t) = h + a\cos t,\ y(t) = k + b\sin t \text{ for } 0 \le t \le 2\pi$$

Example

Find the parametric equations of the ellipse $\frac{x^2}{36}+\frac{y^2}{25}=1$.

Solution

In the given equation of the ellipse, $a^2=36$, $b^2=25$, $h=0$, and $k=0$.

Therefore, the parametric equations are $x(t)=6\cos t$, $y(t)=5\sin t$.

Parametric Form of the Hyperbola

A hyperbola can be parameterized using trigonometric functions.

The equation of a hyperbola centered at (h, k) in standard form is either $\frac{(x-h)^2}{a^2}-\frac{(y-k)^2}{b^2}=1$ or $\frac{(y-k)^2}{a^2}-\frac{(x-h)^2}{b^2}=1$.

To express in parametric form, first solve for $x-h$ in the first equation:

$$x-h=\frac{a}{b}\sqrt{b^2+(y-k)^2}$$

Let $y-k=b\tan t$:

$$x-h=\frac{a}{b}\sqrt{b^2+b^2\tan^2 t}=a\sec t$$

Therefore, the equation of a hyperbola $\frac{(x-h)^2}{a^2}-\frac{(y-k)^2}{b^2}=1$ centered at (h, k) in parametric form is

$$x(t)=h+a\sec t,\ y(t)=k+b\tan t \text{ for } 0\le t\le 2\pi$$

If the hyperbola is defined by the equation $\frac{(y-k)^2}{a^2}-\frac{(x-h)^2}{b^2}=1$, the parametric form is

$$x(t)=h+b\tan t,\ y(t)=k+a\sec t \text{ for } 0\le t\le 2\pi$$

Example

Parameterize the hyperbola $\frac{(x+2)^2}{9}-\frac{(y-3)^2}{4}=1$.

Solution

In the given equation of the hyperbola, $a^2=9$, $b^2=4$, $h=-2$, and $k=3$.

Therefore, the parametric equations are $x(t)=-2+3\sec t$, $y(t)=3+2\tan t$.

Multiple-Choice Questions

1. Which of the following choices are pairs of parametric equations for the ellipse $\frac{(x-3)^2}{16}+\frac{(y+2)^2}{49}=1$?
   (A) $x(t)=3+4\cos t,\ y(t)=-2+7\sin t$
   (B) $x(t)=4\cos t,\ y(t)=7\sin t$
   (C) $x(t)=-3+4\cos t,\ y(t)=2+7\sin t$
   (D) $x(t)=4+3\cos t,\ y(t)=7-2\sin t$

2. Which of the following choices are pairs of parametric equations for the hyperbola $\frac{(x+3)^2}{25}-\frac{(y-1)^2}{9}=1$?
   (A) $x(t)=3+5\cos t,\ y(t)=-1+3\sin t$
   (B) $x(t)=3+5\sec t,\ y(t)=-1+3\tan t$
   (C) $x(t)=-3+5\cos t,\ y(t)=1+3\sin t$
   (D) $x(t)=-3+5\sec t,\ y(t)=1+3\tan t$

3. Find the rectangular equation for the curve given by the parametric equations $x(t)=t-1$, $y(t)=t^2-1$, $-2\le t\le 4$.
   (A) $y=x^2+2x$
   (B) $y=x^2-2x$
   (C) $y=x^2-1$
   (D) $y=x^2-2$

4. Find the rectangular equation for the curve given by the parametric equations $x(t)=\sin t,\ y(t)=1-\cos^2 t,\ 0\le t\le \pi$.
   (A) $y=x$
   (B) $y=x^2+1$
   (C) $y=x^2$
   (D) $y=x^2-1$

5. Which of the following choices represents a set of parametric equations that represents the graph of $y=\sqrt{x+4}$ using the parameter $t=\frac{1}{4}+\frac{1}{4}x$?
   (A) $x(t)=\frac{1}{4}+\frac{1}{4}t,\ y(t)=\sqrt{\frac{1}{4}t+\frac{17}{4}},\ t\ge -\frac{3}{4}$
   (B) $x(t)=4t-1,\ y(t)=\sqrt{4t+3},\ t\ge -\frac{3}{4}$
   (C) $x(t)=t,\ y(t)=\sqrt{t+4},\ t\ge -\frac{3}{4}$
   (D) $x(t)=4t-1,\ y(t)=\sqrt{4t+5},\ t\ge -\frac{3}{4}$

6. Find the rectangular form of the equation of a line represented by the parametric equations $x(t)=1+2t$ and $y(t)=2-t$.
   (A) $y-2=-\frac{1}{2}(x-1)$
   (B) $y-2=2x-2$
   (C) $y+2=-\frac{1}{2}(x-1)$
   (D) $y+2=2x-2$

7. Which of the following equations represents the graph shown below?

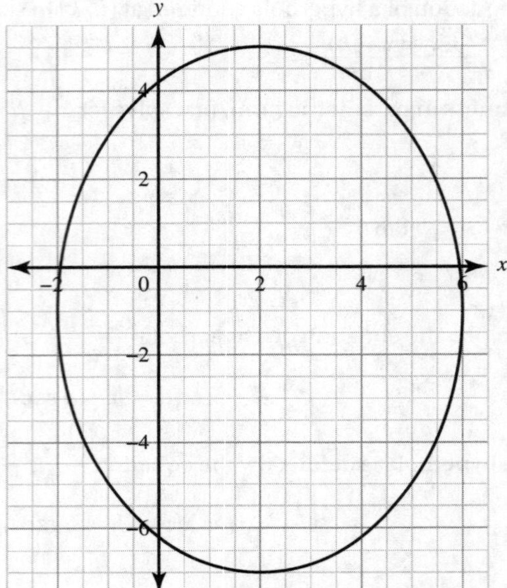

   (A) $\frac{(x+2)^2}{16}+\frac{(y-1)^2}{36}=1$
   (B) $\frac{(x-2)^2}{36}+\frac{(y+1)^2}{16}=1$
   (C) $\frac{(x-2)^2}{16}+\frac{(y+1)^2}{36}=1$
   (D) $\frac{(x+2)^2}{36}+\frac{(y-1)^2}{16}=1$

8. Which of the following represents the parameterization of the line through the points $(-2, 5)$ and $(4, 2)$?
   (A) $x(t)=-2-3t,\ y(t)=5+6t$
   (B) $x(t)=-2+6t,\ y(t)=5-3t$
   (C) $x(t)=5-3t,\ y(t)=-2+6t$
   (D) $x(t)=-2+\frac{1}{6}t,\ y(t)=5-\frac{1}{3}t$

9. Which of the following represents the parametric equations for the graph shown below?

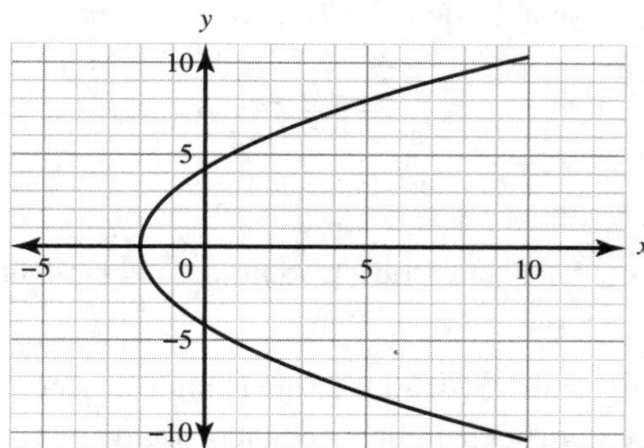

(A) $x(t) = t^2 - 2, y(t) = 3t$
(B) $y^2 = 9(x + 2)$
(C) $x(t) = 3t, y(t) = t^2 - 2$
(D) $x^2 = 9(y + 2)$

10. Given the graph of the closed curve in the xy-plane shown below:

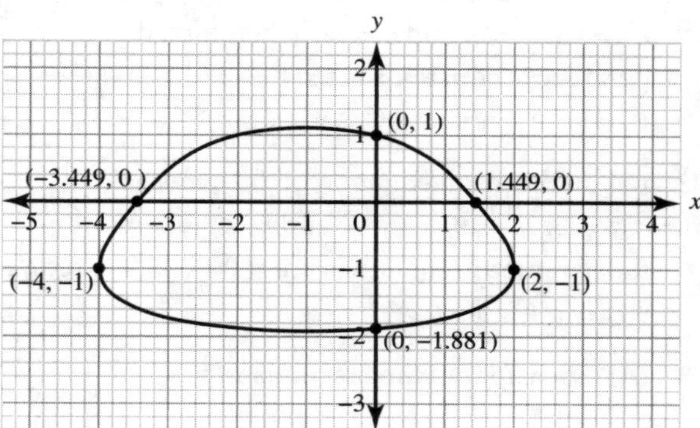

Which of the following choices represents its implicit equation?

(A) $x^2 + 2x + y^4 + 4y = 5$
(B) $x^4 + y^4 = 1$
(C) $\frac{(x+1)^2}{9} + \frac{(y+1)^2}{1} = 1$
(D) $x^4 + 2x + y^2 + 4y = 5$

Answer Explanations

1. **(A)** In the given equation of the ellipse, $a^2 = 16$, $b^2 = 49$, $h = 3$, and $k = -2$. Therefore, the parametric equations are $x(t) = 3 + 4\cos t$, $y(t) = -2 + 7\sin t$.

2. **(D)** In the given equation of the hyperbola, $a^2 = 25$, $b^2 = 9$, $h = -3$, and $k = 1$. Therefore, the parametric equations are $x(t) = -3 + 5\sec t$, $y(t) = 1 + 3\tan t$.

3. **(A)** Solve for t in the first equation:

$$t = x + 1$$

Substitute that expression for t into the second equation:

$$y = (x+1)^2 - 1 = x^2 + 2x + 1 - 1 = x^2 + 2x$$

4. **(C)** Use the Pythagorean identity $\sin^2 t + \cos^2 t = 1$. Substitute x for $\sin t$ and $1 - y$ for $\cos^2 t$:

$$(x)^2 + (1 - y) = 1$$

Solve for y:

$$y = x^2$$

5. **(B)** Using the parameter $t = \frac{1}{4} + \frac{1}{4}x$, solve for x:

$$x = 4t - 1$$

Take the parametric equation for x and substitute it into the equation for y:

$$y = \sqrt{4t - 1 + 4} = \sqrt{4t + 3}$$

6. **(A)** Solve for t in the second equation and substitute for t in the first equation:

$$y = 2 - t \rightarrow t = 2 - y$$

$$x = 1 + 2t \rightarrow x = 1 + 2(2 - y) = 1 + 4 - 2y = 5 - 2y$$

Write y in terms of x: $x = 5 - 2y \rightarrow y = -\frac{1}{2}x + \frac{5}{2}$.

Since all the choices are written in point-slope form, rewrite the choices into slope-intercept form to see which matches.

7. **(C)** The graph provided is an ellipse whose center is $(h, k) = (2, -1)$. The minor axis is horizontal, meaning that $b = 4$. So $b^2 = 16$ will be under the x-term. The major axis is vertical, meaning that $a = 6$. So $a^2 = 36$ will be under the y-term. The equation of the ellipse is $\frac{(x-2)^2}{16} + \frac{(y+1)^2}{36} = 1$.

8. **(B)** Let $t = 0$ represent the point $(-2, 5)$ and $t = 1$ represent the point $(4, 2)$. Then:

$$\frac{\Delta x}{\Delta t} = \frac{4 - (-2)}{1 - 0} = 6$$

$$\frac{\Delta y}{\Delta t} = \frac{2 - 5}{1 - 0} = -3$$

Possible parametric equations are $x = -2 + 6t$ and $y = 5 - 3t$.

9. **(A)** Choices (B) and (D) are eliminated since they are not parametric equations and are instead rectangular equations. A way to choose between Choices (A) and (C) would be to create a table of values with different times to see if the table matches up with the points on the graph. The point $(-2, 0)$ is on the graph. When $t = 0$, substituting into the equations in Choice (A) gives $x = (0)^2 - 2 = -2$, $y = 3(0) = 0$. This matches.

10. **(A)** The correct equation must satisfy all of the points on the graph. Begin with the x-intercepts, where $y = 0$. Substitute 0 for y into all the choices. For Choice (A), the equation turns into the quadratic $x^2 + 2x = 5$. By using the quadratic formula to solve for x, the x-intercepts are $(-1 + \sqrt{6}, 0) \approx (1.449, 0)$ and $(-1 - \sqrt{6}, 0) \approx (-3.449, 0)$.

11

Vectors

Learning Objectives

In this chapter, you will learn:

- ➔ The different properties of vectors
- ➔ How to use vectors to model real-life scenarios

NOTE

These Unit 4 topics are **not** assessed on the AP exam but may be part of your school's course curriculum.

11.1 Algebraic and Graphical Properties of Vectors

The previous chapter demonstrated that particles often need three variables to describe their behavior. For example, when describing the path of an object, its position could move horizontally, $x(t)$, and vertically, $y(t)$, based on time. Changes in the position of a particle affect its velocity and acceleration. This can best be visualized with the paths of planes or projectiles. To organize and graph the paths better, a vector can be used to help describe all the various quantities of its path of motion. A vector allows us to examine the size or magnitude of the quantity being measured as well as its direction. Quantities that only have magnitude are called scalars, and quantities that have both magnitude and direction are called vectors. Velocity is an example of a vector because it describes speed, a scalar, in a particular direction.

Vectors are typically notated using an arrow and are graphically represented using a directed line segment. Consider the vector represented by the directed line segment O to A as shown in Figure 11.1.

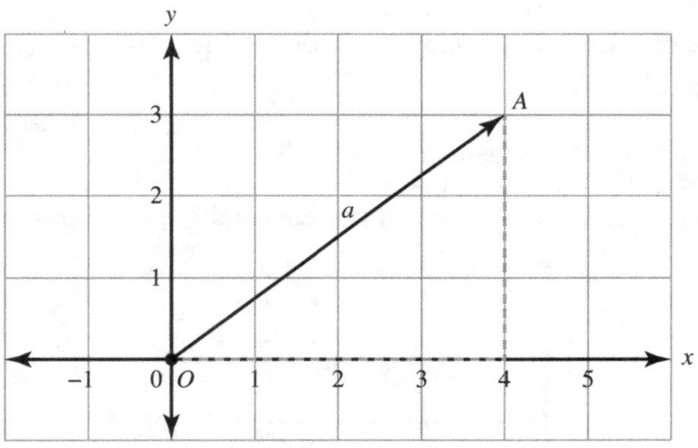

Figure 11.1

This vector can be notated as $\overrightarrow{OA}$ or as $\vec{a}$. We say that $\overrightarrow{OA}$ is the vector that emanates from O and terminates at A. The point O is also referred to as the tail, and the point A is also referred to as the head. $\overrightarrow{OA}$ is the position vector of A relative to O.

The length of the arrow represents the size or magnitude of the vector and can be represented by $|\overrightarrow{OA}| = OA = |\vec{a}|$. The arrowhead shows the direction of travel as shown in Figure 11.2.

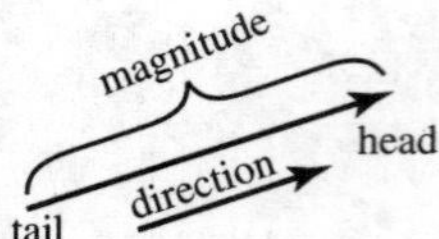

Figure 11.2

Vectors in two dimensions, 2D, have an x-component and a y-component. They can be written in various ways, such as $\binom{x}{y} = \langle x, y \rangle$.

When graphed on the coordinate plane, the x-component can be calculated by counting the boxes horizontally from the tail to the head and the y-component can be calculated by counting the boxes vertically from the tail to the head. As shown in Figure 11.3, the x-component of the vector is 4 and the y-component is 3. Therefore, $\overrightarrow{OA} = \vec{a} = \binom{4}{3} = \langle 4, 3 \rangle$.

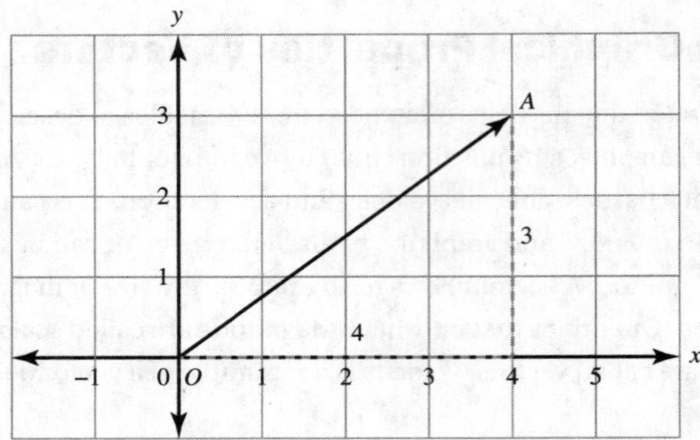

Figure 11.3

To calculate the magnitude, we can use the Pythagorean theorem. The magnitude of the vector can be calculated as:

$$|\overrightarrow{OA}| = \sqrt{4^2 + 3^2} = \sqrt{25} = 5$$

If given the components of the vector, the magnitude of vector $\vec{a} = \langle a_1, a_2 \rangle$ is $|\vec{a}| = \sqrt{(a_1)^2 + (a_2)^2}$.

Not all vectors emanate from the origin. To find the components of the vector between two points, subtract the x- and y-components of the tail from the head.

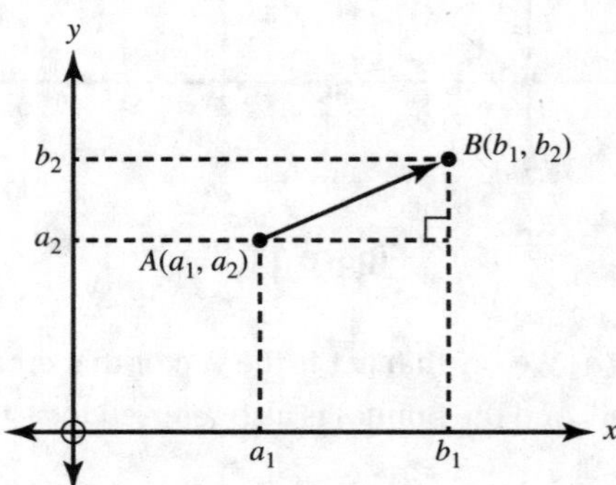

Figure 11.4

As illustrated in Figure 11.4, $\overrightarrow{AB} = B(b_1, b_2) - A(a_1, a_2) = \langle b_1 - a_1, b_2 - a_2 \rangle$. Since the length can be calculated by the distance between the two points, the magnitude of a vector between two points $= \sqrt{(x_2 - x_1)^2 + (y_2 - y_1)^2}$. In this case:

$$|\overrightarrow{AB}| = \sqrt{(b_1 - a_1)^2 + (b_2 - a_2)^2}$$

Example

Find the components and magnitude of $\overrightarrow{CP}$ shown in Figure 11.5.

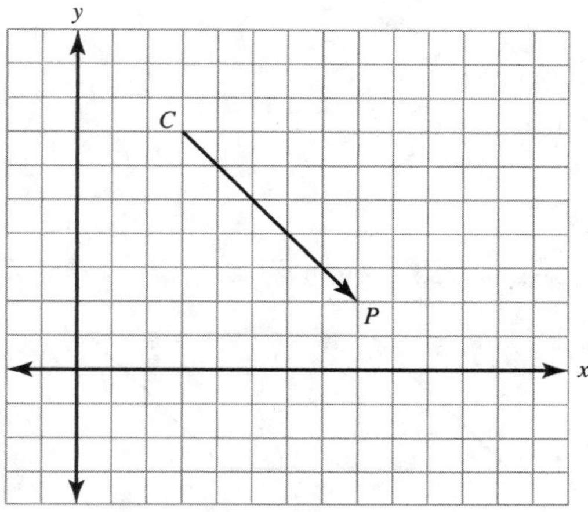

Figure 11.5

Solution

The coordinates of point C are (3, 7), and the coordinates of point P are (8, 2). The components of the vector are $\overrightarrow{CP} = \langle 8 - 3, 2 - 7 \rangle = \langle 5, -5 \rangle$.

Calculate the magnitude of $\overrightarrow{CP}$:

$$|\overrightarrow{CP}| = \sqrt{(5)^2 + (-5)^2} = \sqrt{25 + 25} = \sqrt{50}$$

In the case where the tail and head are the same point, this results in the zero vector, $\langle 0,0 \rangle$.

Vector Equality

Two vectors are equal if they have the same magnitude and direction. In other words, equal vectors are parallel and equal in length. In Figure 11.6, all the vectors are equal since they are translations of each other. Translation preserves length and parallelism.

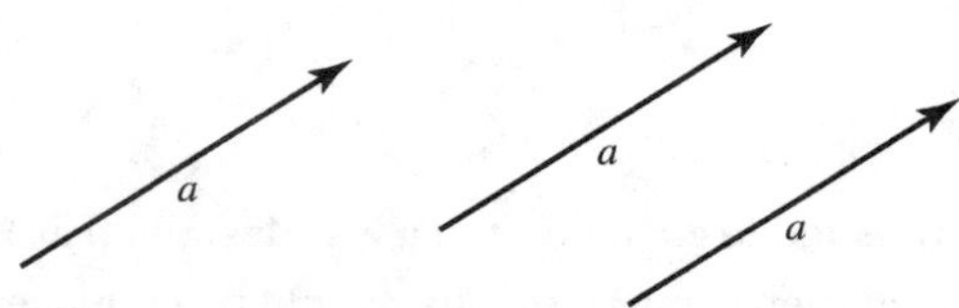

Figure 11.6

Two vectors are equal if and only if their corresponding components are equal. In other words, if $\vec{a} = \langle a_1, a_2 \rangle$ and $\vec{b} = \langle b_1, b_2 \rangle$ then $\vec{a} = \vec{b}$ if and only if $a_1 = b_1$ and $a_2 = b_2$.

Negative Vectors

If two vectors have the same length, are parallel, but have opposite directions, they are negative vectors. As illustrated in Figure 11.7, vectors $\overrightarrow{AB}$ and $\overrightarrow{BA}$ have the same length but run parallel to each other in opposite directions. We would say $\overrightarrow{BA} = -\overrightarrow{AB}$.

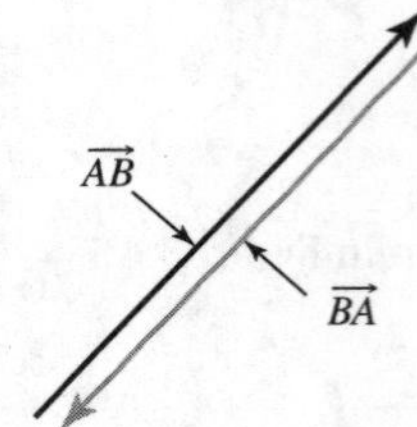

Figure 11.7

Example

Figure 11.8 gives parallelogram *CMJP* where $\overrightarrow{CM} = a$ and $\overrightarrow{MJ} = b$.

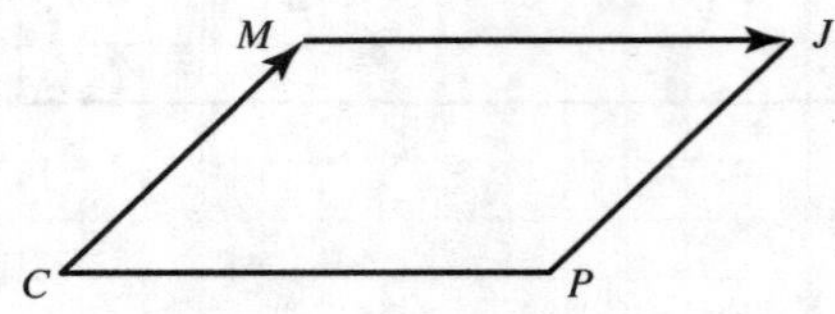

Figure 11.8

Find vector expressions for each of the following.

1. $\overrightarrow{MC}$
2. $\overrightarrow{JM}$
3. $\overrightarrow{PJ}$
4. $\overrightarrow{PC}$

Solution

1. $\overrightarrow{MC} = -\overrightarrow{CM} = -a$
2. $\overrightarrow{JM} = -\overrightarrow{MJ} = -b$
3. $\overrightarrow{PJ} = \overrightarrow{CM} = a$
4. $\overrightarrow{PC} = -\overrightarrow{MJ} = -b$

Geometric Vector Addition

There are times when vectors are connected to each other in a way described as "head to tail." Consider Figure 11.9. If we travel from *C* to *M* and then from *M* to *J*, that would be equivalent to traveling from *C* directly to *J*.

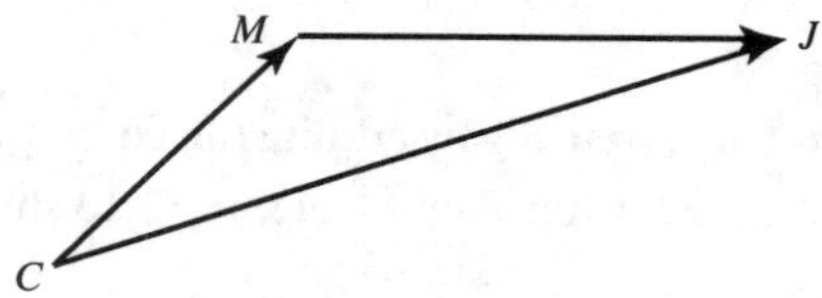

Figure 11.9

This can be expressed in vector form as the sum $\overrightarrow{CM} + \overrightarrow{MJ} = \overrightarrow{CJ}$. If the first vector being added has its head where the tail of the second vector is, then graphically the new vector's tail is the tail of the first vector and the new vector's head is the head of the second vector.

Example

Given $\vec{a}$ and $\vec{b}$ in Figure 11.10, construct $\vec{a} + \vec{b}$.

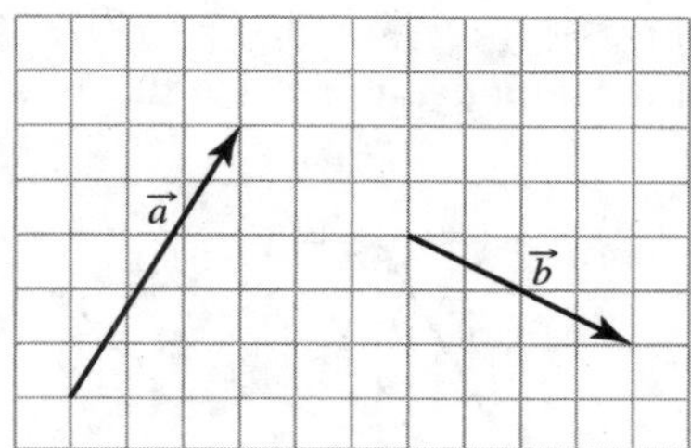

Figure 11.10

Solution

Using $\vec{a}$, draw $\vec{b}$ so that its tail is at the head of $\vec{a}$. This is shown in Figure 11.11.

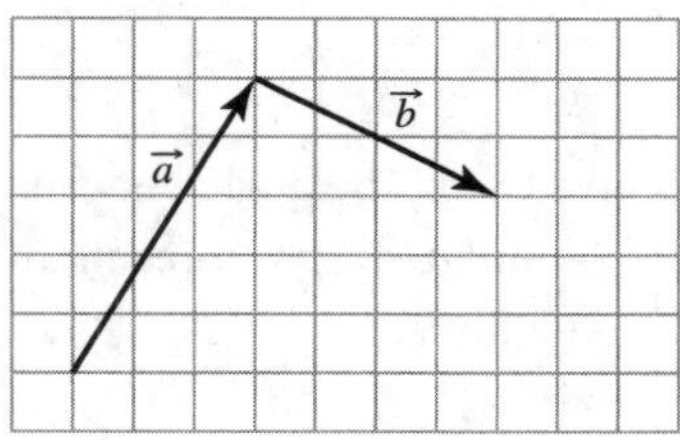

Figure 11.11

To graph the sum, start at the tail of $\vec{a}$ and connect to the head of $\vec{b}$, as shown in Figure 11.12.

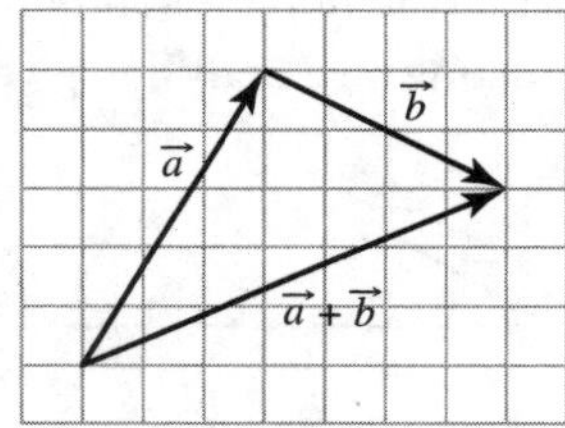

Figure 11.12

Algebraic Vector Addition

The sum of two vectors can also be found algebraically by adding the corresponding components of the two vectors. The previous example can be checked using algebra. Figure 11.13 shows the two vectors.

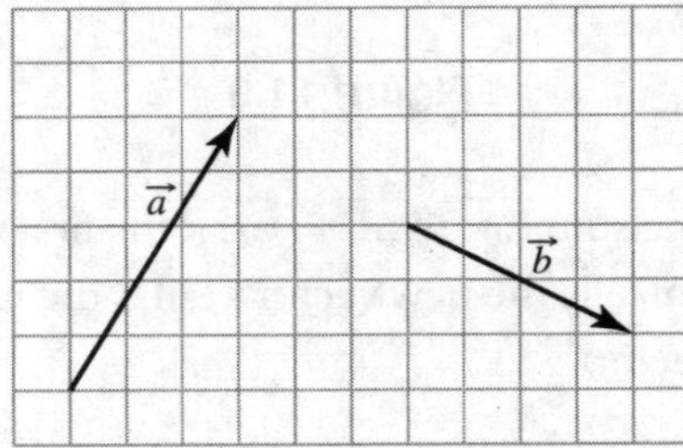

Figure 11.13

To find the vector components, count the boxes from the tail to the head for each vector: $\vec{a} = \langle 3, 5 \rangle$ and $\vec{b} = \langle 4, -2 \rangle$. Adding the corresponding components gives $\vec{a} + \vec{b} = \langle 3 + 4, 5 + (-2) \rangle = \langle 7, 3 \rangle$.

This can be verified by finding the vector components of $\vec{a} + \vec{b}$ that was found graphically.

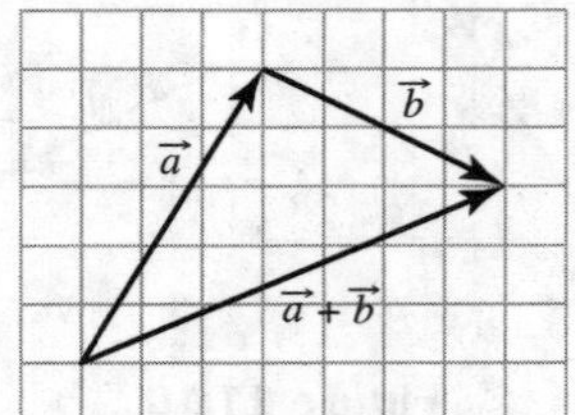

Figure 11.14

Counting the boxes from tail to head in Figure 11.14 shows that $\vec{a} + \vec{b} = \langle 7, 3 \rangle$.

Example

Two forces of 25 pounds and 60 pounds act on a body. The angle between the two forces is 30°. Find the magnitude of the resultant force to the nearest tenth of a pound. Using your resultant force answer, find the angle, to the nearest degree, between the resultant and the smaller force.

Solution

Forces are examples of vectors since they have magnitude and direction. When reading through a word problem, it is a good idea to pause for punctuation as the given information will help you create and mark a model or sketch. This allows you to see what equations are necessary to find the solution.

The first two sentences describe two forces acting on a body and give the magnitudes and angle, which are modeled in Figure 11.15.

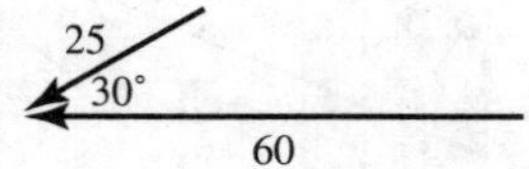

Figure 11.15

The resultant force is found by finding the sum of the two forces. In order to find the sum, translate the 25-pound force along the 60-pound force so that the vectors follow head to tail as modeled in Figure 11.16.

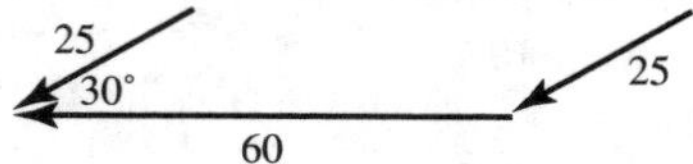

Figure 11.16

Since we translated the vector, the vectors are equal and parallel. Now the angle is supplementary to the given angle and is 150°. We can add this to the model and draw in the resultant force, as shown in Figure 11.17.

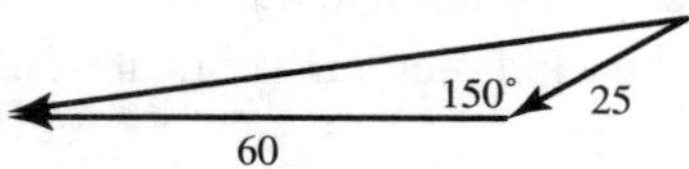

Figure 11.17

This force diagram resembles a triangle with two given sides and an included angle. To find the third side, use the law of cosines:

$$c^2 = a^2 + b^2 - 2ab\cos C$$
$$c^2 = (60)^2 + (25)^2 - 2(60)(25)\cos 150^\circ$$
$$c^2 \approx 6{,}823.076211$$
$$c = 82.6 \text{ pounds}$$

We can add this value to our model to find the measure of the angle between the resultant and the smaller force.

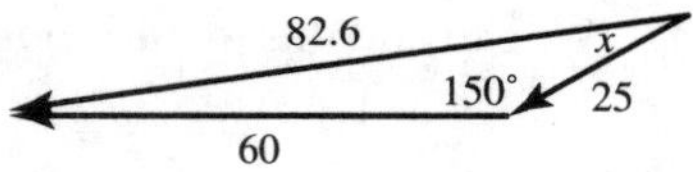

Figure 11.18

In Figure 11.18, we have two marked angles and the sides opposite the angles. To find the measure of the missing angle, use the law of sines:

$$\frac{a}{\sin A} = \frac{b}{\sin B}$$
$$\frac{82.6}{\sin 150^\circ} = \frac{60}{\sin x}$$
$$(82.6)(\sin x) = (60)(\sin 150^\circ)$$
$$\sin x = \frac{(60)(\sin 150^\circ)}{82.6}$$
$$x = \arcsin\left(\frac{(60)(\sin 150^\circ)}{82.6}\right) \approx 21^\circ$$

Example

A plane is flying at a speed of 534 miles per hour. The pilot encounters turbulence due to wind blowing at 47 miles per hour at an angle of 38° as shown in Figure 11.19.

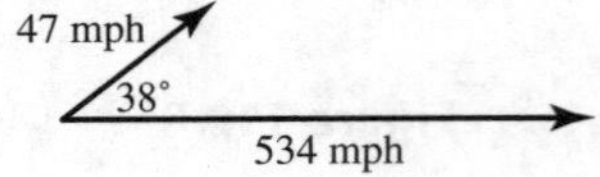

Figure 11.19

Find the resultant speed of the plane to the nearest tenth in miles per hour. Use this answer to find the measure of the angle between the resultant force and the wind vector to the nearest tenth of a degree.

Solution

The resultant force is found by finding the sum of the two forces. In order to find the sum, translate the 534 mph vector along the 47 mph force so that the vectors follow head to tail as modeled in Figure 11.20.

Figure 11.20

The translated vector is equal and parallel to the original vector. Therefore, the angle between the two vectors is supplementary to the given angle and is 142°. We can add this to the model and draw in the resultant force, as shown in Figure 11.21.

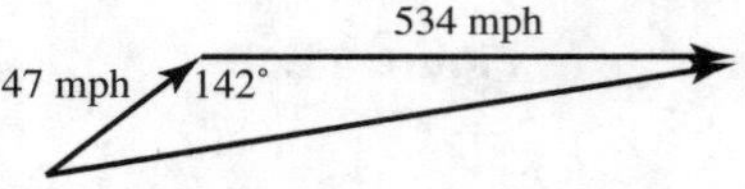

Figure 11.21

This force diagram resembles a triangle with two given sides and an included angle. To find the third side, use the law of cosines:

$$c^2 = a^2 + b^2 - 2ab\cos C$$
$$c^2 = (47)^2 + (534)^2 - 2(47)(534)\cos 142^\circ$$
$$c^2 \approx 326{,}919.9878$$
$$c = 571.8 \text{ mph}$$

We can add this value to our model, as shown in Figure 11.22, to find the measure of the angle between the resultant force and the wind vector.

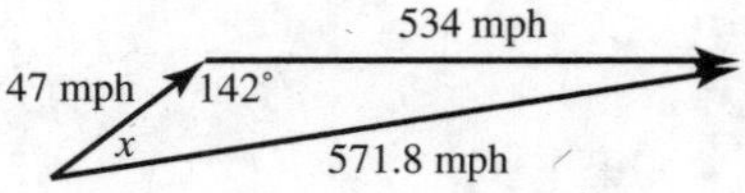

Figure 11.22

In this figure, we have two marked angles and the sides opposite the angles. To find the measure of the missing angle, use the law of sines:

$$\frac{a}{\sin A} = \frac{b}{\sin B}$$

$$\frac{571.8}{\sin 142^\circ} = \frac{534}{\sin x}$$

$$(571.8)(\sin x) = (534)(\sin 142^\circ)$$

$$\sin x = \frac{(534)(\sin 142^\circ)}{571.8}$$

$$x = \arcsin\left(\frac{(534)(\sin 142^\circ)}{571.8}\right) \approx 35.1^\circ$$

Geometric Scalar Multiplication

A scalar is a nonvector quantity; it has magnitude but no direction.

We can use vector addition to help demonstrate scalar multiplication. If $\vec{a} = \langle -4, 5\rangle$ and we multiply it by the scalar quantity of 2:

$$2\vec{a} = \vec{a} + \vec{a} = \langle -4 + -4, 5 + 5\rangle = \langle -8, 10\rangle$$

This is the same as multiplying each vector component by the scalar quantity:

$$2\vec{a} = \langle 2 \cdot -4, 2 \cdot 5\rangle = \langle -8, 10\rangle$$

Geometrically, if $\vec{a}$ is shown by Figure 11.23, then $2\vec{a}$ is shown by Figure 11.24.

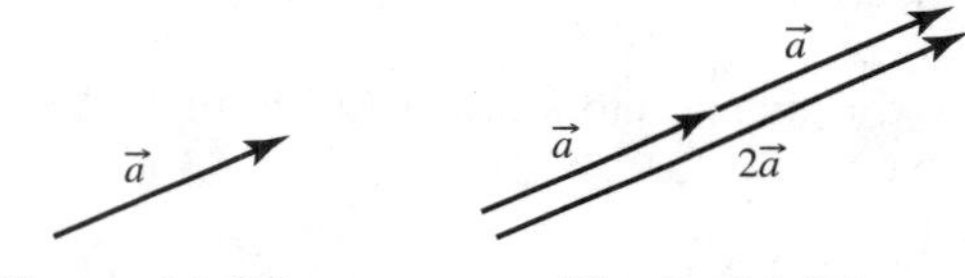

Figure 11.23 Figure 11.24

If $\vec{a} = \langle -4, 5\rangle$ and we multiply it by a scalar quantity of -3 then:

$$-3\vec{a} = 3(-\vec{a}) = -\vec{a} + -\vec{a} + -\vec{a} = \langle 4 + 4 + 4, -5 + -5 + -5\rangle = \langle 12, -15\rangle$$

This is the same as multiplying each vector component by the scalar quantity -3:

$$\vec{a} = \langle -3 \cdot -4, -3 \cdot 5\rangle = \langle 12, -15\rangle$$

Geometrically, if $\vec{a}$ is shown by Figure 11.23, then $-3\vec{a}$ is shown by Figure 11.25.

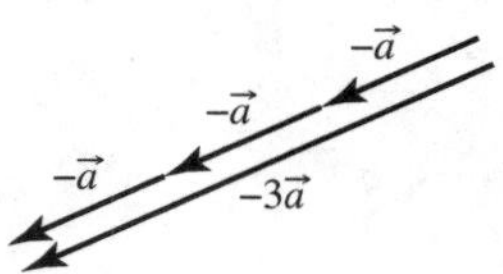

Figure 11.25

Scalar Multiplication and Vector Direction

In general, if $\vec{a}$ is a vector and k is a scalar, then $k\vec{a}$ is also a vector:

- If $k > 0$, $k\vec{a}$ and $\vec{a}$ have the same direction.
- If $k < 0$, $k\vec{a}$ and $\vec{a}$ have opposite directions.
- If $k = 0$, $k\vec{a}$ is the zero vector.

Example

Given $\vec{p} = \langle -6, 9\rangle$ and $\vec{q} = \langle 3, -7\rangle$, find the following.

1. $4\vec{q}$
2. $2\vec{p} + \vec{q}$
3. $\frac{1}{3}\vec{p} - 5\vec{q}$

Solution

1. $4\vec{q} = \langle 4 \cdot 3, 4 \cdot -7\rangle = \langle 12, -28\rangle$
2. $2\vec{p} + \vec{q} = \langle 2 \cdot -6, 2 \cdot 9\rangle + \langle 3, -7\rangle = \langle -12, 18\rangle + \langle 3, -7\rangle = \langle -12 + 3, 18 + -7\rangle = \langle -9, 11\rangle$
3. $\frac{1}{3}\vec{p} - 5\vec{q} = \left\langle \frac{1}{3} \cdot -6, \frac{1}{3} \cdot 9\right\rangle + \langle -5 \cdot 3, -5 \cdot -7\rangle = \langle -2, 3\rangle + \langle -15, 35\rangle = \langle -17, 38\rangle$

Vector Subtraction

The subtraction of two vectors is an application of the addition of two vectors and scalar multiplication. To subtract one vector from another, add its negative scalar multiple.

In general, $\vec{a} - \vec{b} = \vec{a} + (-1)\vec{b}$.

Algebraically, if $\vec{p} = \langle 5, 6\rangle$ and $\vec{q} = \langle -3, 2\rangle$:

$$\vec{p} - \vec{q} = \vec{p} + (-1)\vec{q} = \langle 5, 6\rangle + (-1)\langle -3, 2\rangle = \langle 5, 6\rangle + \langle 3, -2\rangle = \langle 5 + 3, 6 - 2\rangle = \langle 8, 4\rangle$$

Geometrically, to find the difference of the given vectors shown in Figures 11.26 and 11.27, follow the steps listed.

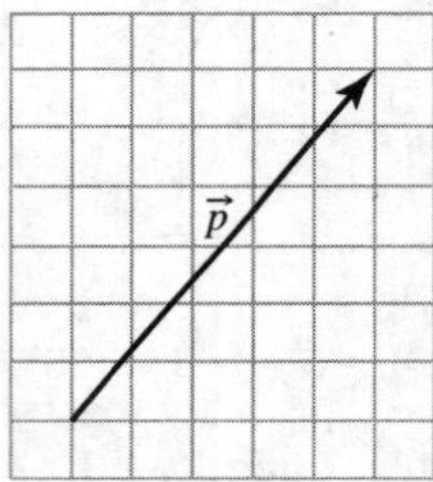

Figure 11.26

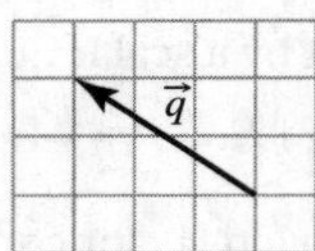

Figure 11.27

1. Since $\vec{p} - \vec{q} = \vec{p} + (-1)\vec{q}$, graph the negative vector of $\vec{q}$. This is shown in Figure 11.28.

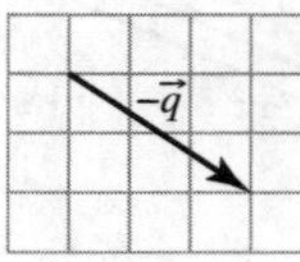

Figure 11.28

2. Connect the head of $\vec{p}$ to the tail of $-\vec{q}$. This is shown in Figure 11.29.

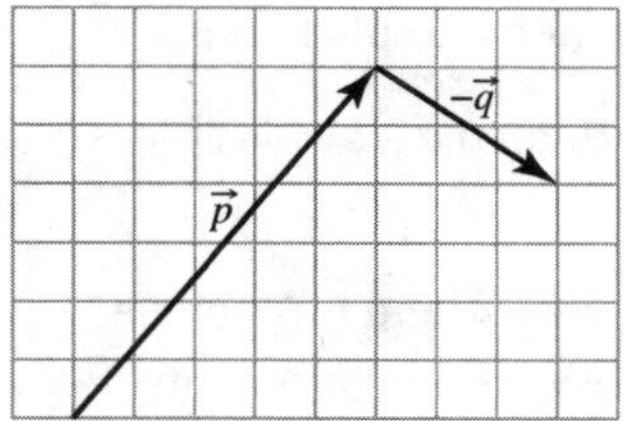

Figure 11.29

3. Graph the sum of $\vec{p} + (-1)\vec{q}$ by connecting the tail of $\vec{p}$ to the head of $-\vec{q}$. This is shown in Figure 11.30.

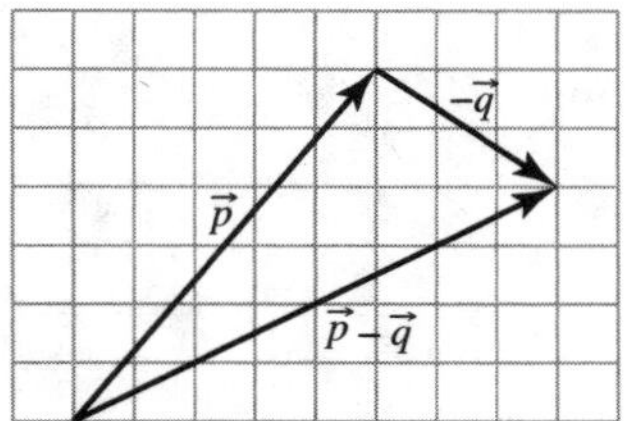

Figure 11.30

Scalar Product of Two Vectors

There are many practical applications for the product of two vectors. One way to find the product of two vectors is known as the scalar product of two vectors or the dot product. If $\vec{a} = \langle a_1, a_2 \rangle$ and $\vec{b} = \langle b_1, b_2 \rangle$, then $\vec{a} \cdot \vec{b}$, which is read as $\vec{a}$ dot $\vec{b}$, can be found using the following formula:

$$\vec{a} \cdot \vec{b} = a_1 \cdot b_1 + a_2 \cdot b_2$$

Since the dot product results in a single scalar, that is why it is called a scalar product. This is different from scalar multiplication when a scalar is multiplied with a vector yielding a new vector.

Properties of the Dot Product

The dot product has the following algebraic properties:

1. $\vec{a} \cdot \vec{b} = \vec{b} \cdot \vec{a}$
2. $\vec{a} \cdot \vec{a} = |\vec{a}|^2$
3. $\vec{a} \cdot (\vec{b} + \vec{c}) = \vec{a} \cdot \vec{b} + \vec{a} \cdot \vec{c}$
4. $(\vec{a} + \vec{b}) \cdot (\vec{c} + \vec{d}) = \vec{a} \cdot \vec{c} + \vec{a} \cdot \vec{d} + \vec{b} \cdot \vec{c} + \vec{b} \cdot \vec{d}$

Example

If $\vec{p} = \langle 2, 3 \rangle$, $\vec{q} = \langle 5, -1 \rangle$, and $\vec{r} = \langle 4, -2 \rangle$, find each of the following.

1. $\vec{p} \cdot \vec{q}$
2. $2\vec{r} \cdot \vec{p}$
3. $\vec{p} \cdot (\vec{q} + \vec{r})$

Solution

1. $\vec{p} \cdot \vec{q} = 2 \cdot 5 + 3 \cdot -1 = 10 - 3 = 7$
2. $2\vec{r} \cdot \vec{p} = \langle 8, -4 \rangle \cdot \langle 2, 3 \rangle = 8 \cdot 2 + -4 \cdot 3 = 16 - 12 = 4$
3. $\vec{p} \cdot (\vec{q} + \vec{r}) = \vec{p} \cdot \vec{q} + \vec{p} \cdot \vec{r} = (2 \cdot 5 + 3 \cdot -1) + (2 \cdot 4 + 3 \cdot -2) = 7 + 2 = 9$

Geometric Properties of the Scalar Product

The dot product of two vectors has a few geometric implications:

1. If θ is the angle between vectors $\vec{a}$ and $\vec{b}$, then $\vec{a} \cdot \vec{b} = |\vec{a}||\vec{b}|\cos\theta$.
2. For nonzero vectors $\vec{a}$ and $\vec{b}$:
   - $\vec{a} \cdot \vec{b} < 0$ if and only if the angle between $\vec{a}$ and $\vec{b}$ is obtuse.
   - $\vec{a} \cdot \vec{b} > 0$ if and only if the angle between $\vec{a}$ and $\vec{b}$ is acute.
   - $\vec{a} \cdot \vec{b} = 0$ if and only if $\vec{a}$ and $\vec{b}$ are perpendicular.
   - $\vec{a} \cdot \vec{b} = \pm|\vec{a}||\vec{b}|\cos\theta$ if and only if $\vec{a}$ and $\vec{b}$ are nonzero, parallel vectors.

Example

If $\vec{a} = \langle -1, 3\rangle$ and $\vec{b} = \langle 0, 2\rangle$, find the following.

(a) $\vec{a} \cdot \vec{b}$

(b) The angle between $\vec{a}$ and $\vec{b}$

Solution

(a) $\vec{a} \cdot \vec{b} = \langle -1, 3\rangle \cdot \langle 0, 2\rangle = (-1)(0) + (3)(2) = 0 + 6 = 6$

(b) Since the dot product is greater than 0, the angle between the two vectors is acute:

$$\begin{aligned}
\vec{a} \cdot \vec{b} &= |\vec{a}||\vec{b}|\cos\theta \\
6 &= \sqrt{(-1)^2 + (3)^2}\sqrt{(0)^2 + (2)^2}\cos\theta \\
6 &= \sqrt{10}\sqrt{4}\cos\theta \\
6 &= 2\sqrt{10}\cos\theta \\
\cos\theta &= \frac{6}{2\sqrt{10}}
\end{aligned}$$

The values of cosine are positive for angles in the intervals $[0°, 90°) \cup (270°, 360°]$. However, we already know the angle is acute and is therefore limited to the interval $[0°, 90°)$:

$$\theta = \arccos\frac{6}{2\sqrt{10}} \approx 18.43°$$

Example

Find x such that $\vec{p} = \langle x, -3\rangle$ and $\vec{q} = \langle 7, -4\rangle$ are perpendicular.

Solution

Two vectors are perpendicular if their dot product is equal to 0. Find the dot product between the two vectors, and set it equal to 0 to solve for x:

$$\begin{aligned}
\vec{p} \cdot \vec{q} = \langle x, -3\rangle \cdot \langle 7, -4\rangle = (x)(7) + (-3)(-4) &= 7x + 12 \\
7x + 12 &= 0 \\
x &= \frac{-12}{7}
\end{aligned}$$

Angle Between Two Vectors

If two nonzero vectors emanate from the same point, an angle is formed between them. We can use the law of cosines to find the measure of the angle.

Imagine you are given two vectors, $\vec{a} = \langle a_1, a_2 \rangle$ and $\vec{b} = \langle b_1, b_2 \rangle$, as shown in Figures 11.31 and 11.32.

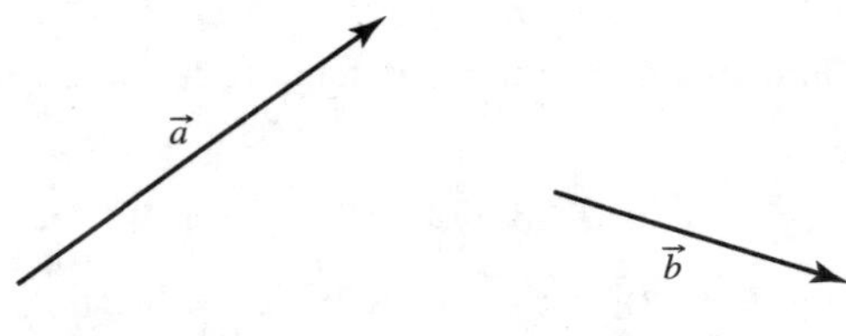

Figure 11.31 **Figure 11.32**

We can translate the two vectors so they both emanate from the same point to form angle θ, as shown in Figure 11.33.

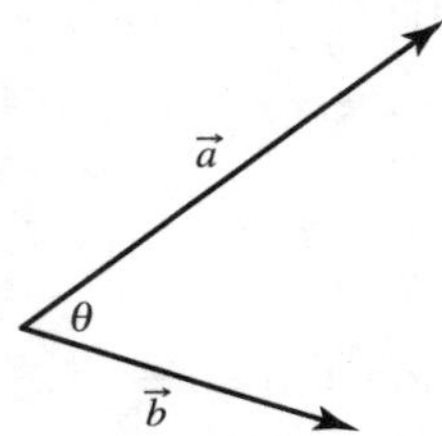

Figure 11.33

Connecting the tails of the two vectors results in vector $-\vec{a} + \vec{b} = \vec{b} - \vec{a}$. This is shown in Figure 11.34.

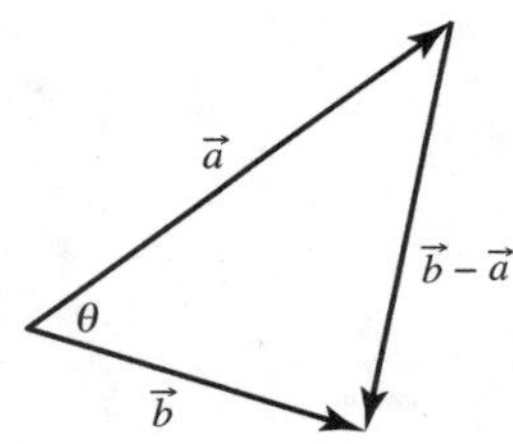

Figure 11.34

The length of each vector can be found by calculating their magnitudes. Use the law of cosines, $c^2 = a^2 + b^2 - 2ab \cos C$:

$$|\vec{b} - \vec{a}|^2 = |\vec{a}|^2 + |\vec{b}|^2 - 2|\vec{a}||\vec{b}|\cos\theta$$

Calculating and simplifying all the vectors' magnitudes yields the following formula to find the angle θ, $0 \leq \theta \leq \pi$, between two vectors:

$$\cos\theta = \frac{a_1 b_1 + a_2 b_2}{|\vec{a}||\vec{b}|} = \frac{\vec{a} \cdot \vec{b}}{|\vec{a}||\vec{b}|}$$

Example

Find the angle θ between $\vec{u} = \langle 3, -5\rangle$ and $\vec{v} = \langle -4, -2\rangle$.

Solution

To find the angle, it is easier to calculate the dot product and magnitudes of the vectors separately and then substitute the values into the formula:

$$\vec{u} \cdot \vec{v} = (3)(-4) + (-5)(-2) = -12 + 10 = -2$$

$$|\vec{u}| = \sqrt{(3)^2 + (-5)^2} = \sqrt{9 + 25} = \sqrt{34}$$

$$|\vec{v}| = \sqrt{(-4)^2 + (-2)^2} = \sqrt{16 + 4} = \sqrt{20}$$

Substitute into the formula:

$$\cos\theta = \frac{\vec{u} \cdot \vec{v}}{|\vec{u}||\vec{v}|} = \frac{-2}{\sqrt{34}\sqrt{20}}$$

$$\theta = \arccos\frac{-2}{\sqrt{34}\sqrt{20}} \approx 94.40°$$

Unit Vectors

A unit vector is any vector that has a length of 1 unit. Examples of unit vectors are shown to the right as their magnitudes are equal to 1.

Examples of Unit Vectors

$\langle 1, 0\rangle$ is a unit vector since its magnitude equals 1:

$$\sqrt{(1)^2 + (0)^2} = \sqrt{1 + 0} = \sqrt{1} = 1$$

$\left\langle -\frac{1}{\sqrt{2}}, \frac{1}{\sqrt{2}}\right\rangle$ is a unit vector since its magnitude equals 1:

$$\sqrt{\left(-\frac{1}{\sqrt{2}}\right)^2 + \left(\frac{1}{\sqrt{2}}\right)^2} = \sqrt{\frac{1}{2} + \frac{1}{2}} = \sqrt{1} = 1$$

Example

Find t for each of the following unit vectors.

1. $\langle 1, t\rangle$
2. $\left\langle t, -\frac{2}{3}\right\rangle$

Solution

1. For $\langle 1, t\rangle$ to be a unit vector, its magnitude must equal 1.

$$\text{Magnitude} = \sqrt{(1)^2 + (t)^2} = \sqrt{1 + t^2}$$

Set the magnitude equal to 1 and solve for t:

$$\sqrt{1 + t^2} = 1$$

$$\left(\sqrt{1 + t^2}\right)^2 = (1)^2$$

$$1 + t^2 = 1$$

$$t^2 = 0$$

$$t = 0$$

2. For $\left\langle t, -\frac{2}{3}\right\rangle$ to be a unit vector, its magnitude must equal 1.

$$\text{Magnitude} = \sqrt{(t)^2 + \left(-\frac{2}{3}\right)^2} = \sqrt{t^2 + \frac{4}{9}}$$

Set the magnitude equal to 1 and solve for t:

$$\sqrt{t^2 + \frac{4}{9}} = 1$$
$$\left(\sqrt{t^2 + \frac{4}{9}}\right)^2 = (1)^2$$
$$t^2 + \frac{4}{9} = 1$$
$$t^2 = \frac{5}{9}$$
$$t = \pm\sqrt{\frac{5}{9}} = \pm\frac{\sqrt{5}}{3}$$

Special unit vectors are of the form $\vec{i} = \langle 1, 0\rangle$ and $\vec{j} = \langle 0, 1\rangle$. They are vectors in the direction of the positive x- and y-axes as graphed in Figure 11.35.

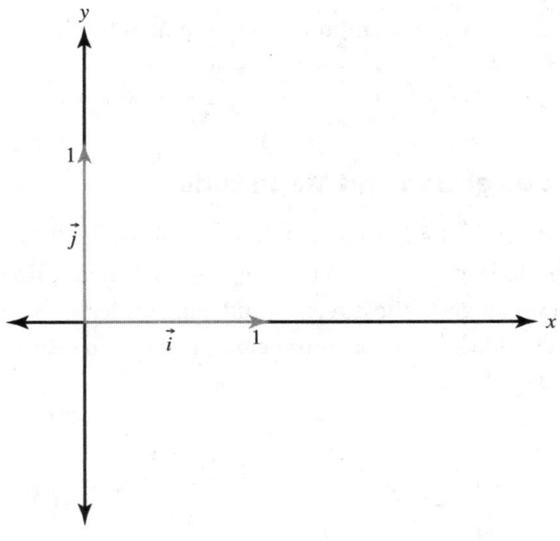

Figure 11.35

This also allows us to write vectors in an additional notation known as unit vector form. If $\vec{a} = \langle a_1, a_2\rangle$, this is written in component form and is also equivalent to $\vec{a} = a_1\vec{i} + a_2\vec{j}$, which is written in unit vector form.

Example

Express the following vectors in component form and find their length.

1. $\vec{p} = -3\vec{i} + 5\vec{j}$
2. $\vec{q} = \frac{1}{2}(\vec{i} - 6\vec{j})$

Solution

1. The component form is $\vec{p} = \langle -3, 5\rangle$. Calculate the length:

$$\vec{p} = |\vec{p}| = \sqrt{(-3)^2 + (5)^2} = \sqrt{9 + 25} = \sqrt{34}$$

2. The component form of $\vec{q} = \left\langle \frac{1}{2}, -3\right\rangle$. Calculate the length:

$$\vec{q} = |\vec{q}| = \sqrt{\left(\frac{1}{2}\right)^2 + (-3)^2} = \sqrt{\frac{1}{4} + 9} = \sqrt{\frac{37}{4}}$$

Unit Vector and a Given Direction

For any nonzero vector $\vec{p}$, unit vector $\vec{u}$ that has the same direction as $\vec{p}$ is equivalent to:

$$\vec{u} = \frac{\vec{p}}{|\vec{p}|}$$

Sometimes it is useful to find a unit vector that has the same direction as a given vector.

Example

Find a unit vector in the same direction as $\vec{p} = 2\vec{i} - 7\vec{j}$.

Solution

First find the magnitude of $\vec{p}$:

$$|\vec{p}| = \sqrt{(2)^2 + (-7)^2} = \sqrt{4 + 49} = \sqrt{53}$$

Multiply $\vec{p}$ by the scalar $\frac{1}{\sqrt{53}}$ to find unit vector $\vec{u}$:

$$\vec{u} = \frac{1}{\sqrt{53}}(2\vec{i} - 7\vec{j}) = \frac{2}{\sqrt{53}}\vec{i} - \frac{7}{\sqrt{53}}\vec{j}$$

To check that $\vec{u}$ is a unit vector, show that the magnitude of $\vec{u}$ is equal to 1:

$$|\vec{u}| = \sqrt{\left(\frac{2}{\sqrt{53}}\right)^2 + \left(-\frac{7}{\sqrt{53}}\right)^2} = \sqrt{1} = 1$$

Finding a Vector from Its Direction and Magnitude

It is more common for vectors to be described using their direction and magnitude rather than their components. However, since many of our operations were calculated using the components of a vector, it is important to learn how to find the components of a vector given the direction and magnitude.

Suppose we are given the magnitude $|\vec{v}|$ of a nonzero vector $\vec{v}$ and the angle θ, between $\vec{v}$ and the unit vector $\vec{i} = \langle 1, 0 \rangle$, as shown in Figure 11.36.

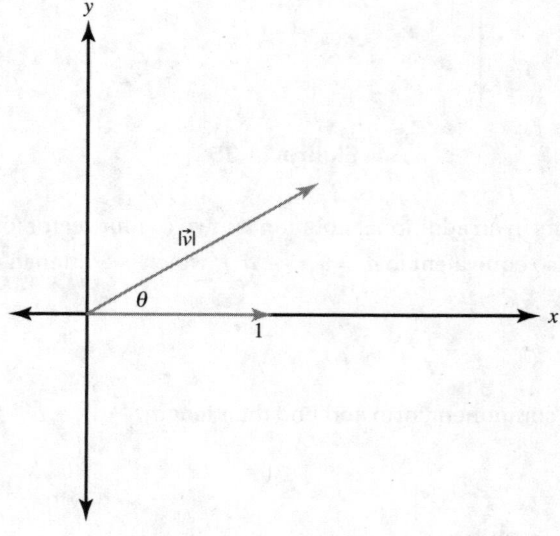

Figure 11.36

In general, to find $\vec{v}$ in terms of $|\vec{v}|$ and θ, we can find the unit vector $\vec{u}$ having the same direction as $\vec{v}$ and right triangle trigonometry to derive the formula $\vec{v} = |\vec{v}|\left(\cos\theta\vec{i} + \sin\theta\vec{j}\right)$.

Example

A ball is thrown with an initial speed of 20 miles per hour in a direction that makes an angle of 45° with the positive x-axis. Express the velocity vector $\vec{v}$ in terms of $\vec{i}$ and $\vec{j}$.

Solution

First find the magnitude of $\vec{v}$. Since the speed was given as 20 miles per hour and is the scalar value of velocity, $|\vec{v}| = 20$. The angle between the vector and the positive x-axis is 45°, so $\theta = 45°$.

Find the velocity vector:

$$\vec{v} = 20\left(\cos 45°\vec{i} + \sin 45°\vec{j}\right) = 20\left(\frac{\sqrt{2}}{2}\vec{i} + \frac{\sqrt{2}}{2}\vec{j}\right) = 10\sqrt{2}\vec{i} + 10\sqrt{2}\vec{j}$$

11.2 Vector-Valued Functions

Planar Motion as Vector-Valued Functions

An important application of vectors relates to the motion of a particle represented by parametric equations.

Position and Magnitude

The position of a particle moving in a plane that is given by the parametric function $f(t) = (x(t), y(t))$ can be expressed as a vector-valued function, $\vec{p}(t) = \langle x(t), y(t)\rangle$ or $\vec{p}(t) = x(t)\vec{i} + y(t)\vec{j}$. At time t, the sign of $x(t)$ indicates if the particle is to the right or left of the origin and the sign of $y(t)$ indicates if the particle is above or below the origin.

The magnitude of the position vector at time t is the distance of the particle from the origin.

Example

An object is moving where the position of the particle is given by the parametric equations $f(t) = (t^2 - t, t^2 + t)$, $-3 \leq t \leq 3$, where distances are measured in meters and time is measured in seconds. State the vector-valued function that represents the position of the particle, and find the distance of the particle from the origin at $t = 2$.

Solution

Write the parametric equations as a vector-valued function:

$$\vec{p}(t) = \langle t^2 - t, t^2 + t\rangle = (t^2 - t)\vec{i} + (t^2 + t)\vec{j}$$

At $t = 2$, the position of the particle can be represented by the following vector:

$$\vec{p}(2) = \langle (2)^2 - (2), (2)^2 + (2)\rangle = \langle 2, 6\rangle$$

The magnitude of the vector is the distance of the particle from the origin:

$$|\vec{p}(t)| = \sqrt{(2)^2 + (6)^2} = \sqrt{4 + 36} = \sqrt{40} \text{ meters}$$

Velocity

The velocity of a particle moving in a plane at time t can be represented by the vector-valued functions $\vec{v}(t) = \langle x(t), y(t)\rangle$ or $\vec{v}(t) = x(t)\vec{i} + y(t)\vec{j}$. At time t, the sign of $x(t)$ indicates if the particle is moving right or left and the sign of $y(t)$ indicates if the particle is moving up or down. The magnitude of the velocity vector at time t gives the speed of the particle.

Example

A particle moves in a plane whose vector velocity is given by $\vec{v}(t) = \langle 2t, 2t + 1\rangle$, where distances are measured in miles and time is measured in hours. At $t = 4$ hours, describe the direction the particle is moving and find the speed of the particle.

Solution

At $t = 4$:

$$\vec{v}(4) = \langle 2(4), 2(4) + 1\rangle = \langle 8, 9\rangle$$

Since $x(4) = 8 > 0$, the particle is moving to the right. Since $y(4) = 9 > 0$, the particle is moving up.

The speed of the particle is found by calculating the magnitude of the velocity vector at $t = 4$:

$$|\vec{v}(4)| = \sqrt{(8)^2 + (9)^2} = \sqrt{64 + 81} = \sqrt{145} \text{ miles per hour}$$

Example

An object moves with a position vector $\vec{s}(t) = (\cos t)\vec{i} + (\sin t)\vec{j}$ and velocity vector $\vec{v}(t) = (-\sin t)\vec{i} + (\cos t)\vec{j}$, where distances are measured in meters and time is in minutes. Find the distance of the object from the origin and its speed at time t.

Solution

Find the distance of the object from the origin:

$$|\vec{s}(t)| = \sqrt{(\cos t)^2 + (\sin t)^2} = \sqrt{\cos^2 t + \sin^2 t} = \sqrt{1} = 1 \text{ meter}$$

This means that at any time t, the distance between the object and the origin is 1. This makes sense because if the components are graphed, they would create a unit circle whose radius is 1.

Find the speed of the object:

$$|\vec{v}(t)| = \sqrt{(-\sin t)^2 + (\cos t)^2} = \sqrt{\sin^2 t + \cos^2 t} = \sqrt{1} = 1 \text{ meter/minute}$$

This means that the object moves at a constant speed.

Multiple-Choice Questions

1. If a particle moves in the xy-plane so that at time $t > 0$ its velocity vector is $\vec{v}(t) = \langle t^3 + 2t, t^2 + 1 \rangle$, what is the speed of the particle at $t = 2$?

   (A) $\sqrt{13}$
   (B) $\sqrt{17}$
   (C) 13
   (D) 17

2. Find the value of k such that $\vec{c} = 3\vec{i} + 10\vec{k}$ and $\vec{p} = k\vec{i} - 6\vec{j}$ are perpendicular.

   (A) -7
   (B) 1.8
   (C) 3
   (D) 20

3. Given a particle whose velocity vector is $\vec{v} = \langle 3 - t, -1 + 2t \rangle$ in meters per minute when $t > 0$, find the time, t, when the speed of the particle is $\sqrt{10}$ meters/minute.

   (A) 0
   (B) 2
   (C) 8
   (D) 10

4. Given the figure below where $\overrightarrow{CP} = a$, $\overrightarrow{JM} = c$, and $\overrightarrow{KB} = d$, which of the following is equivalent to $\overrightarrow{CB}$?

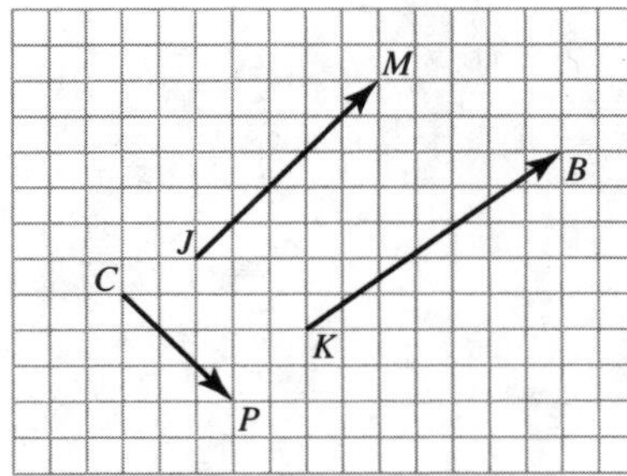

   (A) $a + d$
   (B) $a - d$
   (C) $-a + d$
   (D) $-a - d$

5. A particle moves on a plane curve so that at any time $t > 0$ its x-coordinate is $t^4 - 3t$ and its y-coordinate is $(5t - 2)^2$. Its velocity vector is $\vec{v} = \langle 4t^3 - 3, 10(5t - 2) \rangle$. What is the speed of the particle, to the nearest tenth, when $t = 1$?

   (A) 5.6
   (B) 14
   (C) 30
   (D) 62

6. Find the value of x and y such that $\vec{v} = \vec{z}$ given that $\vec{v} = (3x - 2)\vec{i} + (-y - 3)\vec{j}$ and $\vec{z} = (y + 5)\vec{i} + (2x - 11)\vec{j}$.

   (A) $x = 2, y = 3$
   (B) $x = 3, y = 2$
   (C) $x = -9, y = -4$
   (D) $x = -4, y = -9$

7. Given the vectors below, which of the following are parallel to each other?

   I. $\vec{a} = \langle -10, 3 \rangle$
   II. $\vec{b} = \langle -20, 1.5 \rangle$
   III. $\vec{c} = 2\vec{i} - \frac{3}{5}\vec{j}$

   (A) I and II only
   (B) II and III only
   (C) I and III only
   (D) I, II, and III

8. Given the vectors below, which of the following would form an obtuse angle between the two vectors?

   I. $\vec{x} = \langle 1, -2 \rangle$ and $\vec{y} = \langle 2, 3 \rangle$
   II. $\vec{s} = \vec{i} - \vec{j}$ and $\vec{t} = 2\vec{i} - 3\vec{j}$
   III. $\vec{a} = \langle 3, 4 \rangle$ and $\vec{b} = \langle -4, 3 \rangle$

   (A) I only
   (B) II only
   (C) III only
   (D) None of the above

9. Given $|\vec{p}| = 2$, $|\vec{q}| = 5$, and the angle between the two vectors is $\theta = 60°$, find $\vec{p} \cdot \vec{q}$.

(A) 3.5
(B) 5
(C) $\frac{7\sqrt{3}}{2}$
(D) $\frac{10\sqrt{3}}{2}$

10. Which of the following is a vector that has the opposite direction to $\langle -5, 12 \rangle$ and has a length of 26?

(A) $\langle -10, -24 \rangle$
(B) $\langle -10, 24 \rangle$
(C) $\langle 10, 24 \rangle$
(D) $\langle 10, -24 \rangle$

Answer Explanations

1. **(C)** The speed of a particle is found by calculating the magnitude of the velocity vector at the given time.

   Calculate the velocity vector:

   $$\vec{v}(2) = \langle (2)^3 + 2(2), (2)^2 + 1 \rangle = \langle 12, 5 \rangle$$

   Calculate the speed:

   $$|\vec{v}(2)| = \sqrt{(12)^2 + (5)^2} = 13$$

2. **(D)** Two vectors are perpendicular if their dot product is equal to 0:

   $$\vec{p} \cdot \vec{c} = (3)(k) + (10)(-6) = 3k - 60$$

   Set the dot product equal to 0 and solve for k:

   $$3k - 60 = 0$$
   $$3k = 60$$
   $$k = 20$$

3. **(B)** The speed of a particle is found by calculating the magnitude of the velocity vector at the given time:

   $$\text{Speed} = |\vec{v}(t)|$$
   $$\sqrt{10} = \sqrt{(3-t)^2 + (-1+2t)^2}$$
   $$\sqrt{10} = \sqrt{5t^2 - 10t + 10}$$

   Square both sides of the equation:

   $$10 = 5t^2 - 10t + 10$$

   Set the quadratic equal to 0 and factor:

   $$0 = 5t^2 - 10t$$
   $$0 = 5t(t-2)$$

   Set each factor equal to 0:

   $$t = 0, t = 2$$

   Since $t > 0$, then $t = 2$.

4. **(A)** Based on the figure shown in the question, $\overrightarrow{CB} = \overrightarrow{CP} + \overrightarrow{KB}$ since the vectors can be translated so that the head of $\overrightarrow{CP}$ can meet the tail of $\overrightarrow{KB}$.

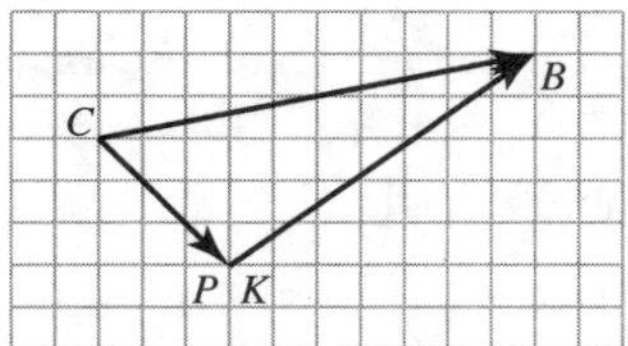

   $$\overrightarrow{CB} = \overrightarrow{CP} + \overrightarrow{KB} = a + d$$

5. **(C)** To find the speed, calculate the magnitude of the velocity vector at $t = 1$:

   $$\vec{v}(1) = \langle 4(1)^3 - 3, 10(5(1) - 2) \rangle = \langle 1, 30 \rangle$$

   Calculate the speed:

   $$|\vec{v}(1)| = \sqrt{(1)^2 + (30)^2} = \sqrt{901} \approx 30$$

6. **(B)** Two vectors are equal if their x- and y-components are equal:

   $$3x - 2 = y + 5 \text{ and } -y - 3 = 2x - 11$$

   Solve the system of equations either graphically or algebraically using the substitution method or the elimination method.

   To use the elimination method, write the variables on one side of the equal sign and add the two equations:

   $$3x - y = 7 \text{ and } 2x + y = 8$$
   $$5x = 15$$
   $$x = 3$$

   Substitute the x-value into either equation and solve for y:

   $$2(3) + y = 8$$
   $$y = 2$$

7. **(C)** Vectors are parallel to each other if they are scalar multiples of each other. Multiplying Choice I by $-\frac{1}{5}$ results in Choice III:

$$2\vec{i} - \frac{3}{5}\vec{j} = -\frac{1}{5}\langle -10, 3\rangle$$
$$\vec{c} = -\frac{1}{5}\vec{a}$$

8. **(A)** Two vectors form an obtuse angle if their dot product is less than zero. The vectors in Choice I form an obtuse angle:

$$\vec{x} \cdot \vec{y} = (1)(2) + (-2)(3) = 2 - 6 = -4 < 0$$

The vectors in Choice II form an acute angle:

$$\vec{s} \cdot \vec{t} = (1)(2) + (-1)(-3) = 2 + 3 = 5 > 0$$

The vectors in Choice III form a right angle:

$$\vec{a} \cdot \vec{b} = (3)(-4) + (4)(3) = -12 + 12 = 0$$

9. **(B)** Based on the given information, to find the dot product between the two vectors it would be easiest to use the formula $\cos\theta = \dfrac{\vec{p} \cdot \vec{q}}{|\vec{p}||\vec{q}|}$:

$$\cos 60° = \frac{\vec{p} \cdot \vec{q}}{(2)(5)}$$
$$\frac{1}{2} = \frac{\vec{p} \cdot \vec{q}}{10}$$
$$\vec{p} \cdot \vec{q} = (10)\left(\frac{1}{2}\right) = 5$$

10. **(D)** First find the unit vector in the direction of the given vector by calculating the magnitude of the given vector:

$$\sqrt{(-5)^2 + (12)^2} = \sqrt{169} = 13$$

The unit vector $= \frac{1}{13}\langle -5, 12\rangle$.

To have the vector go in the opposite direction and have the desired length, multiply the unit vector by -26:

$$-26 \cdot \frac{1}{13}\langle -5, 12\rangle = -2\langle -5, 12\rangle = \langle 10, -24\rangle$$

12

Matrices

Learning Objectives

In this chapter, you will learn:

- ➔ How to organize data in a matrix
- ➔ Properties and operations of matrices
- ➔ How to use matrices as functions
- ➔ How to use matrices to model real-life scenarios

NOTE

These Unit 4 topics are **not** assessed on the AP exam but may be part of your school's course curriculum.

In previous chapters, various math processes were used to solve real-life situations by way of modeling situations and working with data. Being able to organize various equations and data is an important skill. Matrices are a way to do that. A matrix is a set of numbers arranged in rows and columns to form a rectangular array. The numbers are called elements or entries of the matrix. The size of the matrix, or order, is determined by the number of rows by the number of columns, as shown in Table 12.1.

Table 12.1

Matrix	Order
$\begin{bmatrix} 8 \\ 8 \\ 4 \end{bmatrix}$	This matrix has 3 rows and 1 column. We say this is a 3×1 matrix; it is read as "3 by 1" matrix. It can also be called a 3×1 column matrix or column vector.
$\begin{bmatrix} 1 & -2 & 3 \\ 0 & 1 & 4 \\ -7 & \frac{1}{2} & 15 \end{bmatrix}$	This matrix has 3 rows and 3 columns. It can be called a 3×3 matrix. It can also be called a 3×3 square matrix since the number of rows is equal to the number of columns.
$[2 \;\; 10 \;\; 18 \;\; 26]$	This matrix has 1 row and 4 columns. It can be called a 1×4 matrix. It can also be called a 1×4 row matrix or row vector.

When identifying an element in a matrix, you always list the row and then the column. For example, element h is in row 3, column 2.

$$\begin{bmatrix} a & b & c \\ d & e & f \\ g & h & i \end{bmatrix}$$

Matrices are present all around us. A calendar (seen in Figure 12.1) is a form of a matrix. The different dates are elements, the weeks are rows, and the days are columns. Even a recipe (shown in Figure 12.2) can be an example of a matrix with the ingredients as rows and the amounts needed as a column.

July 2001						
M	T	W	T	F	S	S
						1
2	3	4	5	6	7	8
9	10	11	12	13	14	15
16	17	18	19	20	21	22
23	24	25	26	27	28	29
30	31					

Figure 12.1

Ingredients	Amount
sugar	1 tsp.
flour	1 cup
milk	200 mL
salt	1 pinch

Figure 12.2

12.1 Matrix Operations

This section will discuss the various operations that can be performed with matrices. Many are intuitive and follow the same procedures as operations with equations.

Matrix Equality

Two matrices are equal if they have the same order and the elements in corresponding positions are equal. For example, if $\begin{bmatrix} a & b \\ c & d \end{bmatrix} = \begin{bmatrix} w & x \\ y & z \end{bmatrix}$, then because they are both 2×2 matrices, their corresponding elements are equal: $a = w$, $b = x$, $c = y$, and $d = z$.

Matrix Addition

Two matrices of the same order can be added together by adding their corresponding elements.

Example

If $A = \begin{bmatrix} 6 & -2 & 3 \\ -4 & 5 & 1 \end{bmatrix}$, $B = \begin{bmatrix} 0 & 7 & 9 \\ 10 & 2 & 8 \end{bmatrix}$, $C = \begin{bmatrix} 7 & 3 \\ 5 & -2 \end{bmatrix}$, find the following.

1. $A + B$
2. $A + C$

Solution

1. Since A and B are the same order:

$$A + B = \begin{bmatrix} 6+0 & -2+7 & 3+9 \\ -4+10 & 5+2 & 1+8 \end{bmatrix} = \begin{bmatrix} 6 & 5 & 12 \\ 6 & 7 & 9 \end{bmatrix}$$

2. A and C are of different orders, so those matrices cannot be added together.

Matrix Subtraction

Two matrices of the same order can be subtracted by subtracting their corresponding elements.

Example

If $A = \begin{bmatrix} 29 & 31 & 19 \\ 13 & 82 & 23 \\ 4 & 7 & 9 \end{bmatrix}$ and $B = \begin{bmatrix} -1 & 7 & 6 \\ 8 & 20 & 24 \\ 42 & 32 & 21 \end{bmatrix}$, find $A - B$.

Solution

Since A and B are the same order:

$$A - B = \begin{bmatrix} 29-(-1) & 31-7 & 19-6 \\ 13-8 & 82-20 & 23-24 \\ 4-42 & 7-32 & 9-21 \end{bmatrix} = \begin{bmatrix} 30 & 24 & 13 \\ 5 & 62 & -1 \\ -38 & -25 & -12 \end{bmatrix}$$

Matrix Multiplication

To multiply two matrices, the number of columns in the first matrix must equal the number of rows in the second matrix. In other words, the product of an $m \times n$ matrix A with an $n \times p$ matrix B is the $m \times p$ matrix called AB. The new matrix will have the same number of rows as the first matrix and the same number of columns as the second matrix.

For example, if $A = \begin{bmatrix} a & b & c \\ d & e & f \end{bmatrix}$ and $B = \begin{bmatrix} x \\ y \\ z \end{bmatrix}$, then $AB = \begin{bmatrix} ax+by+cz \\ dx+ey+fz \end{bmatrix}$. Matrix A is order 2×3, matrix B is order 3×1, and matrix AB is order 2×1.

Example

If $A = \begin{bmatrix} 2 & 7 & 5 \\ 3 & 8 & 4 \end{bmatrix}$ and $B = \begin{bmatrix} 1 & 3 \\ -5 & 0 \\ 2 & 10 \end{bmatrix}$, find AB.

Solution

Since A is a 2×3 matrix and B is a 3×2 matrix, they can be multiplied together since the number of columns in the first matrix is equal to the number of rows in the second matrix. The order of their product matrix will be 2×2.

To multiply, the corresponding elements in the first row of the first matrix are multiplied with the corresponding elements in the first column of the second matrix and then added together. This will be the entry in row 1, column 1. This process continues, multiplying each row from the first matrix with each column in the second matrix as shown below:

$$A \cdot B = \begin{bmatrix} 2\cdot 1 + 7\cdot -5 + 5\cdot 2 & 2\cdot 3 + 7\cdot 0 + 5\cdot 10 \\ 3\cdot 1 + 8\cdot -5 + 4\cdot 2 & 3\cdot 3 + 8\cdot 0 + 4\cdot 10 \end{bmatrix} = \begin{bmatrix} 2-35+10 & 6+0+50 \\ 3-40+8 & 9+0+40 \end{bmatrix} = \begin{bmatrix} -23 & 56 \\ -29 & 49 \end{bmatrix}$$

Important

Unlike ordinary algebra where multiplication is commutative, matrix multiplication is not commutative since the number of rows and columns might not match if the order of the matrices is reversed.

12.2 The Inverse and Determinant of a Matrix

Identity Matrix

The identity matrix, I, is a square matrix whose elements along the diagonal from the top left to bottom right are 1 and the remaining elements are 0. Examples of two identity matrices of different orders are shown below.

$$I = \begin{bmatrix} 1 & 0 \\ 0 & 1 \end{bmatrix}, I = \begin{bmatrix} 1 & 0 & 0 \\ 0 & 1 & 0 \\ 0 & 0 & 1 \end{bmatrix}$$

When an identity matrix, I, is multiplied by a square matrix, A, that multiplication will be commutative and will result in matrix A. In other words, $AI = IA = A$. This can be demonstrated below.

If $A = \begin{bmatrix} 2 & 5 \\ 7 & 6 \end{bmatrix}$ then:

$$AI = \begin{bmatrix} 2 & 5 \\ 7 & 6 \end{bmatrix} \cdot \begin{bmatrix} 1 & 0 \\ 0 & 1 \end{bmatrix} = \begin{bmatrix} 2+0 & 0+5 \\ 7+0 & 0+6 \end{bmatrix} = \begin{bmatrix} 2 & 5 \\ 7 & 6 \end{bmatrix} = A$$

$$IA = \begin{bmatrix} 1 & 0 \\ 0 & 1 \end{bmatrix} \cdot \begin{bmatrix} 2 & 5 \\ 7 & 6 \end{bmatrix} = \begin{bmatrix} 2+0 & 5+0 \\ 0+7 & 0+6 \end{bmatrix} = \begin{bmatrix} 2 & 5 \\ 7 & 6 \end{bmatrix} = A$$

Inverse Matrix

An $n \times n$ matrix A is said to be invertible if there is an $n \times n$ matrix B such that their product results in the identity matrix, $AB = I_n = BA$. The matrix B is called the inverse of A and can be denoted as A^{-1}.

Example

Verify that matrix $B = \begin{bmatrix} -1 & 1 \\ 3 & -2 \end{bmatrix}$ is the inverse of matrix $A = \begin{bmatrix} 2 & 1 \\ 3 & 1 \end{bmatrix}$.

Solution

To verify that the two matrices are inverses, their product in either order should result in the identity matrix. The product is shown below:

$$AB = \begin{bmatrix} 2 & 1 \\ 3 & 1 \end{bmatrix} \cdot \begin{bmatrix} -1 & 1 \\ 3 & -2 \end{bmatrix} = \begin{bmatrix} -2+3 & 2-2 \\ -3+3 & 3-2 \end{bmatrix} = \begin{bmatrix} 1 & 0 \\ 0 & 1 \end{bmatrix} = I$$

$$BA = \begin{bmatrix} -1 & 1 \\ 3 & -2 \end{bmatrix} \cdot \begin{bmatrix} 2 & 1 \\ 3 & 1 \end{bmatrix} = \begin{bmatrix} -2+3 & -1+1 \\ 6-6 & 3-2 \end{bmatrix} = \begin{bmatrix} 1 & 0 \\ 0 & 1 \end{bmatrix} = I$$

There are two algebraic methods to find the inverse of a matrix and one method using technology. All three methods will be discussed here to find the inverse of the matrix $A = \begin{bmatrix} 2 & 6 \\ 1 & 4 \end{bmatrix}$.

Method 1: System of Equations

Using the property that $AA^{-1} = I$, let $A^{-1} = \begin{bmatrix} w & x \\ y & z \end{bmatrix}$, then:

$$AA^{-1} = \begin{bmatrix} 2 & 6 \\ 1 & 4 \end{bmatrix} \cdot \begin{bmatrix} w & x \\ y & z \end{bmatrix} = \begin{bmatrix} 1 & 0 \\ 0 & 1 \end{bmatrix} = I$$

Multiply the matrices together:

$$\begin{bmatrix} 2w+6y & 2x+6z \\ w+4y & x+4z \end{bmatrix} = \begin{bmatrix} 1 & 0 \\ 0 & 1 \end{bmatrix}$$

Matrix equality creates the following system of equations:

$$2w + 6y = 1 \qquad\qquad 2x + 6z = 0$$
$$w + 4y = 0 \qquad\qquad x + 4z = 1$$

Solve the left system for w and y using the elimination method:

$$2w + 6y = 1 \;\rightarrow\; 2w + 6y = 1$$
$$-2(w + 4y = 0) \rightarrow -2w - 8y = 0$$

Add the equations together:

$$-2y = 1$$
$$y = -\frac{1}{2}$$

Substitute the value for y into one of the two original equations:

$$w + 4\left(-\frac{1}{2}\right) = 0$$
$$w - 2 = 0$$
$$w = 2$$

Solve the right system for x and z using the elimination method:

$$2x + 6z = 0 \;\rightarrow\; 2x + 6z = 0$$
$$-2(x + 4z = 1) \rightarrow -2x - 8z = -2$$

Add the equations together:

$$-2z = -2$$
$$z = 1$$

Substitute the value for z into one of the two original equations:

$$x + 4(1) = 1$$
$$x + 4 = 1$$
$$x = -3$$

Therefore, the inverse is

$$A^{-1} = \begin{bmatrix} 2 & -3 \\ -\frac{1}{2} & 1 \end{bmatrix}$$

Method 2: The Determinant

The determinant of the 2×2 matrix $\begin{bmatrix} a & b \\ c & d \end{bmatrix}$ is $ad - bc$ and is notated $\det(A)$. The value of the determinant can be used to find the inverse of a matrix.

If $\det(A) \neq 0$, then if $A = \begin{bmatrix} a & b \\ c & d \end{bmatrix}$:

$$A^{-1} = \frac{1}{\det(A)}\begin{bmatrix} d & -b \\ -c & a \end{bmatrix}$$

The determinant of $\begin{bmatrix} 2 & 6 \\ 1 & 4 \end{bmatrix}$ is $(2)(4) - (1)(6) = 8 - 6 = 2$.

Then A^{-1}:

$$A^{-1} = \frac{1}{2}\begin{bmatrix} 4 & -6 \\ -1 & 2 \end{bmatrix} = \begin{bmatrix} 2 & -3 \\ -\frac{1}{2} & 1 \end{bmatrix}$$

The square matrix A has an inverse if and only if $\det(A) \neq 0$.

Method 3: Technology

The inverse of a matrix can be found using a graphing calculator. First enter the elements of the matrix by pressing 2ND X^{-1} and moving the right arrow twice to EDIT. Next enter the dimensions, row $\times$ column $= 2 \times 2$. Then enter each element of matrix $A = \begin{bmatrix} 2 & 6 \\ 1 & 4 \end{bmatrix}$. Type 2ND MODE to return to the home screen, and then press 2ND X^{-1} and ENTER. [A] should appear on the screen. Press the X^{-1} button, and press ENTER. The inverse of the matrix will appear on the screen:

$$\begin{bmatrix} 2 & -3 \\ -.5 & 1 \end{bmatrix}$$

This confirms the results of the two previous methods.

Example

Find the inverse of matrix $A = \begin{bmatrix} 3 & -2 \\ -1 & 1 \end{bmatrix}$.

Solution

Using Method 2, $\det(A) = (3)(1) - (-1)(-2) = 1$. Therefore, $A^{-1} = \frac{1}{1}\begin{bmatrix} 1 & 2 \\ 1 & 3 \end{bmatrix} = \begin{bmatrix} 1 & 2 \\ 1 & 3 \end{bmatrix}$.
This can be verified using the calculator.

Application of the Determinant

Matrices can be a useful tool when working with vectors. Vectors $\overrightarrow{v_1} = ai + cj = \langle a, c\rangle = \binom{a}{c}$ and $\overrightarrow{v_2} = bi + dj = \langle b, d\rangle = \binom{b}{d}$ are shown below in Figure 12.3.

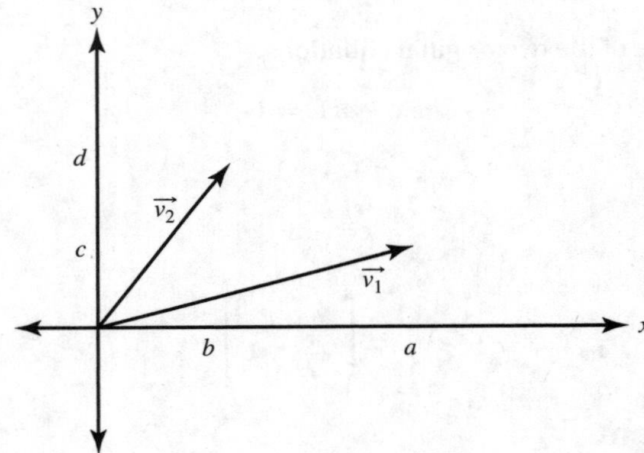

Figure 12.3

The coordinates of the head of $\overrightarrow{v_1}$ are (a, c), and the coordinates of the head of $\overrightarrow{v_2}$ are (b, d). A parallelogram can be created with the two vectors where their tails both intersect at the origin as shown in Figure 12.4.

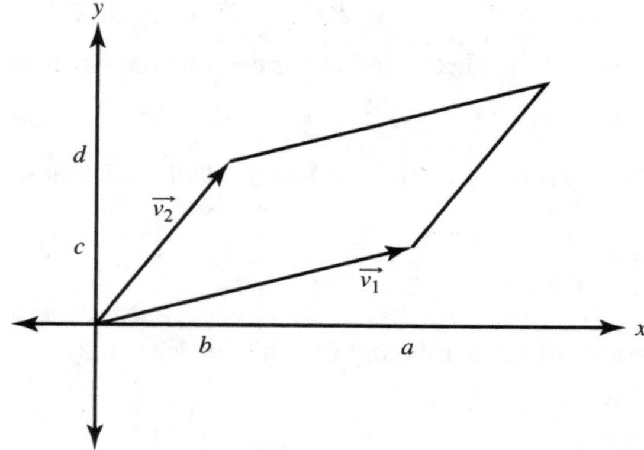

Figure 12.4

Let A represent the 2×2 matrix that consists of vectors $\vec{v_1}$ and $\vec{v_2}$ as two column vectors from R^2, where R^2 represents the collection of ordered pairs in a plane. This means $A = \begin{bmatrix} a & b \\ c & d \end{bmatrix}$.

The area of the parallelogram created by $\vec{v_1}$ and $\vec{v_2} = |\det(A)|$ as long as $\det(A) \neq 0$. If $\det(A) = 0$, the vectors are parallel.

Note that the matrix comprised of the components of the vectors can be written as two column vectors or as two row vectors. The area of the parallelogram will still be $= |\det(A)|$.

Example

Use determinants to calculate the area of the parallelogram with vertices $C(0, 0)$, $A(7, 2)$, $L(12, 11)$, and $Y(5, 9)$.

Solution

Begin by graphing the points of the parallelogram to make a model, as shown in Figure 12.5.

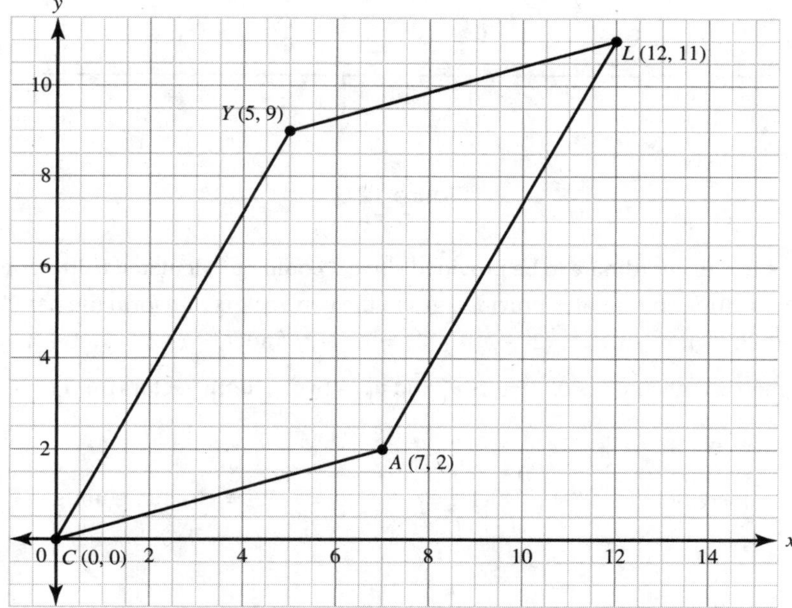

Figure 12.5

Two vectors forming the parallelogram are $\overrightarrow{CA} = \langle 7-0, 2-0\rangle = 7\vec{i} + 2\vec{j}$ and $\overrightarrow{CY} = \langle 5-0, 9-0\rangle = 5\vec{i} + 9\vec{j}$. Let $E = \begin{bmatrix} 7 & 2 \\ 5 & 9 \end{bmatrix}$, a 2×2 matrix that consists of vectors $\overrightarrow{CA}$ and $\overrightarrow{CY}$ as two row vectors from R^2. Calculate the area of the parallelogram:

$$|\det(E)| = |(7)(9) - (5)(2)| = |63 - 10| = |53| = 53 \text{ square units}$$

Example

Use determinants to calculate the area of the parallelogram with vertices (1, 2), (−3, 4), and (2, 4).

Solution

A graph of the given points is provided in Figure 12.6. It is a visual of the vectors that form the parallelogram.

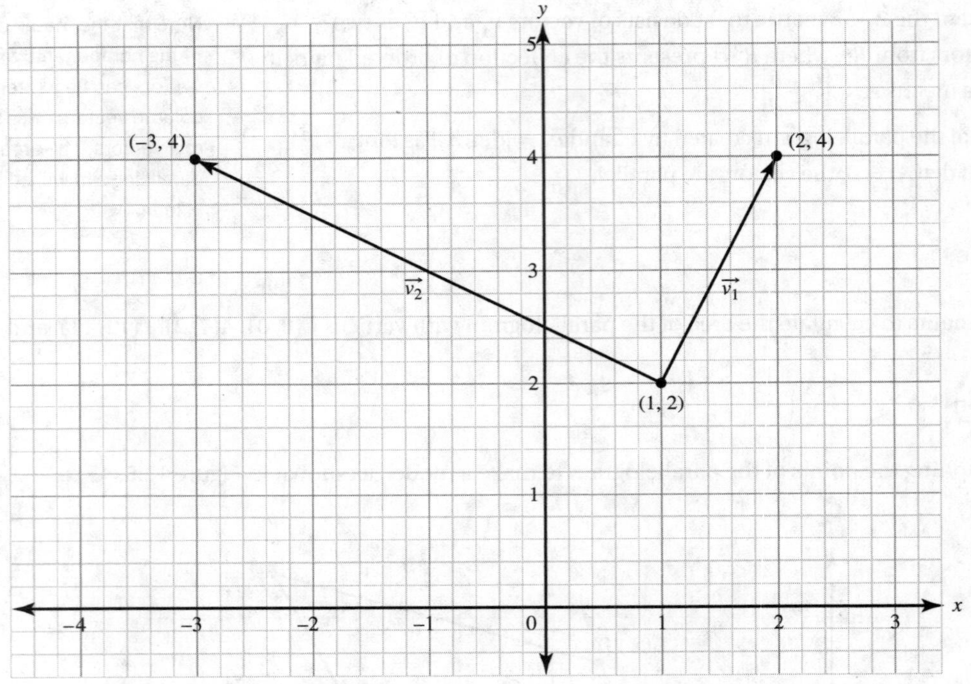

Figure 12.6

There is no need for the fourth coordinate to be provided since it could be calculated using slopes. However, that is not necessary since the three coordinates provided graph the two vectors that form the parallelogram. The two vectors forming the parallelogram are $\vec{v_1} = \langle 2-1, 4-2\rangle = 1\vec{i} + 2\vec{j}$ and $\vec{v_2} = \langle -3-1, 4-2\rangle = -4\vec{i} + 2\vec{j}$. Let $E = \begin{bmatrix} -4 & 1 \\ 2 & 2 \end{bmatrix}$, a 2×2 matrix that consists of vectors $\vec{v_1}$ and $\vec{v_2}$ as two column vectors from R^2. Calculate the area of the parallelogram:

$$|\det(E)| = |(-4)(2) - (2)(1)| = |-8 - 2| = |-10| = 10 \text{ square units}$$

12.3 Linear Transformations and Matrices

Recall that a function is a rule f that associates each element in set A with one and only one element in set B. If f associates the element b with element a, it is notated as $f(a) = b$. That means $f(a)$ is the value of f at a. For the most common functions, A and B are sets of real numbers, in which case f is called a real-valued function of a real variable.

Other common functions occur when B is a set of real numbers and A is a set of vectors in R^2, R^3, or more generally, in R^n.

When $n = 2$ or $n = 3$, it is common to use the terms ordered pair and ordered triple, respectively, instead of saying ordered 2-tuple and 3-tuple. When $n = 1$, each ordered n-tuple consists of one real number. So R^1 may be viewed as the set of real numbers. It is more common to write R or $\mathbb{R}$ rather than R^1 for this set.

> If n is a positive integer, then an ordered n-tuple is a sequence of n real numbers $(a_1, a_2, \ldots, a_n)$. The set of all ordered n-tuples is called n-space and is denoted by R^n.

We are most familiar with the study of 2-space. In 2-space, the symbol (a_1, a_2) has two different geometric interpretations. First, it can be interpreted as a point, in which case a_1 and a_2 are the coordinates as shown in Figure 12.7. Second, it can be interpreted as a vector, in which case a_1 and a_2 are the components as shown in Figure 12.8.

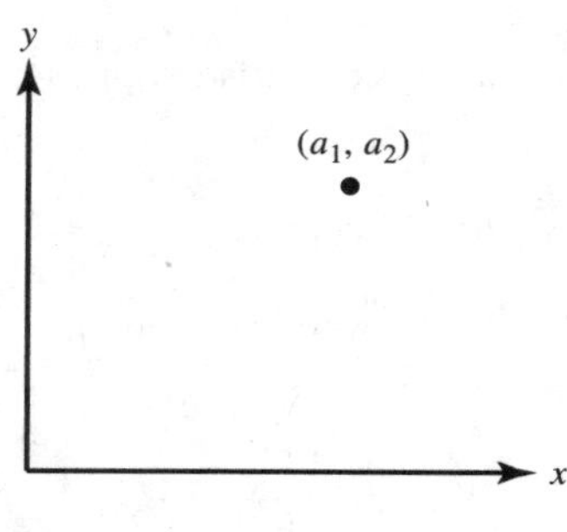

Figure 12.7

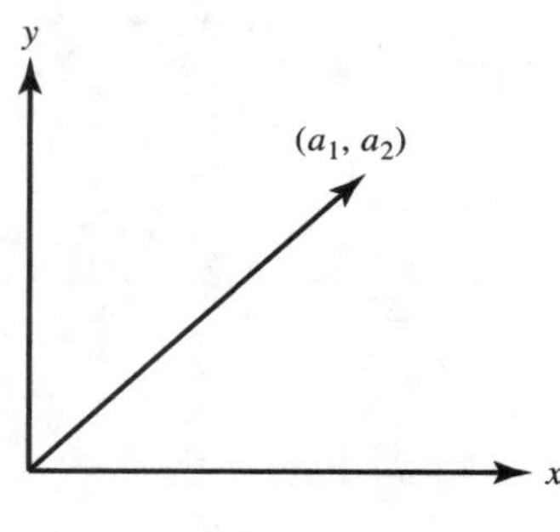

Figure 12.8

It follows, therefore, that an ordered n-tuple $(a_1, a_2, \ldots, a_n)$ can be viewed as a generalized point or as a generalized vector.

Now that we have a better understanding of R^n, we can return to the discussion of common functions in R^n. Table 12.2 outlines some examples.

Table 12.2

Formula	Example	Classification	Description
$f(x)$	$f(x) = x^2$	Real-valued function of a real variable	Function from R to R
$f(x, y)$	$f(x, y) = x^2 + y^2$	Real-valued function of two real variables	Function from R^2 to R
$f(x, y, z)$	$f(x, y, z) = x^2 + y^2 + z^2$	Real-valued function of three real variables	Function from R^3 to R
$f(x_1, x_2, \ldots, x_n)$	$f(x_1, x_2, \ldots, x_n) = {x_1}^2 + {x_2}^2 + {x_3}^2 + \cdots + {x_n}^2$	Real-valued function of n real variables	Function from R^n to R

Functions from R^n to R^m

If the domain of a function f is R^n and the range is R^m, where m and n could be equal, then f is called a map or transformation from R^n to R^m. This is notated as $f: R^n \rightarrow R^m$. The functions shown in Table 12.2 are transformations for which $m = 1$. In the case where $m = n$, the transformation $f: R^n \rightarrow R^n$ is called an operator on R^n. The first entry in Table 12.2 is an operator on R.

To demonstrate one way in which transformations can arise, suppose that $f_1, f_2, \ldots, f_m$ are real-valued functions of n real variables such that:

$$\begin{aligned} w_1 &= f_1(x_1, x_2, \ldots, x_n) \\ w_2 &= f_2(x_1, x_2, \ldots, x_n) \\ &\vdots \\ w_m &= f_m(x_1, x_2, \ldots, x_n) \end{aligned}$$

These m equations assign a unique point $(w_1, w_2, \ldots, w_m)$ in R^m to each point $(x_1, x_2, \ldots, x_n)$ in R^n and so defines a transformation from R^n to R^m. This transformation can be notated by T and so $T: R^n \rightarrow R^m$ and $T(x_1, x_2, \ldots, x_n) = (w_1, w_2, \ldots, w_m)$.

Example

The equations below define a transformation $T: R^2 \rightarrow R^3$:

$$\begin{aligned} w_1 &= x_1 + x_2 \\ w_2 &= 3x_1x_2 \\ w_3 &= {x_1}^2 - {x_2}^2 \end{aligned}$$

Find the image of the point $(1, -2)$.

Solution

In general, with this transformation the image of the point (x_1, x_2) is

$$T(x_1, x_2) = \left(x_1 + x_2, 3x_1x_2, {x_1}^2 - {x_2}^2\right)$$

Therefore:

$$T(1, -2) = (1 + -2, 3(1)(-2), (1)^2 - (-2)^2) = (-1, -6, -3)$$

Linear Transformations from R^n to R^m

In the special case where the following equations are linear:

$$\begin{aligned} w_1 &= f_1(x_1, x_2, \ldots, x_n) \\ w_2 &= f_2(x_1, x_2, \ldots, x_n) \\ &\vdots \qquad\qquad \vdots \\ w_m &= f_m(x_1, x_2, \ldots, x_n) \end{aligned}$$

The transformation $T: R^n \to R^m$ defined by those equations is called a linear transformation, or a linear operator if $m = n$. Thus, a linear transformation $T: R^n \to R^m$ is defined by equations of the following form:

$$\begin{aligned} w_1 &= a_{11}x_1 + a_{12}x_2 + \cdots + a_{1n}x_n \\ w_2 &= a_{21}x_1 + a_{22}x_2 + \cdots + a_{2n}x_n \\ &\vdots \qquad\qquad \vdots \\ w_m &= a_{m1}x_1 + a_{m2}x_2 + \cdots + a_{mn}x_n \end{aligned}$$

or is written in matrix notation:

$$\begin{bmatrix} w_1 \\ w_2 \\ \vdots \\ w_m \end{bmatrix} = \begin{bmatrix} a_{11} & a_{12} & \cdots & a_{1n} \\ a_{21} & a_{22} & \cdots & a_{2n} \\ \vdots & \vdots & & \vdots \\ a_{m1} & a_{m2} & \cdots & a_{mn} \end{bmatrix} \begin{bmatrix} x_1 \\ x_2 \\ \vdots \\ x_n \end{bmatrix}$$

or can be written more generally as:

$$\vec{w} = A\vec{x}$$

The following matrix is called the standard matrix for the linear transformation T, and T is called multiplication by A:

$$A = \begin{bmatrix} a_{11} & a_{12} & \cdots & a_{1n} \\ a_{21} & a_{22} & \cdots & a_{2n} \\ \vdots & \vdots & & \vdots \\ a_{m1} & a_{m2} & \cdots & a_{mn} \end{bmatrix}$$

If $T: R^n \to R^m$ is multiplication by A, this can also be notated as $T_A: R^n \to R^m$ or $T_A(\vec{x}) = A\vec{x}$, where the vector $\vec{x}$ in R^n is expressed as a column matrix.

Geometric Effect of an Operator

Note that depending on whether the inputs are regarded as points or vectors, the geometric effect of an operator $T: R^n \to R^m$ is to transform each point or vector in R^n into some new point or vector. This is shown in Figures 12.9 and 12.10.

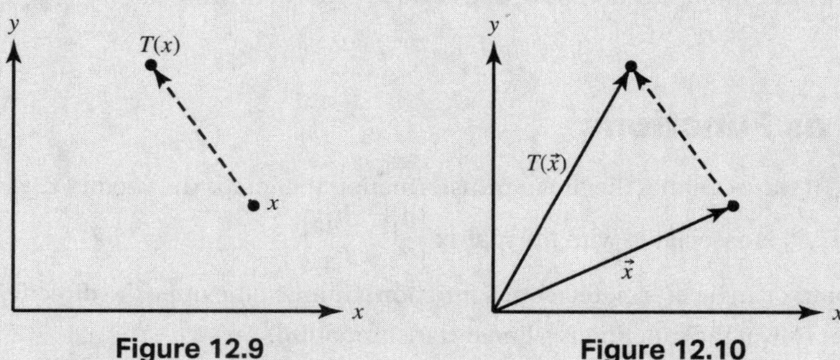

Figure 12.9

Figure 12.10

Let A be a 2×3 matrix such that $A = \begin{bmatrix} 1 & 0 & -1 \\ 3 & 1 & 2 \end{bmatrix}$. If A is multiplied by the vector $\vec{v} = \langle x, y, z \rangle$ then:

$$A\vec{v} = \begin{bmatrix} 1 & 0 & -1 \\ 3 & 1 & 2 \end{bmatrix} \begin{bmatrix} x \\ y \\ z \end{bmatrix} = \begin{bmatrix} (1)(x) + (0)(y) + (-1)(z) \\ (3)(x) + (1)(y) + (2)(z) \end{bmatrix} = \begin{bmatrix} x - z \\ 3x + y + 2z \end{bmatrix} = \langle x - z,\ 3x + y + 2z \rangle$$

What was created was a function of three variables, (x, y, z) whose output is a two-dimensional vector $\langle x - z, 3x + y + 2z \rangle$ that can be defined using function notation as $f = A\vec{v}$ where $f: R^3 \rightarrow R^2$. For example, $f(1, 2, 3) = \langle (1) - (3), 3(1) + (2) + 2(3) \rangle = \langle -2, 11 \rangle$.

Given any $m \times n$ matrix B, a function can be defined such that $g: R^n \rightarrow R^m$ by $g(\vec{x}) = B\vec{x}$, where $\vec{x}$ is an n-dimensional vector.

Example

Find the image of vector $\vec{x} = \langle 1, -3, 0, 2 \rangle$ under the linear transformation $T: R^4 \rightarrow R^3$ defined by the following equations:

$$w_1 = 2x_1 - 3x_2 + x_3 - 5x_4$$
$$w_2 = 4x_1 + x_2 - 2x_3 + x_4$$
$$w_3 = 5x_1 - 2x_2 + 4x_3$$

Solution

The image vector can be expressed in matrix form as:

$$\begin{bmatrix} w_1 \\ w_2 \\ w_3 \end{bmatrix} = \begin{bmatrix} 2 & -3 & 1 & -5 \\ 4 & 1 & -2 & 1 \\ 5 & -2 & 4 & 0 \end{bmatrix} \begin{bmatrix} 1 \\ -3 \\ 0 \\ 2 \end{bmatrix} = \begin{bmatrix} (2)(1) + (-3)(-3) + (1)(0) + (-5)(2) \\ (4)(1) + (1)(-3) + (-2)(0) + (1)(2) \\ (5)(1) + (-2)(-3) + (4)(0) + (0)(2) \end{bmatrix} = \begin{bmatrix} 1 \\ 3 \\ 11 \end{bmatrix}$$

The function $g(\vec{x})$ is a linear transformation if each term of each component of $g(\vec{x})$ is a number multiplied by one of the variables. For example, the functions $f(x, y) = \left(2x + y, \frac{y}{2}\right)$ and $g(x, y, z) = (z, 0, 1.2x)$ are linear transformations. However, none of the following functions are linear transformations: $h(x, y) = (x^2, y, x)$; $m(x, y, z) = (y, xyz)$; or $n(x, y, z) = (x + 1, y, z)$.

A linear transformation is a function that maps an input vector to an output vector such that each component of the output vector is the sum of constant multiples of the input vector components.

Zero Transformation from R^n to R^m

If 0 is the $m \times n$ zero matrix and $\vec{0}$ is the zero vector in R^n, for every vector $\vec{x}$ in R^n, $0\vec{x} = \vec{0}$. In other words, multiplication by 0 maps every vector in R^n into the zero vector in R^m.

A linear transformation will map the zero vector to the zero vector.

12.4 Matrices as Functions

In the previous section, it was seen that the linear transformation that maps the vector $\langle x, y \rangle$ to $\langle a_{11}x + a_{12}y, a_{21}x + a_{22}y \rangle$ is associated with the matrix $\begin{bmatrix} a_{11} & a_{12} \\ a_{21} & a_{22} \end{bmatrix}$.

In this way, every matrix can be associated with a function. Going in the opposite direction, a function can be associated with a matrix only if the function is a linear transformation.

Example

Find the matrix A associated with the function $f(x, y) = (2x + y, y, x - 3y)$, which is a linear transformation from R^2 to R^3.

Solution

The matrix A associated with f is a 3×2 matrix that can be written in general as:

$$A = \begin{bmatrix} a_{11} & a_{12} \\ a_{21} & a_{22} \\ a_{31} & a_{32} \end{bmatrix}$$

Matrix A needs to satisfy $f(\vec{x}) = A\vec{x}$, where $\vec{x} = \langle x, y \rangle$. To aid in finding matrix A, use the unit vectors $\langle 1, 0 \rangle$ and $\langle 0, 1 \rangle$.

Let $\vec{x} = \langle 1, 0 \rangle$. Then $f(\vec{x}) = A\vec{x}$ is the first column of A due to matrix multiplication. Therefore, the first column of A is

$$f(1, 0) = (2(1) + (0), (0), (1) - 3(0)) = (2, 0, 1) = \begin{bmatrix} 2 \\ 0 \\ 1 \end{bmatrix}$$

Similarly, if $\vec{x} = \langle 0, 1 \rangle$, then $f(\vec{x}) = A\vec{x}$ is the second column of A. This can be shown as:

$$f(0, 1) = (2(0) + (1), (1), (0) - 3(1)) = (1, 1, -3) = \begin{bmatrix} 1 \\ 1 \\ -3 \end{bmatrix}$$

Putting these together shows that the linear transformation $f(\vec{x})$ is associated with the following matrix:

$$A = \begin{bmatrix} 2 & 1 \\ 0 & 1 \\ 1 & -3 \end{bmatrix}$$

Rotation Operators

A linear operator that rotates each vector in R^2 with an initial point at (0, 0) through a fixed angle θ is called a rotation operator on R^2. This is shown in Figure 12.11.

The mapping of the unit vectors in a linear transformation provides valuable information for determining the associated matrix.

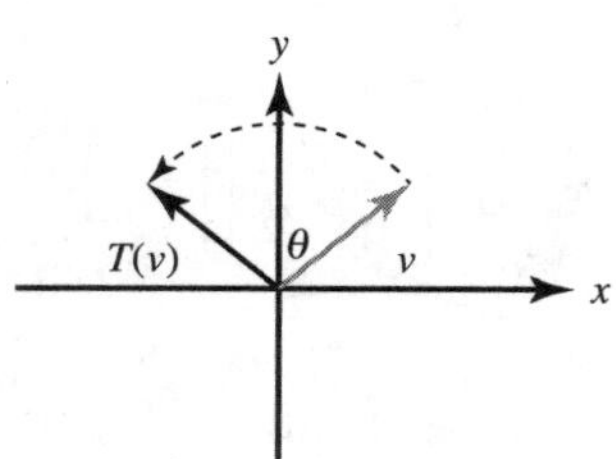

Figure 12.11

It can be shown that the standard matrix $[T]$ of the rotation operator $T: R^2 \to R^2$ that rotates each vector counter-clockwise about the origin through an angle of θ radians is

$$[T] = \begin{bmatrix} \cos\theta & -\sin\theta \\ \sin\theta & \cos\theta \end{bmatrix}$$

Therefore, the image $\vec{w}$ of vector $\vec{x} = \langle x, y \rangle$ can be found by:

$$\vec{w} = \langle w_1, w_2 \rangle = \begin{bmatrix} \cos\theta & -\sin\theta \\ \sin\theta & \cos\theta \end{bmatrix} \begin{bmatrix} x \\ y \end{bmatrix} = \langle x\cos\theta - y\sin\theta, x\sin\theta + y\cos\theta \rangle$$

Example

Find the image of vector $\langle 1, 1\rangle$ in R^2 rotated through an angle of $\frac{\pi}{6}$.

Solution

$$\vec{w} = \begin{bmatrix} \cos\frac{\pi}{6} & -\sin\frac{\pi}{6} \\ \sin\frac{\pi}{6} & \cos\frac{\pi}{6} \end{bmatrix}\begin{bmatrix}1\\1\end{bmatrix} = \begin{bmatrix} \frac{\sqrt{3}}{2} & -\frac{1}{2} \\ \frac{1}{2} & \frac{\sqrt{3}}{2} \end{bmatrix}\begin{bmatrix}1\\1\end{bmatrix} = \begin{bmatrix} \frac{\sqrt{3}-1}{2} \\ \frac{1+\sqrt{3}}{2} \end{bmatrix}$$

Dilations

Let the function $T: R^2 \to R^2$ be a two-dimensional linear transformation of the form $T(\vec{x}) = A\vec{x}$ where $\vec{x} = \begin{bmatrix}x\\y\end{bmatrix}$ and $A = \begin{bmatrix} a & b \\ c & d\end{bmatrix}$. T maps objects from the xy-plane onto an $x'y'$-plane: $(x', y') = T(x, y)$. It will be shown that the geometric properties of this mapping will be reflected in the determinant of the matrix A associated with T.

Consider this linear transformation:

$$T(x, y) = \begin{bmatrix} -2 & 0 \\ 0 & -2\end{bmatrix}\begin{bmatrix}x\\y\end{bmatrix}$$

To understand the behavior of T, consider how it maps the unit vectors $\langle 1, 0\rangle$ and $\langle 0, 1\rangle$.

For the unit vector $\langle 1, 0\rangle$:

$$T(1, 0) = \begin{bmatrix} -2 & 0 \\ 0 & -2\end{bmatrix}\begin{bmatrix}1\\0\end{bmatrix} = \langle -2, 0\rangle$$

For the unit vector $\langle 0, 1\rangle$:

$$T(0, 1) = \begin{bmatrix} -2 & 0 \\ 0 & -2\end{bmatrix}\begin{bmatrix}0\\1\end{bmatrix} = \langle 0, -2\rangle$$

The vectors were both stretched by a factor of 2, and the change in the sign of the components occurred from a rotation of π radians.

Figure 12.12 shows the mapping of T on the unit square $[0, 1] \times [1, 0]$, demonstrating that T did rotate the square by π radians around the origin and stretched each side of the square by a factor of 2.

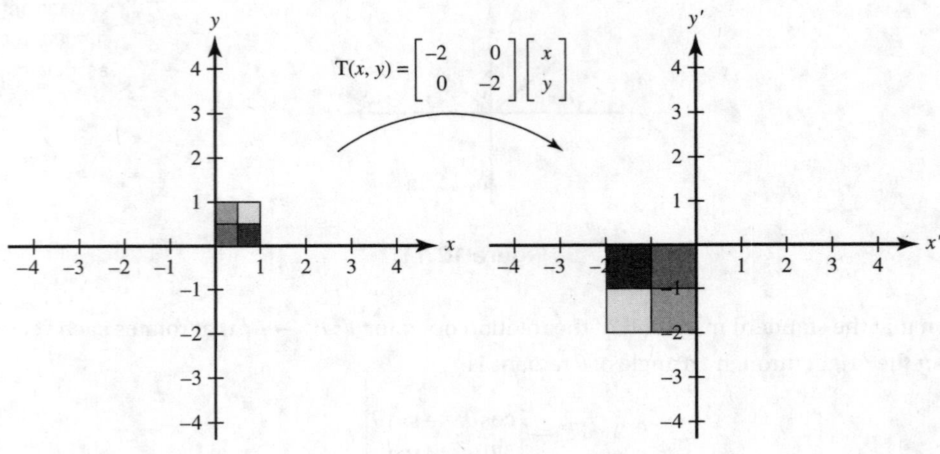

Figure 12.12

The area of the original unit square is (1)(1) = 1 square unit. The area of the transformed square is (2)(2) = 4 square units.

In Section 12.2, it was discovered that the absolute value of the determinant of a 2 × 2 matrix is the area of the parallelogram spanned by the vectors represented in the columns or rows of the matrix. In this case, the area of the square, which is a parallelogram, is 4 and the determinant of the matrix A associated with T is $\det(A) = (-2)(-2) - (0)(0) = 4$.

The absolute value of the determinant of a 2 × 2 transformation matrix gives the magnitude of the dilation of regions in R^2 under the transformation.

Example

Given the linear transformation $T(\vec{x}) = A\vec{x}$, where $A = \begin{bmatrix} -1 & -1 \\ 1 & 3 \end{bmatrix}$, find the magnitude of the dilation of regions in R^2 under the transformation.

Solution

Calculate the absolute value of the determinant:

$$A = |\det(A)| = |(-1)(3) - (1)(-1)| = |-3 - (-1)| = |-2| = 2$$

Compositions of Linear Transformations

If $T_A: R^n \to R^k$ and $T_B: R^k \to R^m$ are linear transformations, for each $\vec{x}$ in R^n you can first compute $T_A(\vec{x})$, which is a vector in R^k, and then compute $T_B(T_A(\vec{x}))$, which is a vector in R^m. Therefore, the composition of T_B with T_A denoted as $(T_B \circ T_A)(\vec{x}) = T_B(T_A(\vec{x}))$ produces a transformation from R^n to R^m.

The composition $T_B \circ T_A$ is linear since:

$$(T_B \circ T_A)(\vec{x}) = T_B(T_A(\vec{x})) = B(A\vec{x}) = (BA)\vec{x}$$

So $T_B \circ T_A$ is multiplication by the matrices BA, which is a linear transformation.

This leads to two big discoveries outlined below.

Two Big Discoveries

1. The composition of two linear transformations is a linear transformation.
2. The matrix associated with the composition of two linear transformations is the product of the matrices associated with each linear transformation.

Example

Given the following linear transformations:

$$T: R^2 \to R^2 \text{ by } T(\vec{u}) = \begin{bmatrix} u_1 + u_2 \\ 3u_1 + 3u_2 \end{bmatrix}$$

$$S: R^2 \to R^2 \text{ by } T(\vec{v}) = \begin{bmatrix} 3v_1 - v_2 \\ -3v_1 + v_2 \end{bmatrix}$$

express $S \circ T$ as a matrix transformation.

Solution

The matrix associated with the composition of the given linear transformations is the product of the matrices associated with each linear transformation.

The matrix associated with T is $\begin{bmatrix} 1 & 1 \\ 3 & 3 \end{bmatrix}$, and the matrix associated with S is $\begin{bmatrix} 3 & -1 \\ -3 & 1 \end{bmatrix}$. The product associated with $S \circ T$ is

$$\begin{bmatrix} 3 & -1 \\ -3 & 1 \end{bmatrix}\begin{bmatrix} 1 & 1 \\ 3 & 3 \end{bmatrix} = \begin{bmatrix} (3)(1)+(-1)(3) & (3)(1)+(-1)(3) \\ (-3)(1)+(1)(3) & (-3)(1)+(1)(3) \end{bmatrix} = \begin{bmatrix} 0 & 0 \\ 0 & 0 \end{bmatrix}$$

This means that $S \circ T$ maps all vectors of R^2 to the zero vector.

Inverse of a Linear Transformation

Inverses of linear transformations are like inverses of functions. If A is an $n \times n$ matrix and if $T_A : R^n \to R^n$ is multiplication by A, the following statements are equivalent:

- A is invertible.
- The range of T_A is R^n.
- T_A is one-to-one.

If $T_A : R^n \to R^n$ is a one-to-one linear operator, the matrix A is invertible. So $T_{A^{-1}} : R^n \to R^n$ is also a linear operator and is called the inverse of T_A. The linear operators T_A and $T_{A^{-1}}$ cancel the effect of one another, which means:

$$T_A\left(T_{A^{-1}}(\vec{x})\right) = AA^{-1}\vec{x} = I\vec{x} = \vec{x}$$

$$T_{A^{-1}}\left(T_A(\vec{x})\right) = A^{-1}A\vec{x} = I\vec{x} = \vec{x}$$

From a more geometric viewpoint, if $\vec{w}$ is the image of $\vec{x}$ under T_A, then $T_{A^{-1}}$ maps $\vec{w}$ back into $\vec{x}$ since:

$$T_{A^{-1}}(\vec{w}) = T_{A^{-1}}\left(T_A(\vec{x})\right) = \vec{x}$$

This is shown in Figure 12.13.

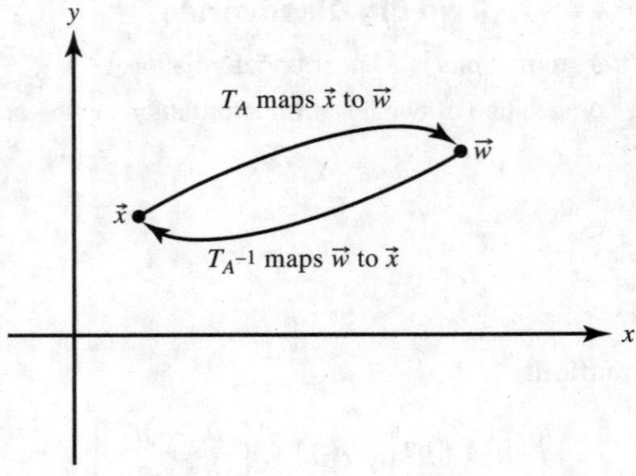

Figure 12.13

When a one-to-one linear operator on R^n is written as $T : R^n \to R^n$ rather than $T_A : R^n \to R^n$, the inverse operator T is denoted by T^{-1} rather than $T_{A^{-1}}$.

Example

Show that the linear operator $T: R^2 \to R^2$ defined by the following equations:

$$w_1 = 2x_1 + x_2$$

$$w_2 = 3x_1 + 4x_2$$

is one-to-one and find $T^{-1}(w_1, w_2)$.

Solution

The matrix form is written as:

$$\begin{bmatrix} w_1 \\ w_2 \end{bmatrix} = \begin{bmatrix} 2 & 1 \\ 3 & 4 \end{bmatrix} \begin{bmatrix} x_1 \\ x_2 \end{bmatrix}$$

So the standard matrix for T is

$$\begin{bmatrix} 2 & 1 \\ 3 & 4 \end{bmatrix}$$

The determinant of this matrix is $(2)(4) - (3)(1) = 8 - 3 = 5$. Since the determinant is not 0, the matrix is invertible. Therefore, T is one-to-one.

The standard matrix for T^{-1} is

$$\frac{1}{5}\begin{bmatrix} 4 & -1 \\ -3 & 2 \end{bmatrix} = \begin{bmatrix} \frac{4}{5} & -\frac{1}{5} \\ -\frac{3}{5} & \frac{2}{5} \end{bmatrix}$$

Therefore:

$$T^{-1}(\vec{w}) = \begin{bmatrix} \frac{4}{5} & -\frac{1}{5} \\ -\frac{3}{5} & \frac{2}{5} \end{bmatrix} \begin{bmatrix} w_1 \\ w_2 \end{bmatrix} = \begin{bmatrix} \frac{4}{5}w_1 - \frac{1}{5}w_2 \\ -\frac{3}{5}w_1 + \frac{2}{5}w_2 \end{bmatrix}$$

This means:

$$T^{-1}(w_1, w_2) = \left(\frac{4}{5}w_1 - \frac{1}{5}w_2, -\frac{3}{5}w_1 + \frac{2}{5}w_2\right)$$

12.5 Matrices Modeling Context

Suppose a physical or mathematical system undergoes a process of change such that at any moment it can occupy one of a finite number of states. For example, the weather in a certain city could be in one of four possible states: sunny, cloudy, rainy, or snowy. Suppose that such a system changes with time from one state to another and that at scheduled times the state of the system is observed. If the state of the system at any observation cannot be predicted by just knowing the state of the system at the preceding observation, the process of change is called a Markov chain.

A Markov Chain

If a Markov chain has k possible states, which are labeled as 1, 2, ..., k, the probability that the system is in state i at any observation after it was in state j at the preceding observation is denoted by p_{ij} and is called the transition probability from state j to state i. The matrix $P = [p_{ij}]$ is called the transition matrix of the Markov chain.

For example, in a three-state Markov chain, the transition matrix has the form shown in Figure 12.14.

$$\begin{array}{c} \text{Preceding State} \\ \begin{array}{ccc} 1 & 2 & 3 \end{array} \\ \begin{bmatrix} p_{11} & p_{12} & p_{13} \\ p_{21} & p_{22} & p_{23} \\ p_{31} & p_{32} & p_{33} \end{bmatrix} \begin{array}{l} 1 \\ 2 \text{ New State} \\ 3 \end{array} \end{array}$$

Figure 12.14

In this matrix, p_{32} is the probability that the system will change from state 2 to state 3, p_{11} is the probability that the system will remain in state 1 if it was previously in state 1, etc.

Example

A library has three locations denoted by 1, 2, and 3. A patron may borrow a book from any of the three locations and return the book to any of the three locations. The head librarian finds that patrons return the books to the various locations according to the probabilities found in Figure 12.15.

$$\begin{array}{c} \text{Borrowed from Location} \\ \begin{array}{ccc} 1 & 2 & 3 \end{array} \\ \begin{bmatrix} 0.8 & 0.3 & 0.2 \\ 0.1 & 0.2 & 0.6 \\ 0.1 & 0.5 & 0.2 \end{bmatrix} \begin{array}{l} 1 \\ 2 \text{ Returned to Location} \\ 3 \end{array} \end{array}$$

Figure 12.15

1. What is the probability that a book borrowed from location 3 will be returned to location 2?
2. Which scenario has the highest probability?

Solution

The given matrix is the transition matrix of the Markov chain. From this matrix, the probabilities of the different scenarios of a patron borrowing and returning books to libraries can be determined.

1. The probability is 0.6 that a book borrowed from location 3 will be returned to location 2.
2. The highest probability is 0.8. This is the probability that a book borrowed from location 1 will be returned to location 1.

A Markov Matrix

In the previous example, the transition matrix of the Markov chain has the property that the entries in any column add to 1. If $P = [p_{ij}]$ is the transition matrix of any Markov chain with k states, for each j it is true that $p_{1j} + p_{2j} + \cdots + p_{kj} = 1$. A matrix with this property is known as a Markov matrix.

Matrices of the Markov chain are to be used to make predictions about future and past states. Although in a Markov chain the state of the system at any observation time cannot generally be determined with certainty, the best that can be done is to specify probabilities for each of the possible states. For example, in a Markov chain with 3 states, the possible state of the system at some observation time can be described by a column vector:

$$\vec{x} = \begin{bmatrix} x_1 \\ x_2 \\ x_3 \end{bmatrix}$$

where x_1 is the probability that the system is in state 1, x_2 is the probability that it is in state 2, and x_3 is the probability that it is in state 3.

Knowing the initial state vector $\vec{x}^{(0)}$ for a Markov chain makes it possible to determine the state vectors $\vec{x}^{(1)}, \vec{x}^{(2)}, \ldots, \vec{x}^{(n)}, \ldots$ at future observation times.

The state vector for an observation of a Markov chain with k states is a column vector $\vec{x}$ whose ith component x_i is the probability that the system is in the ith state at that time.

The entries in any state vector for a Markov chain are nonnegative and have a sum of 1.

If P is the transition matrix of a Markov chain and $\vec{x}^{(n)}$ is the state vector at the nth observation, $\vec{x}^{(n+1)} = P\vec{x}^{(n)}$.

Example

By reviewing its donation records, the alumni office of a college finds that 80% of its alumni who contribute to the annual fund one year will also contribute the next year and that 30% of those who do not contribute one year will contribute the next.

(a) Create a transition matrix for this Markov chain with two states. State 1 corresponds to an alumnus giving a donation in any one year. State 2 corresponds to the alumnus not giving a donation in that year.

(b) Construct the probable future donation record for the next 3 years of a new graduate who did not give a donation in the initial year after graduation.

(c) Using technology, construct the probable future donation record for the next 9, 10, and 11 years for the person described in part (b).

Solution

(a) State 1 corresponds to an alumnus giving a donation in any one year, and state 2 corresponds to the alumnus not giving a donation in that year. So the transition matrix is

$$P = \begin{bmatrix} 0.8 & 0.3 \\ 0.2 & 0.7 \end{bmatrix}$$

(b) For a graduate who did not give a donation in the initial year after graduation, the system is initially in state 2 with certainty. So the initial state vector is

$$\vec{x}^{(0)} = \begin{bmatrix} 0 \\ 1 \end{bmatrix}$$

To predict that person's future donation records, perform repeated multiplication of the transition matrix and the corresponding resultant state vectors:

$$\vec{x}^{(1)} = P\vec{x}^{(0)} = \begin{bmatrix} 0.8 & 0.3 \\ 0.2 & 0.7 \end{bmatrix}\begin{bmatrix} 0 \\ 1 \end{bmatrix} = \begin{bmatrix} 0.3 \\ 0.7 \end{bmatrix}$$

$$\vec{x}^{(2)} = P\vec{x}^{(1)} = \begin{bmatrix} 0.8 & 0.3 \\ 0.2 & 0.7 \end{bmatrix}\begin{bmatrix} 0.3 \\ 0.7 \end{bmatrix} = \begin{bmatrix} 0.45 \\ 0.55 \end{bmatrix}$$

$$\vec{x}^{(3)} = P\vec{x}^{(2)} = \begin{bmatrix} 0.8 & 0.3 \\ 0.2 & 0.7 \end{bmatrix}\begin{bmatrix} 0.45 \\ 0.55 \end{bmatrix} = \begin{bmatrix} 0.525 \\ 0.475 \end{bmatrix}$$

After 3 years, the alumnus can be expected to make a donation with probability 0.525.

(c) To figure out the probable future donation record for the next 9, 10, and 11 years, the multiplication shown in part (b) will be repeated. So using a graphing calculator can prove useful:

$$\vec{x}^{(9)} = P\vec{x}^{(8)} = \begin{bmatrix} 0.8 & 0.3 \\ 0.2 & 0.7 \end{bmatrix}\begin{bmatrix} 0.598 \\ 0.402 \end{bmatrix} = \begin{bmatrix} 0.599 \\ 0.401 \end{bmatrix}$$

$$\vec{x}^{(10)} = P\vec{x}^{(9)} = \begin{bmatrix} 0.8 & 0.3 \\ 0.2 & 0.7 \end{bmatrix} \begin{bmatrix} 0.599 \\ 0.401 \end{bmatrix} = \begin{bmatrix} 0.599 \\ 0.401 \end{bmatrix}$$

$$\vec{x}^{(11)} = P\vec{x}^{(10)} = \begin{bmatrix} 0.8 & 0.3 \\ 0.2 & 0.7 \end{bmatrix} \begin{bmatrix} 0.599 \\ 0.401 \end{bmatrix} = \begin{bmatrix} 0.600 \\ 0.400 \end{bmatrix}$$

After 9 years, the alumnus can be expected to make a donation with probability 0.599. After 10 years, the alumnus can be expected to make a donation with probability 0.599. After 11 years, the alumnus can be expected to make a donation with probability 0.600. In this example, it can be observed that the state vectors converge to a fixed vector as the number of observations increases.

A similar procedure can be performed to predict past states. The product of the inverse of a matrix that models transitions between states and a corresponding state vector can predict past states.

Repeated multiplication of a matrix that models the transitions between states and corresponding resultant state vectors can predict the steady state, which is a distribution between states that does not change from one step to the next.

Example

Use the inverse of the transition matrix to find the state vector that represents the future donation record of the alumnus after 7 years as described in the scenario in the previous example.

Solution

Since $\vec{x}^{(8)} = \begin{bmatrix} 0.598 \\ 0.402 \end{bmatrix}$ and the transition matrix $P = \begin{bmatrix} 0.8 & 0.3 \\ 0.2 & 0.7 \end{bmatrix}$, the determinant can be found:

$$\det(P) = (0.8)(0.7) - (0.2)(0.3) = 0.5$$

Find the inverse of the transition matrix:

$$P^{-1} = \frac{1}{0.5} \begin{bmatrix} 0.7 & -0.3 \\ -0.2 & 0.8 \end{bmatrix} = \begin{bmatrix} 1.4 & -0.6 \\ -0.4 & 1.6 \end{bmatrix}$$

Therefore:

$$\vec{x}^{(7)} = P^{-1}\vec{x}^{(8)} = \begin{bmatrix} 1.4 & -0.6 \\ -0.4 & 1.6 \end{bmatrix} \begin{bmatrix} 0.598 \\ 0.402 \end{bmatrix} = \begin{bmatrix} 0.596 \\ 0.405 \end{bmatrix}$$

This answer can be verified from the work completed in part (c) of the previous example.

Multiple-Choice Questions

1. A certain stock price has been observed to follow a pattern. If the stock price goes up one day, there's a 20% chance of it rising tomorrow, a 30% chance of it falling, and a 50% chance of it remaining the same. If the stock price falls one day, there's a 35% chance of it rising tomorrow, a 50% chance of it falling, and a 15% chance of it remaining the same. Finally, if the price is stable on one day, then it has a 50% chance of rising the next day and a 50% chance of falling. Which matrix is the transition matrix for this Markov chain if the states are listed in order of rising, falling, and constant?

   (A) $\begin{bmatrix} 20 & 30 & 50 \\ 35 & 50 & 15 \\ 50 & 50 & 0 \end{bmatrix}$

   (B) $\begin{bmatrix} 0.2 & 0.35 & 0.5 \\ 0.3 & 0.5 & 0.5 \\ 0.5 & 0.15 & 0 \end{bmatrix}$

   (C) $\begin{bmatrix} 20 & 35 & 50 \\ 30 & 50 & 50 \\ 50 & 15 & 0 \end{bmatrix}$

   (D) $\begin{bmatrix} 0.2 & 0.3 & 0.5 \\ 0.35 & 0.5 & 0.15 \\ 0.5 & 0.5 & 0 \end{bmatrix}$

2. Calculate the area of the parallelogram with the given vertices $C(0, 0)$, $M(2, 6)$, $P(11, 8)$, and $J(9, 2)$.

   (A) 100
   (B) 52
   (C) 50
   (D) 49

3. Which of the following is the inverse matrix of $A = \begin{bmatrix} -5 & 4 \\ 0 & 4 \end{bmatrix}$?

   (A) $\begin{bmatrix} -\frac{1}{5} & -\frac{1}{5} \\ 0 & \frac{1}{4} \end{bmatrix}$

   (B) $\begin{bmatrix} -\frac{1}{5} & \frac{1}{5} \\ 0 & \frac{1}{4} \end{bmatrix}$

   (C) $\begin{bmatrix} \frac{1}{4} & \frac{1}{5} \\ 0 & -\frac{1}{5} \end{bmatrix}$

   (D) $\begin{bmatrix} 0 & \frac{1}{4} \\ -\frac{1}{5} & \frac{1}{5} \end{bmatrix}$

4. Let $A = [-5 \quad 2]$ and $B = [1 \quad 0]$. Find $2A + 3B$.

   (A) $[-10 \quad 4]$
   (B) $[-9 \quad 4]$
   (C) $[-7 \quad 4]$
   (D) $[-2 \quad 2]$

5. Let $A = \begin{bmatrix} 1 & 3 & -3 \\ 3 & 0 & 5 \end{bmatrix}$ and $B = \begin{bmatrix} 3 & 0 \\ -3 & 1 \\ 0 & 5 \end{bmatrix}$. Find AB if it is defined.

   (A) $\begin{bmatrix} -12 & -6 \\ 25 & 9 \end{bmatrix}$

   (B) $\begin{bmatrix} -6 & -12 \\ 9 & 25 \end{bmatrix}$

   (C) $\begin{bmatrix} 3 & -9 & 0 \\ 0 & 0 & 25 \end{bmatrix}$

   (D) AB is undefined

6. Which of the following is the correct geometric interpretation of the associated linear transformations for the standard matrix $A = \begin{bmatrix} 0 & -2 \\ 2 & 0 \end{bmatrix}$?

(A) Rotates counterclockwise through π radians
(B) Rotates counterclockwise through $\frac{\pi}{2}$ radians and doubles the length
(C) Rotates counterclockwise through $\frac{\pi}{2}$ radians and halves the length
(D) Rotates counterclockwise through $\frac{\pi}{2}$ radians and quadruples the length

7. Let T be a linear transformation from R^n to R^m and let $\vec{v} = T(\vec{0})$, where $\vec{0}$ is the zero vector in R^n. Which of the following statements is true?

(A) $\vec{v}$ is a zero vector in R^n.
(B) $\vec{v}$ is a zero vector in R^m if and only if $n = m$.
(C) $\vec{v}$ is a zero vector in R^m.
(D) None of the statements is true.

8. Consider the following three statements about an invertible $n \times n$ matrix A.

I. $A = A^{-1}$
II. $AA = I$
III. $A^{-1}A^{-1} = I$

Which of the statements are equivalent?

(A) I and II only
(B) II and III only
(C) I and III only
(D) I, II, and III

9. Suppose $AB = \begin{bmatrix} 5 & 4 \\ -2 & 3 \end{bmatrix}$ and $B = \begin{bmatrix} 7 & 3 \\ 2 & 1 \end{bmatrix}$. Find A.

(A) $\begin{bmatrix} 1 & -3 \\ -2 & 7 \end{bmatrix}$

(B) $\begin{bmatrix} 11 & -17 \\ -27 & 41 \end{bmatrix}$

(C) $\begin{bmatrix} 31 & 43 \\ 4 & 18 \end{bmatrix}$

(D) $\begin{bmatrix} -3 & 13 \\ -8 & 27 \end{bmatrix}$

10. Given that $T: R^3 \to R^2$, which of the following choices is not a linear transformation?

(A) $T(x, y, z) = (x, x + y + z)$
(B) $T(x, y, z) = (1, 1)$
(C) $T(x, y, z) = (0, 0)$
(D) $T(x, y, z) = (3x - 4y, 2x - 5z)$

Answer Explanations

1. **(B)** The transition matrix of the Markov chain should have columns that sum to 1. This eliminates Choices (A), (C), and (D). The transition matrix in Choice (B) reflects the correct presentation of the given information for each of the days and the given percentages.

2. **(C)** The graph of the parallelogram is provided in the figure below:

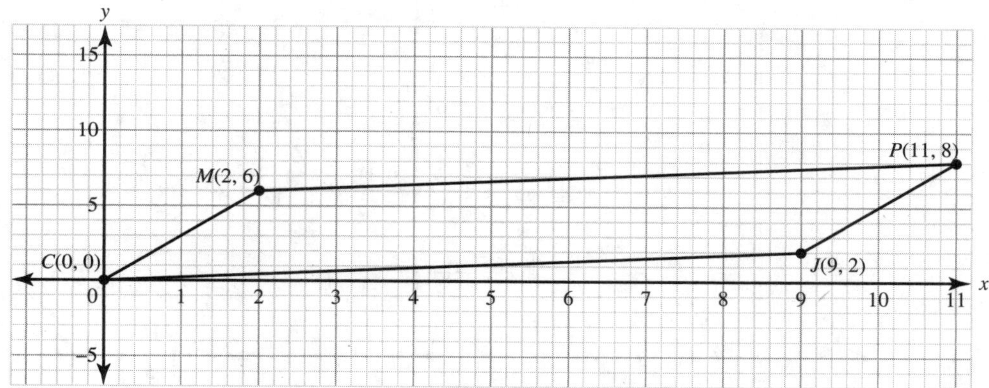

The two vectors forming the parallelogram are $\overrightarrow{CM} = \langle 2-0, 6-0\rangle = 2\vec{i} + 6\vec{j}$ and $\overrightarrow{CJ} = \langle 9-0, 2-0\rangle = 9\vec{i} + 2\vec{j}$. Let $E = \begin{bmatrix} 2 & 6 \\ 9 & 2 \end{bmatrix}$, a 2×2 matrix that consists of vectors $\overrightarrow{CM}$ and $\overrightarrow{CJ}$ as two row vectors from R^2. The area of the parallelogram can be calculated:

$$|\det(E)| = |(2)(2) - (9)(6)| = |4 - 54| = |-50| = 50 \text{ square units}$$

3. **(B)** The $\det(A) = (-5)(4) - (0)(4) = -20$. Since $\det(A) \neq 0$, matrix A is invertible:

$$A^{-1} = -\frac{1}{20}\begin{bmatrix} 4 & -4 \\ 0 & -5 \end{bmatrix} = \begin{bmatrix} -\frac{1}{5} & \frac{1}{5} \\ 0 & \frac{1}{4} \end{bmatrix}$$

4. **(C)** $2A + 3B = 2[-5 \quad 2] + 3[1 \quad 0] = [-10 \quad 4] + [3 \quad 0] = [-7 \quad 4]$.

5. **(B)** Since A is a 2×3 matrix and B is a 3×2 matrix, they can be multiplied together. Their product will be a 2×2 matrix. The product is AB:

$$AB = \begin{bmatrix} 1 & 3 & -3 \\ 3 & 0 & 5 \end{bmatrix}\begin{bmatrix} 3 & 0 \\ -3 & 1 \\ 0 & 5 \end{bmatrix} = \begin{bmatrix} (1)(3)+(3)(-3)+(-3)(0) & (1)(0)+(3)(1)+(-3)(5) \\ (3)(3)+(0)(-3)+(5)(0) & (3)(0)+(0)(1)+(5)(5) \end{bmatrix} = \begin{bmatrix} -6 & -12 \\ 9 & 25 \end{bmatrix}$$

6. **(B)** Let the function $T: R^2 \to R^2$ be a two-dimensional linear transformation of the form $T(\vec{x}) = A\vec{x}$ where $\vec{x} = \begin{bmatrix} x \\ y \end{bmatrix}$ and $A = \begin{bmatrix} a & b \\ c & d \end{bmatrix}$. The linear transformation given was that of $T(x, y) = \begin{bmatrix} 0 & -2 \\ 2 & 0 \end{bmatrix}\begin{bmatrix} x \\ y \end{bmatrix}$. To understand the behavior of T, consider how it maps the unit vectors $\langle 1, 0\rangle$ and $\langle 0, 1\rangle$.

For the unit vector $\langle 1, 0\rangle$:

$$T(1, 0) = \begin{bmatrix} 0 & -2 \\ 2 & 0 \end{bmatrix}\begin{bmatrix} 1 \\ 0 \end{bmatrix} = \langle 0, 2\rangle$$

For the unit vector $\langle 0, 1\rangle$:

$$T(0, 1) = \begin{bmatrix} 0 & -2 \\ 2 & 0 \end{bmatrix}\begin{bmatrix} 0 \\ 1 \end{bmatrix} = \langle -2, 0\rangle$$

The transformation can be illustrated graphically.

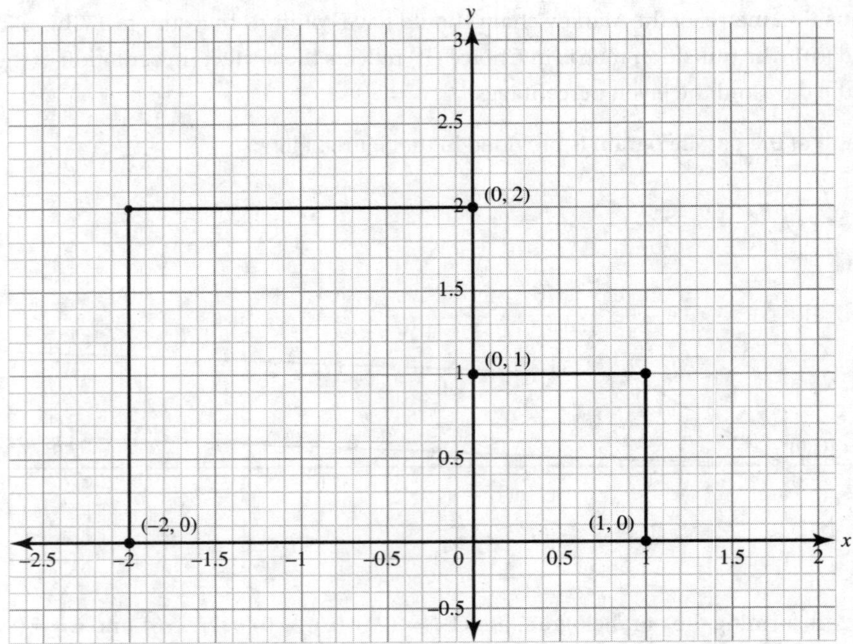

The vectors were both stretched by a factor of 2. The change in the sign of the components occurred from a rotation of $\frac{\pi}{2}$ radians counterclockwise.

7. **(C)** If 0 is the $m \times n$ zero matrix and $\vec{0}$ is the zero vector in R^n, for every vector $\vec{x}$ in R^n $0\vec{x} = \vec{0}$. In other words, multiplication by 0 maps every vector in R^n into the zero vector in R^m.

8. **(D)** Work in order of the statements. If $A = A^{-1}$ is true, by definition $AA^{-1} = A^{-1}A = I$, which is the identity matrix. However since $A = A^{-1}$, each can be substituted for the other. So $A(A) = A^{-1}(A^{-1}) = I$. Therefore, all the statements are equivalent.

9. **(D)** To find A, multiply AB by the inverse matrix of B. So $AB \cdot B^{-1} = AI = A$.

   First find the determinant of B:

$$\det(B) = (7)(1) - (2)(3) = 7 - 6 = 1$$

   Then find the matrix for B^{-1}:

$$B^{-1} = \frac{1}{1}\begin{bmatrix} 1 & -3 \\ -2 & 7 \end{bmatrix} = \begin{bmatrix} 1 & -3 \\ -2 & 7 \end{bmatrix}$$

   Therefore:

$$AB \cdot B^{-1} = \begin{bmatrix} 5 & 4 \\ -2 & 3 \end{bmatrix}\begin{bmatrix} 1 & -3 \\ -2 & 7 \end{bmatrix} = \begin{bmatrix} -3 & 13 \\ -8 & 27 \end{bmatrix} = A$$

10. **(B)** Given that $T: R^3 \rightarrow R^2$, this takes the set of points with 3 coordinates and maps them to a plane with 2 coordinates. A linear transformation is a function that maps an input vector to an output vector such that each component of the output vector is the sum of constant multiples of the input vector components. This can be seen in Choices (A) and (D). A linear transformation will map the zero vector to the zero vector, which is seen in Choice (C). The only choice that does not follow any of the linear transformation definitions or properties is Choice (B).

Practice Tests

ANSWER SHEET
Practice Test 1

Section I

1. Ⓐ Ⓑ Ⓒ Ⓓ
2. Ⓐ Ⓑ Ⓒ Ⓓ
3. Ⓐ Ⓑ Ⓒ Ⓓ
4. Ⓐ Ⓑ Ⓒ Ⓓ
5. Ⓐ Ⓑ Ⓒ Ⓓ
6. Ⓐ Ⓑ Ⓒ Ⓓ
7. Ⓐ Ⓑ Ⓒ Ⓓ
8. Ⓐ Ⓑ Ⓒ Ⓓ
9. Ⓐ Ⓑ Ⓒ Ⓓ
10. Ⓐ Ⓑ Ⓒ Ⓓ
11. Ⓐ Ⓑ Ⓒ Ⓓ
12. Ⓐ Ⓑ Ⓒ Ⓓ
13. Ⓐ Ⓑ Ⓒ Ⓓ
14. Ⓐ Ⓑ Ⓒ Ⓓ
15. Ⓐ Ⓑ Ⓒ Ⓓ
16. Ⓐ Ⓑ Ⓒ Ⓓ
17. Ⓐ Ⓑ Ⓒ Ⓓ
18. Ⓐ Ⓑ Ⓒ Ⓓ
19. Ⓐ Ⓑ Ⓒ Ⓓ
20. Ⓐ Ⓑ Ⓒ Ⓓ
21. Ⓐ Ⓑ Ⓒ Ⓓ
22. Ⓐ Ⓑ Ⓒ Ⓓ
23. Ⓐ Ⓑ Ⓒ Ⓓ
24. Ⓐ Ⓑ Ⓒ Ⓓ
25. Ⓐ Ⓑ Ⓒ Ⓓ
26. Ⓐ Ⓑ Ⓒ Ⓓ
27. Ⓐ Ⓑ Ⓒ Ⓓ
28. Ⓐ Ⓑ Ⓒ Ⓓ
29. Ⓐ Ⓑ Ⓒ Ⓓ
30. Ⓐ Ⓑ Ⓒ Ⓓ
31. Ⓐ Ⓑ Ⓒ Ⓓ
32. Ⓐ Ⓑ Ⓒ Ⓓ
33. Ⓐ Ⓑ Ⓒ Ⓓ
34. Ⓐ Ⓑ Ⓒ Ⓓ
35. Ⓐ Ⓑ Ⓒ Ⓓ
36. Ⓐ Ⓑ Ⓒ Ⓓ
37. Ⓐ Ⓑ Ⓒ Ⓓ
38. Ⓐ Ⓑ Ⓒ Ⓓ
39. Ⓐ Ⓑ Ⓒ Ⓓ
40. Ⓐ Ⓑ Ⓒ Ⓓ

Practice Test 1

Section I

Part A: Graphing calculator not permitted

TIME: 80 MINUTES

1. What is the numerical coefficient of the term containing x^3y^2 in the expansion of $(x + 2y)^5$?

   (A) 10
   (B) 20
   (C) 40
   (D) 80

2. If $f(x) = x^3 + Ax^2 + Bx - 3$ and if $f(1) = 4$ and $f(-1) = -6$, what is the value of $2A + B$?

   (A) -2
   (B) 0
   (C) 8
   (D) 12

3. If $f(x) = x^3 - 2x - 1$, then $f(-2) =$

   (A) -17
   (B) -13
   (C) -5
   (D) -1

4. What is the domain of $f(x) = \dfrac{x-5}{x^2+25}$?

   (A) All reals, $x \neq 5$
   (B) All reals, $x \neq \pm 5$
   (C) All reals, $x \neq -5$
   (D) All reals

5. What is the domain of $g(x) = \dfrac{\sqrt{x-3}}{x^2-2x}$?

   (A) All reals, $x \neq 0, 2$
   (B) All reals, $x \geq -3, x \neq 0, 2$
   (C) All reals, $x \geq 3$
   (D) All reals

6. Which of the following is equivalent to the expression $\log \sqrt[3]{\dfrac{a^2b}{c}}$?

   (A) $\sqrt[3]{\dfrac{2\log a + \log b}{\log c}}$
   (B) $\dfrac{1}{3}\left(\dfrac{2\log a + \log b}{\log c}\right)$
   (C) $\dfrac{1}{3}(2\log a + \log b - \log c)$
   (D) $\sqrt[3]{2\log a + \log b - \log c}$

7. Identify the vertical asymptote(s) for the function $f(x) = \dfrac{x^2-4}{x^3+2x^2}$.

   (A) $x = 0$
   (B) $x = -2, x = 2$
   (C) $x = 0, x = 2$
   (D) $x = -2, x = 0$

8. Given $f(x) = \begin{cases} -\frac{1}{3}x^2 + 9, x \leq 0 \\ (x-3)^2, x > 0 \end{cases}$, find $f(-3)$.

(A) 36
(B) 10
(C) 8
(D) 6

9. Find the slant asymptote of $f(x) = \dfrac{x^2 + 2x - 1}{x - 1}$.

(A) $y = 1$
(B) $y = x + 3$
(C) $y = x + 1$
(D) $y = x - 1$

10. In polar coordinates, which of the following choices is *not* equivalent to $\left(2, \frac{5\pi}{6}\right)$?

(A) $\left(-2, -\frac{\pi}{6}\right)$
(B) $\left(-2, \frac{11\pi}{6}\right)$
(C) $\left(2, -\frac{11\pi}{6}\right)$
(D) $\left(2, -\frac{7\pi}{6}\right)$

11. Which of the following represents zeros of $r = 2 - 4\sin 2\theta$?

(A) $\frac{\pi}{6}, \frac{5\pi}{6}$
(B) $0, \frac{\pi}{2}, \pi, \frac{3\pi}{2}$
(C) $\frac{\pi}{4}, \frac{3\pi}{4}, \frac{5\pi}{4}, \frac{7\pi}{4}$
(D) $\frac{\pi}{12}, \frac{5\pi}{12}, \frac{13\pi}{12}, \frac{17\pi}{12}$

12. Evaluate: $\cos\left[\arctan\left(-\frac{2}{3}\right)\right]$, given that $\frac{\pi}{2} \leq \theta \leq \pi$

(A) $-\frac{3\sqrt{13}}{13}$
(B) $-\frac{2\sqrt{13}}{13}$
(C) $\frac{2\sqrt{13}}{13}$
(D) $\frac{3\sqrt{13}}{13}$

13. Given that $\sin\theta = -\frac{1}{5}$ and $\tan\theta > 0$, find $\cos\theta$.

(A) $\frac{\sqrt{26}}{5}$
(B) $\frac{2\sqrt{6}}{5}$
(C) $-\frac{2\sqrt{6}}{5}$
(D) $-\frac{\sqrt{26}}{5}$

14. Which of the following is a possible equation for the sinusoidal graph shown below?

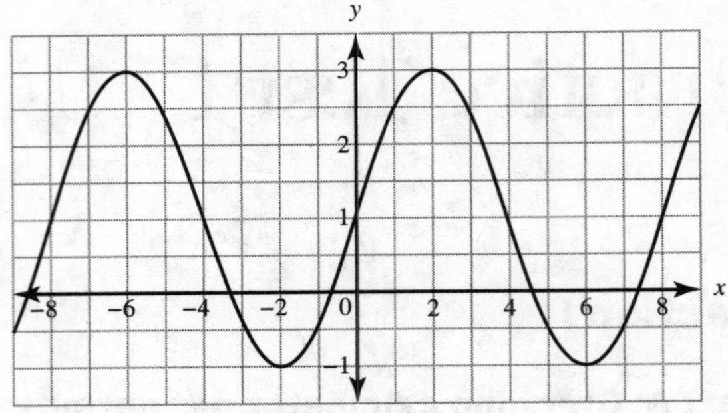

(A) $y = 2\sin(x) + 1$
(B) $y = 2\cos\left(\frac{\pi}{4}x\right) + 1$
(C) $y = 2\sin\left(\frac{\pi}{4}x + 1\right)$
(D) $y = 2\sin\left(\frac{\pi}{4}x\right) + 1$

15. Which of the following is equivalent to $\sin 5x \cos 3x - \cos 5x \sin 3x$?

(A) $\cos 8x$
(B) $\sin 8x$
(C) $\cos 2x$
(D) $\sin 2x$

16. Given $\sin x = -\frac{1}{8}$ and $\tan x < 0$, find $\sin 2x$.

(A) $-\frac{\sqrt{65}}{32}$
(B) $-\frac{1}{4}$
(C) $-\frac{3\sqrt{7}}{32}$
(D) $\frac{3\sqrt{7}}{32}$

17. Determine the period of the function $y = \frac{1}{2}\sin\left(\frac{1}{3}x - \pi\right)$.

(A) $\frac{1}{2}$
(B) $\frac{2\pi}{3}$
(C) 3π
(D) 6π

18. Solve the equation $\log_a 3 - \log_a b = c$ for b.

(A) $\frac{c}{3a}$
(B) $\frac{3}{a^c}$
(C) $\frac{a}{3c}$
(D) a^{3c}

19. Which of the following statements is true about the function $g(x) = x^7 + x^3 + \sin x$?

(A) The function is even and symmetric with respect to the origin.
(B) The function is even and symmetric with respect to the y-axis.
(C) The function is odd and symmetric with respect to the origin.
(D) The function is odd and symmetric with respect to the y-axis.

20. Which is an equation for the linear function f that satisfies the conditions $f(-3) = -7$ and $f(5) = -11$?

(A) $y + 7 = -\frac{1}{2}(x + 3)$
(B) $y + 11 = \frac{1}{2}(x - 5)$
(C) $y - 7 = -\frac{1}{2}(x - 3)$
(D) $y - 11 = -\frac{1}{2}(x + 5)$

21. Given the functions $f(x) = x^2 - 4$ and $g(x) = \sqrt{x} + 4$, determine the composition $(g \circ f)(x)$.

(A) $x^4 - 8x^2 + 12$
(B) $\sqrt{x^2 - 4} + 4$
(C) $x + 8\sqrt{x} + 12$
(D) $x + 12$

22. Given the function $g(x) = -(12x - 7)^2(34x + 89)^3$, determine $\lim_{x \to -\infty} g(x)$.

(A) $-\infty$
(B) 0
(C) 1
(D) ∞

23. Evaluate $\lim_{x \to \infty} -5^{-x} - 2$.

(A) $-\infty$
(B) -2
(C) 0
(D) ∞

24. What is the function whose graph is a reflection over the y-axis of the graph of $f(x) = 1 - 3^x$?

(A) $g(x) = 1 - 3^{-x}$
(B) $g(x) = 1 + 3^x$
(C) $g(x) = 3^x - 1$
(D) $g(x) = \log_3(x - 1)$

25. Which of the following functions does *not* have an inverse function?

(A) $y = \sin x\left(-\frac{\pi}{2} \leq x \leq \frac{\pi}{2}\right)$
(B) $y = x^3 + 2$
(C) $y = \frac{x^2}{x + 2}$
(D) $y = \frac{1}{2}e^x$

26. The table shows the predicted growth of a particular bacteria after various numbers of hours. Write an explicit formula for the sequence of the number of bacteria.

Hours (n)	1	2	3	4	5
Number of Bacteria (a_n)	19	38	57	76	95

(A) $a_n = 19n + 19$
(B) $a_n = n + 19$
(C) $a_n = \frac{1}{19}n$
(D) $a_n = 19n$

27. What are the points where the graph of the polynomial $f(x) = 7(x - 5)(x + 5)^2$ passes through the x-axis?

(A) $x = 5$ only
(B) $x = -5$ only
(C) $x = -5$ and $x = 5$
(D) Nowhere

28. Evaluate: $\lim_{x \to \infty} \frac{3x^2 - 9x - 4}{5x^2 - 4x + 8}$

(A) Does not exist
(B) 0
(C) $\frac{3}{5}$
(D) $\frac{9}{4}$

PRACTICE TEST 1

PRACTICE TEST 1

Part B: Graphing calculator required

TIME: 40 MINUTES

29. Given the graph of the function $y = f(x)$ shown below with points a, b, c, d, e, f, g, and h, which of the following pairs of points below has the greatest average rate of change?

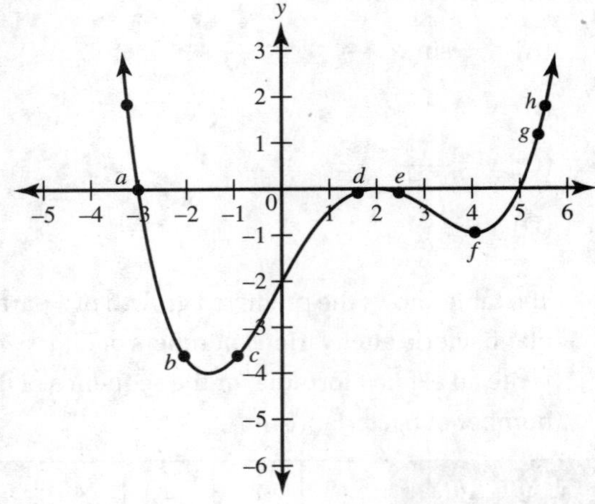

(A) a and c
(B) b and c
(C) c and d
(D) f and h

30. Given the function $f(x) = (x^2 - 5)(x - 3)(3x - 2)$, at what value of x is the absolute maximum of $f(x)$ over the interval $[-2.28, 3.25]$?

(A) -2.28
(B) -1.27
(C) 1.35
(D) 3.25

31. Approximate $\log_a 24$, given that $\log_a 2 \approx 0.4307$ and $\log_a 3 \approx 0.6826$.

(A) 1.9747
(B) 1.1133
(C) 0.8820
(D) 0.2940

32. Solve for x: $3^{2x} = 5^{x-1}$

(A) -2.7381
(B) -1
(C) -0.5563
(D) 15.2755

33. Which of the following choices is equivalent to the complex number $3 - 8i$?

(A) $\sqrt{73}\left(\cos\frac{37\pi}{60} + i\sin\frac{37\pi}{60}\right)$
(B) $\sqrt{73}\left(\cos\frac{97\pi}{60} + i\sin\frac{97\pi}{60}\right)$
(C) $\sqrt{55}\left(\cos\frac{37\pi}{60} + i\sin\frac{37\pi}{60}\right)$
(D) $\sqrt{55}\left(\cos\frac{97\pi}{60} + i\sin\frac{97\pi}{60}\right)$

34. Find all solutions in the interval $[0, 2\pi]$ for the trigonometric equation $6\cos^2 x - 5\sin x - 2 = 0$.

(A) $-1.333333, -4.4749, \frac{\pi}{6}, \frac{5\pi}{6}$
(B) $\frac{\pi}{6}, \frac{5\pi}{6}$
(C) 2, 5.1416
(D) $\frac{7\pi}{6}, \frac{11\pi}{6}$

35. Give an algebraic expression for $\cos(\sin^{-1} x)$.

(A) $\sqrt{1 - x^2}$
(B) $\sqrt{1 + x^2}$
(C) $\sqrt{x^2 - 1}$
(D) $1 - x^2$

36. How long will it take for $1,000 to double in an investment when interest is compounded continuously at the rate of 5.8% per annum? Round your answer to the nearest year.

(A) 11 years
(B) 12 years
(C) 13 years
(D) 14 years

37. An object is launched straight up from ground level with an initial velocity of 50 feet per second. The height, h (in feet above ground level), of the object t seconds after the launch is given by the function $h(t) = -16t^2 + 50t$. At approximately what value of t will the object have a height of 28 feet and be traveling downward?

(A) 0.73 seconds
(B) 1.56 seconds
(C) 1.84 seconds
(D) 2.39 seconds

38. Suppose you drop a tennis ball from a height of 15 feet. After the ball hits the floor, it rebounds to 85% of its previous height. How high, to the nearest tenth, will the ball rebound after its third bounce?

(A) 1.9 feet
(B) 7.8 feet
(C) 9.2 feet
(D) 10.8 feet

39. Let $f(x) = -2x^6 + 15x^5$. For what values of x does the graph of f have a point of inflection?

I. $x = 0$
II. $x = 6.5$
III. $x = \frac{15}{2}$

(A) I only
(B) I and II only
(C) I, II, and III
(D) none of these

40. After performing analyses on a set of data, Joseph examined the scatterplot of the residual values for each analysis. Which scatterplot indicates the best linear fit for the data?

(A)
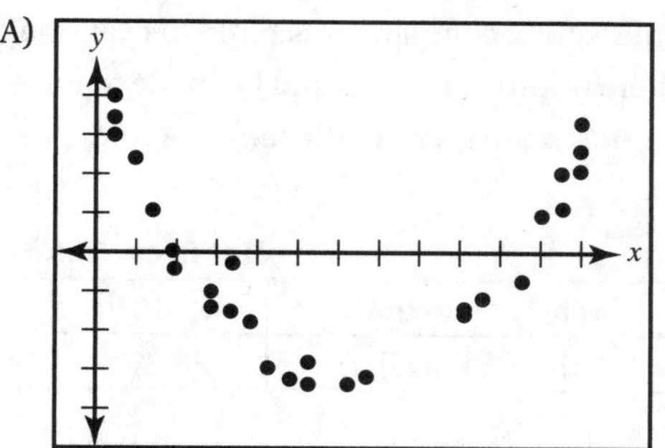

(B)
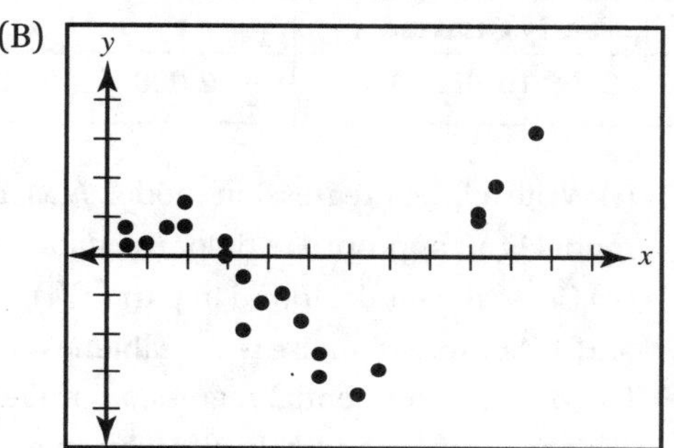

(C)
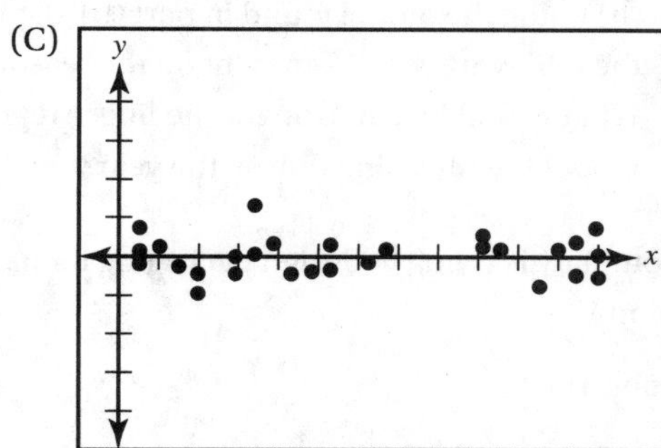

(D)
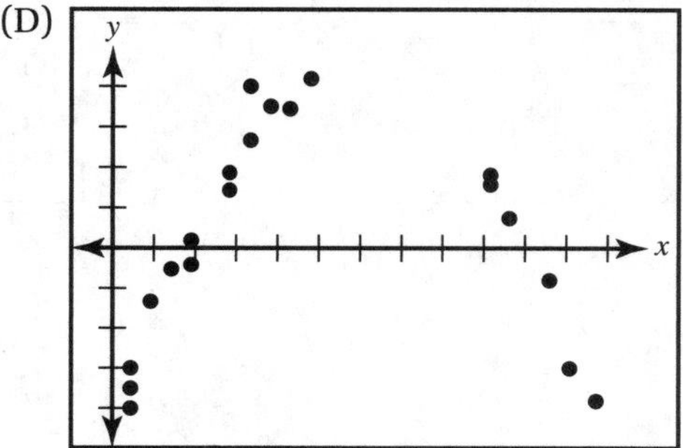

Section II

Part A: Graphing calculator required

TIME: 30 MINUTES

1. A marketing company was hired to examine the data of the sales in a single year of a local coffee shop to see where improvements could be made to increase profits. The data of their most popular hot and cold brew beverages are given in the tables below.

Hot Brew Coffee Beverages

Yearly Quarter, x	1	2	3	4
Sales (dollars), h	36,208	37,620	39,088	40,594

Cold Brew Coffee Beverages

Yearly Quarter, x	1	2	3	4
Sales (dollars), c	42,000	37,800	34,020	27,556

(a) (i) Write a linear regression model, $h(x)$, for the hot brew coffee sales. Justify why a linear regression model is an appropriate fit for the data.
(ii) Using the model found in part (a)(i), predict the sales, *to the nearest dollar,* halfway between quarters 3 and 4. Comment on the reasonableness of the prediction.

(b) (i) Write an exponential regression model, $c(x)$, for the cold brew coffee sales, rounding all coefficients *to the nearest thousandth.* Justify why an exponential regression model is an appropriate fit for the data.
(ii) Using the model found in part (b)(i), predict the sales, *to the nearest dollar*, for the second quarter of the following year. Comment on the reasonableness of the prediction.

(c) What would be a reason for the linear regression model to be increasing while the exponential regression model is decreasing during the year?

2. A company is trying to design packaging for its new product and is deciding between two different design options.

Option 1:

An open box with the dimensions labeled in the figure below. All units are inches.

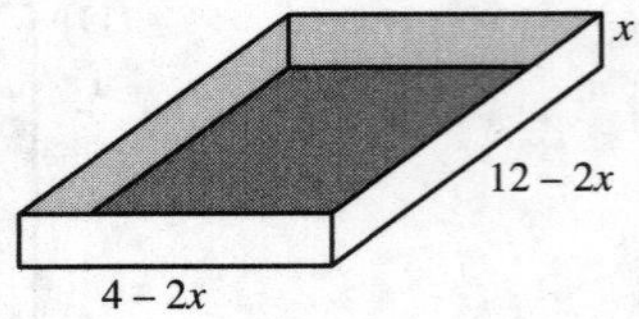

Option 2:

An open box that is made from a square sheet of cardboard by cutting out 4-inch squares from each corner as shown below and then folding along the dotted lines.

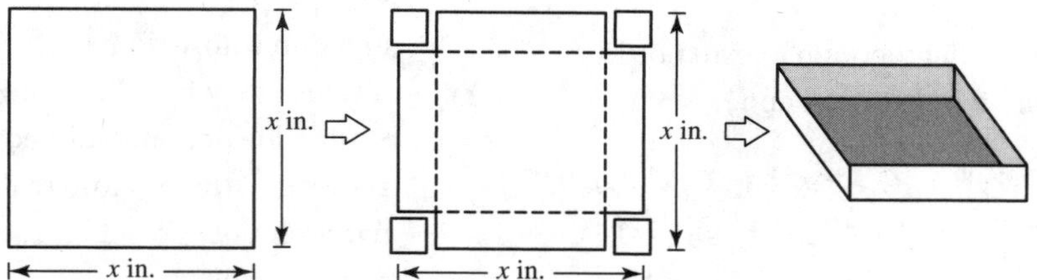

The company wants to ensure that both options can hold a product that has a volume of 36 cubic inches.

(a) If possible, find the dimensions of the boxes in Options 1 and 2. If it is not possible, explain why.

(b) The company has decided that it is going to build a rectangular section to hold all the products in the warehouse. If the company has 400 feet of fencing, what is the area of the largest rectangular section it can enclose if it uses the building as one side of the section?

(c) (i) The company examines the costs involved with shipping the product and derives the following piecewise function that determines the cost of shipping, $c(p)$, per pound, p:

$$c(p) = \begin{cases} ap + 3, & p < 100 \\ p^2 + 2p, & 100 \le p \le 2000 \end{cases}$$

Find the cost of shipping if the weight is 100 pounds.

(ii) Using the piecewise function from part (c)(i), for what value of a would the cost of shipping be a continuous function when $p = 100$?

Part B: Graphing calculator not permitted

TIME: 30 MINUTES

3. Given the function $f(x) = -(7x - 14)^2(10x + 80)^3$, do the following.

(a) State the degree of $f(x)$, and evaluate $\lim_{x\to-\infty} f(x)$ and $\lim_{x\to\infty} f(x)$.

(b) (i) Find the zeros of $f(x)$, and state their corresponding multiplicities.

(ii) Using part (b)(i), state the value(s) of x for which the graph of $f(x)$ is tangent to the x-axis. Justify your answer.

(c) Create a sketch of the graph of $f(x)$. The axes need not be scaled.

4. The function f is defined by $f(x) = \dfrac{3x + 6}{2 - x}$, where $x \neq 2$.

(a) Find the equations of the vertical and horizontal asymptotes of the graph of $f(x)$.

(b) Find the coordinates where the graph of $f(x)$ crosses the x-axis and the y-axis.

(c) The function g is defined by $g(x) = \dfrac{ax + 6}{2 - x}$, where $x \neq 2$. Given that $g(x) = g^{-1}(x)$, find the value of a.

Answer Explanations

Section I

Part A

1. **(C)** Use the binomial theorem to expand $(x+2y)^5$, where $a=x$, $b=2y$, and $n=5$:

$$(x+2y)^5 = {}_5C_0x^5(2y)^0 + {}_5C_1x^4(2y)^1 + {}_5C_2x^3(2y)^2 + {}_5C_3x^2(2y)^3 + {}_5C_4x^1(2y)^4 + {}_5C_5x^0(2y)^5$$

Evaluate the coefficients in front of each term either by using the formula ${}_nC_r = \frac{n!}{r!(n-r)!}$ or by copying the values from row 5 of Pascal's triangle:

$$(x+2y)^5 = 1x^5 + 5x^4(2y) + 10x^3(4y^2) + 10x^2(8y^3) + 5x(16y^4) + 1(32y^5)$$
$$= x^5 + 10x^4y + 40x^3y^2 + 80x^2y^3 + 80xy^4 + 32y^5$$

The coefficient in front of x^3y^2 is 40.

2. **(C)** Given $f(1)=4$, when $x=1$ the function value at 1 is 4. Substitute these values in for x and $f(x)$, respectively: $1+A+B-3=4$. Given $f(-1)=-6$, when $x=-1$ the function value at -1 is -6. Substitute these values in for x and $f(x)$, respectively: $-1+A-B-3=-6$. This is a system of two equations with two variables, A and B. Solve the system by the elimination method by adding the two equations together to eliminate B:

$$2A-6=-2$$
$$A=2$$

Substitute $A=2$ into either of the two equations, and solve for B:

$$1+2+B-3=4$$
$$B=4$$

Substituting $A=2$ and $B=4$ gives:

$$2A+B=2(2)+4=8$$

3. **(C)** If $f(x)=x^3-2x-1$, then $f(-2)=(-2)^3-2(-2)-1=-5$.

4. **(D)** The given function $f(x)=\frac{x-5}{x^2+25}$ is a rational function whose domain is all reals excluding any x-values for which the denominator is equal to 0. Setting the denominator equal to 0, $x^2+25=0$, shows that there are no real solutions, only imaginary solutions, $x=\pm 5i$. Therefore, the domain is all real numbers.

5. **(C)** The given function $g(x)=\frac{\sqrt{x-3}}{x^2-2x}$ is a rational function with a radical function. Its domain is all real numbers, excluding any x-values for which the denominator is equal to 0 or the radicand is less than 0. Set the denominator equal to 0 and solve for x:

$$x^2-2x=0$$
$$x(x-2)=0$$
$$x=0 \text{ and } x=2$$

Therefore, the domain needs to exclude $x=0, 2$. The radicand cannot be less than 0; otherwise the function values are imaginary. Set the radicand greater than or equal to 0:

$$x-3\geq 0$$
$$x\geq 3$$

Therefore, the domain is all real numbers such that $x\geq 3$.

6. **(C)** Use log laws:

$$\log\sqrt[3]{\frac{a^2b}{c}} = \log\left(\frac{a^2b}{c}\right)^{\frac{1}{3}} = \frac{1}{3}\log\left(\frac{a^2b}{c}\right) =$$
$$\frac{1}{3}(\log a^2 + \log b - \log c) = \frac{1}{3}(2\log a + \log b - \log c)$$

7. **(A)** To find the vertical asymptote(s) for the function $f(x)=\frac{x^2-4}{x^3+2x^2}$, first factor the numerator and denominator, $f(x)=\frac{(x+2)(x-2)}{x^2(x+2)}$. Remove the common factor $(x+2)$; $x=-2$ represents a hole. The remaining factor in the denominator, x^2, indicates that there is a vertical asymptote at $x=0$.

8. **(D)** To evaluate $f(-3)$, first identify which of the equations to use by referring to the inequalities. The input -3 is less than 0, so the first equation is used:

$$f(-3) = -\frac{1}{3}(-3)^2 + 9 = -3 + 9 = 6$$

9. **(B)** The slant asymptote of $f(x) = \dfrac{x^2 + 2x - 1}{x - 1}$ can be found by using long division:

$$\begin{array}{r} x + 3 \\ x - 1 \overline{)\, x^2 + 2x - 1} \\ -(x^2 - 1x) \\ 3x - 1 \\ -(3x - 3) \\ 2 \end{array}$$

An equivalent representation of the function is $f(x) = \dfrac{x^2 + 2x - 1}{x - 1} = x + 3 + \dfrac{2}{x - 1}$.

Therefore, the equation of the slant asymptote is $y = x + 3$.

10. **(C)** Given $\left(2, \frac{5\pi}{6}\right)$, $r = 2$ and $\theta = \frac{5\pi}{6}$. This point lies in Quadrant II. Choice (A) is equivalent because even though $\theta = -\frac{\pi}{6}$ is found in Quadrant IV, since $r = -2$, the point is drawn in the reverse direction and is in Quadrant II. Choice (B) is equivalent because like Choice (A), $\theta = \frac{11\pi}{6}$ is found in Quadrant IV. Since $r = -2$, the point is drawn in the reverse direction and is in Quadrant II. Choice (C) is not equivalent because $\theta = -\frac{11\pi}{6}$ is found in Quadrant I, so Choice (C) is equivalent to $\left(2, \frac{\pi}{6}\right)$. Choice (D) is equivalent to the given point since it is found in Quadrant II.

11. **(D)** To find the zeros of $r = 2 - 4\sin 2\theta$, set $r = 0$ and solve for θ:

$$0 = 2 - 4\sin 2\theta$$

$$\sin 2\theta = \frac{1}{2}$$

Using the unit circle, $2\theta = \frac{\pi}{6}$ and $2\theta = \frac{5\pi}{6}$. Therefore, $\theta = \frac{\pi}{12}$ and $\theta = \frac{5\pi}{12}$ along with $\theta = \frac{13\pi}{12}$ and $\theta = \frac{17\pi}{12}$.

12. **(A)** Given $\cos\left[\arctan\left(-\frac{2}{3}\right)\right]$, first evaluate the inner expression. So $\arctan\left(-\frac{2}{3}\right)$ is the angle whose legs of a right triangle drawn in Quadrant II are 2 and 3. Using the Pythagorean theorem, the hypotenuse of the triangle can be found:

$$(2)^2 + (3)^2 = r^2 \rightarrow r = \sqrt{13}$$

Therefore, $\cos\theta = -\frac{3}{\sqrt{13}}$. The cosine value of this angle is negative since it is in Quadrant II. Rationalizing the denominator gives:

$$\cos\theta = -\frac{3}{\sqrt{13}} \cdot \frac{\sqrt{13}}{\sqrt{13}} = -\frac{3\sqrt{13}}{13}$$

13. **(C)** Given that $\sin\theta = -\frac{1}{5}$ and $\tan\theta > 0$, since $\tan\theta = \frac{\sin\theta}{\cos\theta}$ then $\cos\theta < 0$. Use the Pythagorean trigonometric identity:

$$\cos^2\theta + \sin^2\theta = 1$$

Substitute $\sin\theta = -\frac{1}{5}$ and solve for $\cos\theta$:

$$\cos^2\theta + \left(-\frac{1}{5}\right)^2 = 1$$

$$\cos^2\theta = \frac{24}{25}$$

$$\cos\theta = -\frac{\sqrt{24}}{5} = -\frac{\sqrt{4}\sqrt{6}}{5} = -\frac{2\sqrt{6}}{5}$$

14. **(D)** The general equation of a sinusoidal curve is $y = a\sin b(x - c) + d$. The midline is $d = 1$. The amplitude is $a = 2$. Half of a sinusoidal curve is seen from 0 to 4, meaning a full curve is from 0 to 8. So the period is $8 = \frac{2\pi}{b}$ and $b = \frac{\pi}{4}$. A possible equation for the graph is $y = 2\sin\frac{\pi}{4}x + 1$.

15. **(D)** Using the difference identity, $\sin 5x\cos 3x - \cos 5x\sin 3x = \sin(5x - 3x) = \sin 2x$.

16. **(C)** Given $\sin x = -\frac{1}{8}$ and $\tan x < 0$, since $\tan x = \frac{\sin x}{\cos x}$ then $\cos x > 0$. To find $\cos x$, use the Pythagorean trigonometric identity:

$$\cos^2\theta + \sin^2\theta = 1$$

Substitute $\sin\theta = -\frac{1}{8}$ and solve for $\cos\theta$:

$$\cos^2\theta + \left(-\frac{1}{8}\right)^2 = 1$$

$$\cos^2\theta = \frac{63}{64}$$

$$\cos\theta = \frac{\sqrt{63}}{8}$$

Then use the double-angle identity:

$$\sin 2x = 2\sin x\cos x = 2\left(-\frac{1}{8}\right)\left(\frac{\sqrt{63}}{8}\right)$$

$$= -\frac{2\sqrt{63}}{64} = -\frac{\sqrt{63}}{32} = -\frac{\sqrt{9}\sqrt{7}}{32} = -\frac{3\sqrt{7}}{32}$$

17. **(D)** The period $= \frac{2\pi}{b}$, where $b = \frac{1}{3}$. Therefore, the period $= \frac{2\pi}{\frac{1}{3}} = 6\pi$.

18. **(B)** Condense the logarithmic equation:

$$\log_a 3 - \log_a b = c$$

$$\log_a \frac{3}{b} = c$$

Rewrite in exponential form:

$$a^c = \frac{3}{b}$$

Solve for b:

$$b = \frac{3}{a^c}$$

19. **(C)** To determine if a function is even or odd, substitute $-x$ for x:

$$g(-x) = (-x)^7 + (-x)^3 + \sin(-x)$$
$$= -x^7 - x^3 - \sin x = -g(x)$$

Since $g(-x) = -g(x)$, the function is odd and is symmetric with respect to the origin.

20. **(A)** To write the equation of the line, determine the slope by calculating the average rate of change:

$$\frac{\Delta y}{\Delta x} = \frac{-11-(-7)}{5-(-3)} = \frac{-4}{8} = -\frac{1}{2} = m$$

Substitute either input-output pair and the slope into the point-slope formula:

$$y - y_1 = m(x - x_1)$$

Using the first condition gives the equation:

$$y - (-7) = -\frac{1}{2}(x - (-3))$$

$$y + 7 = -\frac{1}{2}(x + 3)$$

21. **(B)** Determine the composition:

$$(g \circ f)(x) = g(f(x)) = g(x^2 - 4) = \sqrt{x^2 - 4} + 4$$

22. **(D)** To determine $\lim_{x\to-\infty} g(x)$, find the end behavior of the function on the left. Since the leading coefficient is negative and the degree is the sum of the multiplicities, 5, which is odd, the function rises on the left and falls on the right. In other words, $\lim_{x\to-\infty} g(x) = +\infty$.

23. **(B)** The graph of $g(x) = -5^{-x} - 2$ is the transformation of the parent function $f(x) = 5^x$. The graph of $f(x)$ is reflected over the x- and y-axes and shifted down 2 units. After the transformation, $\lim_{x\to\infty} f(x) = -2$.

24. **(A)** A reflection over the y-axis transforms the equation of a function by replacing x with $-x$. Therefore, the transformed equation is $g(x) = 1 - 3^{-x}$.

25. **(C)** For a function to have an inverse function, it must be one-to-one. The function in Choice (C) is not one-to-one since it fails the horizontal line test.

26. **(D)** The given sequence is arithmetic since there is a common difference, $d = 19$. Use the general arithmetic sequence formula

$$a_n = a_1 + d(n-1) = 19 + 19(n-1)$$
$$= 19 + 19n - 19 = 19n$$

27. **(A)** The graphs of polynomial functions pass through the x-axis at x-intercepts whose factors have odd multiplicity. The only factor with an odd multiplicity is $x = 5$. Therefore, the graph of the polynomial function passes through the x-intercept at $(5, 0)$ and is tangent to the x-axis at $(-5, 0)$.

28. **(C)** To evaluate infinite limits of rational functions, determine their end behavior by considering the degree of both the numerator and the denominator of the expression. The degree of the numerator and denominator are both 2. Since they are equal, the limit is the ratio of their leading coefficients, $\frac{3}{5}$.

Part B

29. **(D)** Graphically, the greatest average rate of change is equal to the steepest slope of a secant line. The lines can be drawn in to see the steepest slope, or estimations of the coordinates can be used to calculate the slope of the secant for each choice as shown in the table below.

Points	Slope of Secant Line
$a(-3, 0)$ and $c(-1, -3.5)$	$m = \frac{-3.5-0}{-1-(-3)} = \frac{-3.5}{2} = -1.75$
$b(-2, -3.5)$ and $c(-1, -3.5)$	$m = \frac{-3.5-(-3.5)}{-2-(-1)} = \frac{0}{-1} = 0$
$c(-1, -3.5)$ and $d(1.75, 0)$	$m = \frac{0-(-3.5)}{1.75-(-1)} = \frac{4}{2.75} \approx 1.2727$
$f(4, -1)$ and $h(5.5, 2)$	$m = \frac{2-(-1)}{5.5-4} = \frac{3}{1.5} = 2$

The sign indicates direction of the secant line. So the slope that has the greatest absolute value is between f and h.

30. **(D)** The absolute maximum value of a function is the highest y-value in the range of a given function and can occur at relative maxima or at endpoints of the interval. Below is the graph of $f(x)$.

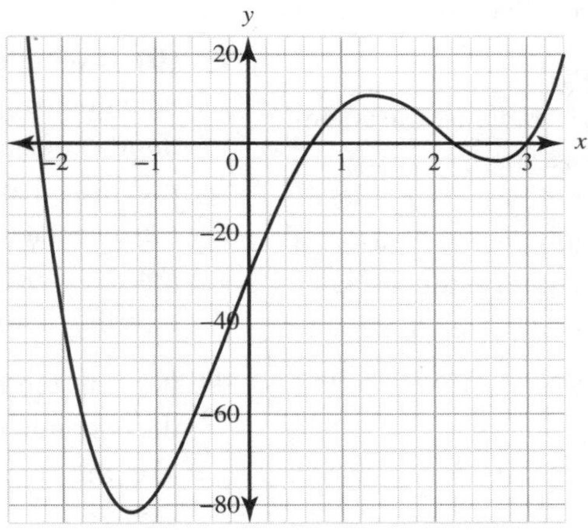

From the graph it is hard to tell if the absolute maximum value occurs at Choice (A), (C), or (D). In this case, finding the y-values for each of the x-values is a way to choose the correct answer. The table of values is given below.

x	$f(x)$
−2.28	9.2604
−1.27	−84.03
1.35	10.748
3.25	10.777

The highest y-value is 10.777, which occurs at the endpoint of the interval and would be the x-value with the absolute maximum value for $f(x)$.

31. **(A)** Using log laws:

$$\log_a 24 = \log_a(3 \cdot 2 \cdot 2 \cdot 2)$$
$$= \log_a 3 + \log_a 2 + \log_a 2 + \log_a 2$$
$$= \log_a 3 + 3\log_a 2$$
$$= 0.6826 + 3(0.4307) = 1.9747$$

32. **(A)** To solve for x given the equation $3^{2x} = 5^{x-1}$, use a graphing calculator. Let $Y1 = 3^{2x}$ and $Y2 = 5^{x-1}$. To find their point of intersection, press 2ND TRACE and select 5:intersect. The point of intersection is $(-2.738153, 0.00243869)$.

33. **(B)** When graphed, the complex number $3 - 8i$ is in Quadrant IV. To find r, use the Pythagorean theorem:

$$(3)^2 + (8)^2 = r^2$$
$$r = \sqrt{73}$$

This eliminates Choices (C) and (D). To find the angle, use the calculator to find the reference angle:

$$\tan^{-1}\frac{8}{3} \approx 1.2120256 \text{ radians}$$

Since the angle is in Quadrant IV, $\theta \approx 2\pi - 1.2120256 \approx 5.071159$ radians. The angle in Choice (B), $\frac{97\pi}{60} \approx 5.071159$ radians, is correct.

34. **(B)** Use the Pythagorean trigonometric identity $\cos^2 x + \sin^2 x = 1$ to substitute in for $\cos^2 x$. Then the given equation $6\cos^2 x - 5\sin x - 2 = 0$ becomes $6(1 - \sin^2 x) - 5\sin x - 2 = 0$. Write the equation in standard form:

$$6\sin^2 x + 5\sin x - 4 = 0$$

Factor the trinomial:

$$(3\sin x + 4)(2\sin x - 1) = 0$$

Set each factor equal to 0 and solve for $\sin x$:

$$\sin x = -\frac{4}{3} \text{ and } \sin x = \frac{1}{2}$$

There is no value of x for which the first equation can be true since the range of $\sin x$ is $[-1, 1]$. Therefore, the only solution comes from $\sin x = \frac{1}{2}$. Using the unit circle, $x = \frac{\pi}{6}, \frac{5\pi}{6}$.

35. **(A)** $\cos(\sin^{-1} x)$ means that $\theta = \sin^{-1} x$, so $\sin\theta = x$. Using the Pythagorean identity, solve for $\cos\theta$:

$$x^2 + \cos^2\theta = 1$$

Therefore:

$$\cos\theta = \sqrt{1 - x^2}$$

36. **(B)** When an investment is compounded continuously, it is modeled by the equation $f(t) = a \cdot e^{rt}$. In this case, $a = 1{,}000$ $f(t) = 2{,}000$ and $r = 0.058$. Substitute the values:

$$2{,}000 = 1{,}000\, e^{(0.058)t}$$

To solve for t, you can use technology to graph the left and right sides of the equation and find their point of intersection. Algebraically, isolate the exponential expression and use natural logs to solve for t:

$$2 = e^{0.058t}$$
$$t = \frac{\ln 2}{0.058} \approx 11.9508$$

37. **(D)** Given the function $h(t) = -16t^2 + 50t$, set the equation equal to 28 and solve for t. This can be done graphically by finding the point of intersection using technology. There are two places where the object reaches a height of 28 feet. The second point is chosen as this is also where the object is traveling downward, (2.394, 28).

38. **(C)** This represents a geometric sequence with the equation $a_n = 15(0.85)^n$. To find the height after the third bounce, substitute $n = 3$ into the equation:

$$a_3 = 15(0.85)^{(3)} = 9.2$$

39. **(B)** Using technology, graph the function and look at where the graph changes concavity.

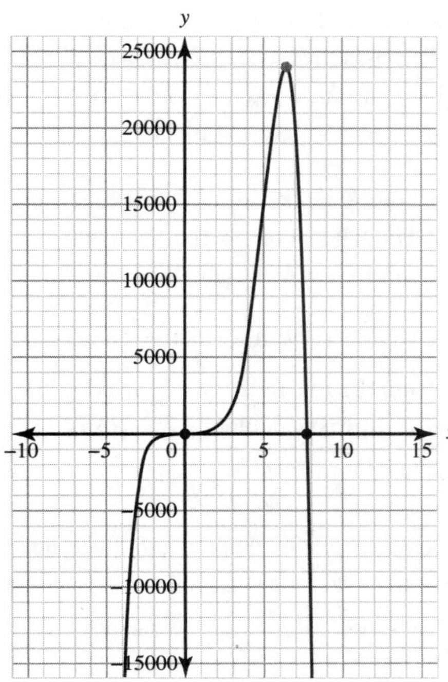

The graph changes from concave down to concave up at $x = 0$ and then changes from concave up to concave down at $x = 6.5$.

40. **(C)** For a residual plot, there should be no observable pattern and a similar distribution of residuals above and below the x-axis.

Section II

Part A

1. (a)(i) Using the calculator, the linear regression model is $h(x) = 1462.6x + 34721$. Since the correlation coefficient is $r = 0.999895996$, which is close to 1, this indicates that the linear regression model is an appropriate fit for the data.

   (a)(ii) Halfway between quarters 3 and 4 is $x = 3.5$. Substitute $x = 3.5$ into the linear regression model:

$$h(3.5) = 1462.6(3.5) + 34721 = \$39{,}840.10 \approx \$39{,}840$$

   This seems reasonable since this value falls between the sales of quarters 3 and 4.

   (b)(i) Using the calculator, the exponential regression model is $c(x) = 49191.164 \cdot 0.872^x$. Since the coefficient of determination is $r^2 = 0.9657107767$, which is close to 1, this indicates that the exponential regression model is an appropriate fit for the data.

   (b)(ii) The second quarter of the following year is $x = 6$. Substitute $x = 6$ into the exponential regression model:

$$c(6) = 49191.164 \cdot 0.872^6 = \$21626.48851 \approx \$21{,}626$$

   This does not seem reasonable because you would expect that the sales would be similar to that from quarter 2 from the previous year. Instead, the sales continue to decrease. This is a result of the x-value being outside of the provided domain. Most likely, the domain should be restricted to only include values within the year since the pattern of the exponential model will continue to decrease as x increases.

   (c) Since the data are comparing hot brew coffee to cold brew coffee, one possible explanation is that quarter 1 is during the summer, so more people want cold beverages, and then as the year continues and it approaches the winter months, more people want hot beverages.

2. (a) Volume of Box 1 = (length)(width)(height) = $(4-2x)(12-2x)(x) = 4x^3 - 32x^2 + 48x$. Set the volume equal to 36, and use technology to graph the two sides of the equation and find their point of intersection.

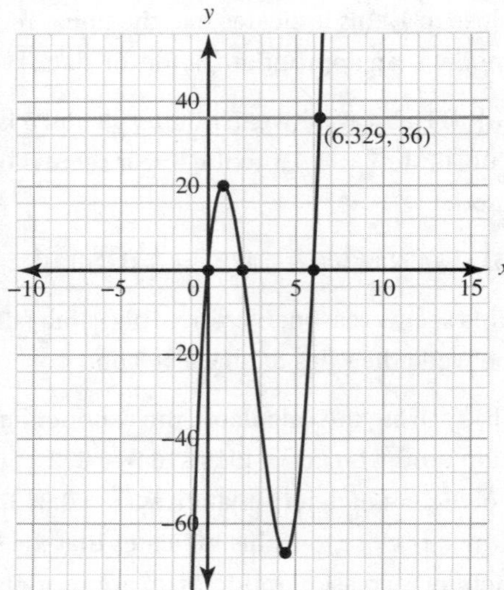

Therefore, $x = 6.3$ inches. The dimensions of the Option 1 box are $4 - 2(6.3) = -8.6$ inches, $12 - 2(6.3) = -0.6$ inches, $x = 6.3$ inches. It is not possible to have a box with these dimensions since the length and width are negative numbers.

Volume of Box 2 = (length)(width)(height) = $(x-8)(x-8)(4) = 4x^2$. Set the volume equal to 36, and use technology to graph the two sides of the equation and find their point of intersection.

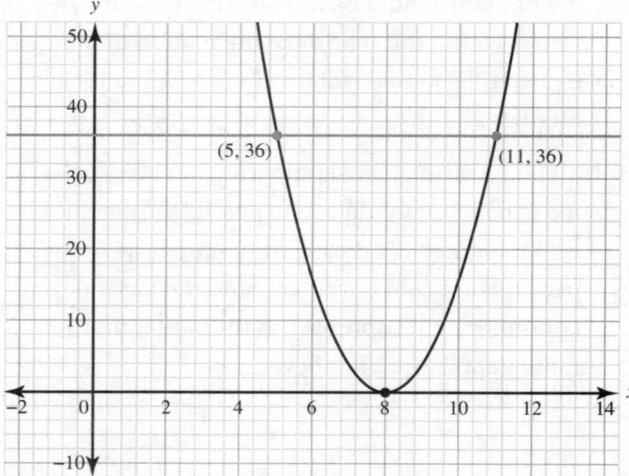

The x-values are $x = 5$, $x = 11$. Using $x = 5$, the dimensions of the Option 2 box are $(5) - 8 = -3$ inches, $(5) - 8 = -3$ inches, 4 inches. Since there cannot be negative lengths, this solution is rejected. Using $x = 11$, the dimensions of the Option 2 box are $(11) - 8 = 3$ inches, $(11) - 8 = 3$ inches, 4 inches.

(b) Let x = width and y = length. If there is a total of 400 feet of fencing, then $400 = 2x + y$. Therefore, $y = 400 - 2x$. The area of a rectangle is:

$$\text{Area} = (\text{length})(\text{width}) = (400 - 2x)(x) = 400x - 2x^2$$

To find the maximum, find the coordinates of the vertex by first calculating the axis of symmetry:

$$x = -\frac{b}{2a} = -\frac{400}{2(-2)} = 100$$

The width is $x = 100$ feet, and the length is $y = 400 - 2(100) = 200$ feet. The largest possible area is $(200)(100) = 20{,}000$ square feet.

(c)(i) The cost of shipping with a weight of 100 pounds is equivalent to evaluating $c(100)$. For this piecewise function, the second equation should be used:

$$c(100) = (100)^2 + 2(100) = \$10,200$$

(c)(ii) For the function to be continuous, there cannot be any breaks when $p = 100$. Both values of $c(100)$ for the two functions must be equal:

$$a(100) + 3 = (100)^2 + 2(100) = \$10,200$$

Solve for a:

$$a = 101.97$$

Part B

3. (a) The degree is equal to the sum of the multiplicities, which equals $2 + 3 = 5$. Evaluating limits as x approaches infinity means determining the end behavior of a function. Since this is a polynomial, the end behavior is determined by the degree and the leading coefficient. Since the degree is odd and the leading coefficient is negative, the graph will rise on the left and fall on the right. Therefore, $\lim_{x\to-\infty} f(x) = \infty$ and $\lim_{x\to\infty} f(x) = -\infty$.

(b)(i) The zeros are the x-values for which the function is equal to 0. Since the function is already factored, set each factor equal to 0 and solve for x:

$$(7x - 14)^2 = 0 \qquad (10x + 80)^3 = 0$$
$$x = 2 \qquad x = -8$$

The multiplicity for $x = 2$ is 2, and the multiplicity for $x = -8$ is 3.

(b)(ii) The graph of f is tangent to the x-axis at $x = 2$ since $(7x - 14)$ has an even multiplicity. The graph of f will pass through $x = -8$ since $(10x + 80)$ has an odd multiplicity.

(c)

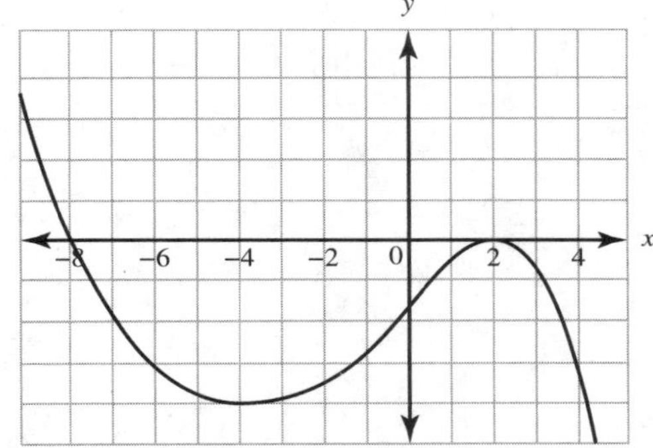

4. (a) The vertical asymptote is the x-value for which the denominator is equal to 0: $x = 2$. The horizontal asymptote is determined by the degree of the numerator and the degree of the denominator. Since both degrees are equal to 1, the equation of the horizontal asymptote is the ratio of the leading coefficients: $y = \frac{3}{-1} = -3$.

(b) The graph crosses the x-axis when $y = 0$:

$$0 = \frac{3x + 6}{2 - x}$$
$$3x + 6 = 0$$
$$x = -2$$

The x-intercept is $(-2, 0)$.

The graph crosses the y-axis when $x = 0$:

$$f(0) = \frac{3(0) + 6}{2 - (0)} = \frac{6}{2} = 3$$

The y-intercept is $(0, 3)$.

(c) The inverse is found by switching x with y and solving for y. Therefore, $x = \frac{ay + 6}{2 - y}$. Cross multiply and solve for y:

$$g^{-1}(x) = \frac{2x - 6}{a + x}$$

The given equation for $g(x)$ has an equivalent form if -1 is factored from the denominator:

$$g(x) = \frac{ax + 6}{2 - x} = \frac{ax + 6}{-1(-2 + x)} = -\frac{ax + 6}{-2 + x} = \frac{-ax - 6}{-2 + x}$$

Set $g(x) = g^{-1}(x)$:

$$\frac{-ax - 6}{-2 + x} = \frac{2x - 6}{a + x}$$

By looking at each term and matching coefficients, $a = -2$.

ANSWER SHEET
Practice Test 2

Section I

1. Ⓐ Ⓑ Ⓒ Ⓓ	11. Ⓐ Ⓑ Ⓒ Ⓓ	21. Ⓐ Ⓑ Ⓒ Ⓓ	31. Ⓐ Ⓑ Ⓒ Ⓓ
2. Ⓐ Ⓑ Ⓒ Ⓓ	12. Ⓐ Ⓑ Ⓒ Ⓓ	22. Ⓐ Ⓑ Ⓒ Ⓓ	32. Ⓐ Ⓑ Ⓒ Ⓓ
3. Ⓐ Ⓑ Ⓒ Ⓓ	13. Ⓐ Ⓑ Ⓒ Ⓓ	23. Ⓐ Ⓑ Ⓒ Ⓓ	33. Ⓐ Ⓑ Ⓒ Ⓓ
4. Ⓐ Ⓑ Ⓒ Ⓓ	14. Ⓐ Ⓑ Ⓒ Ⓓ	24. Ⓐ Ⓑ Ⓒ Ⓓ	34. Ⓐ Ⓑ Ⓒ Ⓓ
5. Ⓐ Ⓑ Ⓒ Ⓓ	15. Ⓐ Ⓑ Ⓒ Ⓓ	25. Ⓐ Ⓑ Ⓒ Ⓓ	35. Ⓐ Ⓑ Ⓒ Ⓓ
6. Ⓐ Ⓑ Ⓒ Ⓓ	16. Ⓐ Ⓑ Ⓒ Ⓓ	26. Ⓐ Ⓑ Ⓒ Ⓓ	36. Ⓐ Ⓑ Ⓒ Ⓓ
7. Ⓐ Ⓑ Ⓒ Ⓓ	17. Ⓐ Ⓑ Ⓒ Ⓓ	27. Ⓐ Ⓑ Ⓒ Ⓓ	37. Ⓐ Ⓑ Ⓒ Ⓓ
8. Ⓐ Ⓑ Ⓒ Ⓓ	18. Ⓐ Ⓑ Ⓒ Ⓓ	28. Ⓐ Ⓑ Ⓒ Ⓓ	38. Ⓐ Ⓑ Ⓒ Ⓓ
9. Ⓐ Ⓑ Ⓒ Ⓓ	19. Ⓐ Ⓑ Ⓒ Ⓓ	29. Ⓐ Ⓑ Ⓒ Ⓓ	39. Ⓐ Ⓑ Ⓒ Ⓓ
10. Ⓐ Ⓑ Ⓒ Ⓓ	20. Ⓐ Ⓑ Ⓒ Ⓓ	30. Ⓐ Ⓑ Ⓒ Ⓓ	40. Ⓐ Ⓑ Ⓒ Ⓓ

Practice Test 2

Section I

Part A: Graphing calculator not permitted

TIME: 80 MINUTES

1. What is the fifth term in the expansion of $(3a - b)^6$?

   (A) $135a^2b^4$
   (B) $540a^3b^3$
   (C) $-18ab^5$
   (D) $-135a^2b^4$

2. Simplify: $\ln\sqrt[3]{e^2x}$

   (A) $\frac{2e}{3} + \frac{1}{3}\ln x$
   (B) $\frac{2}{3} + \ln\frac{x}{3}$
   (C) $\frac{2}{3} + \frac{1}{3}\ln x$
   (D) $\frac{2e}{3} + \ln\frac{x}{3}$

3. If $\csc x \neq -1$, which of the following is equivalent to $\frac{\cot^2 x}{1 + \csc x}$?

   (A) $1 - \csc x$
   (B) $\csc x - 1$
   (C) $-\csc x$
   (D) $\csc x - \cot x$

4. If $\sin\theta = b$, then what is $\sin\theta \cdot \cos\theta \cdot \tan\theta$?

   (A) 1
   (B) $\frac{1}{b}$
   (C) b
   (D) b^2

5. If $f(\theta) = a\cos\theta + b$, what is the minimum value of $f(\theta)$?

   (A) $b - a$
   (B) $\frac{b-a}{2}$
   (C) $b + a$
   (D) $\frac{b+a}{2}$

6. Which of the following equations could represent the graph shown below?

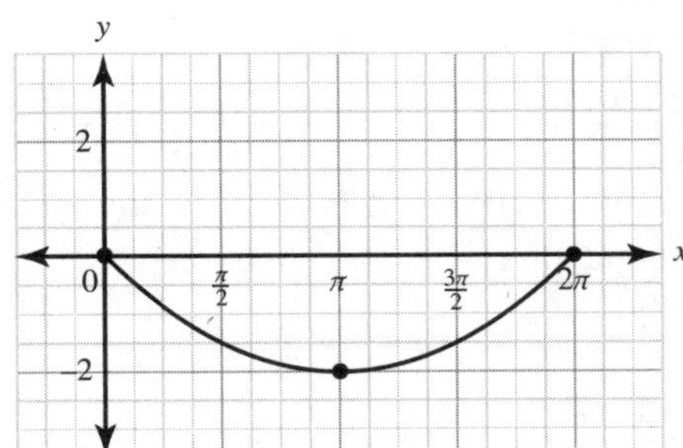

   (A) $y = 2\sin(2x)$
   (B) $y = -2\sin\left(\frac{1}{2}x\right)$
   (C) $y = -\frac{1}{2}\sin(2x)$
   (D) $y = \frac{1}{2}\sin(2x)$

7. Which of the following expressions is equivalent to $\csc\left(-\frac{\pi}{2}\right)$?

   (A) $2\cos\frac{\pi}{3}$
   (B) $\tan\pi$
   (C) $2\sin\frac{5\pi}{6}$
   (D) $\cos\pi$

8. In which quadrant is the terminal side of angle θ located if the graphs of $y = \sin\theta$ and $y = \cos\theta$ are both decreasing when angle θ is increasing?

(A) I
(B) II
(C) III
(D) IV

9. Evaluate: $\csc\left(\tan^{-1}\left(\frac{4}{3}\right)\right)$

(A) $\frac{5}{3}$
(B) $\frac{5}{4}$
(C) $\frac{4}{5}$
(D) $\frac{3}{5}$

10. For the expression $k - \frac{1}{\sec^2 x} = \sin^2 x$ to be an identity, what does k equal?

(A) 1
(B) 0
(C) $\cos^2 x$
(D) $\csc^2 x$

11. What is the expression $\frac{\sin 2\theta}{2\tan\theta}$ equivalent to?

(A) $\cot^2\theta$
(B) $\sin^2\theta$
(C) $\cos^2\theta$
(D) $\sec^2\theta$

12. In the interval $0 \leq x < 2\pi$, what are the solutions of the equation $\sin^2 x = \sin x$?

(A) $0, \frac{\pi}{2}, \pi$
(B) $\frac{\pi}{2}, \frac{3\pi}{2}$
(C) $0, \frac{\pi}{2}, \frac{3\pi}{2}$
(D) $\frac{\pi}{2}, \pi, \frac{3\pi}{2}$

13. What is the expression $\frac{\sin\left(x + \frac{\pi}{2}\right)}{-\sin x}$ equivalent to?

(A) -1
(B) 1
(C) $-\cot x$
(D) $\cot x$

14. Which of the following choices represents the graph of $r = 6\cos\theta$?

(A)

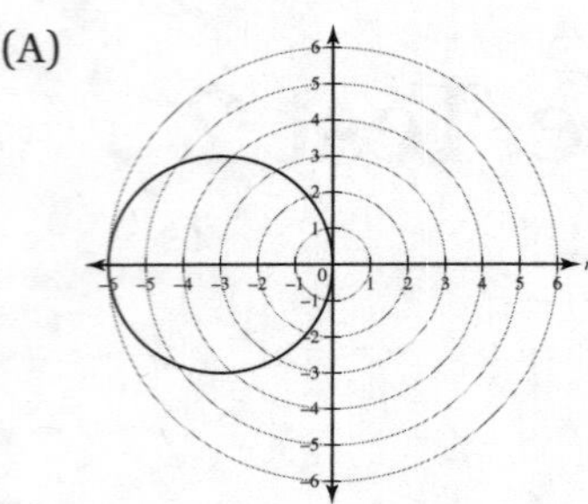

(B)

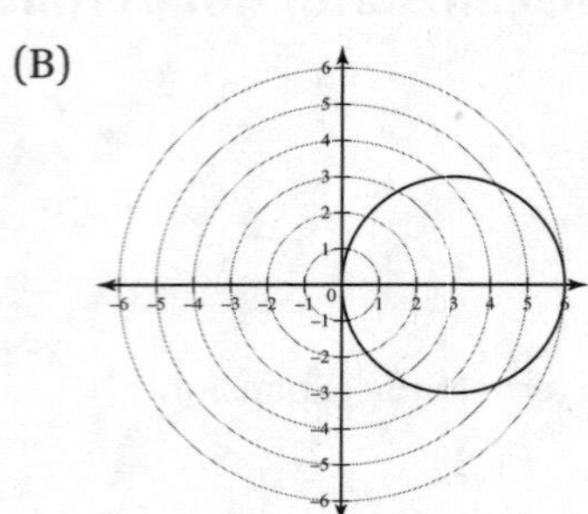

(C)

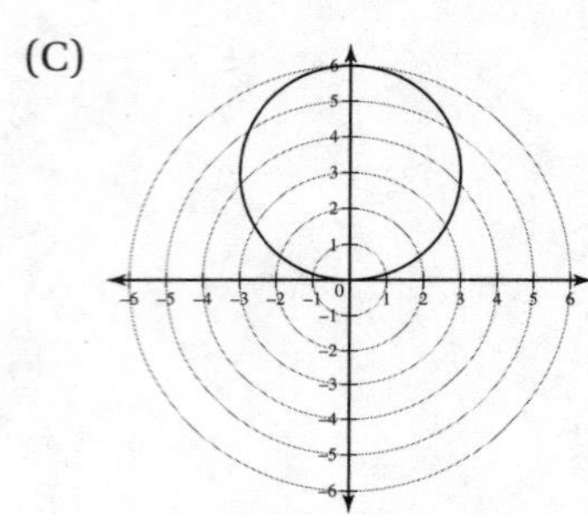

(D)

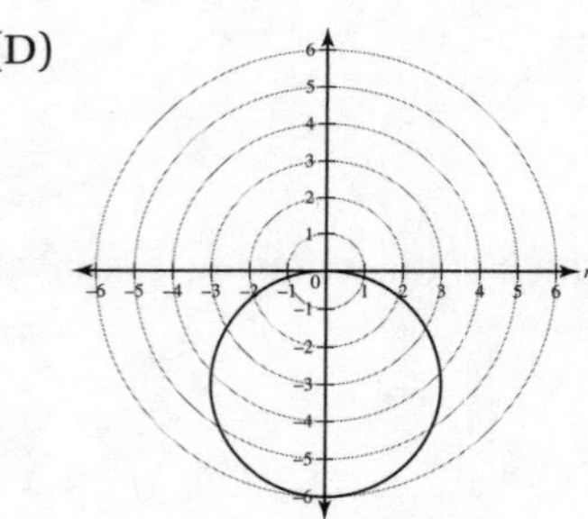

15. Which of the following points does *not* change the location of the point $\left(4, \frac{11\pi}{18}\right)$ in polar coordinates?

(A) $\left(4, \frac{29\pi}{18}\right)$
(B) $\left(-4, \frac{10\pi}{9}\right)$
(C) $\left(-4, \frac{47\pi}{18}\right)$
(D) $\left(4, \frac{47\pi}{18}\right)$

PRACTICE TEST 2

16. Given the polar coordinates $\left(3, -\frac{3\pi}{4}\right)$, find the rectangular coordinates of this point.

(A) $\left(\frac{3\sqrt{2}}{2}, -\frac{3\sqrt{2}}{2}\right)$
(B) $\left(\frac{3\sqrt{2}}{2}, \frac{3\sqrt{2}}{2}\right)$
(C) $\left(-\frac{3\sqrt{2}}{2}, -\frac{3\sqrt{2}}{2}\right)$
(D) $\left(-\frac{3\sqrt{2}}{2}, \frac{3\sqrt{2}}{2}\right)$

17. If $f(x) = x^3 - 3x^2 - 2x + 5$ and $g(x) = 2$, then what does $g(f(x))$ equal?

(A) $2x^3 - 6x^2 - 2x + 10$
(B) $2x^2 - 6x + 1$
(C) -3
(D) 2

18. Which of the following functions is *not* odd?

(A) $f(x) = \sin(x)$
(B) $g(x) = \sin(2x)$
(C) $h(x) = x^5 + 1$
(D) $j(x) = \frac{x}{x^2 + 3}$

19. At what value(s) of x do the graphs of $y = x + 2$ and $y^2 = 4x$ intersect?

(A) -2 and 2
(B) -2
(C) 2
(D) none of these

20. If $\log_b(3^b) = \frac{b}{2}$, then what does b equal?

(A) $\frac{1}{9}$
(B) $\frac{1}{3}$
(C) 3
(D) 9

21. If $f^{-1}(x)$ is the inverse of $f(x) = 2e^{-x}$, then what does $f^{-1}(x)$ equal?

(A) $\ln\left(\frac{2}{x}\right)$
(B) $\ln\left(\frac{x}{2}\right)$
(C) $\frac{1}{2}\ln(x)$
(D) $\ln(2 - x)$

22. Find $f(x + h)$ when $f(x) = 2x^2 + 4x - 1$.

(A) $2x^2 + 2h^2 + 8x + 8h - 1$
(B) $2x^2 + 2h^2 + 4x + 4h - 1$
(C) $2x^2 + 2xh + 2h^2 + 4x + 4h - 1$
(D) $2x^2 + 4xh + 2h^2 + 4x + 4h - 1$

23. For $f(x) = \begin{cases} 8x + 1, x < 1 \\ 4x, 4 \le x \le 6 \\ 4 - 2x, x > 6 \end{cases}$, evaluate $f(6)$.

(A) -8
(B) 6
(C) 24
(D) 49

24. Use the graph of $h(x)$ below to determine the intervals where $h(x)$ is increasing and where $h(x)$ is decreasing.

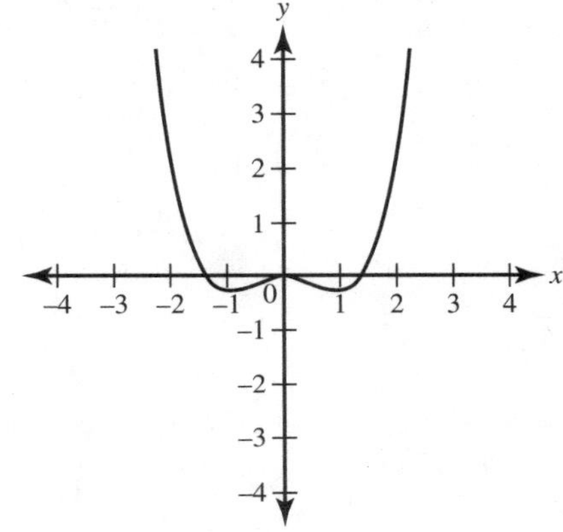

(A) Increasing: $(-1, 1)$; decreasing: $(-\infty, -1) \cup (1, \infty)$
(B) Increasing: $(-1, 0) \cup (1, \infty)$; decreasing: $(-\infty, -1) \cup (0, 1)$
(C) Increasing: $(0, \infty)$; decreasing: $(-\infty, 0)$
(D) Increasing: $(-1, \infty)$; decreasing: $(-\infty, -1)$

25. Using the tables below, evaluate $(f \circ g)(4)$.

x	13	9	5	7
$f(x)$	26	18	10	14

x	6	4	7	5
$g(x)$	11	7	13	9

(A) 4
(B) 7
(C) 14
(D) 18

PRACTICE TEST 2

26. Find the horizontal asymptote, if any, for the rational function $g(x) = \frac{7x^2 - 3x - 6}{6 - 8x + 3x^2}$.

(A) $y = 0$

(B) $y = \frac{7}{3}$

(C) $y = \frac{3}{8}$

(D) no horizontal asymptote

27. A farmer has 1,400 yards of fencing to enclose a rectangular garden. Express the area, A, of the rectangle as a function of the width, x, of the rectangle. What is the domain of A?

(A) $A(x) = -x^2 + 700x;\quad 0 < x < 700$

(B) $A(x) = -x^2 + 700x;\quad 0 < x < 1{,}400$

(C) $A(x) = x^2 + 700x;\quad 0 < x < 700$

(D) $A(x) = -x^2 + 1{,}400x;\quad 0 < x < 1{,}400$

28. If $x = -2$ is a real zero of the polynomial $f(x) = x^3 + 12x^2 + 46x + 52$, write $f(x)$ as a product of linear factors.

(A) $f(x) = (x + 1)(x + 5 + i\sqrt{3})(x - 2 - i\sqrt{3})$

(B) $f(x) = (x - 1)(x + 5 + i\sqrt{3})(x + 5 - i\sqrt{3})$

(C) $f(x) = (x + 2)(x + 5 + i)(x + 5 - i)$

(D) $f(x) = (x + 2)(x + 5 + i)(x - 5 - i)$

Part B: Graphing calculator required

TIME: 40 MINUTES

29. A deposit of $12,000 is made in an account that earns 7.6% interest compounded quarterly. The balance in the account after n quarters is given by the sequence $a_n = 12{,}000\left(1 + \frac{0.076}{4}\right)^n$, $n = 1, 2, 3, \ldots$ Find the balance in the account after 5 years.

    (A) $6,051.40
    (B) $7,285.40
    (C) $13,184.15
    (D) $17,484.97

30. If $\log_{0.3}(x - 1) < \log_{0.09}(x - 1)$, then x lies in which of the following intervals?

    (A) $(2, \infty)$
    (B) $(-2, -1)$
    (C) $(-\infty, 2)$
    (D) $(1, 2)$

31. If θ is an angle in standard position and its terminal side passes through the point $P(-0.8, 0.6)$ on the unit circle, which of the following is a possible radian value of θ to the nearest hundredth?

    (A) 2.22
    (B) 2.50
    (C) 5.36
    (D) 5.64

32. The owner of a video store has determined that the cost, C, in dollars, of operating the store is approximately modeled by the function $C(x) = 2x^2 - 22x + 760$, where x is the number of videos rented daily. Find the lowest cost to the nearest dollar.

    (A) $518
    (B) $639
    (C) $700
    (D) $821

33. The height, h, in meters of a rock thrown straight up at t seconds is described by the equation $h(t) = 2 + 20t - 4.9t^2$. Find the average rate of change of the height of the rock over the first 2 seconds of flight.

    (A) -10.2
    (B) -0.0980
    (C) 0.0980
    (D) 10.2

34. Find the solution set for the rational inequality $\frac{x+2}{2x-1} > 5$.

    (A) $\left(-\infty, \frac{1}{2}\right) \cup \left(\frac{7}{9}, \infty\right)$
    (B) $(-\infty, 2)$
    (C) $\left(\frac{1}{2}, 2\right)$
    (D) $\left(\frac{1}{2}, \frac{7}{9}\right)$

35. The table below shows selected values of (x, y) for a continuous polynomial relation. Based on the table, within which interval of x must there exist a zero of the polynomial?

x	-3	-1	1	3	5	7
y	4.24	-10.18	-16.6	-15.02	-5.44	6.35

    (A) $(-1, 1)$
    (B) $(1, 3)$
    (C) $(3, 5)$
    (D) $(5, 7)$

36. Find the domain of the function $f(x) = \sqrt{x-a} - \frac{1}{\sqrt{b-x}}$, $0 < a < b$.

    (A) $[a, b)$
    (B) $(a, b]$
    (C) (a, b)
    (D) $(-\infty, \infty)$

37. Find the difference of the largest and smallest solutions of the equation $5x^4 + x^2 - 1 = 2^x$ rounded to 2 decimal places.

    (A) 0
    (B) 1.13
    (C) 1.34
    (D) 1.50

PRACTICE TEST 2

38. The population of Old Bethpage is given in the table below.

Year	1980	1990	2000	2010
Population (in thousands)	2	4	8	17

If the same pattern of population growth continues, what will be the population of Old Bethpage in the year 2024, to the nearest thousand?

(A) 20,000
(B) 32,000
(C) 45,000
(D) 64,000

39. A box with an open top is to be constructed from a rectangular piece of cardboard with dimensions 11 inches by 27 inches by cutting out equal squares of side x at each corner and then folding up the sides as seen in the figure below.

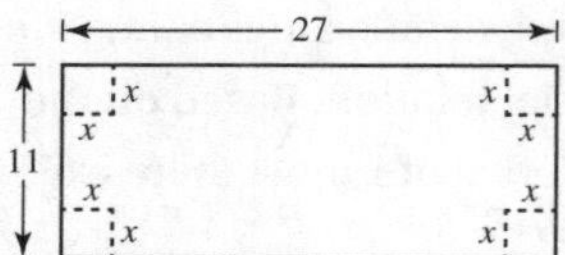

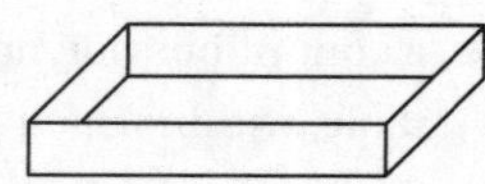

What would be the length of the side of the square for the volume of the box to be an absolute maximum?

(A) 2.41
(B) 5.5
(C) 5.8
(D) 330.35

40. Given the rectangular coordinates $(4, -4\sqrt{3})$, find the polar coordinates of this point.

(A) $\left(4, \frac{5\pi}{3}\right)$
(B) $\left(4, \frac{11\pi}{6}\right)$
(C) $\left(8, \frac{11\pi}{6}\right)$
(D) $\left(8, \frac{5\pi}{3}\right)$

Section II

Part A: Graphing calculator required

TIME: 30 MINUTES

1. A study was performed in the high school of a small town where the population data was collected and organized in the table below.

t, years since 1980	5	7	9	10	12	14	18	20	23	27	32	38
$P(t)$, population	50	160	270	310	370	440	510	580	640	700	770	830

   (a) Write a logarithmic regression model for the given data, rounding all coefficients *to the nearest thousandth*. Explain why a logarithmic regression model is appropriate to use in the context of this problem.

   (b) Using the regression model from part (a), approximate the population in the year 1995, *to the nearest tenth*. Comment on the reasonableness of this prediction.

   (c) Using the regression model from part (a), approximate the population in the year 2024, *to the nearest tenth*. Comment on the reasonableness of this prediction.

2. A spacecraft landed on a planet whose temperature fluctuates periodically. It was recorded, for the first time, that 30 hours after the landing, the minimum temperature was 4°F. It was recorded, for the first time, that 72 hours after the landing, the maximum temperature was 120°F.

   (a) (i) Find a sinusoidal function $f(t) = a\sin[b(t - c)] + d$ that models the temperature t hours after the landing.

   (ii) Using the model from part (a)(i), approximate the temperature 42 hours after the landing.

   (b) Within the first 100 hours after the landing, find all the values of t when the temperature is 80°F. Round all answers to *two decimal places*.

   (c) (i) The gravity on this new planet is such that when the astronauts drop a ball from a height of 1.8 meters, it bounces so that the maximum height reached by the ball, after each bounce, is 85% of the previous maximum height. Show that the maximum height reached by the ball after it has bounced for the sixth time is 0.68 m.

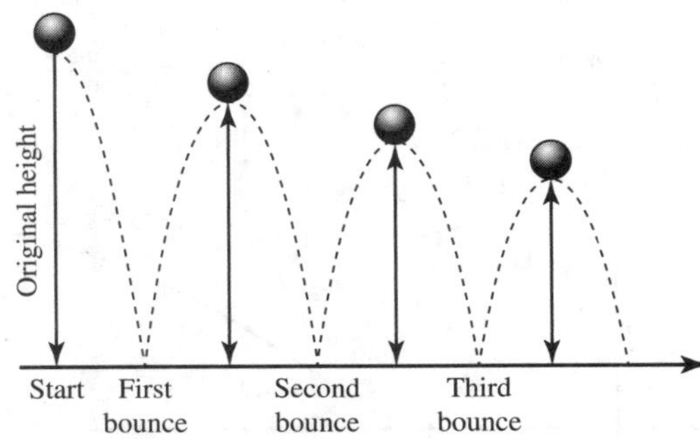

   (ii) Find the number of times, after the first bounce, that the maximum height reached is greater than 0.1 m.

Part B: Graphing calculator not permitted

TIME: 30 MINUTES

3. Given the rational function $f(x) = \dfrac{4x - 6}{x - 2}$:

   (a) (i) State the coordinates of the x-intercept and y-intercept of $f(x)$.

   (ii) Write the equation of the vertical asymptote and horizontal asymptote of $f(x)$.

   (iii) Evaluate $f(3)$.

   (b) Using the values found in part (a), sketch the graph of $f(x)$ on the axes provided below.

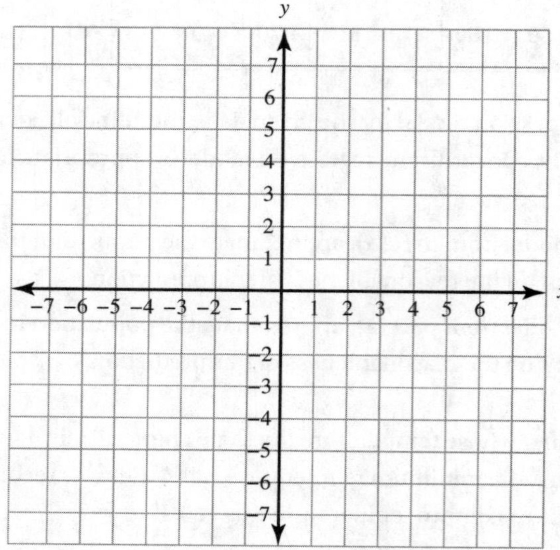

   (c) (i) Sketch the graph of $g(x) = |x + 1|$ on the axes from part (b).

   (ii) How many solutions does the equation $f(x) = g(x)$ have?

4. (a) (i) Find the solution of $\log_2 x - \log_2 5 = 2 + \log_2 3$.

   (ii) Find the solutions of $(\ln y)^2 - (\ln 2)(\ln y) = 2(\ln 2)^2$.

   (b) Find integer values of m and n for which $m - n\log_3 2 = 10\log_9 6$.

   (c) The graph of $y = \dfrac{(\ln x)^2}{x}$, $x > 0$ is shown below. Given that the curve passes through the point $(a, 0)$, find the value of a.

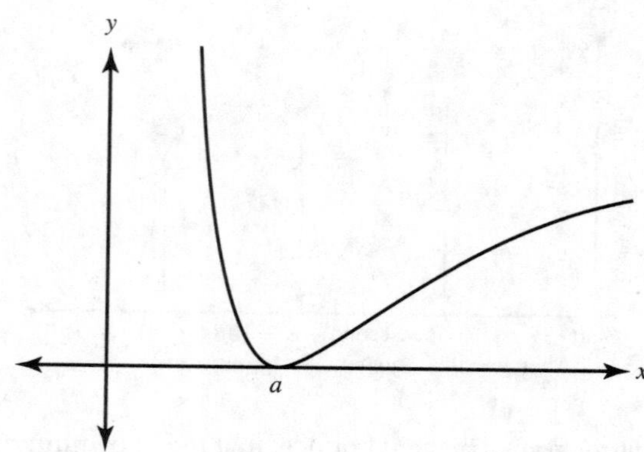

Answer Explanations

Section I

Part A

1. **(A)** Use the binomial theorem to expand $(3a - b)^6$, where $a = 3a$, $b = -b$, and $n = 6$:

$$(3a - b)^6 = {}_6C_0\,(3a)^6(-b)^0 + {}_6C_1\,(3a)^5(-b)^1 + {}_6C_2\,(3a)^4(-b)^2 + {}_6C_3\,(3a)^3(-b)^3 + {}_6C_4\,(3a)^2(-b)^4 + {}_6C_5\,(3a)^1(-b)^5 + {}_6C_6\,(3a)^0(-b)^6$$

Evaluate the coefficients in front of each term either by using the formula ${}_nC_r = \frac{n!}{r!(n-r)!}$ or by using the values from row 6 of Pascal's triangle:

$$(3a - b)^6 = 1\,(3a)^6 - 6\,(3a)^5 b + 15\,(3a)^4 b^2 - 20\,(3a)^3 b^3 + 15\,(3a)^2 b^4 - 6(3a)\,b^5 + 1b^6$$

Expand the fifth term:

$$15(3a)^2 b^4 = 15 \cdot 9a^2 b^4 = 135a^2 b^4$$

2. **(C)** Use log laws:

$$\ln\sqrt[3]{e^2 x} = \ln\left(e^2 x\right)^{\frac{1}{3}} = \tfrac{1}{3}[\ln e^2 + \ln x] = \tfrac{1}{3}[2 + \ln x] = \tfrac{2}{3} + \tfrac{1}{3}\ln x$$

3. **(B)** Rewrite the expression in terms of sine and cosine:

$$\frac{\cot^2 x}{1 + \csc x} = \frac{\frac{\cos^2 x}{\sin^2 x}}{1 + \frac{1}{\sin x}} = \frac{\frac{\cos^2 x}{\sin^2 x}}{\frac{\sin x + 1}{\sin x}} = \frac{\cos^2 x}{\sin^2 x} \cdot \frac{\sin x}{\sin x + 1}$$

$$= \frac{1 - \sin^2 x}{\sin^2 x} \cdot \frac{\sin x}{\sin x + 1}$$

$$= \frac{(1 + \sin x)(1 - \sin x)}{\sin^2 x} \cdot \frac{\sin x}{\sin x + 1} = \frac{1 - \sin x}{\sin x}$$

$$= \frac{1}{\sin x} - \frac{\sin x}{\sin x} = \csc x - 1$$

4. **(D)** Use trigonometric substitutions:

$$\sin\theta \cdot \cos\theta \cdot \tan\theta = \sin\theta \cdot \cos\theta \cdot \frac{\sin\theta}{\cos\theta} = \sin^2\theta$$

Since $\sin\theta = b$, $\sin^2\theta = (\sin\theta)^2 = b^2$.

5. **(A)** Given $f(\theta) = a\cos\theta + b$, the amplitude $= a$ and the midline $= b$. Therefore, the maximum value is $b + a$ and the minimum value is $b - a$.

6. **(B)** Since the graph is starting at the origin, the parent function is $y = \sin x$. Since it decreases as x increases from 0, it is a reflection over the x-axis, making the parent function $y = -\sin x$. The amplitude is $2 = a$, and there is half of the curve from $(0, 2\pi)$. So $b = \frac{1}{2}$. Therefore, the transformed equation is $y = -2\sin\left(\frac{1}{2}x\right)$.

7. **(D)** The coterminal angle of $-\frac{\pi}{2}$ is $\frac{3\pi}{2}$ so:

$$\csc\left(-\frac{\pi}{2}\right) = \csc\left(\frac{3\pi}{2}\right) = \frac{1}{\sin\left(\frac{3\pi}{2}\right)} = \frac{1}{-1} = -1$$

The only choice that is equivalent to -1 is $\cos\pi$.

8. **(B)** The graphs of the two functions are shown below. Both are decreasing in the interval $\left(\frac{\pi}{2}, \pi\right)$, which represents Quadrant II.

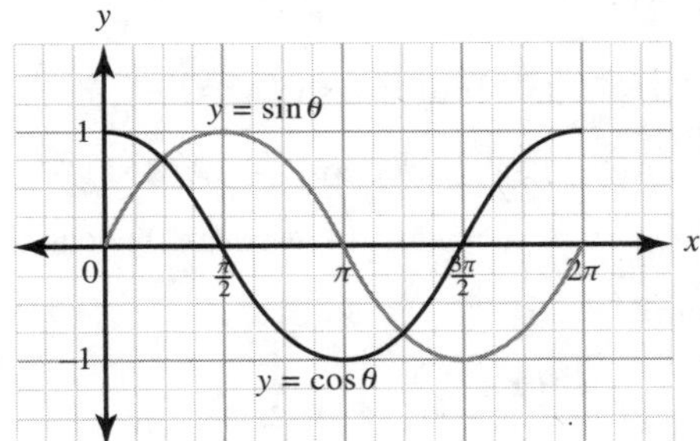

9. **(B)** Starting with the inner function of the composition, we know that $\tan\theta = \frac{4}{3}$. This creates a right triangle whose side opposite angle θ is 4 and whose side adjacent to angle θ is 3. Using the Pythagorean theorem, the hypotenuse would be 5. Then evaluate:

$$\csc(\theta) = \frac{1}{\sin\theta} = \frac{1}{\frac{4}{5}} = \frac{5}{4}$$

10. **(A)** Starting with the left side of the equation, $k - \frac{1}{\sec^2 x} = k - \cos^2 x$. For this to equal $\sin^2 x$, then $k = 1$ so that it will match the Pythagorean trigonometric identity $\sin^2 x + \cos^2 x = 1$.

11. **(C)** Use the double-angle trigonometric identities:

$$\frac{\sin 2\theta}{2\tan\theta} = \frac{2\sin\theta\cos\theta}{2\tan\theta}$$

Rewrite $\tan\theta$ in terms of $\sin\theta$ and $\cos\theta$:

$$\frac{2\sin\theta\cos\theta}{2\tan\theta} = \frac{2\sin\theta\cos\theta}{2\frac{\sin\theta}{\cos\theta}} = \frac{2\sin\theta\cos\theta}{2\sin\theta}\cdot\cos\theta$$
$$= \cos^2\theta$$

12. **(A)** Set the equation equal to 0:

$$\sin^2 x - \sin x = 0$$

Factor and set each factor equal to 0:

$$\sin x(\sin x - 1) = 0$$
$$\sin x = 0 \text{ and } \sin x - 1 = 0$$

Isolate $\sin x$ and find angle x using the unit circle. For $\sin x = 0$, then $x = 0, \pi$. For $\sin x = 1$, then $x = \frac{\pi}{2}$.

13. **(C)** Use the identity for $\sin(A + B)$:

$$\sin\left(x + \frac{\pi}{2}\right) = \sin x \cos\frac{\pi}{2} + \cos x \sin\frac{\pi}{2}$$
$$= \sin x(0) + \cos x(1) = \cos x$$

Then:

$$\frac{\sin\left(x + \frac{\pi}{2}\right)}{-\sin x} = \frac{\cos x}{-\sin x} = -\cot x$$

14. **(C)** The parent polar equation for $r = 6\cos\theta$ can be found in Table 9.2 in Section 9.2. A table of values can also be created that inputs different angle measures for θ from the unit circle, and then the points can be plotted.

15. **(D)** In the polar plane, the point $\left(4, \frac{11\pi}{18}\right)$ lies in Quadrant II with $r = 4$ and $\theta = \frac{11\pi}{18}$. The point in Choice (D) has an angle that is coterminal to $\theta = \frac{11\pi}{18}$ with the same $r = 4$. Therefore, the two points would be in the same location.

16. **(C)** To convert $\left(3, -\frac{3\pi}{4}\right)$ into rectangular coordinates (x, y), use the conversion formulas with $r = 3$ and $\theta = -\frac{3\pi}{4}$. Therefore:

$$x = r\cos\theta = 3\cos\left(-\frac{3\pi}{4}\right) = 3\left(-\cos\left(\frac{5\pi}{4}\right)\right)$$
$$= 3\left(-\cos\left(\frac{\pi}{4}\right)\right) = 3\left(-\frac{\sqrt{2}}{2}\right) = -\frac{3\sqrt{2}}{2}$$

Similarly:

$$y = r\sin\theta = 3\sin\left(-\frac{3\pi}{4}\right) = 3\left(-\sin\left(\frac{5\pi}{4}\right)\right)$$
$$= 3\left(-\sin\left(\frac{\pi}{4}\right)\right) = 3\left(-\frac{\sqrt{2}}{2}\right) = -\frac{3\sqrt{2}}{2}$$

The equivalent rectangular coordinates are $(x, y) = \left(-\frac{3\sqrt{2}}{2}, -\frac{3\sqrt{2}}{2}\right)$.

17. **(D)** You are given $g(f(x)) = g(x^3 - 3x^2 - 2x + 5) = 2$. Since $g(x) = 2$, it is a constant function. This means that no matter the input, the output will always be 2.

18. **(C)** For a function to be odd, graphically, it is symmetric with respect to the origin and algebraically $f(-x) = -f(x)$. For the function h, $h(-x) = (-x)^5 + 1 = -x^5 + 1 \neq -h(x)$ since all the signs were not negated.

19. **(D)** The solutions are the points of intersection of the two graphs, which can be algebraically found by substituting. Substitute $y = x + 2$ into the second equation for y:

$$(x + 2)^2 = 4x$$
$$x^2 + 4x + 4 = 4x$$

Set the equation equal to 0 and solve for x:

$$x^2 + 4 = 0$$

The solutions to this equation are imaginary, specifically $x = \pm 2i$. There are no real solutions.

PRACTICE TEST 2

20. **(D)** Given that $\log_b(3^b) = \frac{b}{2}$, using log laws lets the left side of the equation be rewritten as $b \cdot \log_b 3$. Set this equal to the right side of the equation and divide out a common factor of b:

$$b \cdot \log_b 3 = \frac{b}{2}$$
$$\log_b 3 = \frac{1}{2}$$

Rewrite in exponential form:

$$b^{\frac{1}{2}} = 3$$

Square both sides of the equation to solve for b:

$$b = 3^2 = 9$$

21. **(A)** To find the inverse equation, switch x and y and solve for y:

$$x = 2e^{-y}$$
$$\frac{x}{2} = e^{-y}$$

Rewrite as a natural log equation:

$$-y = \ln\left(\frac{x}{2}\right)$$
$$y = -\ln\left(\frac{x}{2}\right)$$

Therefore, the inverse equation is $f^{-1}(x) = -\ln\left(\frac{x}{2}\right)$. Using log laws this is equivalent to:

$$f^{-1}(x) = -\ln\left(\frac{x}{2}\right) = \ln\left(\frac{x}{2}\right)^{-1} = \ln\left(\frac{2}{x}\right)$$

22. **(D)** If $f(x) = 2x^2 + 4x - 1$:

$$\begin{aligned} f(x+h) &= 2(x+h)^2 + 4(x+h) - 1 \\ &= 2(x^2 + 2xh + h^2) + 4(x+h) - 1 \\ &= 2x^2 + 4xh + 2h^2 + 4x + 4h - 1 \end{aligned}$$

23. **(C)** To evaluate $f(6)$, first identify which of the functions to use in the piecewise function. Since $x = 6$, it fits the interval of the second equation. So $f(6) = 4(6) = 24$.

24. **(B)** Starting at the left of the graph, it decreases from $(-\infty, -1)$ then increases from $(-1, 0)$ then decreases from $(0, 1)$ and then increases from $(1, \infty)$.

25. **(C)** The composition $(f \circ g)(4) = f(g(4)) = f(7) = 14$.

26. **(B)** To find the horizontal asymptote, examine the degree of the polynomial in both the numerator and the denominator. Since both are degree 2, the horizontal asymptote is the ratio of the leading coefficients, $y = \frac{7}{3}$.

27. **(A)** If x = length and y = width, the perimeter of the rectangular garden is $1{,}400 = 2x + 2y$. Therefore, the width is $y = 700 - x$. Since the area of a rectangle is the product of the length and the width, $A(x) = x(700 - x) = 700x - x^2$. Since no one side can be all the fencing, the domain must be restricted to $(0, 700)$ so that there is enough fencing for the garden to exist with all four sides.

28. **(C)** Since it is given that $x = -2$ is a real zero of $f(x)$, use long division to divide the factor $(x + 2)$ into $f(x)$. An equivalent representation of $f(x) = (x + 2)(x^2 + 10x + 26)$ after long division. Since the second factor is a quadratic, the zeros can be solved for by using the quadratic formula:

$$x = \frac{-(10) \pm \sqrt{(10)^2 - 4(1)(26)}}{2(1)} = \frac{-10 \pm \sqrt{-4}}{2}$$
$$= \frac{-10 \pm 2i}{2} = -5 \pm i$$

The imaginary solutions are in conjugate pairs and can be written as factors as:

$$(x - (-5 - i))(x - (-5 + i)) = (x + 5 + i)(x + 5 - i)$$

So:

$$f(x) = (x + 2)(x + 5 + i)(x + 5 - i)$$

Part B

29. **(C)** Substitute $n = 5$ into the sequence:

$$a_5 = 12{,}000\left(1 + \frac{0.076}{4}\right)^{(5)} = \$13{,}184.15$$

30. **(A)** Since both log expressions are of different bases that do not have a relationship to each other, use the change of base formula to rewrite each of the logarithmic expressions:

$$\log_{0.3}(x-1) < \log_{0.09}(x-1)$$
$$\frac{\log(x-1)}{\log 0.3} < \frac{\log(x-1)}{\log 0.09}$$

Since $\log 0.3 > \log 0.09$, for this inequality to be true then both fractions must be negative, meaning that $\log(x-1)$ must be negative.

This leads to the following inequality:

$$\log(x-1) < 0$$

Change the inequality to an equal sign and rewrite in exponential form:

$$10^0 = x - 1$$
$$x - 1 = 1$$
$$x = 2$$

To determine if $x < 2$ or $x > 2$, choose x-values in those intervals and test those values in the original inequality.

$x = 1.5$	$x = 3$
$\log_{0.3}((1.5)-1) < \log_{0.09}((1.5)-1)$ $0.5757 < 0.287$	$\log_{0.3}((3)-1) < \log_{0.09}((3)-1)$ $-0.5757 < -0.287$
This is not a true statement, so x cannot be less than 2.	This is a true statement, so x is greater than 2.

For the final solution, it is important to consider the domain of a logarithmic function, which is all input values greater than 0. If $x - 1 > 0$, then $x > 1$.

Therefore, the final solution is $x > 2$.

31. **(B)** Point $P(-0.8, 0.6)$ lies in Quadrant II. Since P lies on the unit circle, the y-coordinate is $0.6 = \sin\theta$. The reference angle is $\sin^{-1}(0.6) \approx 0.6435011$ radians. The angle in Quadrant II is $\theta \approx \pi - 0.6435011 \approx 2.4981$, which is 2.50 when rounded to the nearest hundredth.

32. **(C)** Finding the minimum value of $C(x) = 2x^2 - 22x + 760$ can be done graphically using technology. It can also be done algebraically by finding the axis of symmetry, $x = -\frac{b}{2a}$, and substituting that x-value into the function $C(x)$:

$$x = -\frac{-22}{2(2)} = 5.5$$
$$C(5.5) = 699.50 \approx 700$$

33. **(D)** Calculate the average rate of change:

$$\frac{\Delta y}{\Delta x} = \frac{h(2) - h(0)}{2 - 0} = \frac{22.4 - 2}{2 - 0} = 10.2$$

34. **(D)** To solve the rational inequality, first get it in the form $f(x) > 0$ by subtracting the 5 from both sides and combining the expressions:

$$\frac{x+2}{2x-1} - 5 > 0$$

$$\frac{x + 2 - 5(2x - 1)}{5(2x - 1)} > 0$$

$$\frac{x + 2 - 10x + 5}{5(2x - 1)} > 0$$

$$\frac{7 - 9x}{5(2x - 1)} > 0$$

Find the critical values by setting the numerator and denominator equal to 0 and solving for x:

$$7 - 9x = 0 \qquad\qquad 5(2x - 1) = 0$$

$$x = \frac{7}{9} \qquad\qquad x = \frac{1}{2}$$

Place these critical values on a number line, and choose test points within each interval to find the sign of $f(x)$.

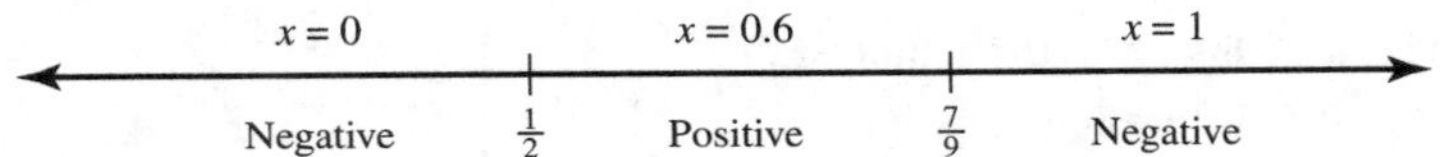

The function $f(x) > 0$ in the interval where the values are positive. Since $f(x) \neq 0$, the endpoints of the intervals are contained by parentheses. The solution is $\left(\frac{1}{2}, \frac{7}{9}\right)$.

35. **(D)** Since the relation is said to be continuous, there are no breaks in the output values. Therefore, a y-value of 0 must exist in an interval where the y-values switch from positive to negative or from negative to positive. This happens in the intervals $(-3, -1) \cup (5, 7)$.

36. **(A)** You are given the function $f(x) = \sqrt{x - a} - \frac{1}{\sqrt{b - x}}$, $0 < a < b$. To find the domain, there are two different sets of restrictions. The first expression of the function is a radical expression where the radicand must be greater than or equal to 0. Therefore, $x - a \geq 0 \rightarrow x \geq a$. For the second expression, the function is a rational expression where the denominator has a radical expression. Therefore, $b - x > 0 \rightarrow b > x \rightarrow x < b$. Since it was given that $0 < a < b$, the domain of $f(x)$ is $a \leq x < b$.

37. **(D)** To find the solutions of the equation $5x^4 + x^2 - 1 = 2^x$, use technology to graph the two different sides of the equation and find the points of intersection.

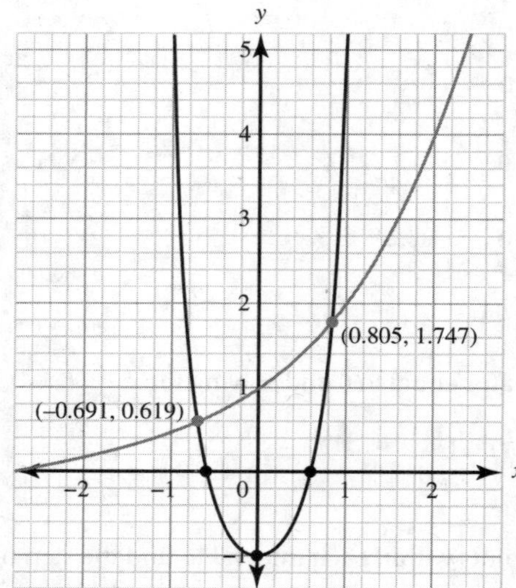

The largest solution is $x = 0.805$, and the smallest solution is $x = -0.691$. The difference between the largest and smallest solutions is $0.805 - (-0.691) = 1.496 \approx 1.50$.

38. **(C)** Looking at the growth of the population, the data resemble an exponential pattern. Using technology, enter the data into the lists of a graphing calculator where L1 = number of years since 1980 and L2 = population (in thousands). Calculate the exponential regression model:

$$y = 1.975896573(1.0737245)^x$$

Substitute $x = 44$ for the year 2024:

$$y = 1.975896573(1.0737245)^{(44)}$$
$$= 45.19081499 \approx 45{,}000 \text{ people}$$

39. **(A)** Let x = length of the side of the square. This will also represent the height of the box. Since 2 squares are being cut on either side of the original piece of cardboard, the length of the box is $27 - 2x$ and the width of the box is $11 - 2x$. It is necessary to restrict the domain here. Since we want the box to exist, $x > 0$. The width of the box needs to be restricted such that:

$$11 - 2x > 0$$
$$x < 5.5$$

The domain is $0 < x < 5.5$. A function to represent the volume is $V(x) = (27 - 2x)(11 - 2x)(x)$. Using technology, this function can be graphed within the restricted domain and the calculator can determine the coordinates of the maximum.

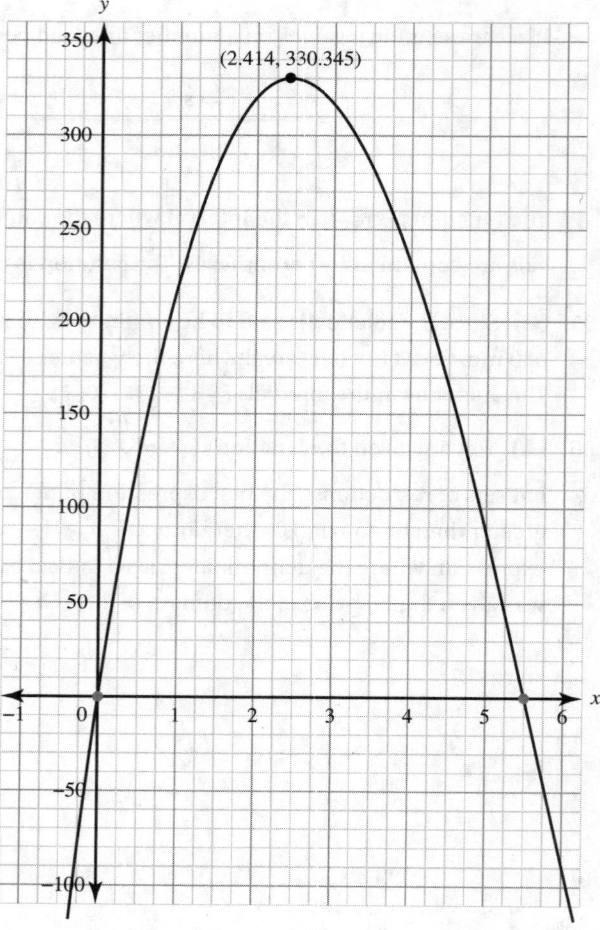

Then $x = 2.414$, $V = 330.345$. This means that there will be a maximum volume of 330.345 cubic inches when the length of the squares cut out are $2.414 \approx 2.41$ inches.

40. **(D)** To convert $(4, -4\sqrt{3})$ into polar coordinates (r, θ), find θ and r when $x = 4$ and $y = -4\sqrt{3}$.

To find r:

$$r = \sqrt{x^2 + y^2} = \sqrt{(4)^2 + (-4\sqrt{3})^2} = \sqrt{16 + (16 \cdot 3)} = \sqrt{16 + 48} = \sqrt{64} = 8$$

To find θ:

$$\theta = \arctan\left(\frac{y}{x}\right) \text{ for } x > 0,\ y < 0$$

$$\theta = -\arctan\left|\frac{-4\sqrt{3}}{4}\right| = -\arctan\left(\frac{4\sqrt{3}}{4}\right) = -\arctan(\sqrt{3}) = -\frac{\pi}{3} = \frac{5\pi}{3}$$

The equivalent polar coordinates are $(r, \theta) = \left(8, \frac{5\pi}{3}\right)$.

Section II

Part A

1. (a) Use technology to enter the data into L1 and L2, and calculate the logarithmic regression model using STAT CALC 9:LnReg. The regression model is $y = -591.427 + 390.404 \ln x$. The coefficient of determination is $r^2 \approx 0.99816$. Using a logarithmic regression model is appropriate since the population seems to increase rather rapidly in the beginning but then begins to level off although still increasing.

   (b) To approximate the population in the year 1995, evaluate $P(15) = -591.427 + 390.404 \ln(15) \approx 465.8$. This seems reasonable as it fits within the data values between $t = 14$ and $t = 18$.

   (c) To approximate the population in the year 2024, evaluate $P(44) = -591.427 + 390.404 \ln(44) \approx 885.9$. It is difficult to comment on the reasonableness of data outside of the provided domain. There are limiting factors on population and no guarantee that the population will continue to increase as it does with the provided data.

2. (a)(i) The amplitude $= |a| = \dfrac{\text{maximum} - \text{minimum}}{2} = \dfrac{120 - 4}{2} = 58$. Since the sinusoidal curve hits the minimum before the maximum, it is reflected over the x-axis; therefore, $a = -58$.

   The midline $= d = \dfrac{\text{maximum} + \text{minimum}}{2} = \dfrac{120 + 4}{2} = 62$.

   For a basic parent sine curve, the length of time between the minimum and maximum values is half of a full cycle. Since it is known that the first minimum value occurs 30 hours after the landing and the first maximum value occurs 72 hours after the landing, the difference is 42 and half of that interval is 21. This value represents the x-scale of the graph. Adding 21 to 72 gives an upper bound of the sinusoidal cycle of 93, and subtracting 21 from 30 gives a lower bound of the sinusoidal cycle of 9. So to complete one full cycle of sine, the period is $93 - 9 = 84$. Therefore, to find the b-value, the period $= \frac{2\pi}{b} = 84 \rightarrow b = \frac{2\pi}{84} = \frac{\pi}{42}$.

A phase shift exists where the graph is shifted to the right 9 units; therefore, $c = 9$.

Thus, the function is represented as $f(t) = -58 \sin\left[\frac{\pi}{42}(t - 9)\right] + 62$.

(a)(ii) $f(42) = -58 \sin\left[\frac{\pi}{42}(42 - 9)\right] + 62 \approx 25.8883759$. The temperature is 25.89°F.

(b) With the restricted domain of [0, 100], using technology, graph the equation found in part (a)(i), graph the line $y = 80$, and find their points of intersection.

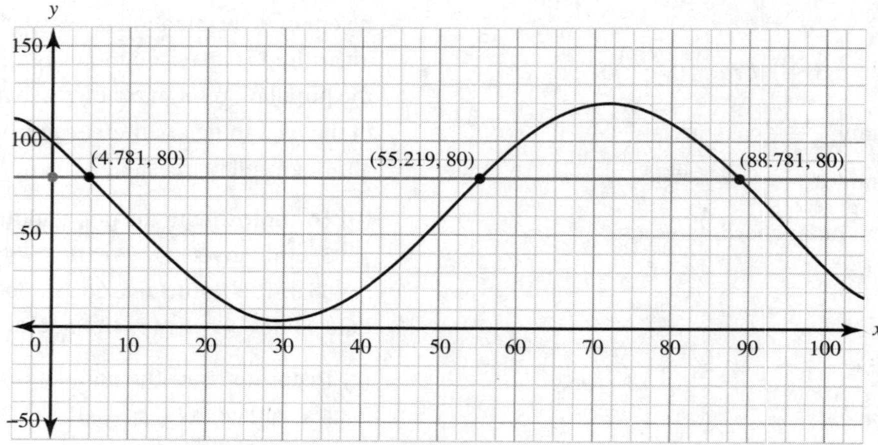

The times are $t = 4.78$ hours after the landing, $t = 55.22$ hours after the landing, and $t = 88.78$ hours after the landing.

(c)(i) This scenario represents a geometric sequence with $g_0 = 1.8$ and $r = 0.85$. Using $g_n = g_0 r^n$, on the sixth bounce, $n = 6$, so $g_6 = (1.8)(0.85)^6 \approx 0.678869 = 0.68$ m.

(c)(ii) Set the general equation for the sequence greater than 0.1: $(1.8)(0.85)^n > 0.1$. Use technology to graph the two sides of the inequality.

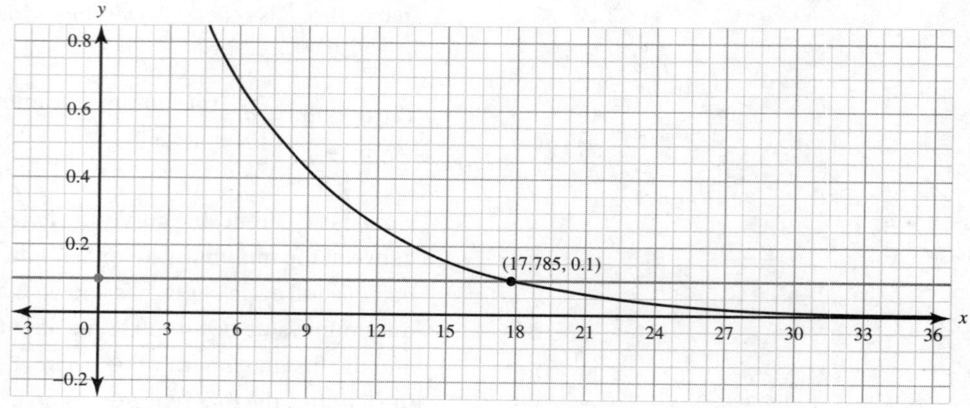

The graph of $y = (1.8)(0.85)^n$ and $y = 0.1$ intersect at the point (17.785, 0.1). Since the domain of a sequence is the set of whole numbers, $n = 17$.

Part B

3. (a)(i) To find the x-intercept of $f(x)$, set the equation equal to 0 and solve for x:

$$\frac{4x-6}{x-2}=0 \rightarrow 4x-6=0 \rightarrow x=\frac{6}{4}=\frac{3}{2}$$

The coordinates of the x-intercept are $\left(\frac{3}{2},0\right)$.

To find the y-intercept of $f(x)$, evaluate $f(0)$:

$$f(0)=\frac{4(0)-6}{(0)-2}=3$$

The coordinates of the y-intercept are (0, 3).

(a)(ii) To find the equation of the vertical asymptote of $f(x)$, set the denominator equal to 0 and solve for x:

$$x-2=0 \rightarrow x=2$$

To find the equation of the horizontal asymptote of $f(x)$, consider the degree of the numerator and denominator; since they are both equal to 1, the horizontal asymptote is the ratio of the leading coefficients, 4 and 1, so $y=4$.

(a)(iii)

$$f(3)=\frac{4(3)-6}{(3)-2}=\frac{12-6}{1}=6$$

This represents the point (3, 6) on the graph of $f(x)$.

(b) The graph of $f(x)$ is provided below.

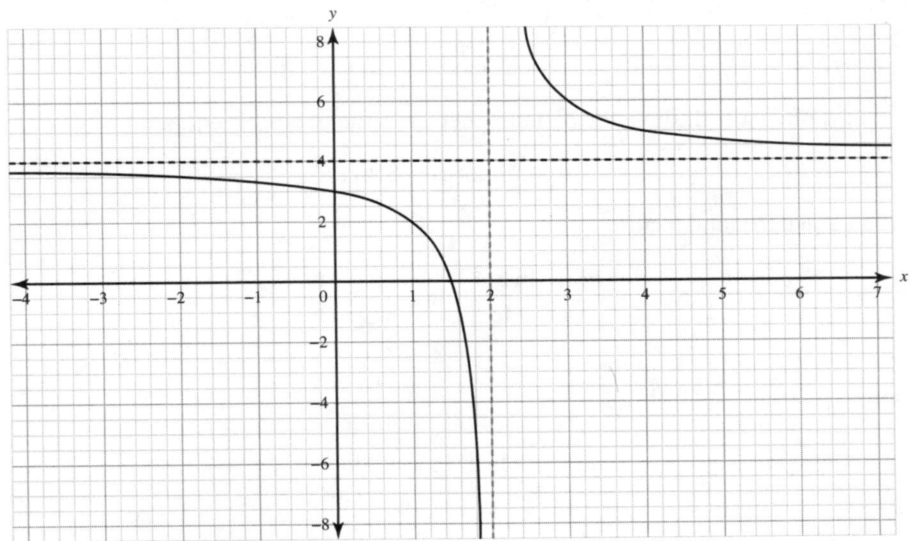

(c)(i) The graph of $g(x)=|x+1|$, shown on the following page, is the graph of the parent function $p(x)=|x|$ shifted to the left 1 unit.

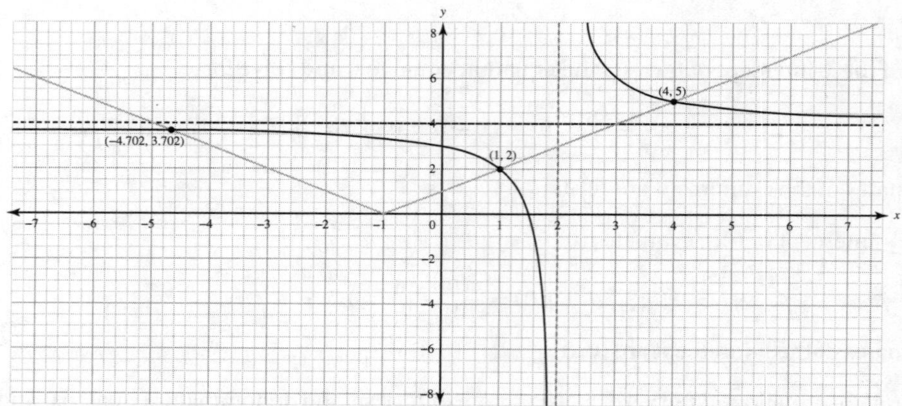

(c)(ii) The solutions to the equation $f(x) = g(x)$ are where the graphs intersect. They intersect at three points, $(-4.702, 3.702)$, $(1, 2)$, and $(4, 5)$, so there are three solutions to the equation.

4. (a)(i) Isolate the logarithms to one side of the equation:

$$\log_2 x - \log_2 5 - \log_2 3 = 2$$

Condense the left side of the equation using log laws:

$$\log_2\left(\frac{x}{5\cdot 3}\right) = 2$$

Rewrite in exponential form:

$$2^2 = \frac{x}{15}$$

Solve for x:

$$x = 60$$

(a)(ii) Let $u = \ln y$. After substituting, the equation becomes the quadratic equation:

$$(u)^2 - (\ln 2)(u) = 2(\ln 2)^2$$

Set the equation equal to 0 and factor:

$$u^2 - (\ln 2)u - 2(\ln 2)^2 = 0$$
$$(u - 2\ln 2)(u + \ln 2) = 0$$
$$u = 2\ln 2,\ u = -\ln 2$$

Replace u with $\ln y$:

$$\ln y = 2\ln 2,\ \ln y = -\ln 2$$

Use log laws:

$$\ln y = \ln 2^2,\ \ln y = \ln 2^{-1}$$
$$y = 4,\ y = \frac{1}{2}$$

(b) Since $9 = 3^2$, it can be written in logarithmic form: $\log_3 9 = 2$. Using the change of base formula, all the logarithms can be written as base 3:

$$\log_9 6 = \frac{\log_3 6}{\log_3 9} = \frac{\log_3 6}{2}$$

So $m - n\log_3 2 = 10\log_9 6$ can be rewritten as:

$$m - n\log_3 2 = 10\left(\frac{\log_3 6}{2}\right) = 5\log_3 6$$

Using log laws:

$$\begin{aligned}
m - n\log_3 2 &= 5\log_3 6 = 5\log_3(3 \cdot 2) = 5(\log_3 3 + \log_3 2)\\
m - n\log_3 2 &= 5(\log_3 3 + \log_3 2)\\
m - n\log_3 2 &= 5(1 + \log_3 2)\\
m - n\log_3 2 &= 5 + 5\log_3 2
\end{aligned}$$

Matching up the coefficients on the left side and right side of the equation, $m = 5$ and $-n = 5 \rightarrow n = -5$.

(c) Substitute the coordinates of the point into the equation:

$$\begin{aligned}
0 &= \frac{(\ln a)^2}{a}\\
0 &= (\ln a)^2
\end{aligned}$$

Extract the square root, and rewrite the resulting equation in exponential form:

$$\begin{aligned}
\ln a &= 0\\
e^0 &= a\\
a &= 1
\end{aligned}$$

Glossary of Key Terms

Amplitude Half the difference between the maximum and minimum values of a sinusoidal function.

Arithmetic sequence A sequence in which consecutive terms have a common difference, d, or a constant rate of change.

Asymptote A line whose distance from a curve approaches 0 as the curve tends to infinity.

Complex zero For a function, $f(x)$, a complex zero is a value of x that makes $f(x) = 0$ and is of the form $a \pm bi,\ b \neq 0$.

Composition Links two functions to form a new function by using the output of one function as the input to the other function.

Concavity Describes the curvature of the graph of a function; a function is concave up if the rate of change is increasing; a function is concave down if the rate of change is decreasing.

Cosecant function Notated as $f(\theta) = \csc\theta$; it is the reciprocal of the sine function, where $\sin\theta \neq 0$.

Cotangent function Notated as $f(\theta) = \cot\theta$; it is the reciprocal of the tangent function, where $\tan\theta \neq 0$.

Decreasing A function with a domain and range of real numbers such that the dependent variable decreases as the independent variable increases.

Dependent variable The variable representing output values; typically notated using the letter y.

Determinant The determinant of matrix $A = \begin{bmatrix} a & b \\ c & d \end{bmatrix}$ is $ad - bc$ and is notated as $\det(A)$.

Domain The set of values of a function that can be assumed by the independent variable of the function; typically notated using the letter x.

Dot product The dot product of two vectors is the sum of the products of their corresponding components; that is, $\langle a_1, b_1\rangle \cdot \langle a_2, b_2\rangle = a_1 a_2 + b_1 b_2$.

Error The difference between the predicted and actual values of a model.

Exponential function A function of the general form $f(x) = ab^x$, with the initial value a, where $a \neq 0$, and the base b, where $b > 0$, and $b \neq 1$; when $a > 0$ and $b > 1$, the exponential function is said to demonstrate exponential growth; when $a > 0$ and $0 < b < 1$, the exponential function is said to demonstrate exponential decay.

Factor A number or polynomial that divides a given number or polynomial exactly; for example, 1, 2, 3, and 6 are all factors of 6; $x - 1$ and $x + 2$ are factors of $x^2 + x - 2$ since $(x - 1)(x + 2) = x^2 + x - 2$.

Function A mathematical relation that maps a set of input values to a set of output values such that each input value is mapped to exactly one output value.

Geometric sequence A sequence in which consecutive terms have a common ratio, r, or a constant proportional change.

Holes Exist if there is a common factor in the numerator and denominator of a rational function that can be removed and then graphed using an open circle; holes are also known as removable discontinuities.

Horizontal asymptote Occurs on the graph of a rational function when the degree of the numerator is less than or equal to the degree of the denominator.

Identity function $f(x) = x$.

Identity matrix A square matrix consisting of 1's on the diagonal from the top left to bottom right and 0's everywhere else; notated as I.

Increasing A function with a domain and range of real numbers such that the dependent variable increases as the independent variable increases.

Independent variable The variable representing input values; typically notated using the letter x.

Inverse function A reverse mapping of a function; an inverse function, f^{-1}, maps the output values of a function, f, on its invertible domain to their corresponding input values; that is, if $f(a) = b$, then $f^{-1}(b) = a$.

Limit A value of a function that can be approached arbitrarily closely by the dependent variable when some restriction is placed on the independent variable of the function.

Linear function May be represented as an equation that has the form $y = ax + b,\ a \neq 0$; the graph of a linear function is a line with a constant slope.

Linear transformation A function that maps an input vector to an output vector such that each component of the output vector is the sum of constant multiples of the input vector components.

Logarithmic expression Notated as $\log_b c$; is equal to, or represents, the value that the base b must be exponentially raised to in order to obtain the value c; that is $\log_b c = a$ if and only if $b^a = c$, where a and c are constants, $b > 0$, and $b \neq 1$.

Logarithmic function A function of the general form $f(x) = a\log_b x$, with base b, where $b > 0$, $b \neq 1$, and $a \neq 0$.

Magnitude The magnitude of a vector is the length of the line segment of the vector $\langle a,\ b \rangle$, which is found by evaluating $\sqrt{a^2 + b^2}$.

Matrix A set of quantities (called elements or entries) arranged in a rectangular array with certain rules governing their combination; the array is enclosed in square brackets; the horizontal lines of elements are rows, and the vertical lines are columns.

Midline Determined by the average of the maximum and minimum values of a sinusoidal function.

Multiplicity Refers to a multiple root or a repeated root of an equation; in general, if $(x - r)^n$ is a factor of a polynomial equation, then r is a root of multiplicity n of the equation.

Parabola A U-shaped curve that is the graph of a quadratic function; the slope of a parabola changes along the curve.

Parametric function In $\mathbb{R}^2$, the set of all ordered pairs of two real numbers, consists of a set of two parametric equations in which two dependent variables, x and y, are dependent on a single independent variable, t, called the parameter.

Period The period of a function can be defined as the smallest positive value k such that $f(x + k) = f(x)$ for all x in the domain.

Periodic relationship Identified between two aspects of a context if, as the input values increase, the output values demonstrate a repeating pattern over successive equal-length intervals.

Piecewise-defined function Consists of a set of functions defined over nonoverlapping domain intervals.

Polar coordinates An ordered pair, (r, θ), such that $|r|$ represents the radius of the circle on which the point lies and θ represents the measure of an angle in standard position whose terminal ray includes the point.

Polynomial function A function that can be written in the form $p(x) = a_n x^n + a_{n-1}x^{n-1} + \ldots + a_1 x + a_0$, where n is a positive integer called the degree of the polynomial and $a_n \neq 0$.

Quadratic function A second-degree equation that has the form $y = ax^2 + bx + c,\ a \neq 0$.

Radian measure The radian measure of an angle in standard position is the ratio of the length of the arc of a circle centered at the origin subtended by the angle to the radius of that same circle.

Range The set of output values of a function.

Rate of change A comparison that describes how one quantity changes in relation to another quantity; rates of change can be either positive or negative.

Rational function A function of the form $f(x) = \dfrac{P(x)}{Q(x)}$, where $P(x)$ and $Q(x)$ are polynomial functions and $Q(x) \neq 0$; graphs of rational functions may have discontinuities, such as asymptotes or "holes."

Real zero For a function, $f(x)$, a real zero is a real value of x that makes $f(x) = 0$.

Regression A model that describes the dependence of the mean value of one random variable on one or more other variables.

Residuals In statistics, residuals are the differences between observed values and values predicted by a model; residuals are sometimes called errors.

Secant function Notated as $f(\theta) = \sec\theta$; it is the reciprocal of the cosine function, where $\cos\theta \neq 0$.

Semi-log plot A plot that has one of the axes logarithmically scaled; when the y-axis of a semi-log plot is logarithmically scaled, data or functions that demonstrate exponential characteristics will appear linear.

Sequence A function from the whole numbers to the real numbers; the graph of a sequence consists of discrete points instead of a curve.

Sinusoidal function Any function that involves additive and multiplicative transformations of $f(\theta) = \sin\theta$; the sine and cosine functions are both sinusoidal functions with $\cos\theta = \sin\left(\theta + \frac{\pi}{2}\right)$.

Standard position The standard position of an angle in the coordinate plane occurs when the vertex coincides with the origin and one ray coincides with the positive x-axis; the other ray is called the terminal ray.

Tangent function Notated as $f(\theta) = \tan\theta$; it is the ratio of the sine function and the cosine function, where $\cos\theta \neq 0$.

Transformation Transformation of a graph moves each point of the graph according to a given rule such as a reflection, translation, or dilation.

Unit circle A circle that is centered at the point (0, 0) with a radius length of 1.

Unit vector A vector of magnitude 1.

Vector A directed line segment; when a vector is placed in the plane, the point at the beginning of the line segment is called the tail, and the point at the end of the line segment is called the head.

Vertical asymptote Occurs on the graph of a rational function at a value of x for which the denominator evaluates to 0 but the numerator does not evaluate to 0; it is written as an equation of a vertical line $x = a$ and is typically graphed as a dashed line.

***x*-intercept** The coordinates of the intersection of a graph with the x-axis.

***y*-intercept** The coordinates of the intersection of a graph with the y-axis.

Index